WJEC Physics For A2 Level

2nd Edition

Gareth Kelly
Nigel Wood

Illuminate Publishing

Published in 2021 by Illuminate Publishing, an imprint of Hodder Education (a trading division of Hodder & Stoughton Limited), an Hachette UK Company, Carmelite House, 50 Victoria Embankment, London EC4Y 0DZ

Orders: please contact Hachette UK Distribution, Hely Hutchinson Centre, Milton Road, Didcot, Oxfordshire, OX11 7HH.
Telephone: +44 (0)1235 827827. Email: education@hachette.co.uk.
Lines are open from 9 a.m. to 5 p.m., Monday to Friday. You can also order through our website: www.hoddereducation.co.uk

British Library Cataloguing in Publication Data

A catalogue record for this book is available from the British Library

ISBN 978-1-912820-72-6

Printed by Ashford Colour Ltd, UK

The authorised representative in the EEA is Hachette Ireland,
8 Castlecourt Centre, Dublin 15, D15 XTP3, Ireland (email: info@hbgi.ie)

Impression 5
Year 2026

Hachette UK's policy is to use papers that are natural, renewable and recyclable products and made from wood grown in well-managed forests and other controlled sources. The logging and manufacturing processes are expected to conform to the environment regulations of the country of origin.

Editor: Geoff Tuttle
Design: Nigel Harriss
Layout: John Dickinson Graphic Design

Acknowledgements

The authors would like to thank Jim Newall and Tony Reeves for their eagle eyes in spotting mistakes and inconsistencies and for their many insightful suggestions. We are also indebted to the Illuminate team of Beth Hutchins, Geoff Tuttle, Nigel Harris and John Dickinson for their patience and meticulous attention to detail.

Major thanks to the long-suffering wives of both authors who had been labouring under the delusion that their partners had retired!

Cover Image: © Pavel L Photo and Video/Shutterstock

Image credits:
p.1 Pavel L Photo and Video/Shutterstock.com; **p.7** and **12** NASA and the Hubble Heritage Team; **p.8** (top) © Christian Bertrand/Shutterstock.com; **p.8** (middle) © Abdul Momin Photographer/Shutterstock.com; **p.8** (bottom) ESA, NASA, and Felix Mirabel (French Atomic Energy Commission and Institute for Astronomy and Space Physics/ Conicet of Argentina); **p.13** (top) Mim Friday/Alamy Stock Photo; **p.13** (bottom) © deviyanthi79/Shutterstock.com; **p.16** © Tristan3D/Shutterstock.com; **p.31** (top) © Trinity Mirror/Mirrorpix/ Alamy Stock Photo; **p.31** (bottom) © XtravaganT/stock.adobe.com; **p.49** (left) Chris Allen; **p.49** (middle) © Industryviews/Shutterstock.com; **p.49** (right) © Northallertonman/Shutterstock.com; **p.54** © Adam Hart-Davis/Science Photo Library; **p.63** (top) © Photo Researchers/Science History Images/ Alamy Stock Photo; **p.63** (bottom) Courtesy of the Cavendish Laboratory, University of Cambridge; **p.79** © Dmitr86/stock.adobe.com; **p.97** CERN, Licence: CC-BY-4.0; **p.98** © Surasak/stock.adobe.com; **p.107** © miha de/ Shutterstock.com; **p.112** © Nigel Wood; **p.115** © ART Collection/Alamy Stock Photo; **p.118** © Adam Hart-Davis/Science Photo Library; **p.123** © Denisk0/iStock/Getty Images Plus; **p.131** © Boris15/Shutterstock.com; **p.136** (top) NASA/JPL/University of Arizona; **p.136** (bottom) © NASA Image Collection/Alamy Stock Photo; **p.137** Artist's impression of the Milky Way. jpg: NASA/JPL-Caltech/ESO/R. Hurt derivative work: Cmglee, Public domain, via Wikimedia Commons; **p.138** (top) © Tryfonov/stock.adobe.com; **p.138** (bottom) NASA/JPL-Caltech/L. Jenkins (GSFC) NASA original work – pubic domain; **p.144** Javier Lorenzo, University of Alicante; **p.150** © Nigel Wood; **p.152** © Imago Europe Collection/Alamy Stock Photo; **p.154** (top) KLINGER; **p.154** (bottom) © Shin Okamoto/Shutterstock.com; **p.158** © Mark Duffin; **p.159** © Nigel Wood; **p.167** © Nigel Wood; **p.175** 3B Scientific Limited; **p.184** © Giphotostock/Science Photo Library; **p.199** © Samunella/stock. adobe.com; **p.200** (top left) © Suttha Burawonk/Shutterstock.com; **p.200** (right) © Xray Computer/Shutterstock.com; **p.200** (bottom left) © Xray Computer/Shutterstock.com; **p.201** (top) © Image Supply/Shutterstock. com; **p.201** (bottom) © Creative Endeavors/stock.adobe.com; **p.203** © faustasyan/Shutterstock.com; **p.204** (left) © Donna Beeler/Shutterstock. com; **p.204** (right) © Nejron Photo/Shutterstock.com; **p.206** (top left) © Pnmb/ISM/Science Photo Library; **p.206** (bottom left) © SMK4pix/ Shutterstock.com; **p.206** (bottom right) © LIVING ART ENTERPRISES/ SCIENCE PHOTO LIBRARY; **p.209** © Philippe Plailly/Science Photo Library; **p.210** (right) © Centre Jean Perrin, ISM/Science Photo Library; **p.210** (left) © Brian Bell/Science Photo Library; **p.213** © Mark Duffin; **p.214** (top) © wavebreakmedia/Shutterstock.com; **p.214** (middle left) © stihii/Shutterstock. com; **p.214** (bottom) © studioloco/Shutterstock.com; **p.220** Laszlo Veres/ Beehive Illustration; **p.221** © David J. Phillip/Associated Press/Alamy Stock Photo; **p.223** © DR. Gary Settles/Science Photo Library; **p.224** (top) © hiloi/ Shutterstock.com; **p.224** (middle) © ARENA Creative/Shutterstock.com; **p.224** (bottom) © Ryan/stock.adobe.com; **p.228** © 1059SHU/Shutterstock. com; **p.234** © Ken Edwards/Alamy Stock Photo; **p.235** © Feature Eye Images/Alamy Stock Photo; **p.237** © Science History Images/Alamy Stock Photo; **p.252** © Pukhov K/Shutterstock.com.

Contents

How to use this book

This book has been written to support the second half (A2) of the WJEC A level Physics specification. The layout of the book in the first two sections matches that of Units 3 and 4 respectively of the A level Physics specification. It provides you with information which covers the content requirements of the course as well as plenty of practice questions to allow you to keep track of your progress and to prepare for your A level examinations.

The main chapters in the book are the A level Units 3 and 4.

- Unit 3 covers Oscillations and nuclei
- Unit 4 covers Fields and options

There is an additional chapter on

- Mathematical and data-handling skills

Unit 3 covers six topics of the Physics specification. Unit 4 covers 5 core topics and 4 option topics, of which you should study one. These two chapters cover the specified practical work and mathematical discussion of the topics. The mathematical and data-handling demands of the A2 course are higher than those of the AS; Chapter 3 builds on the Practical and Mathematical skills chapters of the AS book and provides background material for the demands in the topics of Unit 3 and 4. Please note that material covered in Chapters 3 and 4 of the AS Student book is not repeated in this A2 book.

Level of coverage

This book contains material which is examined only in the A2 units of the course. The level of coverage in both the main body of the text and the practice questions includes concepts and treatment designed to stretch and challenge students.

Practice questions

As well as the Knowledge check (KC) questions in the margin of the main text, each section in Units 3 and 4 ends with a Test yourself (TY) exercise. The mathematical and data-handling skills chapter also includes both KC questions and a TY exercise, which are not tied to specific subject content but are designed to allow you to practise these essential skills.

As well as containing material relating to the content of the sections, the exercises in Units 3 and 4 contain data-analysis questions around the specified practical work for the unit.

Some questions also relate to the content of more than one unit: the A level examination requires candidates to answer such synoptic questions, which bring together a range of ideas from the physics course. The solutions to these exercises as well as the KC questions are to be found at the end of the book. Unless the KC or TY question asks for an explanation or reasoning to be shown, the solutions only give a final answer.

Caution: Questions which ask for explanations or other working can often be answered in various ways, all of which gain full credit. The solutions given in this book cannot cover all the possibilities.

Examination practice questions

At the end of each of Units 3 and 4, you will find a set of Exam practice questions. These are all taken from past WJEC examinations. They do not cover all the topics but do reflect a range of styles of question; all the option topics are included. A set of sample answers to these questions is given at the end of this book. Again it should be noted that there is often more than one legitimate way of answering a question – this especially applies to QER questions.

Section B of Unit 3 of the A level examination consists of a structured set of questions based upon a passage, i.e. a comprehension question. Unit 5 of the examination consists of a practical examination. Whilst this book gives important background for your preparation for these units, they are not discussed in examination terms.

Margin features

As in the AS Student book, the margin of each page holds a variety of features to support your learning. These are:

Key term

Avogadro constant: N_A: the number of particles per mole ($N_A = 6.02 \times 10^{23}$ mol^{-1})

These are definitions or laws which you should be able to state. Physics exam papers often contain a few marks for stating key terms.

Study point

ω always has a positive value, so we don't need to write $\omega = \pm 20$ s^{-1}

These are provided to help you understand and use the knowledge content. In this feature, factual information may be emphasised or restated to enhance your understanding.

3.1.4 Knowledge check

Derive the unit of volume by considering a cube.

Knowledge check (KC) questions are short questions to check your understanding of the subject material as you learn it or revise it. You are strongly advised to do these questions *as you work through the book*. Many of them ask for calculations related directly to the text alongside. The answers to all the KC questions are in the back of the book.

Stretch & challenge

Given the mean molecular energy at room temperature and considering the typical separation of the atomic energy levels (see AS Student book), explain why molecular collisions in monatomic gases are elastic.

These are designed to make you think more deeply about the subject. They usually include questions for you to consider and discuss. Unlike the KC questions, the answers are not provided.

Top tip

The equation $C = \frac{C_1 C_2}{C_1 + C_1}$ is not in the WJEC Data Booklet.

These are to provide extra help in preparing for your examination.

Maths tip

Make sure you can use these relationships:

$\ln(e^x) = x \qquad e^{\ln x} = x$

$\ln(2^x) = x \ln 2 \qquad e^{x \ln 2} = 2^x$

These refer to particular techniques and often direct you to the Mathematics and data-handling chapter.

Link

See Section 1.4.3(d) of the AS textbook for more on potential energy.

Links to other sections of the course are highlighted in the margin, near the relevant text. They are accompanied by a reference to areas where sections relate to one another. It may be useful for you to use these Links to recap a topic before beginning to study the current topic.

Units and calculations

Apart from the determination of a ratio, the results of all calculations in physics are values with units. If you leave out the units in the final answer, you will be penalised. Usually one or two questions in each paper have a unit penalty attached. Lack of units is not penalised in intermediate stages of a calculation.

How should you deal with units during the calculation? The authors strongly suggest that you include the units and have usually done so in the worked examples. On page 13, you'll see the following line in the **Example**:

$$v_{max} = \sqrt{20 \text{ m} \times 9.81 \text{ m s}^{-2}} = 14 \text{ m s}^{-1}$$

Using the units m and m s^{-2} provides a check that the method is correct because the unit of $20 \text{ m} \times 9.81 \text{ m s}^{-2}$ is $(\text{m} \times \text{m s}^{-2}) = \text{m}^2 \text{ s}^{-2}$. So taking the square root gives m s^{-1} which agrees with the unit of the answer.

We haven't kept to this 100% because we'd run out of line length. This is particularly the case in the answers to the Examination practice papers. An example is the answer to 3(c) at the top of page 285. Putting in the units would be ideal but it would require two extra lines:

$$\Delta T = \frac{m_{magnet} g}{A \rho_{copper} c_{copper}}$$

$$= \frac{0.300 \text{ kg} \times 9.81 \text{ N kg}^{-1}}{7.85 \times 10^{-6} \text{ m}^2 \times 8960 \text{ kg m}^{-3} \times 385 \text{ J kg}^{-1} \text{ K}^{-1}}$$

$$= 0.11 \text{ K}$$

Why would this be ideal? Because (remembering that J = N m) it lets us check that the units work out.

Units in exponential, logarithmic and trigonometrical functions are also awkward. In all of these functions the *argument*, e.g. $(-\lambda t)$ in $e^{(-\lambda t)}$, has no units, so we just need to be sure that the units of λ and t are compatible. This means that if t is expressed in s, then λ must be in s^{-1}, if you are expressing t in years, then λ has to be in year^{-1}.

The A2 examination

In writing the examination papers, the examiners take care to reflect the range of skills which A level candidates are expected to show. So they include: questions from all areas of the specification; a balance of mathematical and non-mathematical questions; theoretical and practical contexts; questions with definite answers and those seeking reasoned argument. An overarching requirement is for them to set questions which reflect the following three **Assessment objectives (AOs)**.

Assessment objective 1

Learners must: Demonstrate knowledge and understanding of scientific ideas, processes, techniques and procedures

30% of the marks for questions set on the examination papers are for AO1. As well as pure recall, such as stating laws and definitions, this includes knowing which equations to use, substituting into equations and describing experimental techniques.

Assessment objective 2

Learners must: Apply knowledge and understanding of scientific ideas, processes, techniques and procedures:

- in a theoretical context
- in a practical context
- when handling qualitative data
- when handling quantitative data.

45% of the marks for questions set on the examination papers are for AO2. Bringing together ideas to explain phenomena, solving mathematical problems and performing calculations using experimental results and graphs are categorised as AO2. Application involves using the skills you have acquired in situations you have not previously encountered, e.g. in synoptic questions.

Assessment objective 3

Learners must: Analyse, interpret and evaluate scientific information, ideas and evidence, including in relation to issues, to:

- make judgements and reach conclusions
- develop and refine practical design and procedures.

25% of the marks for questions set on the examination papers are for AO3.These marks include determining quantities using experimental results and also responding to data to draw conclusions.

A2 Physics written papers

There are three units in the A2 Physics examination: Unit 3 and Unit 4, which have a weighting of 25% each in the A level and Unit 5, Practical Physics, which is worth 10%.

Unit 3 (2 hours 15 minutes)

The paper has two sections:

Section A consists of approximately six structured questions, with a total of 80 marks. The questions comprise a mixture of short and extended writing.

Section B consists of one 20-mark structured comprehension question based upon a passage.

Unit 4 (2 hours)

The paper has two sections:

Section A consists of approximately five structured questions, with a total of 80 marks. The questions comprise a mixture of short and extended writing.

Section B consists of four structured 20-mark questions, one on each of the four options.

Time and mark allocation

The time allocation is 1¼ minutes per mark with an additional 15 minutes allowed for reading the passage in Unit 3. Allowing yourself one minute per mark gives you 20 minutes for checking over your work, for making sure you have not missed out any question parts.

Synoptic questions

Each A2 unit contains some question parts which examine knowledge from the AS units and the other A2 unit. In many areas of the subject, the A2 course builds naturally on AS content, e.g. capacitor circuits in A2 builds upon DC circuits in AS. Hence revision using the AS Student book is advised.

Question types and command words

The question styles and command words used in A2 Physics are exactly as in AS Physics. For a discussion of these, see pages 6 and 7 of the AS Physics Student book.

Unit 3

Oscillations and nuclei

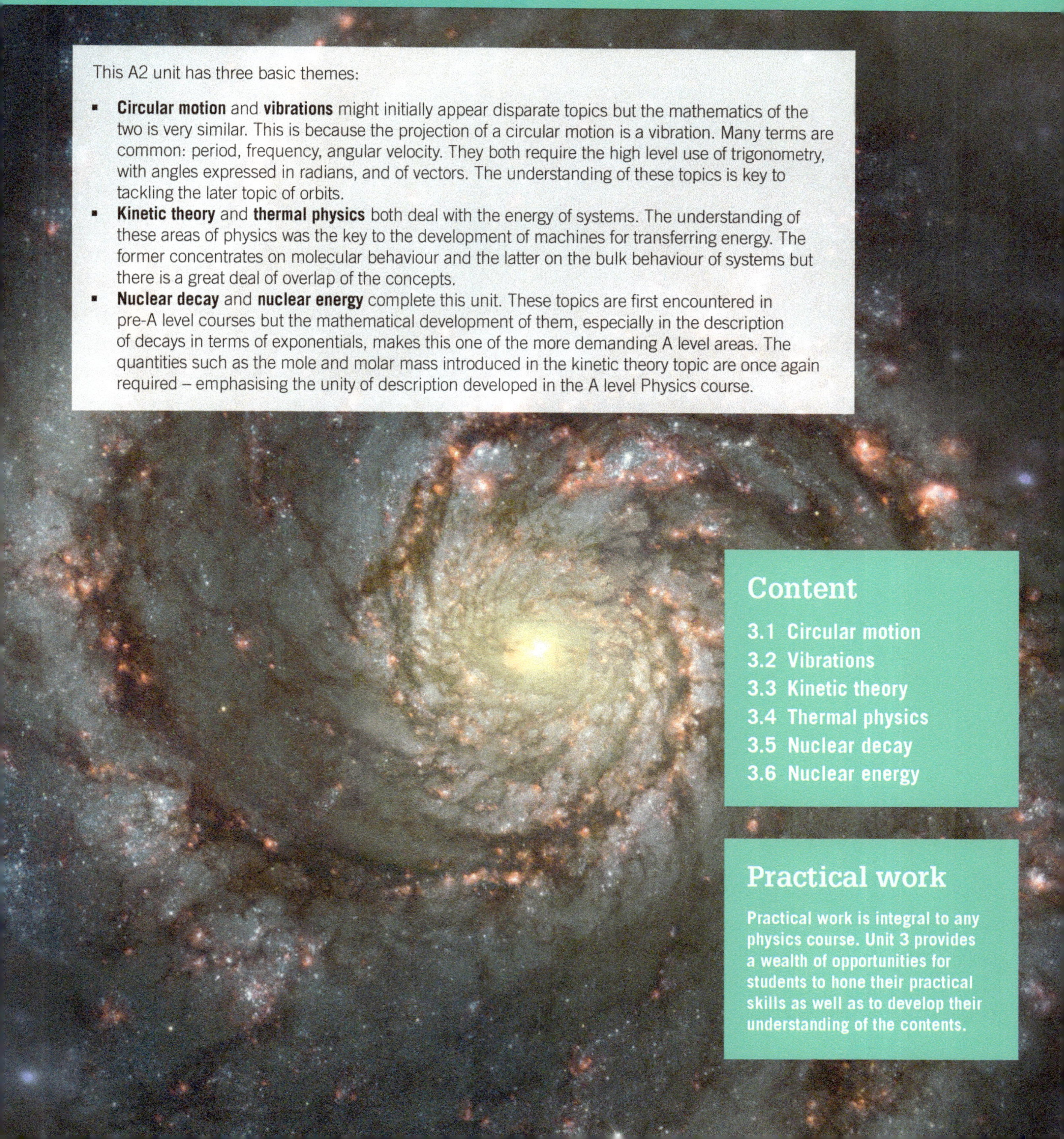

This A2 unit has three basic themes:

- **Circular motion** and **vibrations** might initially appear disparate topics but the mathematics of the two is very similar. This is because the projection of a circular motion is a vibration. Many terms are common: period, frequency, angular velocity. They both require the high level use of trigonometry, with angles expressed in radians, and of vectors. The understanding of these topics is key to tackling the later topic of orbits.
- **Kinetic theory** and **thermal physics** both deal with the energy of systems. The understanding of these areas of physics was the key to the development of machines for transferring energy. The former concentrates on molecular behaviour and the latter on the bulk behaviour of systems but there is a great deal of overlap of the concepts.
- **Nuclear decay** and **nuclear energy** complete this unit. These topics are first encountered in pre-A level courses but the mathematical development of them, especially in the description of decays in terms of exponentials, makes this one of the more demanding A level areas. The quantities such as the mole and molar mass introduced in the kinetic theory topic are once again required – emphasising the unity of description developed in the A level Physics course.

Content

Practical work

Practical work is integral to any physics course. Unit 3 provides a wealth of opportunities for students to hone their practical skills as well as to develop their understanding of the contents.

3.1 Circular motion

Fig. 3.1.1 Circular motion in Barcelona

Fig. 3.1.2 Wall of death rider

Every six seconds or so, the excited riders on the swing carousel in Barcelona travel along a horizontal circle about 20 m in diameter. Nor are they alone in this behaviour as the wall of death driver shows. We, along with other denizens of the Earth, execute a multiplicity of motions that, if not exactly, are close to circular:

- Daily rotation around the axis of the Earth
- Annual path of the Earth around the Sun
- 240 million-year orbit of the Solar System around the galactic centre.

As this section is being written, astronomers are celebrating the detection of the gravitational waves which were emitted when two black holes, previously in mutual orbit, coalesced into a single body. This one observation simultaneously confirmed the outstanding prediction of General Relativity theory and provided the first direct observational evidence for black holes themselves (rather than the matter which spirals into them).

To get back to Earth, this section introduces the mathematics needed to describe and understand the behaviour of circling objects – their motion and the forces which produce it. The mathematics is not difficult but the physics concepts contain some traps for the unwary. You have been warned!

3.1.1 The kinematics of 'uniform' circular motion[1]

For the most part we shall limit our study to objects which are moving at a constant speed in circles. This is not unduly restrictive – we shall still allow the swing carousel to spin up and down but only consider the forces and motion when it is at a steady speed. Option C, The physics of sports, considers angular accelerations and decelerations.

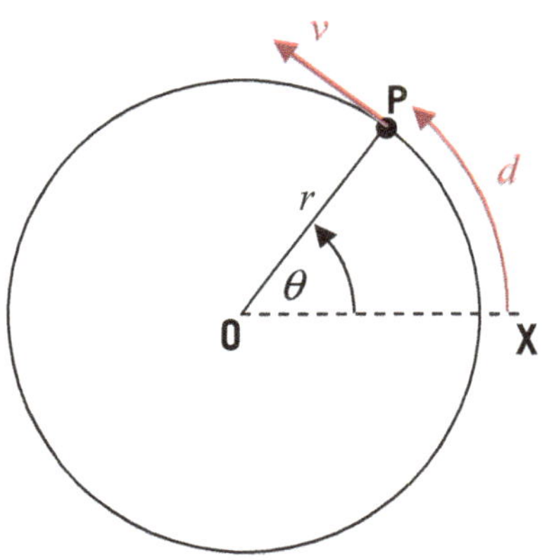

Fig. 3.1.3 Angular position

(a) Angular position and speed

Consider an object, **P**, which is moving at a constant speed, v, in a circular path of radius r about a central point, **O**. We measure its linear motion along the circumference from the point **X.** The angular position, θ, as shown is measured from the line **OX**.

We define the **angular speed,** ω, as the rate of change of θ, i.e. the angle swept out by the radius per unit time.

Mathematically: $\omega = \dfrac{\Delta\theta}{\Delta t}$

Units

Scientists often express angles in degrees (°) for reasons of convenience and the familiarity of the audience with these units. Astronomers often subdivide degrees into minutes (') and seconds (") of arc. However, for rotational and vibrational purposes we shall always use radians. The definition of radian measure is covered in the Year 1 / AS textbook, Section 4.2.4. Hence:

Unit of angular position, θ: rad.

Unit of angular speed, ω: rad s^{-1}.

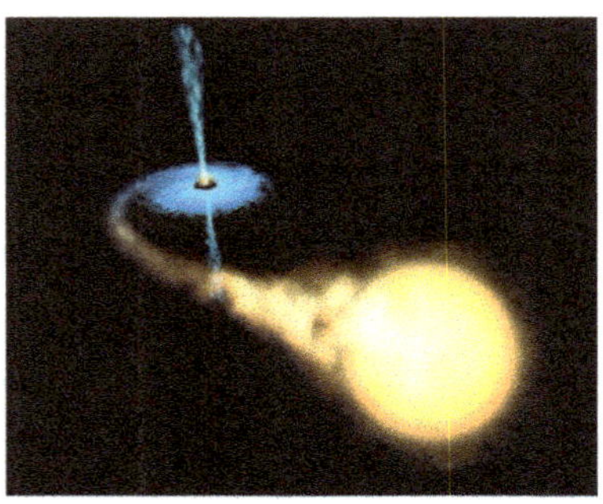

Fig. 3.1.4 Multiple orbits in a star / black hole system

1 The quotation marks around 'uniform' are because an object which is travelling in a circle has not got a uniform velocity because the direction is changing – something to be aware of.

Example

A vinyl record disc spins at a constant $33\frac{1}{3}$ revolutions per minute. Calculate the angular speed in rad s^{-1}.

Answer

In 60 s, the angle swept out $= 33\frac{1}{3} \times 2\pi = 66.7\pi$ rad

$\therefore \omega = \frac{\Delta\theta}{\Delta t} = \frac{66.7\pi \text{ rad}}{60 \text{ s}} = 3.49 \text{ rad s}^{-1}$.

(b) Relationships between angular and linear speed

The linear quantities in circular motion are indicated in red in Fig. 3.1.3; we can use this diagram to relate them to the rotational quantities.

By definition $\theta = \frac{d}{r}$

If **P** is moving with speed v, the increase in distance, Δd, in time Δt is given by $\Delta d = v\Delta t$

So the angular position increase, $\Delta\theta = \frac{v\Delta t}{r}$

$\therefore$ Dividing by Δt and using $\omega = \frac{\Delta\theta}{\Delta t}$ gives $\omega = \frac{v}{r}$

Example

Calculate the speed of the Earth's motion around the Sun. [1 AU = 1.50×10^{11} m]

Answer

The angular velocity $\omega = \frac{\Delta\theta}{\Delta t} = \frac{2\pi}{(3600 \times 24 \times 365.25) \text{ s}} = 1.99 \times 10^{-7} \text{ rad s}^{-1}$.

$\therefore$ The speed, $v = r\omega = 1.50 \times 10^{11} \text{ m} \times 1.99 \times 10^{-7} \text{ rad s}^{-1}$
$= 29.9 \text{ km s}^{-1}$.

Angular speed or angular velocity?

Motion in a circle is directional so it has vector properties. About any axis, we can have clockwise or anticlockwise motion. And the direction of the axis of rotation could change. This is explored in greater detail in Option C, The physics of sports. Although we are only dealing with the scalar aspects, we shall use the terms *angular speed* and *angular velocity* interchangeably.

Study point

It is often convenient to leave π explicitly in an angular speed, e.g. in the example $\omega = 1.1\pi$ rad s^{-1}.

Knowledge check 3.1.1

A rotating reel of diameter 15 cm winds in thread at a speed of 9.5 cm s^{-1}. Calculate the angular velocity of the reel.

Knowledge check 3.1.2

The sound waves on a $33\frac{1}{3}$ rpm vinyl record are produced using wavy grooves.

(a) Calculate the wavelength of the grooves for a 1 kHz note at the outside of the 30 cm diameter disc.

(b) How would a 1 kHz groove be different near the centre of the disc?

Study point

You'll have noticed that you have to calculate the number of seconds in a year quite often in A level physics. You could remember that 1 year = 3.16×10^7 s (to 3 sf). The author finds this too difficult but remembers that there are 86 400 seconds in a day (3600×24) and multiplies by 365.25 for the number of days in a mean year.

(c) Period, T, and frequency, f, of circular motion

The definitions of **period** and **frequency** in the context of circular motion are essentially the same as for waves or any other periodic phenomena. The relationship between them is:

$$\text{frequency} = \frac{1}{\text{period}} \qquad f = \frac{1}{T}$$

In one revolution, the angular displacement is 2π. We can use this to relate the period and frequency to the angular velocity, ω.

$$\omega = \frac{2\pi}{T} \quad \text{and} \quad \omega = 2\pi f$$

These relationships will occur in the same form for vibrational motion. We shall explore the connection at the end of this section.

Key terms

The **period**, T, of circular motion is the time taken for one complete revolution.

UNIT: s

The **frequency**, f, is the number of revolutions per unit time.

UNIT: Hz ≡ s^{-1}

3.1.2 The existence of a resultant force on an orbiting object

Consider the forces on one of the fairground riders in Fig. 3.1.1. You may find it easier to use Fig. 3.1.5. Once the carousel has settled down to a steady angular velocity, the riders move in horizontal circles so they have no vertical motion. The vertical components of the forces on the riders therefore sum to zero. If we ignore air resistance, there are just two forces on the rider: the tension, F, in the support wire and the gravitational force, mg, on the rider.

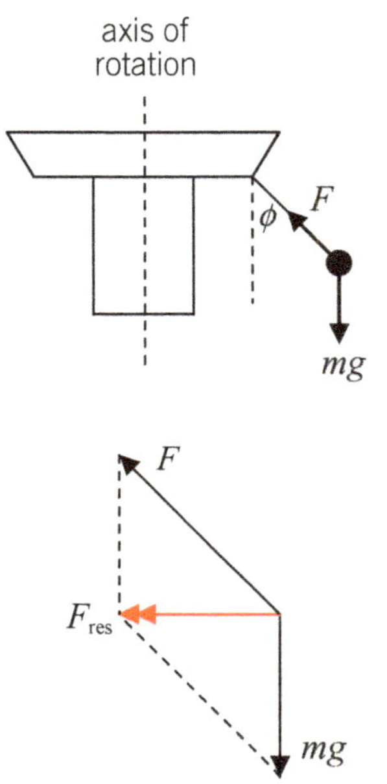

Fig. 3.1.5 Forces on a swing carousel rider

The vertical component of the tension is $F\cos\phi$

$\therefore$ Hence $F\cos\phi = mg$ i.e. $F = \frac{mg}{\cos\phi}$

But F also has a horizontal component, $F\sin\phi$ to the left. As there is no resultant vertical force on the rider and this is the only horizontal force on the rider, we conclude that the resultant force, F_{res} on the rider is, at the instant of Fig. 3.1.6 $F\sin\phi$ horizontally to the left.

Now consider the situation half a revolution later. The rider is on the left and F_{res} is now a force to the right: there is always a resultant force on the rider which is in the direction of the centre of the circle. In the other examples too:

- The wall of death in Fig. 3.1.2 exerts an inward force on the driver as well as a frictional force to hold him up against gravity.
- The black hole and star in Fig. 3.1.4 exert attractive gravitational forces on each other.

We'll now consider the necessity for such a resultant force and how it depends upon the characteristics of the motion.

We will also look at why there is always a resultant force towards the centre for an object in a circular orbit. We'll derive equations to show how the force depends upon the object's speed and the radius of its orbit.

Note that you do not need to be able to derive these equations but you are advised to study Section 3.1.3 carefully.

3.1.3 Knowledge check

The mass of the carousel rider is 75 kg and the angle ϕ of the support cords is 50°.

Calculate:

(a) the tension force, F, in the cord, and

(b) the resultant force on the rider.

3.1.3 Centripetal acceleration

Consider the object in Fig. 3.1.6. It is moving at a constant speed of $20\ \text{m s}^{-1}$ in a circle of radius $5\ \text{m}$. We'll calculate the mean acceleration between points **A** and **B**, i.e. $\frac{\pi}{4}$ either side of the vertical line **OY**.

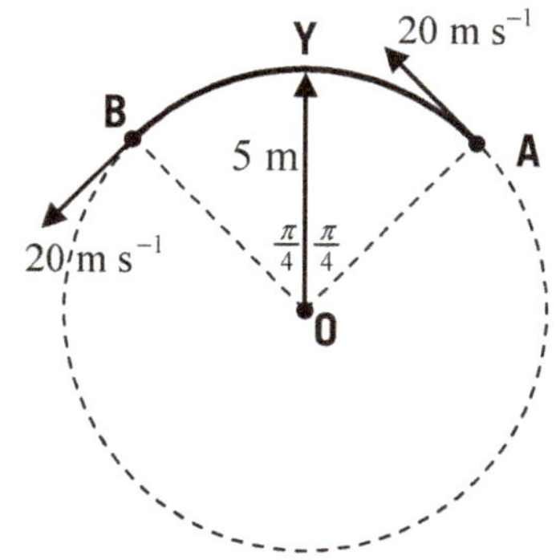

Fig. 3.1.6 What's the acceleration?

The horizontal component of the velocity is unchanged between **A** and **B** at $20\cos\frac{\pi}{4}\ \text{m s}^{-1}$ to the left.

The vertical component of velocity at **A** $= 20\sin\frac{\pi}{4}\ \text{m s}^{-1}$ in direction **OY.**

The vertical component of velocity at **B** $= 20\sin\frac{\pi}{4}\ \text{m s}^{-1}$ in direction **YO.**

$= -20\sin\frac{\pi}{4}\ \text{m s}^{-1}$ in direction **OY**

$\therefore\ \Delta v = v_B - v_A = -20\sin\frac{\pi}{4} - 20\sin\frac{\pi}{4} = -40\sin\frac{\pi}{4}\ \text{m s}^{-1}$ in direction **OY**.

To calculate the mean acceleration we need to know the time interval between **A** and **B**. From the result of KC 3.1.4, we can now write that the mean acceleration $\langle a\rangle$ is given by:

$$\langle a\rangle = \frac{\Delta v}{\Delta t} = \frac{-40\sin\frac{\pi}{4}}{\frac{\pi}{8}} = -72\ \text{m s}^{-2}\ \text{(2 sf) in the direction } \mathbf{OY}.$$

You should check this result. The figure of $72\ \text{m s}^{-2}$ is not so important but the direction is: the mean acceleration is towards the centre of the circle. If you do KC 3.1.5, you will find a result with a very similar magnitude (slightly greater) but in the same direction: there is always a mean acceleration around the point **Y** in the direction **YO**.

We'll now build on this result and calculate the instantaneous acceleration at point **Y** itself. To do this we'll need to use the result:

$$\lim_{\theta\to 0}\frac{\sin\theta}{\theta} = 1,$$

which follows from the definitions of θ (in rad) and $\sin\theta$. Fig. 3.1.7 is just a redrawn Fig. 3.1.6 with general values of the variables.

We could use components again but, for flexibility, we'll use the vector diagram, Fig. 3.1.8, to calculate Δv.

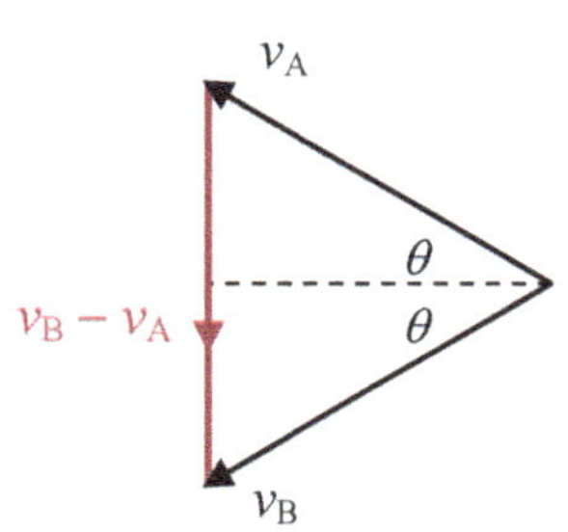

Fig. 3.1.8 $v_B - v_A$

The magnitudes of v_A and v_B are equal and they make the same angles to the horizontal.

$\therefore$ by symmetry, $v_B - v_A$ is vertical

and $\Delta v = v_B - v_A = 2v\sin\theta$

The arc length AB is given by: arc $\text{AB} = 2r\theta$

$\therefore$ The time Δt, between **A** and **B** is given by:

$$\Delta t = \frac{\text{arc AB}}{v} = \frac{2r\theta}{v}$$

Hence the mean acceleration between **A** and **B** is given by:

$$\langle a\rangle = \frac{\Delta v}{\Delta t} = 2v\sin\theta \times \frac{v}{2r\theta} = \frac{v^2}{r} \times \frac{\sin\theta}{\theta}$$

and the direction is downwards in Fig. 3.1.7, i.e. always towards the centre of the circle.

Knowledge check 3.1.4

Use time $= \frac{\text{distance}}{\text{speed}}$ to show that the time interval between **A** and **B** in Fig. 3.1.6 is $\frac{\pi}{8}$ s.

Knowledge check 3.1.5

Repeat the calculation of the mean acceleration for two points which are $\frac{\pi}{6}$ on either side of the vertical.

Maths tip

Use your calculator to determine the ratio $\frac{\sin\theta}{\theta}$, for θ (in rad) $= 1.0, 0.1, 0.01, 0.001$ and 0.0001 and show that the ratio approaches 1.

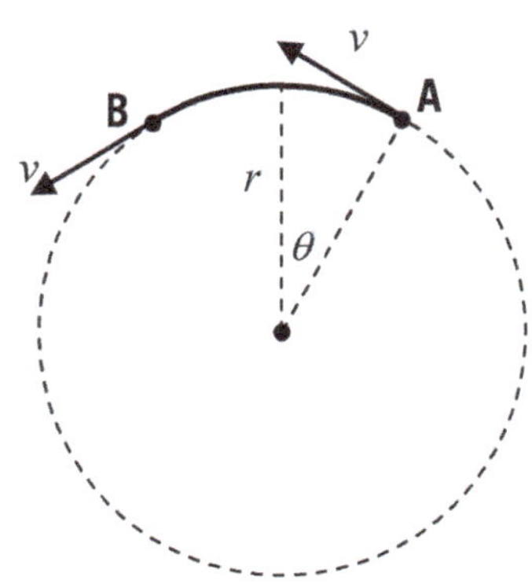

Fig. 3.1.7 Centripetal acceleration

Study point

You might like to check that the formula

$$\langle a \rangle = \frac{v^2 \sin\theta}{r\theta}$$

gives the same answers for the mean acceleration at the beginning of the section.

We can use this result and $\lim_{\theta \to 0} \frac{\sin\theta}{\theta} = 1$ to calculate the instantaneous acceleration at the top of the arc, i.e. when $\theta = 0$. The answer is the very simple equation $a = \frac{v^2}{r}$ which, applying $v = r\omega$, we can write as $a = r\omega^2$

Because this acceleration is directed towards the centre of the circle it is referred to as the **centripetal acceleration** (from the Latin, to seek the centre).

3.1.6 Knowledge check

Use the formula $a = \frac{v^2}{r}$ to calculate the acceleration at point **Y** in Fig. 3.1.7.

Example

The international space station (ISS) orbits at a mean height, h, of 412 km with an orbital period, T, of 92.69 minutes. Calculate its acceleration.

[Mean radius of the Earth, $R_E = 6371$ km.]

Answer

The angular velocity of the ISS $= \frac{2\pi}{T}$ $\therefore$ $a = r\left(\frac{2\pi}{T}\right)^2 = \frac{4\pi^2}{T^2}(R_E + h)$

$$= \frac{4\pi^2}{(92.69 \times 60 \text{ s})^2}(6371 + 412) \times 10^3 \text{ m}$$

$$= 8.66 \text{ m s}^{-2} \text{ (3 sf). (See Study point)}$$

Note that this acceleration is the value of the acceleration due to gravity at the height of the ISS orbit.

Study point

It is no accident that the ISS acceleration is just a little smaller than g at the surface of the Earth. This is because of the inverse square law of gravitation and will be taken up in Section 4.2.

3.1.4 Centripetal force

We can now use Newton's laws to move from kinematics to dynamics; we can explore what makes a body move in a circle at a steady speed. Newton's first law tells us that, if an object is accelerating, the resultant of all the forces on it cannot be zero; there must be a resultant force in the direction of the acceleration, i.e. towards the centre of the circular path. We can calculate the magnitude of this **centripetal (resultant) force** using the N2 equation, $F = ma$.

From above: The centripetal force, $F = \frac{mv^2}{r}$ or $F = mr\omega^2$.

Some examples of centripetal forces

1. Every one of the HI regions (the pink clouds in M51 – the spiral galaxy in Fig. 3.1.9) is acted on by the gravitational fields of all the other objects in the galaxy. The resultant force on each HI region (as well as on the stars) is towards the centre of the galaxy.
2. The resultant of the tension and gravitational forces on the riders on the swing carousel is towards the axis of the carousel.
3. The sideways grip of a car's tyres on a horizontal road as it negotiates a bend (Fig. 3.1.10) is towards the centre of the bend.

It is worth examining that last one. Tyres have a maximum grip, F_{max}, The maximum grip of tyres depends on how heavy the vehicle is and the state of the road surface. Typically, the magnitude of the maximum grip is about equal to the car's weight (mg) on a dry road. It can fall to about one fifth of this value in wet conditions. Worn tyres can reduce the grip even more 'in the wet'.

Fig. 3.1.9 M51 – the Whirlpool galaxy

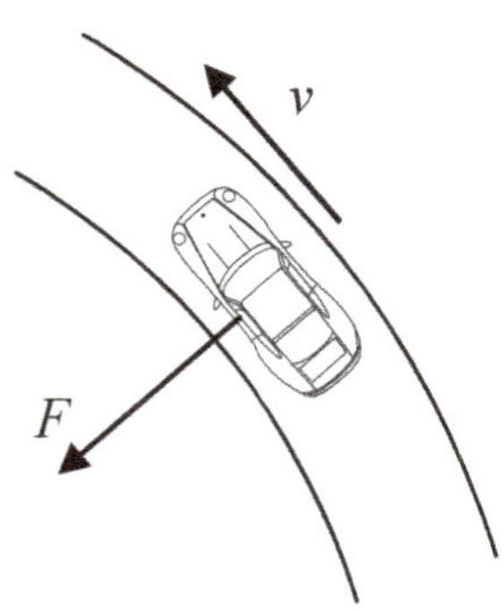

Fig. 3.1.10 Car on a bend

Example

Estimate the maximum safe speed for a car to drive around a tight bend on a country lane, which has a radius of curvature of 20 m, if the road is dry.

Answer

Assuming that the maximum grip is equal to the car's weight:

The only horizontal force on the car with a component along the radius of the bend is provided by the grip on the tyres. So this provides the necessary centripetal force.

The maximum grip $F_{max} = mg$

With a speed of v, the centripetal force $= \frac{mv^2}{r} = \frac{mv^2}{20\text{ m}}$

∴ At the maximum speed $\frac{mv^2}{20\text{ m}} = m \times 9.81\text{ m s}^{-2}$

∴ Cancelling and rearranging: $v_{max} = \sqrt{20\text{ m} \times 9.81\text{ m s}^{-2}} = 14\text{ m s}^{-1}$

Knowledge check 3.1.7

Repeat the calculation in the example for a wet road for which the maximum grip is 0.2 × the car's weight

Knowledge check 3.1.8

Why does the answer to the example and KC 3.1.7 not depend upon the mass of the car?

3.1.5 Two common misconceptions

(a) Centrifugal force

When a car goes quickly round a bend, passengers 'feel' a force throwing them to the outside of the bend. Similarly in the fairground ride the 'cage' (Fig. 3.1.11), the punters 'feel' a force pushing them onto the wall. This sensation is interpreted as indicating the existence of an outward acting force, hence *centrifugal* [from the latin: 'fleeing the centre']. This is an understandable interpretation but it is false – there is no outward force, but there **is** an inward force, as we have seen, which somehow the passengers are ignoring. Suppose (for the moment) that there were an outwards pushing force: what external body is exerting this force on the people and what is its N3 partner force?

Fig. 3.1.11 The 'cage'

This misinterpretation is not confined to rotations: Fig. 3.1.12 is of a (cartoon) airline passenger, but it could equally be a car passenger. The aeroplane is accelerating to the left for takeoff. For twenty seconds or so the seat exerts a forward force on the passenger which accelerates him to take-off speed. But what does the passenger 'feel'? He 'feels' pressed backwards into the seat.

Forward acceleration

Fig. 3.1.12 Force on an accelerating passenger

The whole problem is trying to apply Newton's laws of motion in an accelerated frame of reference (accelerating plane or rotating car). It is *possible* to do so by inventing additional, *'fictitious'* forces.[2] However at A level, we'll stick to inertial (non-accelerating) frames of reference.

There is a similar effect with gravity and weight. There is no sensation of weight if gravity is the only force on us, as in the ISS or during a dive from a high board. We know that gravity is exerting its ineluctable tentacles on us but we are only aware of it when we are stopped from falling freely by the upward force of the floor or a chair. So the *downward* sensation of weight arises from the compression of our tissues between the normal contact force (which is upwards) and the sensationless downward force of gravity.

So the apparently outwards force on the cage passengers is a psychological misinterpretation.

2 Another fictitious force, beloved of meteorologists, is the Coriolis force, which they use to explain the circular motion of air around low and high pressure regions.

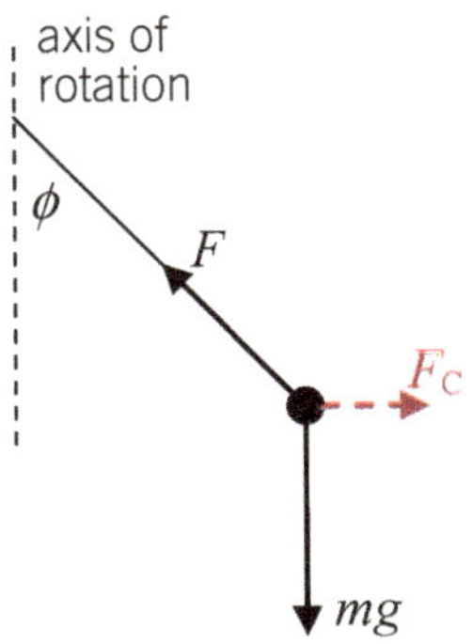

Fig. 3.1.13 Swing carousel rider with fictitious force

Unfortunately, adding this fictitious force can lead to correct results even if the physics is wrong. The red dotted force in Fig. 3.1.13 is this 'centrifugal force'. If a student argues that the three forces, F, mg and F_C are in equilibrium, the triangle of forces can be used to calculate F_C. By a sleight of hand this then becomes the centripetal force which can be equated to $mr\omega^2$: two errors leading to a 'correct' answer.

(b) An additional centripetal force

When solving physics problems, e.g. the swing carousel, students are often tempted to include an *additional* centripetal force over and above the actual forces. The important thing to remember is that there is no separate centripetal force: it is the resultant of all the forces on the rotating object.

Study point

In simple questions, the loop-the-loop car in Fig. 3.1.14 is unpowered and frictionless. However, at the top and the bottom the motion is horizontal, so there is momentarily no changing speed and the acceleration is just the centripetal acceleration. Hence we can use the same equations as for horizontal circles **at these points**.

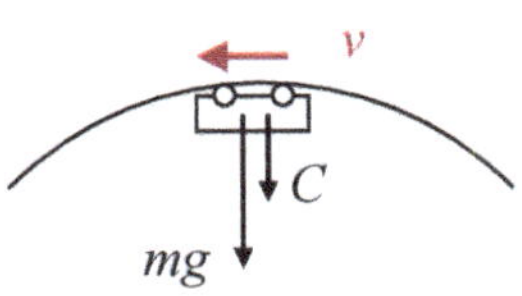

Fig. 3.1.15 Forces at the top of the loop

3.1.9 Knowledge check

A model car track has a radius of 20 cm. Calculate the minimum speed needed at the top of the loop so that the car maintains contact with the track.

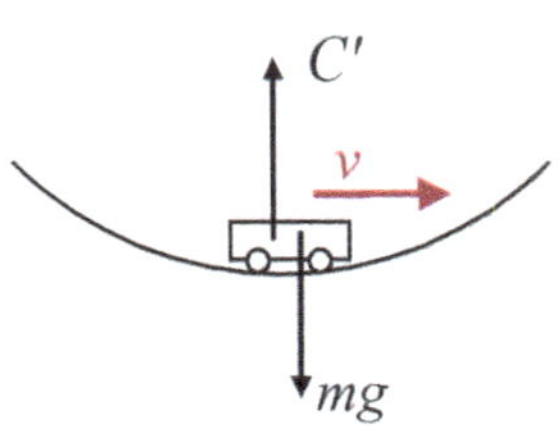

Fig. 3.1.16 Forces at the bottom of the loop

3.1.6 Some nice problems

These 'nice problems' are not strictly on the specification, which only requires horizontal motion in a circle. However, we can solve them using only concepts which **are** on the specification. See the Study point and questions 5 and 6 on page 17.

(a) Loop-the-loop

Whether the question is about toy cars, marbles or an aircraft in flight, the principle of the question is exactly the same.

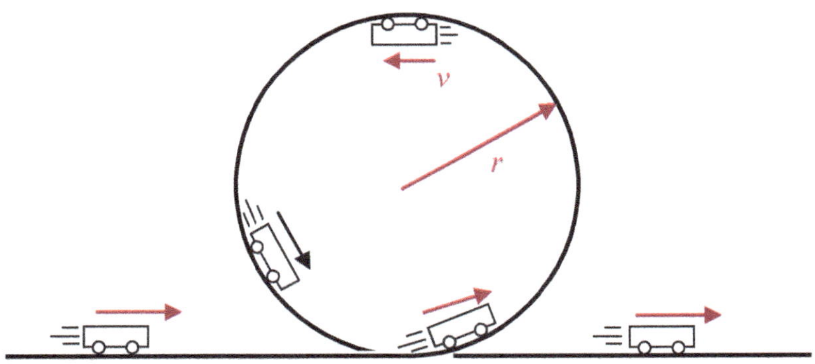

Fig. 3.1.14 Loop-the-loop cars on a track

How fast does the car (or marble, or plane) have to go to stay in the loop?

We'll examine the critical situation at the top of the curve. Supposing for the moment that the car remains on the track, what forces are acting?

- Gravity, mg, downwards
- The normal contact force, C, downwards.

As these are the only forces acting and they are both in the direction of the centre of the circle then the sum of them is the centripetal force.

$$\therefore \quad \frac{mv^2}{r} = mg + C$$

C cannot be less than zero (the track doesn't hold on to the car), so for the car to maintain contact with the track,

$$\frac{mv^2}{r} \geq mg \quad \therefore \quad v \geq \sqrt{rg}$$

At the bottom of the loop (Fig. 3.1.16) the contact force, C', and mg now oppose each other so the resultant towards the centre is $C' - mg$.

Hence $\frac{mv^2}{r} = C' - mg$,

$$\therefore \quad C' = \frac{mv^2}{r} + mg$$

When we derive an equation like this, it is always worth doing a quick check to see if it is reasonable. Looking at the equation, we notice two things:

1. If $v = 0$, then $C' = mg$ as expected, because the car is in equilibrium, so the resultant force on it is zero.
2. As v increases, so does C' as expected, because the resultant force towards the centre (i.e. upwards) must increase.

Example

Find an expression for the speed with which a car enters a loop-the-loop or radius r at the bottom, if it just remains in contact with the loop at the top.

Answer

The speed at the top is given by $v^2 = rg$. So its kinetic energy at the top is $\frac{1}{2}mrg$.

The gain in potential energy as the car climbs the loop = $2mgr$

$\therefore$ The loss in kinetic energy as the car climbs the loop = $2mgr$

$\therefore$ Initial kinetic energy = $\frac{1}{2}mrg + 2mgr = \frac{5}{2}mrg$

$\therefore \frac{1}{2}mv^2 = \frac{5}{2}mrg$, $\therefore$ The initial velocity $v = \sqrt{5rg}$

(b) Newton's law of gravitation

Until the 16th century it was a widely held belief that the scientific laws of the *sublunary sphere* (the geocentric universe below the sphere of the Moon, i.e. the Earth and its atmosphere[3]) were different from those of the rest of creation, which was made of a heavenly material called *quintessence*. Newton showed that one and the same inverse square law related the acceleration due to gravity on the surface of the Earth (i.e. 9.81 m s^{-2}) with the centripetal acceleration of the Moon in orbit around the Earth.

KCs 3.1.10 and 3.1.11 take you through this calculation. Note that the calculation is not 100% accurate; because the Moon doesn't orbit the centre of the Earth but the centre of mass of the Earth–Moon system (see Section 4.3).

Study point

The same principles hold for the roller coaster (Fig. 3.1.17). This time, however, the questions is, 'What is the maximum speed to stay in contact with the track?'

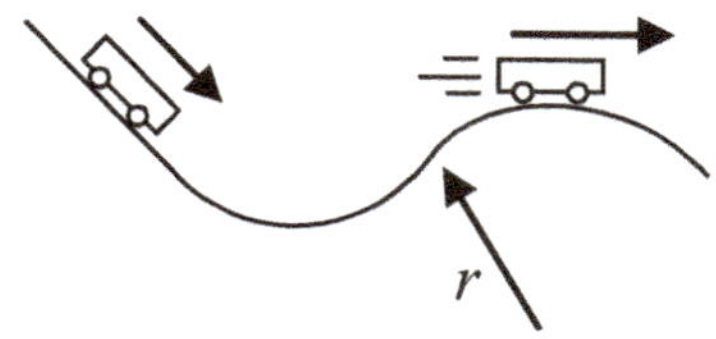

Fig. 3.1.17 Roller coaster

Knowledge check 3.1.10

The mean radius of the Earth is 6371 km. g at its surface in 9.81 m s^{-2}.

Use the inverse square law to estimate the acceleration due to the Earth's gravity at 385 000 km, the orbital radius of the Moon.

Knowledge check 3.1.11

The orbital period of the Moon is 27.3 days. Using the orbital radius from KC 3.1.10, calculate the Moon's centripetal acceleration and compare it with the answer to KC 3.1.10.

3 ... also comets, which were long held to be atmospheric phenomena.

Stretch & challenge

Fig. 3.1.18 The shape of Jupiter

The shape of a rotating planet

What is the shape of the Earth? Apart from a few wrinkles (mountains. continents, oceanic basins…) it is spherical. Well not quite. If you examine the Hubble Space Telescope image of Jupiter in Fig. 3.1.18 and compare its outline with the red circle, you'll notice that it is slightly flattened at the poles. The Earth is similarly shaped, but the deviation from a sphere is less. Why is this? We're not going to allow ourselves to say, 'Centrifugal force,' and swiftly pass on. So, what's the story?

A non-rotating planet would be spherical. This is the minimum potential energy shape. Even dwarf planets – Ceres and Pluto – have sufficient mass to make solid rocks flow into this form. So we'll start by looking at the forces on a hypothetical spherical rotating planet – it might as well be the Earth.

We'll consider the forces on a an object of mass m on the surface of the (spherical) Earth at a latitude, λ, as in Fig. 3.1.19.

Notice that the 'normal' contact force is not directly away from the centre of the Earth. It cannot be because the resultant of C and mg is not zero but points towards the axis of rotation – it is centripetal.However we'll anticipate a result from it.

For $\lambda = \frac{\pi}{4}$, the value of C is almost identical to mg but the angle of C is 0.0017 rad (about 0.1°) away from the 'vertical'.

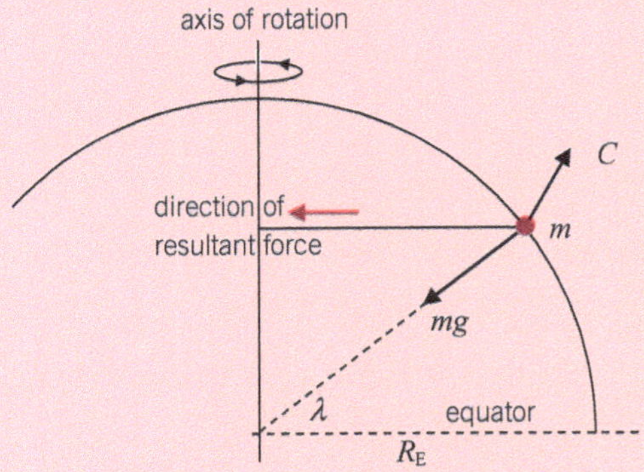

Fig. 3.1.19 Forces on an object at latitude λ

3.1.12 Knowledge check

Given that the rotational period of the Earth is 24 hours and the mean radius is 6731 km. Show that the centripetal acceleration of a point on the equator is approximately 0.04 m s^{-2} (i.e. about 0.4% of g).

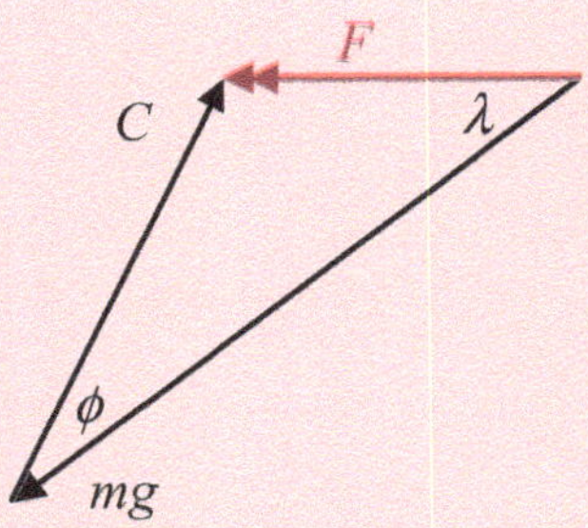

3.1.20 Resultant force

The effect of this is that the horizontal line at the position of the mass is not at right angles to mg but at right angles to C – with the effect that the Earth bulges outwards towards the equator and is flattened at the poles. Note that this undermines (by a small amount) our initial assumption of a spherical Earth! From Fig. 3.1.20 we see that the magnitude of C (i.e. the object's apparent weight) is less than mg: this effect gets smaller towards the pole. There is a secondary effect, which we haven't taken account of: because the Earth now bulges at the equator, actual value of mg is less because the surface is further away from the centre.

Test yourself 3.1

1. A $8.9\ \mathrm{cm}$ diameter hard disk drive rotates at $7200\ \mathrm{rpm}$ (revolutions per minute). Calculate: (a) the angular velocity of the disk, and (b) centripetal acceleration of a point on the rim.

2. A car of mass $1000\ \mathrm{kg}$ travels on a horizontal track at $20\ \mathrm{m\ s^{-1}}$ around a circular bend of radius $80\ \mathrm{m}$.
   (a) Calculate the sideways grip exerted by the tyres.
   (b) The maximum grip of the tyres is $9000\ \mathrm{N}$. Calculate the maximum safe speed of the car without skidding.

3. A planet travels at constant speed in a circular orbit about a star. Ignore any forces from outside the star-planet system.
   (a) Explain why the acceleration of the planet is not zero.
   (b) State the direction of the acceleration of the planet.
   (c) State the direction of the resultant force on the planet.
   (d) The only force on the planet is the gravitational force from the star. State how this relates to your answer to part (c).

4. A load of mass $100\ \mathrm{g}$ is whirled in a horizontal circle on a $50\ \mathrm{cm}$ long horizontal thread, of breaking tension $20\ \mathrm{N}$. Calculate the maximum angular velocity.

5. A teacher uses the apparatus shown to introduce the idea of centripetal force. A small object, of mass m, is whirled in a horizontal circle, of radius r, on the end of a thin string (the red line). Some slotted masses, of total mass M, are hung at the other end of the string. The string passes through a smooth plastic tube. Assume that the tension in the string is constant throughout its length. The speed of rotation, v, is controlled so that the slotted masses remain stationary.

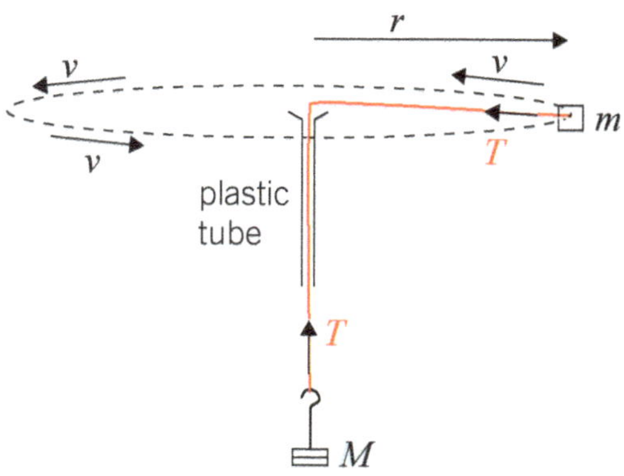

   The string attached to the small mass makes a small angle, θ, to the horizontal.

   (a) Explain why each of the equations (i)–(iii) is valid. [Hint: consider equilibrium]
   (i) $T = Mg$ (1)
   (ii) $\frac{mv^2}{r} = T\cos\theta$ (2)
   (iii) $T\sin\theta = mg$ (3)
   (b) The small object has a mass of $120\ \mathrm{g}$. It is whirled around with a period of $0.80\ \mathrm{s}$ in a circle of radius $1.5\ \mathrm{m}$. Use equations (1) and (2), with the approximation $\cos\theta = 1$, to calculate mass of the slotted masses which need to be used.
   (c) Evaluate whether the approximation in part (b) is valid. [Hint: use the result of part (b) and equation (3) to find a value of θ].
   (d)

   Stretch & challenge

   Answer part (b) without making the assumption that $\cos\theta = 1$. Comment on the value of θ.

6. A metal block of mass $0.50\ \mathrm{kg}$ rests on a turntable at a distance of $15\ \mathrm{cm}$ from the centre.

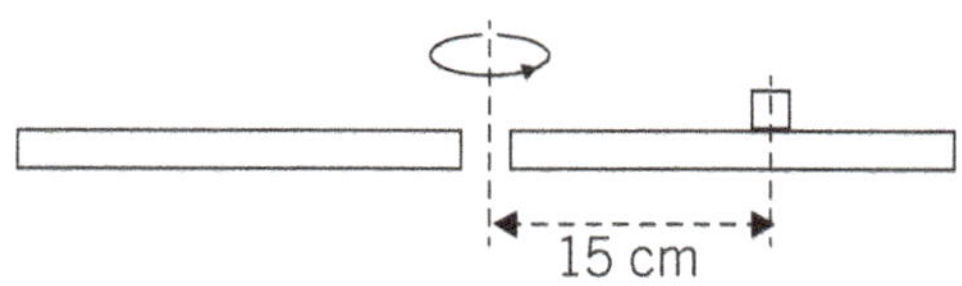

   (a) The turntable rotates at a constant angular speed of $6.0\ \mathrm{rad\ s^{-1}}$. The diagram shows this from the side. The metal block is held in position by a frictional force.
   (i) Explain in which direction this force acts.
   (ii) Calculate the magnitude of the force.
   (b) A small pulley (assumed frictionless) is now added to the turntable and an object of mass $1.00\ \mathrm{kg}$ is attached to the metal block by a thread as shown. The turntable rotates at the same angular speed as before. Calculate the new magnitude and direction of the frictional force.

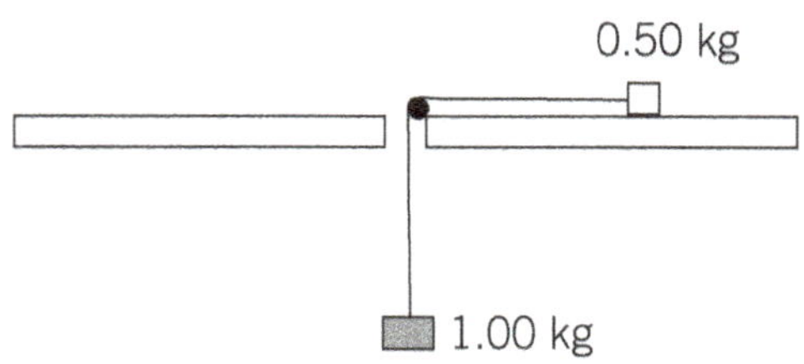

7. The planet Venus orbits the Sun in an almost circular orbit of radius of 1.082×10^{11} km and a period of 224.7 Earth days.

Anwen says that because the distance of Venus from the Sun is constant, its speed must be constant, so there is no resultant force on it. Rhodri says the Sun exerts a force on the planet but doesn't understand why the speed is constant. Discuss the two statements.

Note: Questions 8, 9 and 10 are synoptic questions involving motion in a vertical circle. The centripetal acceleration equations are applicable at the top and bottom of the circles. See Section 3.1.6(a).

8. The vertical loop in a model car set has a radius of 30 cm. A 40 g car enters the bottom of the loop with just enough speed to remain in contact with the loop at the top. Calculate:

(a) The speed at the top of the loop.
(b) The kinetic energy at the top of the loop and the gain in gravitational potential energy from the bottom to the top.
(c) The speed with which the car enters the loop.

9. A model glider executes a loop. Two forces act on the glider: the weight, mg, and the aerodynamic lift, L. The lift depends upon the speed of the glider. These directions of these forces are shown at two places, **A** and **B**, in the loop. In answering this question, you should assume that other forces, e.g. air resistance, are negligible.

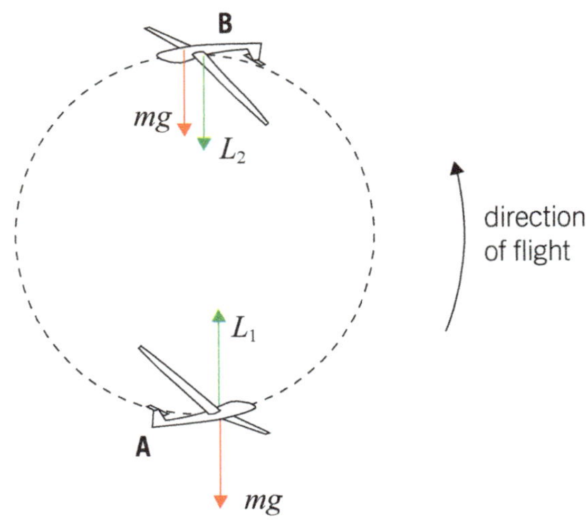

(a) State the direction of the resultant force on the glider at **A**.
(b) State the direction of the resultant force on the glider at **B**.
(c) In terms of L_1 and mg, state the magnitude of the centripetal force on the glider at **A**.
(d) In terms of L_2 and mg, state the magnitude of the centripetal force on the glider at **B**.
(e) Explain, in terms of the principle of conservation of energy, why the speed of the glider at **B** is less than that at **A**.
(f) Discuss, using your answers to (c), (d) and (e) whether the magnitude of L_1 is less than, equal to or greater than that of L_2.

10. A small glass sphere rolls backwards and forwards between points **A**, **B** and **C** in a spherical basin as shown. Assume that no energy is dissipated in the process. **X** is the centre of curvature of the basin.

State the direction of the resultant force on the sphere at each of points **A**, **B** and **C**. Explain your answers.

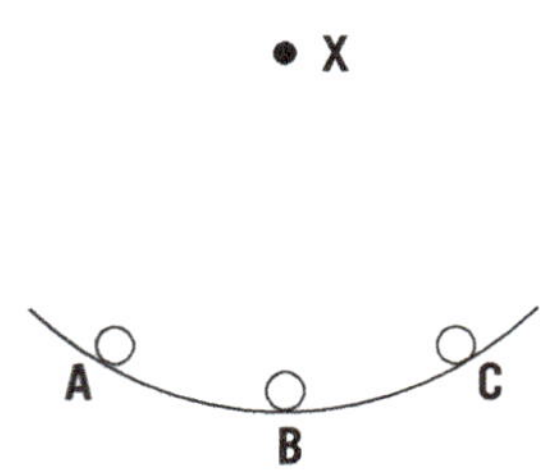

3.2 Vibrations

What is the difference between vibration and oscillations? Not a lot really. An object *oscillates* if it moves with a regular **period**; we tend to use the word *vibrations* for short period oscillations but there is no hard and fast rule. In any case, the study of periodic changes – their characteristics, causes and effects – is an important focus of study in physics. They can cause disasters, often because of resonance effects such as in the collapses of Ferrybridge, Yorkshire, cooling towers (1965) or the bridge over the Tacoma Narrows, Washington State, USA; they can transmit information by acoustic, seismic, electromagnetic and gravitational waves; they form the basis of accurate timekeeping from the pendulum clock to the atomic clock and the pulsar.

A level Physics concentrates on one particular sort of oscillation called **simple harmonic motion**, usually abbreviated to SHM. We shall consider this first.

Key terms

The **period**, T, of an oscillation is the time for one complete cycle.

UNIT: s

Simple harmonic motion occurs when an object moves such that its acceleration is always directed toward a fixed point and is proportional to its distance from the fixed point.

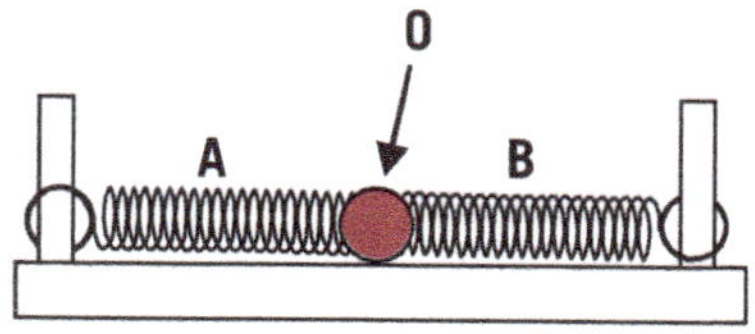

Fig. 3.2.1 Stable equilibrium

3.2.1 Introduction to simple harmonic motion

The red sphere in Fig. 3.2.1 is held in stable equilibrium at point **O** between two springs under tension. We shall consider the horizontal surface to be frictionless. If we move the sphere slightly to the left, the tension in spring **A** will decrease; the tension in spring **B** will increase. If we let the sphere go there will be a resultant force on it to the right and it will therefore move back to its equilibrium position. That's what we mean by stable. What happens when it gets back to **O**? The two tensions are now equal but the sphere is moving, so it carries on to the right – building up the tension in **A** and lowering the tension in **B**. Hence decelerating until it stops and accelerates back to the left...

The tension in the springs in Fig 3.2.1 varies linearly with extension, so the restoring force, F, on the mass is directly proportional to the displacement, x, and in the opposite direction. By N2, so is the acceleration, a, i.e.

$$F = -kx \qquad \text{and} \qquad a = -\frac{k}{m}x$$

where k is the stiffness constant of the arrangement of springs (in fact in Fig. 3.2.1 it is the sum of the spring constants of **A** and **B**). Note that $k > 0$.

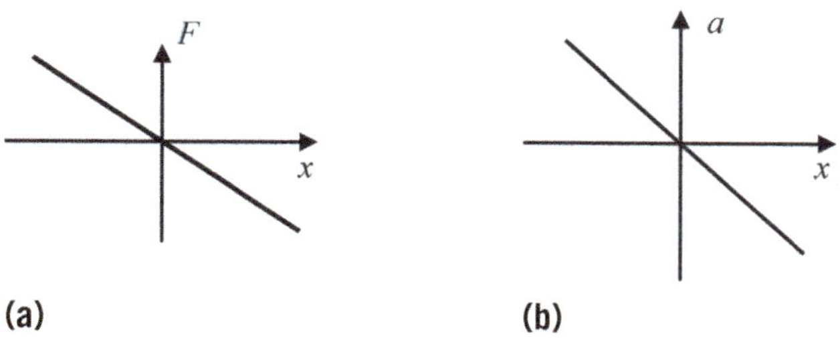

Fig. 3.2.3 (a) F-x and (b) a-x graphs for SHM

The graphs in Fig. 3.2.3 illustrate the definition of simple harmonic motion.

Study point

For vibrational purposes we can consider the molecules of a (crystalline) solid to be small masses linked by springs (which represent the bonds), so our model in Fig. 3.2.1 is not too artificial. See Fig. 3.2.2.

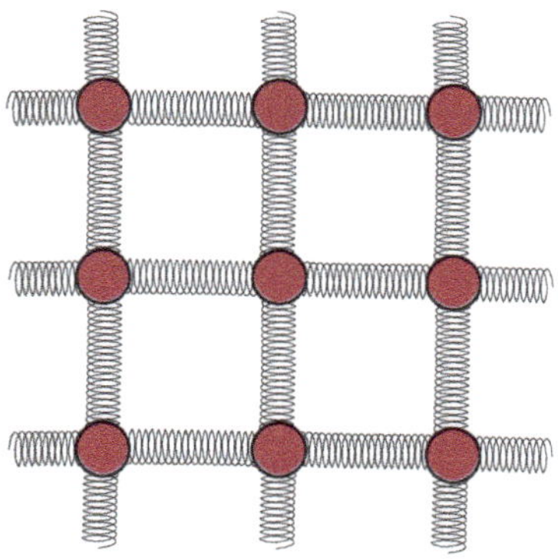

Fig. 3.2.2 Molecules as masses on springs

Knowledge check 3.2.1

In Fig. 3.2.1 the spring constants of **A** and **B** are 200 N m^{-1} and 300 N m^{-1} respectively. The mass is in equilibrium at **O**. The sphere is displaced 2.0 cm to the right and released. Calculate:

(a) the change in the tension, ΔF_A, of spring **A**,

(b) ΔF_B and

(c) the resultant force, F_{res}, on the sphere.

3.2.2 Equations of simple harmonic motion

In this section, we're going to produce the equations for simple harmonic motion, relying heavily on A level mathematics. If you like, you can skip this and just move to the results at the end – **you need to know how to use the equations, but not how to derive them.**

Before we start, we're going to introduce a new quantity, ω, called the **angular frequency**, defined by

$$\omega^2 = \frac{k}{m}.$$

Note that this is the same symbol as **angular velocity** which we met in Section 3.1. The reason that the same symbol is used is that it obeys many of the same equations (see Section 3.2.3).

We can now re-write the equation $a = \frac{k}{m}x$ as

$$a = -\omega^2 x.$$

To help us solve this equation we recall the following two facts from GCSE physics:

- the gradient of a displacement–time graph is the velocity
- the gradient of a velocity–time graph is the acceleration.

Rewriting these in the language of A level maths:

- $v = \frac{dx}{dt}$ and
- $a = \frac{dv}{dt} = \frac{d^2x}{dt^2}$

Hence $x(t)$ must satisfy the equation $\frac{d^2x}{dt^2} = -\omega^2 x$. There are two functions which do this:

$$x = A\cos(\omega t + \varepsilon) \qquad \text{and} \qquad x = A\sin(\omega t + \varepsilon)$$

where A and ε are constants. It is just a matter of convenience which one of these two equations we use because we can always convert one into the other using the fact that $\sin\omega t = \cos\left(\omega t + \frac{\pi}{2}\right)$. We shall always use $x = A\cos(\omega t + \varepsilon)$.

If $\quad x = A\cos(\omega t + \varepsilon)$ [1]

Then $\quad v = \frac{dx}{dt}$, and remembering the rules for differentiating

$\quad v = -A\omega\sin(\omega t + \varepsilon)$ [2]

and $\quad a = \frac{dv}{dt}$ so $a = -A\omega^2\cos(\omega t + \varepsilon)$ [3]

So, comparing [1] and [3], $a = -\omega^2 x$, [4]

which is what we wanted!

Maximum values of displacement, velocity and acceleration

We know that the extreme values of the sine and cosine functions are ±1, so the maximum values of x, v and a are:

$$x_{max} = A \qquad v_{max} = \omega A \qquad a_{max} = \omega^2 A$$

The symbol A is called the **amplitude** of the oscillation.

3.2.2 Knowledge check

The mass of the sphere in KC 3.2.1 is 200 g. Sketch graphs of (a) F against x and (b) a against x.

3.2.3 Knowledge check

Show that the unit of $\sqrt{\frac{k}{m}}$ can be written as s^{-1}.

Key terms

The quantity ω is called the **angular frequency** of the oscillation.

UNIT: s^{-1} [or rad s^{-1}]

Study point

The WJEC specification doesn't give a name for ω, so you will not be asked for it. We shall use the term angular frequency.

Key terms

The **amplitude** of the oscillations of an object is the maximum value of the object's displacement from its equilibrium position.

Example

The resultant force on a body of mass $0.10\ \text{kg}$ is given by $F_{res} = -kx$, where x is the displacement and $k = 40\ \text{N m}^{-1}$. The body is pulled $20\ \text{cm}$ from its equilibrium position and released. Calculate the maximum speed and acceleration.

Answer

The maximum displacement, A, is $20\ \text{cm}$.

$$\omega = \sqrt{\frac{k}{m}} = \sqrt{\frac{40}{0.10}} = \sqrt{400} = 20\ \text{s}^{-1} \text{ (see Study point)}$$

$$\therefore v_{max} = \omega A = 20\ \text{s}^{-1} \times 20\ \text{cm} = 400\ \text{cm s}^{-1}\ [4.0\ \text{m s}^{-1}]$$

$$\text{and } a_{max} = \omega^2 A = (20\ \text{s}^{-1})^2 \times 20\ \text{cm} = 8\,000\ \text{cm s}^{-2}\ [80\ \text{m s}^{-2}]$$

In the next section, we'll examine how the values of the constants, ω, A and ε, in equations [1] – [4] relate to the characteristics of an oscillatory motion. Usually, you will encounter questions in which $\varepsilon = 0$, which makes life simpler. First we'll just present the graphs of x, v and a against t for comparison (with $\varepsilon = 0$).

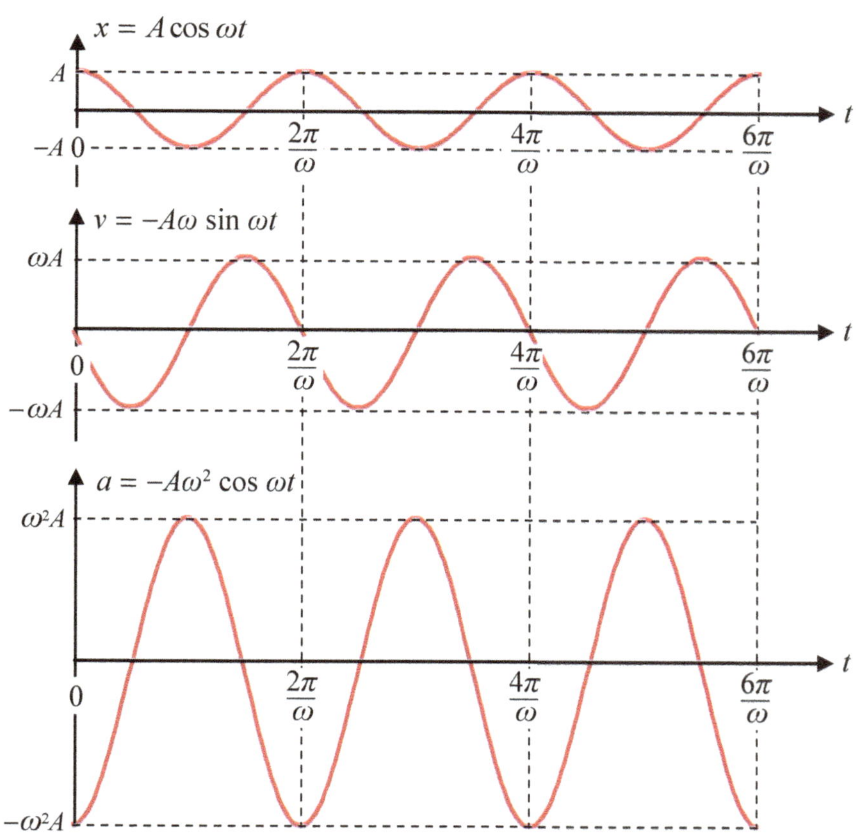

Fig. 3.2.4 Graphs of x, v and a against t

Study point

ω always has a positive value so we don't need to write $\omega = \pm 20\ \text{s}^{-1}$.

Knowledge check 3.2.4

An object oscillates with $x = 2.0 \cos \pi t$, with x in m and t in s. Use Fig. 3.2.4 to help answer the following questions.

(a) Find v at $t = 0$, 0.5 s, 1.0 s, 1.5 s and 2.0 s.

(b) Find a at $t = 0$, 0.5 s, 1.0 s, 1.5 s and 2.0 s.

(c) State the period of the oscillation.

Study point

Note that vertical axes of the graphs in Fig. 3.2.4 are expressing different quantities so the 'heights' of the graphs shown are arbitrary.

Maths tip

When you use trigonometrical functions to analyse vibrations, always make sure that the calculator is in radian mode.

To check this, enter $\cos \pi$. If the calculator returns –1 it is in the correct mode.

3.2.3 Characteristics of simple harmonic motion

Key terms

The angle $\omega t + \varepsilon$ in the equation $x = A\cos(\omega t + \varepsilon)$ is called the **phase** of the oscillation.

The angle ε is called the **phase constant**.

3.2.5 Knowledge check

For the oscillation in Fig. 3.2.5, write down all the times between 0 and 50 μs for which:

(a) the displacement is (i) 0, (ii) 2.0 mm and (iii) −2.0 mm

(b) the velocity is 0

(c) the acceleration is 0.

How does the equation $x = A\cos(\omega t + \varepsilon)$ relate to the amplitude, frequency, period and **phase** of the oscillation? As an example, consider the oscillation represented by the graph in Fig. 3.2.5. We'll see how to write the equation of the motion in the form $x = A\cos(\omega t + \varepsilon)$.

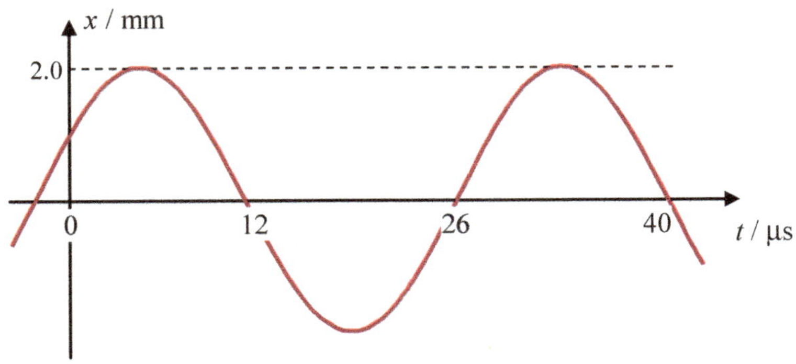

Fig. 3.2.5 Example oscillation for finding the equation

The amplitude and period are clearly 2.0 mm and 28 μs respectively. So the value of A in $x = A\cos(\omega t + \varepsilon)$ is 2.0 mm. In order to establish the other physical characteristics, we need to examine the phase of the oscillation.

(a) The period, T, and frequency, f

Key terms

Frequency: the number of oscillations per unit time (or per second).

UNIT: Hz

The oscillation repeats itself with every increase by 2π in the value of the phase. In other words, if we add T (the period) to any time t, we add 2π to the phase:

So, for any value of t $\quad \omega(t + T) + \varepsilon = \omega t + \varepsilon + 2\pi$

∴ Simplifying $\quad \omega T = 2\pi$

So the period, T, is given by $\quad T = \dfrac{2\pi}{\omega}$ [5]

In this case, the period is 28 μs, so the angular frequency, ω, is given by:

$$\omega = \frac{2\pi}{T} = \frac{2\pi}{28\ \mu s} = 2.2 \times 10^5\ s^{-1} \text{ (to 2 sf)}$$

We can also relate ω to the **frequency** using $f = \dfrac{1}{T}$.

This leads to $\omega = 2\pi f$. [6]

3.2.6 Knowledge check

Calculate the maximum values of (a) the velocity and (b) the acceleration, for the motion in Fig. 3.2.5.

(b) The phase constant, ε

If $\varepsilon = 0$ the graph of displacement against time is a cosine curve, i.e. it has its maximum value at time, $t = 0$. This is clearly not the case in Fig. 3.2.5. To calculate ε, we need to insert known values of the other quantities into the equation $x = A\cos(\omega t + \varepsilon)$. For example, when $t = 5$ μs, $x = 2.0$ mm. Inserting these with the already known values of A and ω gives:

$$2.0 = 2.0\cos\left(\frac{2\pi}{28\ \mu s} \times 5\ \mu s + \varepsilon\right)$$

[Notice we have reverted to the more precise value of ω.]

Simplifying $\quad 1 = \cos(1.12\text{ rad} + \varepsilon)$

So $\quad 1.12\text{ rad} + \varepsilon = \cos^{-1} 1 = \ldots -2\pi, 0, 2\pi, 4\pi \ldots$

3.2.7 Knowledge check

Repeat the calculation for ε by using the values, $x = 0$ when $t = -2.0$ μs. In this case you will get 4 apparent answers in the range $\pm 2\pi$ but you can rule some out, e.g. by considering the value of x at $t = 0$.

It makes sense to choose a value of ε in the range $\pm 2\pi$. In this case this leaves us with

$$\varepsilon = 0 - 1.12 \text{ rad or } 2\pi - 1.12 \text{ rad}$$

i.e. $$\varepsilon = -1.12 \text{ rad or } 5.16 \text{ rad}$$

Combining the results of this and the previous section gives the equation:

$$(x / \text{mm}) = 2.0 \cos(2.2 \times 10^5 (t / \text{s}) - 1.12)$$

This is perhaps an overly fussy way of writing the equation. We could write:

$$x = 2.0 \cos(2.2 \times 10^5 t - 1.12 \text{ rad}) \quad \text{with } x \text{ in mm and } t \text{ in s}$$

However, inserting the units in this has the advantage of clarity and allows us to express the time in other units, e.g.

$$(x / \text{mm}) = 2.0 \cos(0.22(t / \mu\text{s}) - 1.12)$$

(c) The equation for v

We have already seen that, if $x = A\cos(\omega t + \varepsilon)$, then $v(t)$ can be written as $v = -\omega A \sin(\omega t + \varepsilon)$. From the answer to KC 3.2.6, to 2 sf, $\omega A = 450 \text{ m s}^{-1}$, so we can write

$$v = -450 \sin(2.2 \times 10^5 t - 1.12).$$

This is a perfectly acceptable way of writing the velocity equation but we can get rid of the '–' sign. If we add (or subtract) π to the phase, this shifts the graph along to the left (or right) by half a cycle, which turns a positive value of v to the same negative value. This gives us

$$v = 450 \sin(2.2 \times 10^5 t + 2.02)$$

(d) The relationship between velocity and displacement

Equations [1] – [3] in Section 3.2.2 give the time variations of displacement, velocity and acceleration. Equation [4], which is effectively the definition of simple harmonic motion, relates acceleration to displacement. Another useful relationship is the connection between displacement and velocity,

$$v = \pm\omega\sqrt{A^2 - x^2}, \qquad [7]$$

This useful equation is not included in the WJEC equations booklet but its use can simplify many questions and you are advised to learn it. It is derived in the mathematical skills chapter, Section M1.2(c). Equation [7] is clearly consistent with the equation for v_{max}. The speed is maximum when the displacement, x, is zero.

$$\text{maximum speed} = \omega\sqrt{A^2 - 0^2} = \omega\sqrt{A^2} = \omega A$$

Study point

The cosine term could also be $\cos(\omega t + 5.16)$. The difference in the two values of ε is 2π.

Knowledge check 3.2.8

Rewrite the equation for x using the units cm and ms.

Knowledge check 3.2.9

Sketch a v–t graph for the oscillation, labelling significant points.

Hint: rather than starting from the equation, consider the amplitude, period and the phase difference between the x and v graphs (Fig. 3.2.4).

Study point

In $v = \pm\omega\sqrt{A^2 - x^2}$, the maximum possible value of x is A (and the minimum is $-A$). At these points $v = 0$.

Example

An object oscillates with a frequency of 12 Hz and an amplitude of 5.0 cm.

Calculate: (a) the maximum speed of the object and (b) the speed when the displacement is −3.0 cm.

Answer

(a) $\omega = 2\pi f = 2\pi \times 12 = 75.4\ \text{s}^{-1}$

$\therefore\ v_{\text{max}} = A\omega = 5.0\ \text{cm} \times 75.4\ \text{s}^{-1} = 377\ \text{cm s}^{-1}$ [= 380 cm s^{-1} to 2 sf]

(b) $v = \pm\omega\sqrt{A^2 - x^2}$

$\therefore$ When $x = -3.0$ cm, $v = \pm 75.4\sqrt{(5.0)^2 - (-3.0)^2} = \pm 300\ \text{cm s}^{-1}$ to 2 sf

So the speed at −3.0 cm is 300 cm s^{-1}

3.2.4 Examples of simple harmonic motion

Any object for which the resultant force and the displacement are related by $F = -kx$ can oscillate with simple harmonic motion. We shall consider two simple systems. These derivations are not required to be learned for the examination but you will be expected to use the results. The equations for the period of these systems are given in the WJEC Data Booklet.

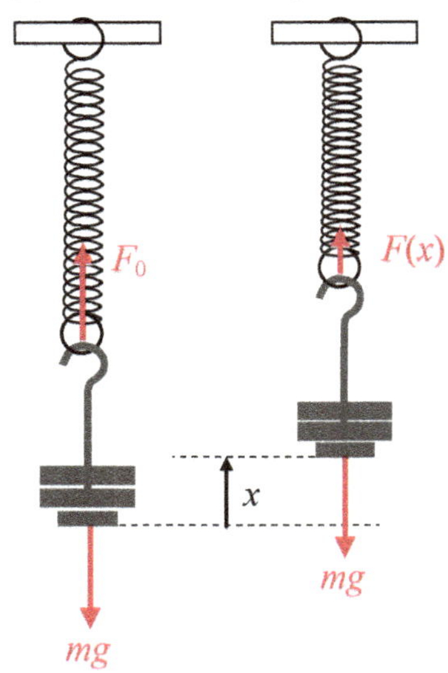

Fig. 3.2.6 Mass on a spring

(a) Oscillating mass on a spring

The spring in Fig. 3.2.6 has a constant k. The spring applies a force F_0 to the mass. The mass, m, in part (a) is in equilibrium, so we can write:

$$F_0 = mg$$

Consider raising the mass and releasing it. It will oscillate vertically. Consider the moment in the oscillation, shown in part (b) of Fig. 3.2.6. The mass is at a displacement x above its equilibrium position. The extension of the spring from its natural length is reduced by x, so the force, $F(x)$, exerted by the spring on the mass is now given by:

$$F(x) = F_0 - kx$$

The resultant upward force, F_{res}, on the mass is given by $F_{\text{res}} = F(x) - mg$

$$\therefore\quad F_{\text{res}} = [F_0 - kx] - mg = [mg - kx] - mg$$

$$\therefore\quad F_{\text{res}} = -kx$$

So the spring oscillates with simple harmonic motion. The values of ω, f and T are given by:

$$\omega = \sqrt{\frac{k}{m}} \qquad f = \frac{1}{2\pi}\sqrt{\frac{k}{m}} \qquad T = 2\pi\sqrt{\frac{m}{k}}$$

The same principle holds for the two-spring system in Fig. 3.2.1. The stiffness constant of the two springs is the sum of the spring constants (see Knowledge check 3.2.1). If friction is negligible, the sphere will execute SHM with $\omega = \sqrt{\frac{k_1 + k_2}{m}}$.

Study point

Notice that the frequency of a mass on a spring is independent of g.

3.2.10 Knowledge check

A 300 g mass oscillates on a spring with a spring constant 30 N m^{-1}. Calculate the angular frequency, frequency and period of the oscillation.

Example

A body of mass 0.5 kg is suspended from a Hookean elastic cord with stiffness constant 50 N m^{-1}. It is raised from its equilibrium point by 5.0 cm and released at time $t = 0$. Find the variation of the body's position, x, as a function of time.

Answer

The body executes simple harmonic motion. The amplitude is 5.0 cm.

$$\omega = \sqrt{\frac{k}{m}} = \sqrt{\frac{50 \text{ N m}^{-1}}{0.5 \text{ kg}}} = 10 \text{ s}^{-1}$$

As the displacement is a maximum at $t = 0$, the equation of motion is

$x = 5.0 \cos 10t$ [with x in cm and t in seconds] .

(b) Simple pendulum

A pendulum bob, of mass, m, on a light inextensible thread of length l (measured to the centre of mass) executes SHM as long as the amplitude of the oscillations is small, e.g. a maximum angle to the vertical of 5°.

Ignoring air resistance, there are two forces on the pendulum bob: the weight, mg, and the tension, F, in the thread. The bob is constrained to travel along the arc of a circle of radius l. The displacement from the centre along this arc is x. Considering tangential components:

Component of F along the arc = 0 [the thread is at 90° to the arc].

Component of mg along the arc = $-mg \sin \theta$ [see Study point].

∴ From N2: $ma = -mg \sin \theta$

For small θ (in radians): $\sin\theta \approx \theta$

Also: by definition, $\theta = \frac{x}{l}$

Combining these and dividing by m: $a = -\frac{g}{l}x$

Comparing this to $a = -\omega^2 x$, this is an SHM equation with $\omega = \sqrt{\frac{g}{l}}$

So the period, T, of a simple pendulum is given by $T = 2\pi\sqrt{\frac{l}{g}}$

We can therefore write $x = A\cos(\omega t + \varepsilon)$ or $\theta = \theta_{max} \cos(\omega t + \varepsilon)$, with the above value of ω. If the oscillation starts from its maximum positive position (e.g. the pendulum is pulled to the side and released) then $\varepsilon = 0$.

Question: What angle of swing is small?

Answer: $\sin (0.2 \text{ rad}) = 0.1987$ which is a difference of 0.65%. So even at this quite large swing (11.5° either side) the difference in period between an actual pendulum and the small-angle approximation is tiny. (See Stretch & challenge on page 26).

Knowledge check 3.2.11

A body of mass 0.20 kg is suspended from two identical springs each with $k = 25 \text{ N m}^{-1}$, arranged vertically one below the other. Assume that both springs remain under tension.

(a) Calculate the distance of the equilibrium point below the midpoint of the two anchor points.

(b) Calculate the frequency of oscillation if the mass is pulled down slightly and released.

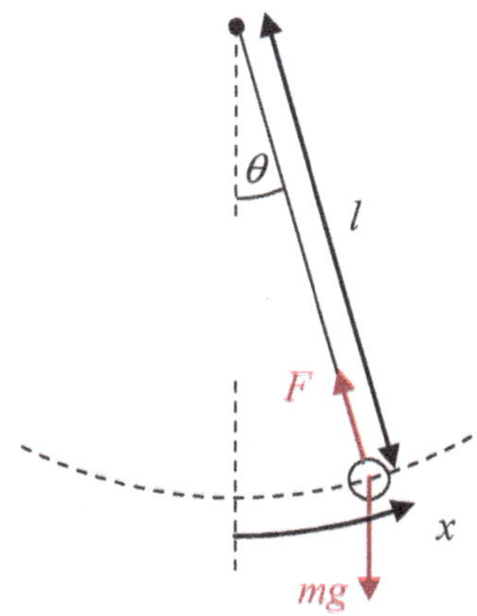

Fig. 3.2.7 Simple pendulum

Study point

The '–' sign in $-mg\sin\theta$ is because the component points in the opposite direction to x.

Knowledge check 3.2.12

A *seconds pendulum* is one of period 2 seconds, i.e. 1 second each way. Historically it was used in clocks.

Calculate the length of the pendulum arm required.

Knowledge check 3.2.13

If the pendulum thread were light but could stretch elastically, why and how would the length of the thread vary throughout the swing? [Qualitative answer]

Stretch & challenge

The period, T, of a simple pendulum for larger angles is related to that of the small-angled pendulum, T_0, by an infinite series:

$$T = T_0(1 + \frac{1}{6}\theta^2 + \frac{11}{3072}\theta^4 + \ldots)$$

Use the second and fourth order approximations to compare the period of a pendulum with $\theta = 0.2$ rad and 0.5 rad with the small-angled pendulum.

Fig. 3.2.8 Loaded cantilever

3.2.14 Knowledge check

Calculate the maximum value of the resultant force on the load on the cantilever in Fig. 3.2.8. State when this first occurs after the load is first released.

(c) General system

As we've seen, all we need to do to work out the period of a system's oscillation is to find out its stiffness, k, and the mass, m, to be oscillated. As long as the oscillating mass can be approximated as a point mass then the period is given by $T = 2\pi\sqrt{\frac{m}{k}}$.

Example

When a load of mass 200 g is attached to the end of the cantilever in Fig. 3.2.8 the latter is depressed by 1.5 cm. The load is then pulled down by another 1.0 cm. Describe in detail the subsequent motion, stating any assumptions.

Answer

Assumptions: 1. The depression is proportional to the load. 2. The mass of the cantilever itself is insignificant. 3. There are no frictional losses.

Calculate k: $k = \frac{F}{x}$. $F = mg$, $\therefore k = \frac{mg}{x} = \frac{0.20 \times 9.81\text{ N}}{0.015\text{ m}} = 131\text{ N m}^{-1}$.

Then $\omega = \sqrt{\frac{k}{m}} = \sqrt{\frac{131\text{ N m}^{-1}}{0.20\text{ kg}}} = 25.6\text{ s}^{-1}$.

So the frequency of the oscillation $= \frac{\omega}{2\pi} = 4.1\text{ Hz}$.

Thus the cantilever oscillates with a frequency of 4.1 Hz and an amplitude of 1.0 cm about its equilibrium position.

The depression d from the equilibrium position at any time is given (in cm) by:
$d = 1.0\cos(25.6t)$

3.2.5 Energy of oscillation

If a system is oscillating without friction or other dissipative forces, the sum of the kinetic and potential energies remains constant. When the displacement is maximum (i.e. $x = \pm A$), the speed is zero, so the kinetic energy, E_k, is zero and the potential energy, E_p, is maximum. When the displacement is zero, the speed is maximum, so E_k is maximum and E_p is minimum. We'll examine how E_k and E_p vary with displacement and time throughout the oscillation.

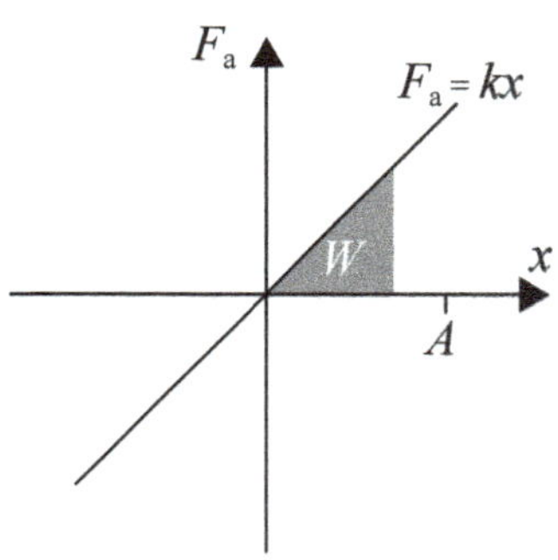

Fig. 3.2.9 Work done in displacing an oscillating system

(a) Variation of E_k and E_p with displacement

To investigate this, we'll consider first the variation of potential energy, E_p, with displacement. To displace the system from its equilibrium position we have to apply a force, F_a, in opposition to the restoring force. The potential energy stored is the work done by F_a, which is the area between the graph and the x-axis. (see Fig. 3.2.9).

Hence, as we saw in your AS studies, the potential energy, E_p, is given by

$$E_p = \frac{1}{2}kx^2$$

with a maximum value of $\frac{1}{2}kA^2$

Link

See Section 1.4.3(d) of the AS textbook for more on potential energy.

At the maximum displacement, A, there is no kinetic energy, so the total energy of vibration must be $\frac{1}{2}kA^2$. This means that the kinetic energy, E_k, varies with displacement according to the equation

$$E_k = \frac{1}{2}kA^2 - \frac{1}{2}kx^2 \quad \text{i.e.} \quad E_k = \frac{1}{2}k(A^2 - x^2)$$

and the graphs of E_k and E_p against x are as shown in Fig. 3.2.10.

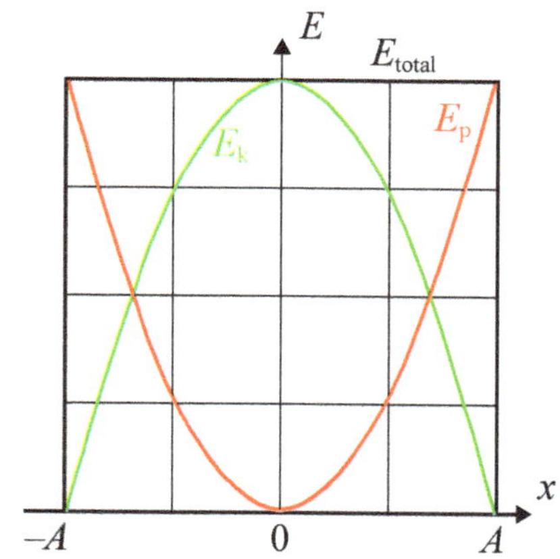

Fig. 3.2.10 Graphs of E_k and E_p against x for SHM

(b) Variation of E_k and E_p with time

Again, we'll start by thinking about the potential energy, E_p. From above:

$$E_p = \frac{1}{2}kx^2$$

To simplify the appearance of the equations a little, we'll assume the phase constant is zero, i.e. the system starts from rest at time zero. So $x = A\cos\omega t$ and we can write

$$E_p = \frac{1}{2}k(A\cos\omega t)^2 = \frac{1}{2}kA^2\cos^2\omega t$$

and $$E_k = \frac{1}{2}mv^2 = \frac{1}{2}m(-A\omega\sin\omega t)^2 = \frac{1}{2}Am\omega^2\sin^2\omega t$$

but $$\omega^2 = \frac{k}{m}, \text{ so } E_k = \frac{1}{2}kA^2\sin^2\omega t$$

and adding the two: $$E_{tot} = E_p + E_k = \frac{1}{2}kA^2(\cos^2\omega t + \sin^2\omega t) = \frac{1}{2}kA^2$$

where, again, we have used the trig form of Pythagoras' theorem.

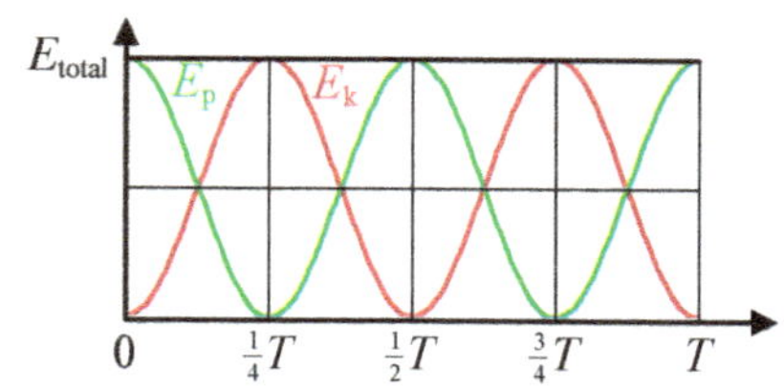

Fig. 3.2.11 Graphs of E_k and E_p against t for SHM

Link

See Section M1.2(c) on page 258 for the trig form of Pythagoras' theorem.

Example

The total energy of an oscillating system is 1.0 J. The period is 2.0 s, the amplitude is 0.25 m and the displacement is maximum at time zero. Sketch graphs of

(a) E_k and E_p against displacement

(b) E_k and E_p against time, over an interval of 2.0 s.

Answer

We can just use Figs. 3.2.10 and 3.2.11 with the appropriate scales.

(a)

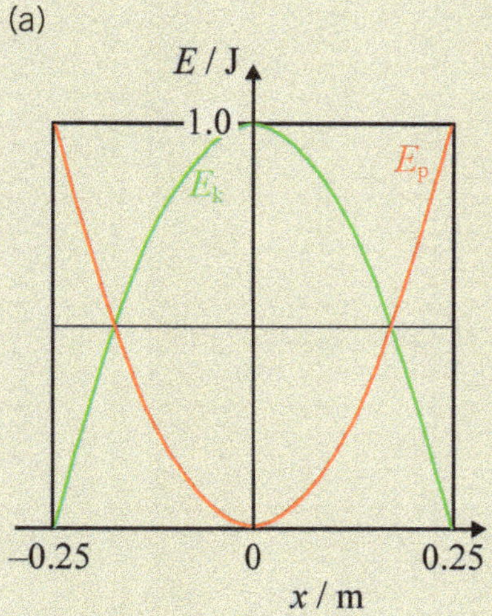

(b)

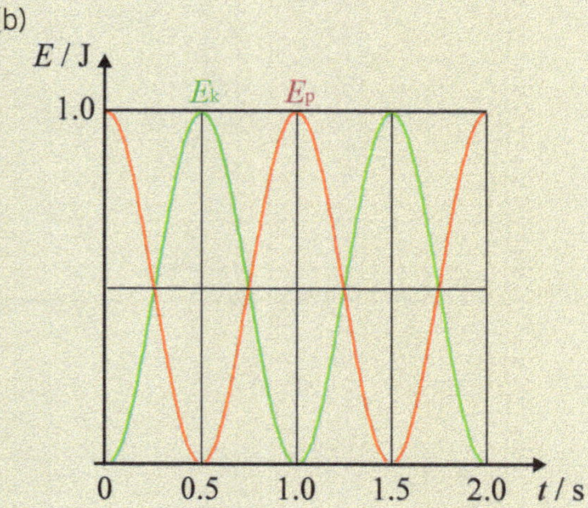

Knowledge check 3.2.15

The displacement, x (in cm), of an object of mass 5 kg object varies with time according to $x = 10\cos 100t$. Calculate:

(a) the total energy,

(b) the PE and KE at $t = 0$

(c) the PE and KE at $t = \frac{\pi}{200}$s.

(c) Where is the point where $E_p = 0$ in an oscillating system?

In this section, we have assumed that $E_p = 0$ at the equilibrium point. Can we justify this? The short answer is 'Yes!' This is true even if there are two or more different forms of potential energy involved, e.g. elastic potential energy and gravitational potential energy in the oscillating mass on a spring system in Fig. 3.2.6. Just as when we look at objects falling under gravity, it is only the **change** in potential energy that matters. Hence we can just decide that the equilibrium point is the zero for potential energy.

As far as A level questions are concerned it is overwhelmingly likely that you can take the zero of E_p at the centre of the oscillation but in principle it could be taken to be anywhere.

Key terms

Free oscillations: These occur when an oscillatory system is displaced and released.

Damping: The dying away, due to resistive forces, of the amplitude of free oscillations.

3.2.6 Damped oscillations

We shall continue for the moment to focus on **free oscillations**. In all real-life situations, systems do not carry on vibrating indefinitely but gradually wind down due to the effect of dissipative forces called **damping**. For example, if a car drives over a small obstacle, the compression in suspension springs is increased. The springs then expand, releasing the excess elastic energy and in the absence of damping forces the car would continue oscillating – with emetic effects. Many damping forces, e.g. air resistance and electromagnetic damping due to eddy currents (see Section 4.5.5), are proportional to the speed of motion. Engineers build in a required degree of damping to a system which can oscillate, e.g. door closures, car suspensions, the Millennium Bridge (in London). We recognise three categories: under-damping, over-damping and critical damping, the last of which is the aim in designing many structures.

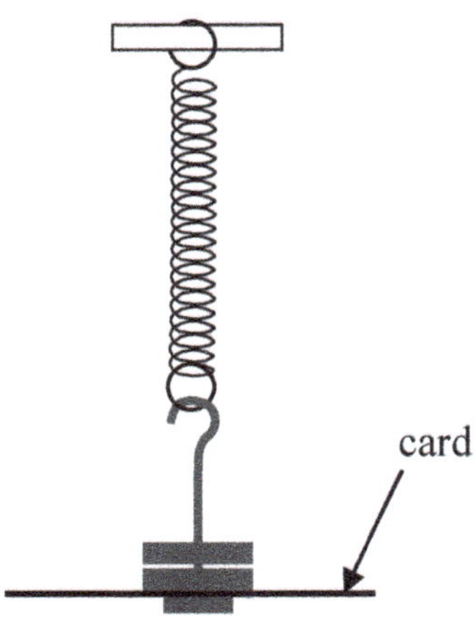

Fig. 3.2.12 Under-damped system

(a) Under-damping

As the name implies, the degree of damping here is light. The mass on a spring system in Fig. 3.2.12 is an example. In these cases, the equation of motion is still $A\cos(\omega t + \varepsilon)$ but the amplitude, A, decreases exponentially as $A_0e^{-\lambda t}$, as in Fig. 3.2.13. The resulting equation is thus

$$x = A_0e^{-\lambda t}\cos(\omega t + \varepsilon)$$

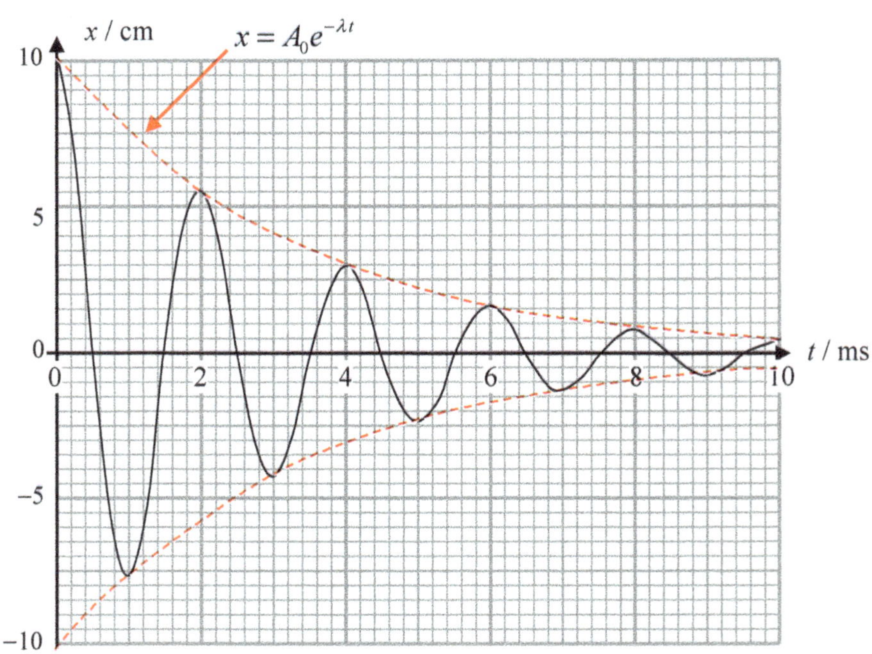

Fig. 3.2.13 Damped oscillations

3.2.16 Knowledge check

The value of A_0 in the graphs in Fig. 3.2.13 is pretty obvious. Use the graphs to calculate (a) ω and (b) λ.

3.2.17 Knowledge check

(a) Use the 0.55 figure from 'things to notice' on the following page to state the fraction of the potential energy (PE) lost in each cycle.

(b) Without considering the PE, how could you measure the fraction of the kinetic energy lost per cycle?

The value of ω is slightly less than its value in the absence of damping, i.e. $\sqrt{\frac{k}{m}}$, but it is very close. A car whose suspension behaved in this may would still be rather sick making.

Things to notice about damped oscillation curves

1. The period of the oscillation is constant – the time for the oscillations doesn't become shorter as the amplitude decays. In Fig. 3.2.13 the period of the oscillations remains at 2 ms even when the amplitude has dropped to less than 10% of the original.
2. The amplitude goes down by the same fraction in each cycle, e.g. in Fig. 3.2.13 the amplitude after 1 cycle is 0.55 of the initial, after 2 cycles it is 0.30 (= 0.55^2) of the initial.
3. The greater the value of λ, the more rapidly the amplitude of the oscillations decreases.

(b) Over-damping and critical damping

As the degree of damping increases, the system initially returns to equilibrium more quickly. The dashed line in Fig. 3.2.14 is a similar degree of damping to that in Fig. 3.2.13. The **green graph** is for a higher degree of damping (about 5 times the dashed graph). It still shows a tendency to oscillate (with a slightly increased period). The **red** and **blue** graphs are for still greater damping (about 10 times and 20 times) respectively.

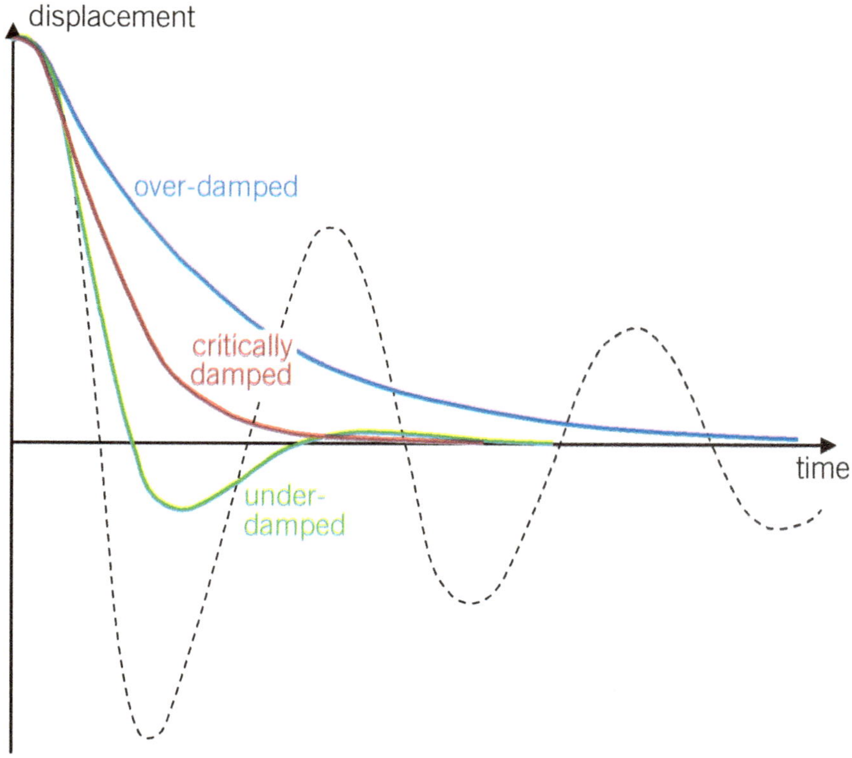

Fig. 3.2.14 Degrees of damping

For many purposes, engineers attempt to design systems which are **critically damped**. For example, consider a bike's suspension system:

- If the damping is too light, the bike will oscillate after going over a pothole in the road.
- If the damping is too heavy, jolts (e.g. from potholes) are transmitted strongly from the wheels to the rest of the bike and the rider.

Fig. 3.2.15 is a schematic diagram of a typical bicycle shock absorber which is coupled to a spring suspension. The damping effect is produced by the hydraulic oil being forced through the holes in the main piston.

Stretch & challenge

What relationship does $\sqrt{0.55}$ have to the graph in Fig. 3.2.13?

Knowledge check 3.2.18

Sketch a graph of the damped oscillation with the same frequency and initial amplitude but with a lighter degree of damping, with λ one half of the value in Fig. 3.2.14.

Study point

In the critically damped graph in Fig. 3.2.14, the displacement decreases monotonically to zero without oscillating. This is not necessarily the case for critical damping but it is so if the system is set into motion by displacing it and releasing with zero velocity.

Knowledge check 3.2.19

What feature of the critically damped and over-damped curves in Fig. 3.2.14 shows that they are not pure exponential decays of the form $x = A_0 e^{-\lambda t}$?

Knowledge check 3.2.20

Automatic door closers are normally designed to be critically damped. Why is this preferable to under-damping or over-damping?

Key terms

A **critically damped** system returns to equilibrium as rapidly as possible without oscillating.

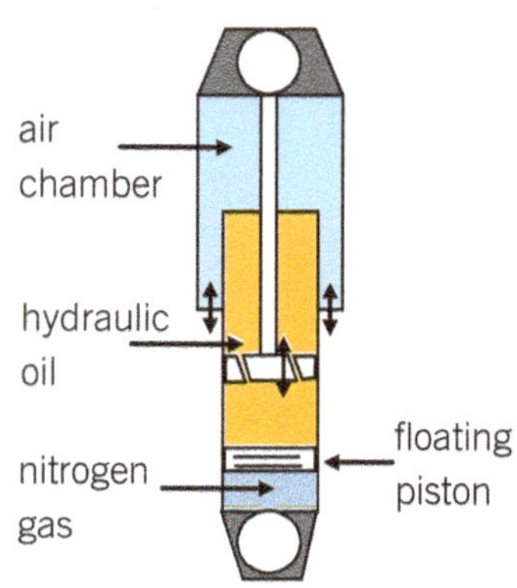

Fig. 3.2.15 Bicycle shock absorber

3.2.7 Forced oscillations – resonance

An oscillatory system can be set into oscillation by the application of a force, be it a one-off impact such as in a car or bike suspension, a periodic driving force or an irregular external stimulus such as a moist finger rubbing around the top of a wine glass. These last two situations are not entirely dissimilar because any driving force can be considered to be composed of the sum of a number of sinusoidal oscillations of different frequencies. For the moment we'll look at **forced oscillations** with a periodic driving force.

Key terms

A **forced oscillation** occurs when a sinusoidal driving force is applied to an oscillatory system, causing it to oscillate with the frequency of the applied force.

Natural frequency: The frequency of the free oscillations.

Study point

The mathematics of breaking down a non-sinusoidal variation into the sum of sinusoidal variations is called Fourier analysis and is covered in university engineering courses.

(a) Resonance curves

The easiest way to investigate this (you are advised to try this yourself) is to use the simple arrangement in Fig. 3.2.16, i.e. our old friend the mass on a spring but supported by an oscillating finger – yours! Try very low frequency oscillations (i.e. $f << f_0$, where f_0 is the **natural frequency** of oscillation of the spring mass system), very high frequency ($f >> f_0$) and then (more difficult) sweeping the frequencies between these two extremes. You should find that:

- For $f << f_0$, the amplitude and phase of the oscillations are the same as that of your finger.
- For $f >> f_0$, the amplitude of the oscillations is very small and the phase difference (more difficult to see) is π (i.e. $180°$).
- For $f \sim f_0$ the amplitude of the oscillations is very large (in fact there is a strong possibility that the masses and spring will part company) and the phase is $\frac{\pi}{2}$ behind that of your finger.

Using more sophisticated set ups, the resonance curves in Fig. 3.2.17 can be obtained.

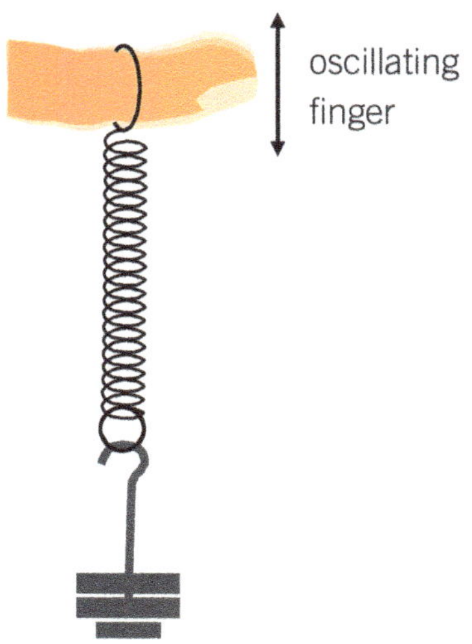

Fig. 3.2.16 Investigating forced oscillations

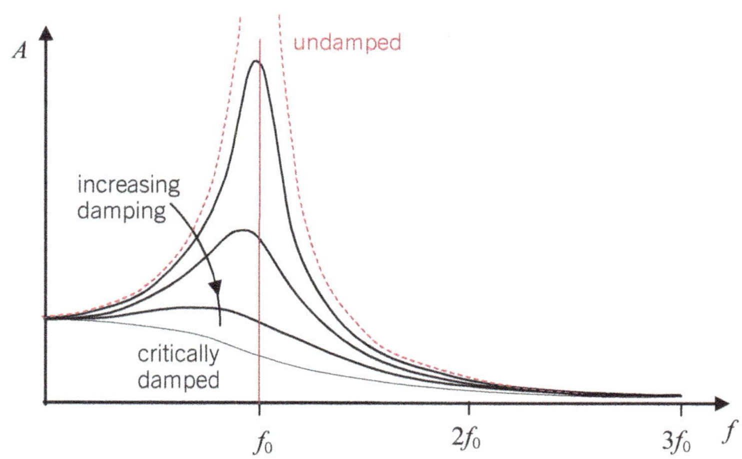

Fig. 3.2.17 Resonance curves

Stretch & challenge

It is not only the amplitude of forced oscillations that vary with the frequency of the driving force.
As Fig. 3.2.18 shows, the forced oscillations are only in phase with the driving force at very low frequencies. At f_0, the forced oscillations are $\frac{\pi}{2}$ (90°) behind the driving force and at much higher frequencies they are almost half a cycle behind.

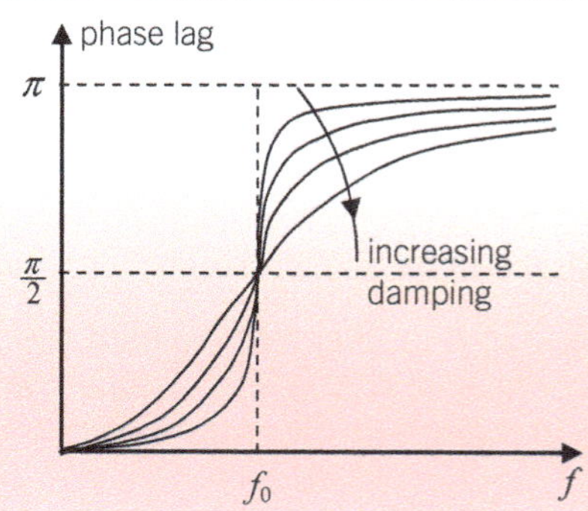

Fig. 3.2.18 Phase lags for forced oscillations

(b) Resonance without periodic driving force

The Tacoma Narrows bridge failure is perhaps the most famous example of resonance causing a major structure to collapse without an apparent periodic driving force. On 1 November 1965 the cooling towers of the newly built power station at Ferrybridge in Yorkshire also underwent catastrophic failure due to resonance. The fitful gusts of wind had a range of frequencies. Unfortunately one of the natural modes of oscillation of the air around the regular array of cooling towers, see Fig. 3.2.19, was within this range and so the air resonated.[1] This would not have mattered but for the fact that this frequency also coincided with one of the natural resonances of the cooling towers.

Woodwind and brass instruments work on a similar (but not catastrophic) principle. The lips of the trumpeter, or the reed of the clarinettist, produce 'white noise' or sound of a wide range of frequencies. Most of these are of no consequence but the ones which coincide with frequencies of stationary waves in the air column inside the instrument will resonate to produce the loud sound. Generally there will be one main note together with several harmonics which are characteristic of the instrument. These sound waves (as well as escaping to the audience) feed back to lips and reed of the instrumentalist to reinforce the same frequencies. The operation of a flute is even more akin to the power station: the air from the flautist's lips rapidly switches between the two paths shown in Fig. 3.2.20 with a wide range of frequencies. Resonance occurs in the air inside the flute which is fed back to the stream of air and reinforces an oscillation at the resonant frequency.

Fig. 3.2.19 Ferrybridge power station, November 1965

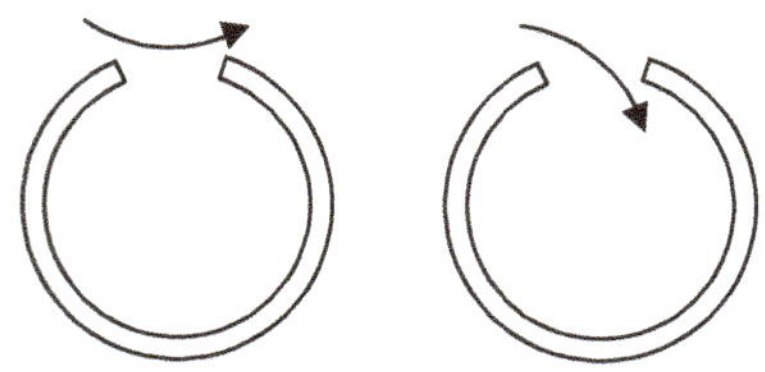

Fig. 3.2.20 Paths of the air stream across a flute mouthpiece

(c) Designing resonance out of systems

During the spin cycle of domestic washing machines, the drum rotates at up to 1800 revolutions per minute. A slight asymmetry in the position of the various items in the washing load, which is inevitable, will result in a periodic driving force of significant amplitude. The frequency of this driving force will vary from 0 up to $30\ \text{Hz}$. Hence, if any of the electrical or mechanical components of the washing machine have a natural frequency of vibration in this range, they will resonate at the appropriate point as the machine spins up and down. Eventually any such components will fail as a result. Hence, in designing washing machines, engineers try and ensure that this does not happen.

Designing resonance out is not always easy. The human eyeball has a natural rotational resonance at $18\ \text{Hz}$. This has been implicated in several helicopter accidents in which pilots have missed seeing power lines because of vibrations from the helicopter engine and rotors.

Engineers must also take account of human factors in designing out resonance as the story of the Millennium Bridge in London shows.

When we walk, as well as exerting vertical forces on the ground, we push slightly outwards with each step, which creates a roughly $1\ \text{Hz}$ sideways forcing function on any structure we are walking on. The central span of the Millennium Bridge has a few transverse resonances in the $1\ \text{Hz}$ range. With several hundred people on the bridge it was not anticipated that they would co-ordinate their footsteps but once slight sideways swaying occurred this is precisely what they subconsciously did, which increased the effect. The bridge was closed until dampers could be fitted!

Fig. 3.2.21 London's Millennium Bridge

Study point

One well-known anti-resonance measure is for marching troops to be ordered to break step when crossing a bridge.

Knowledge check 3.2.21

Two oscillations with frequencies / amplitudes of $0.8\ \text{Hz}$ / $50\ \text{mm}$ and $1.0\ \text{Hz}$ / $75\ \text{mm}$ were recorded on the Millennium Bridge. Calculate the maximum accelerations on pedestrians as a result.

1 The Yorkshire word for this is 'wuthering' (pronounced to rhyme with smotherin') as in *Wuthering Heights*.

(d) Useful resonance

Resonance is hardly ever a good thing in mechanical systems. There are a few notable exceptions, e.g.:

- The basilar membrane in the ear has varying dimensions and stiffness so that different parts of it respond to different input frequencies of sound, enabling us to discriminate between sounds.
- Playground swings are only satisfying once you apply the principle that the forward push be applied from the moment you start moving forwards.

Radio devices are required to respond to a restricted range of frequencies. Audio information is restricted to the range 0–20 kHz. This means that a radio station broadcasting at a frequency of, say, 92.4 MHz has a signal which ranges over 92.40 ± 0.02 MHz, so the tuned detector circuit needs to have a resonance curve which includes this range and drops to zero outside this (see Study point). Television has much more information so greater bandwidth requirements, from about 5 MHz for standard definition TV up to several GHz for some HDTV formats, so the tuned circuit needs to be designed to accommodate this.

Other examples of useful resonance are:

- Medical MRI scans (the clue is in the name – Magnetic Resonance Imaging).
- Laser operation – the incident photon must have the same energy as the difference in the energy levels to stimulate emission.

Study point

In fact the channel separation for FM radio stations is 200 kHz to allow for a variation in frequency of the central *carrier* wave of the transmitter.

Link

See page 204 for more on MRI scans in Option B and the operation of lasers was covered in the AS textbook.

3.2.22 Knowledge check

Compare the frequency of the microwave oven with that of the absorption spectrum peaks of water.

(e) Microwave cooking

This is an example of forced oscillations but not of resonance. Water molecules in the food are dipoles (with the oxygen being slightly negative and the hydrogen atoms slightly positive) so they are able to pick up energy from the electromagnetic microwaves, so the internal energy of the food (see Section 3.4) increases. The reason it is not resonance is that, whereas the peaks in the absorption spectrum of water are at 1.5 μm, 2.0 μm and 3.0 μm, microwave ovens operate at 2.45 GHz which is a long way below the resonance.

Study point

If the microwave resonated with water molecules, we'd expect virtually all the heating to take place in the outer few mm of the food – which it clearly doesn't.

3.2.8 Specified practical work

(a) Measuring *g* by simple pendulum

This is, in principle, the most straightforward of the practicals that A level students are asked to do. However, it does provide a useful introduction to some basic experimental techniques and uncertainty analysis.

The arrangement is as in Fig. 3.2.22. The split cork arrangement is to allow the length of the pendulum to be adjusted easily. The cork should be orientated so that the slit is at right angles to the plane of oscillation of the pendulum. In this way the position of the top of the thread is clear. The cork should be clamped at such a height that lengths, ℓ, of at least 1 m are possible. The bob should be a small heavy sphere, e.g. lead or brass.

To obtain the greatest accuracy, the oscillations should be timed between centres of oscillation, i.e. when the thread is vertical. This is because:

1. The thread is moving most quickly here and hence it is easy to judge the instant it moves past. At the extremes of the oscillation, the bob is moving very slowly and the instant of the extreme difficult to judge.
2. If the pendulum loses energy, it will still pass though the vertical line; the position of maximum displacement will gradually change. Hence a fiducial marker can be used against which to time the oscillations.

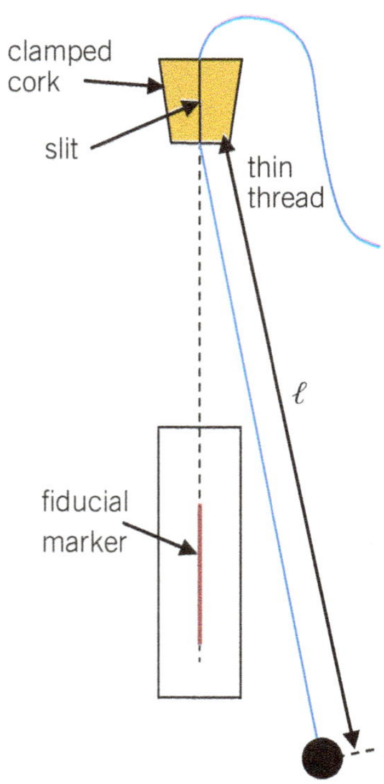

Fig. 3.2.22 Pendulum set up

Procedure

1. Adjust the length of the pendulum to approximately (just less than) 1 m and measure length ℓ using a metre rule. Note that this is to the centre of the bob.
2. Pull the bob to one side a short distance (a few cm) and release.
3. Use a stop watch to measure the time for multiple oscillations. (How many? See Study point.) The uncertainty should be quite low as you can anticipate the thread's crossing the fiducial maker at both the beginning and the end.
4. Take some repeat readings. Not many are needed – it is more a check on miscounting.
5. Repeat for a range of values of ℓ.

Analysis

The period T is given by: $T = 2\pi\sqrt{\frac{\ell}{g}}$ $\quad \therefore T^2 = \frac{4\pi^2}{g}\ell$

A graph of T^2 against ℓ is drawn. It should be a straight line (through the origin) of gradient $\frac{4\pi^2}{g}$. So $g = \frac{4\pi^2}{\text{gradient}}$.

(b) Investigating the damped oscillations of a loaded spring system

The arrangement in Fig. 3.2.23 can be used to investigate the way in which the oscillations decay under the influence of the air resistance on the sheet of thin card. At very low speeds, the magnitude of the drag is approximately proportional to the velocity, v, so the amplitude of the oscillations is expected to decay according to

$$A = A_0 e^{-\lambda t}.$$

Hence a graph of $\ln A$ against t should be a straight line of gradient $-\lambda$ and an intercept of $\ln A_0$ on the vertical axis.

The procedure is as follows:

1. Set up the apparatus as shown, selecting the mass to give as large an amplitude of oscillation as possible without exceeding the elastic limit of the spring.
2. Note the position of the card at equilibrium.
3. Displace the card vertically by lifting or lowering the load by as large a distance as possible so that the spring is always under tension in the subsequent oscillations. Note the initial displacement, i.e. the initial amplitude.
4. Release the load and start a timer. Record the amplitude of the oscillations at suitable intervals, e.g. every 30 s (see Study point) for a suitable period, e.g. 5 minutes.
5. The procedure should be repeated several times.

How to handle the repeat readings

In this investigation, it is worth doing at least one repeat to allow the uncertainty in amplitude to be estimated. There is no need to ensure that the starting amplitude is the same for each repeat. Even with different starting amplitudes, the results can just be added or, if preferred, the means taken.

Some points to consider

A typical spring used in school experiments has a spring constant (k) of about 25 N m^{-1}. With a load of 0.5 kg this gives an equilibrium extension of about 20 cm and a period of oscillation of about 0.9 s. An initial amplitude of about 10 cm is thus possible and the period is long enough to allow the amplitude to be monitored with an uncertainty of about 1 mm [initial percentage uncertainty 1%]. Greater amplitudes can be achieved by connecting two springs in series. This also has the advantage of giving a longer period (1.3 s).

Study point

The fiducial marker doesn't have to be made specially. The vertical post of the clamp stand (or even a convenient window frame) could serve!

Study point

Rather than thinking about the number of oscillations, think about the time. Aiming for at least 10 s duration (for however many oscillations), with an uncertainty of 0.4 s will give a percentage uncertainty of 4% in the period; 20 s would give 2%.

Knowledge check 3.2.23

If the length of the pendulum thread above the bob (rather than to the C of G) is measured, how would the analysis be different?

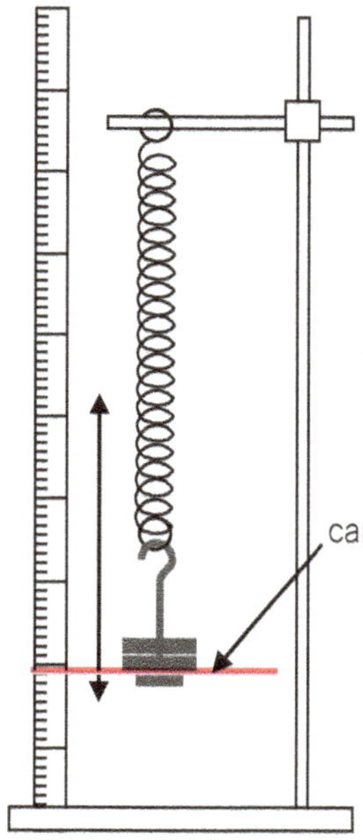

Fig. 3.2.23 Investigating damping

Study point

The oscillation is unlikely to be at maximum displacement at the notional times but the decay time is much longer than the period of oscillation so that an uncertainty of a second is unimportant.

Test yourself 3.2

1. Distinguish between free oscillations and forced oscillations.

2. Samantha defines simple harmonic motion as occurring when an object's acceleration is directly proportional to its distance from a fixed point. What is missing from her definition?

3. Aled notices that the amplitude of the vertical oscillations of a mass on a spring gradually decreases.

   (a) What name is given to this effect? Explain why it occurs in this case.
   (b) Rhian hangs an object with the same mass on to a rubber band. She notices that the amplitude of oscillations decreases more rapidly than in Aled's experiment. State the name of the property of rubber that accounts for this.

4. This question refers to the oscillation shown in Fig. 3.2.5.

   (a) Sketch the a-t curve for the oscillation, labelling significant values.
   (b) Calculate the values of displacement, velocity and acceleration at $t = 0$ for the oscillation.
   (c) Calculate the velocity in the oscillation when the displacement is $+ 1.5$ mm. Why are there two values?

5. A pendulum bob is set into oscillation by applying a nudge to give it a speed of 10 cm s^{-1}. The subsequent period of oscillation is 1.5 s.

   (a) Express the displacement, x, of the bob at time t in the form $x = A\cos(\omega t + \varepsilon)$. [Hint: $v_{max} = \omega A$.]
   (b) Find the position, velocity and acceleration of the bob after 1.0 s.

6. A person on the unmodified Millennium Bridge found herself oscillating laterally with amplitude of 7.5 mm and a period of 1.0 Hz.

   (a) Calculate her maximum velocity and acceleration.
   (b) If her position at time $t = 0$ was 5.0 mm and her velocity was negative, express her displacement in the form $x = A\cos(\omega t + \varepsilon)$.

7. The diagram shows a pendulum consisting of a bob of mass 2.0 kg on a light thread. The pendulum is held at **A** and released at time $t = 0$. It oscillates repeatedly between **A** and **C**. The angle between the threads at **A** and **C** is 0.20 rad and the length of the pendulum (measured to the centre of mass of the bob) is 120 cm.

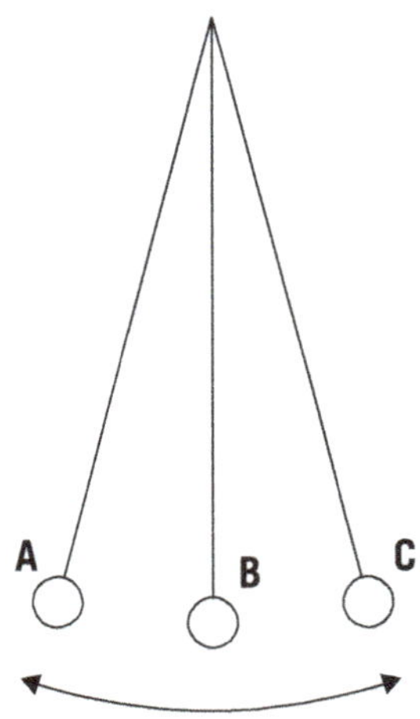

   (a) Identify the positions at which
      (i) the potential energy is maximum
      (ii) the kinetic energy is maximum
   (b) (i) State the amplitude of the oscillations in radians.
      (ii) Calculate the amplitude of the oscillations in cm measured along the arc of the path.
   (c) Calculate the period, T, and angular frequency, ω, of the oscillation.
   (d) Calculate the maximum speed and acceleration of the bob.
   (e) Use your answers to (b)(ii), (c) and (d) to sketch graphs of displacement, velocity and acceleration against time over two complete oscillations.

8. In the 'pin and pendulum' set up in the diagram, a pin, **P**, is placed $\frac{\ell}{2}$ vertically below the pendulum support, so that the pendulum rotates about **P** on the left part of the swing.

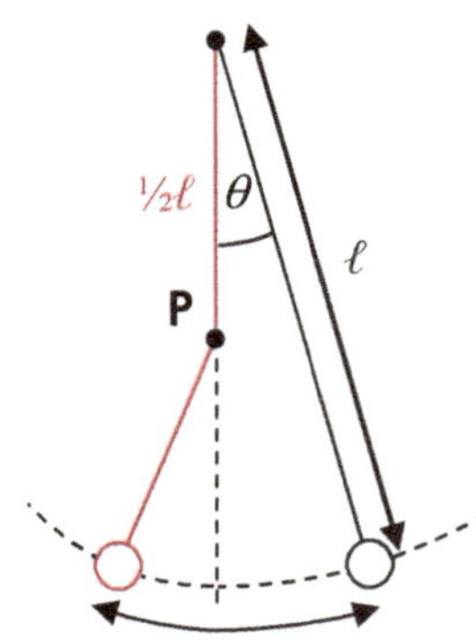

   (a) Write an expression for the period of the pendulum in terms of ℓ.
   (b) Sketch a graph of displacement against time for the pendulum bob, starting from a maximum displacement to the right.

9. A large pendulum is set up in a public building. The mass of the bob is 5.0 kg, the length of the pendulum is 50 m and the horizontal amplitude of the oscillations 5.0 m.
Sketch graphs of the variation of potential energy, kinetic energy and total energy with time over a period of two cycles, starting from a point of maximum positive displacement. Use the lowest position as the point of zero PE.

10. The amplitude of damped oscillation decays from 8.0 cm to 6.0 cm in a period of 25 s. Assuming that the decay is exponential, calculate the amplitude at (a) 50 s (b) 75 s and (c) 120 s.

11. A damped oscillating system has an initial amplitude of 20 cm and a period of 0.50 s.
After 2.5 s the amplitude has become 5 cm. The initial displacement is 20 cm.

(a) Write the equation of motion for the system in the form $x = A_0e^{-\lambda t}\cos(\omega t + \varepsilon)$.
(b) Sketch a graph of displacement against time from 0 to 2.5 s.
(c) Calculate the displacement at $t = 0.9\text{ s}$.

12. A student timed the period of a simple pendulum to measure the acceleration due to gravity.
Results: length = $0.850 \pm 0.002\text{ m}$ Time for 20 oscillations = $36.95 \pm 0.05\text{ s}$
Explain whether these results are consistent with the accepted value of 9.81 m s^{-2}.

13. Bryn and Iestyn each determined the period of a pendulum. Bryn made four measurements, each of five oscillations. Iestyn made one measurement of 20 oscillations. Discuss which method is better from an uncertainty point of view.

14. A load on a spring oscillates. Because of air resistance, the amplitude of oscillations steadily decreases and the following results are obtained:

Time /s	0	30	60	90	120	150
Amplitude / cm	15.0	12.8	9.5	7.5	6.3	5.3

Use these results to show that the decay is approximately exponential and determine the value of the decay constant, λ.

15. Helen determined the period of a pendulum in the following way: She pulled it to one side and let it go, starting a stopwatch at the same time. She waited until it had swung back and forth 10 times and then stopped the stopwatch. She divided the measured time by 10. Eleri said that it would be better to let the pendulum swing a few times and start the stopwatch when the string is vertical. Why is Eleri's idea a good one?

16. A group of students used a simple pendulum, of lengths up to 100 cm to determine a value for the acceleration due to gravity. Their results were as follows:

length, ℓ/cm	20.0	40.0	60.0	80.0	100.00
period, T/s	0.87 ± 0.04	1.22 ± 0.04	1.55 ± 0.04	1.80 ± 0.04	1.97 ± 0.04
$(T/\text{s})^2$	0.76 ± 0.07		2.40 ± 0.20		3.89 ± 0.31

They estimated the uncertainty in the lengths to be $\pm\ 0.1\text{ cm}$.

(a) They expect the relationship between T and ℓ to be $T = 2\pi\sqrt{\dfrac{\ell}{g}}$.
   (i) Explain why they expect a graph of ℓ against T^2 to be a straight line, though the origin.
   (ii) State how a value for g can be found from the graph.
(b) Calculate the missing values of T^2 and their absolute uncertainties.
(c) (i) Plot of graph of ℓ against T^2 including error bars.
   (ii) Draw the lines of maximum and minimum gradient through the error bars.
   (iii) Explain whether the results are consistent with (a)(i).
   (iv) Determine the best value for the gradient together with its percentage uncertainty.
(d) Discuss whether the results are consistent with the accepted value of g (981 cm s^{-2}).
(e) Later, one of the students realised that they had measured the lengths to the bottom of the pendulum bob. Discuss how this has affected the experiment.

3.3 Kinetic theory

Study point

The only properties of gases that you need to know for the examination are those associated with the ideal gas laws and the energy of the gas molecules. The others are given to give an idea of the range of gas properties that can be accounted for by the kinetic theory model. Like all good theories, its strength is in the small number of assumptions related to the large number of observations that it can explain.

Until the early 20th century there was no direct evidence for the existence of atoms and molecules. However, physicists and chemists had developed a particle model of matter that explained a whole series of gas properties:

- The Ideal gas laws of Boyle and Charles: $\frac{pV}{T}$ = constant. See Section 3.3.3.
- Dalton's law: the pressure of mixtures of gases is equal to the sum of the pressures of the individual gas components.
- Brownian motion: the random motion of smoke particles in a gas.
- Graham's law (the rate of diffusion of a gas is inversely proportional to the square root of the molecular mass).

In 1905, Einstein's mathematical explanation of the observed detail of Brownian motion finally provided the direct evidence.

This chapter does not attempt to follow the historical process of the development of the ideas in the 19th century. For example, Section 3.3.1 draws on 20th-century techniques.

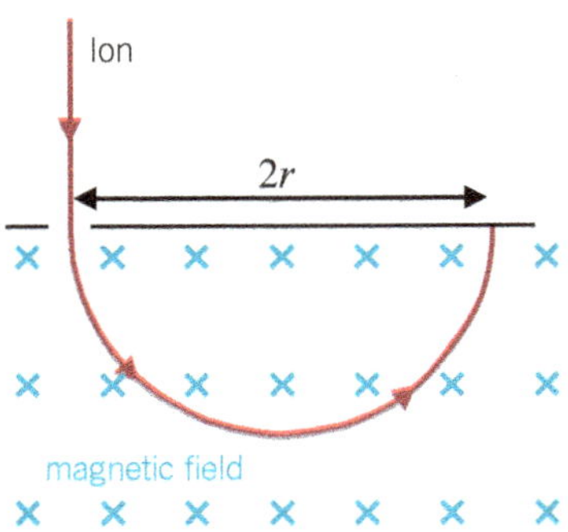

Fig. 3.3.1 Mass spectrometer

3.3.1 Atomic properties

By careful measurements of reacting masses, 19th-century chemists were able to establish the relative masses of the putative atoms of the different elements. For example, relative to hydrogen the masses of carbon, oxygen and iron are approximately 12, 16 and 56. But what are their actual masses? Also atoms are very small – but how small? And how big are the forces which they exert on one another?

Note: This section is not for examination. It is here to give an introduction to some of the basic physical properties of atoms, to provide background for the material that follows. The mass spectrometer, Fig. 3.3.1, is not directly mentioned in the specification but all the concepts involved are examinable and so examiners could use it in a context question in the Unit 4 paper.

Top tip

Remember that the mass of a typical atom is of the order of 10^{-26} to 10^{-25} kg.

(a) Atomic masses

If ions (e.g. H^+, He^+, Na^+) are accelerated and injected at right angles into a magnetic field, they follow a circular path. The reason for this is that, as explained in Section 4.4, they experience a force at right angles to their direction of motion, which is proportional to their charge and velocity. This is the principle of the *mass spectrometer*, which is depicted schematically in Fig. 3.3.1.

The greater the mass of the ions, the larger the radius of their path. By measuring the radius and knowing the energy of the ions and the magnetic field strength, scientists can measure the ionic masses. In this way they have shown that the masses of naturally occurring atoms range from 1.67×10^{-27} kg (hydrogen) to 3.95×10^{-25} kg (uranium 238).

(b) Atomic sizes

Atoms and molecules have no 'hard' surfaces, so there is no exact meaning to the *size* of an atom. However, atoms do not penetrate each other (very far) when they collide or stick together. If they are close to each other in a crystal, we can use the distance between their centres as the sum of their radii. We can measure this by making use of the regular arrangement of particles in a crystal and treating it as a sort of diffraction grating. We then fire radiation of suitable wavelength at it and observe the interference pattern. This was pioneered by Laurence and Henry Bragg from 1913 onwards using X-rays. Beams of neutrons and electrons can also be used, using their wave properties.

Such images, obtained using radiation of wavelength ~10 pm, reveal typical values of lattice spacing in crystals of around 150–200 pm. Hence this may be taken as a typical atomic diameter.

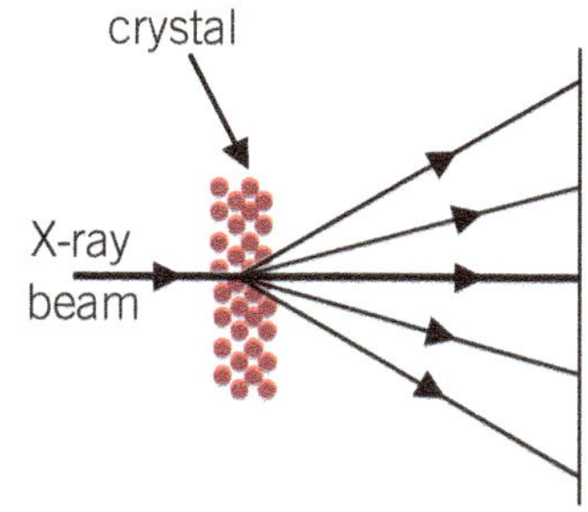

Fig. 3.3.2 X-ray diffraction

Knowledge check 3.3.1

An X-ray photon has a wavelength of 10 pm.

(a) Calculate its energy.

(b) Calculate the kinetic energy of a neutron with a de Broglie wavelength of 10 pm (mass of neutron = 1.67×10^{-27} kg).

(c) Interatomic forces

Neutral atoms and molecules exert very short-range forces on one another that are known as van der Waal's forces.[1] A typical graph of force against separation is shown in Fig. 3.3.3. The separation axis is shown in terms of the equilibrium separation, σ i.e. typically 150–200 pm. There will be no questions on this graph in the A level examination but you should note two things about the intermolecular forces:

1. For $r < \sigma$ there is effectively an infinite repulsion, i.e. the particles behave like hard spheres.
2. The attractive force is very short-range: for $r > 2\sigma$ it is effectively zero.

We shall now use these ideas to develop our treatment on the behaviour of gases.

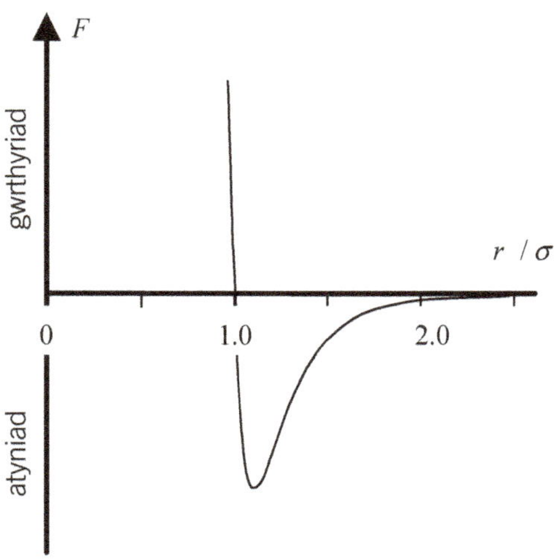

Fig. 3.3.3 Atomic force-separation curve

1 Strictly, the van der Waal's force is only the attractive section of the F-r curve.

3.3.2 Expressing the amount of a substance

Consider a sample of a material, say 47 kg of propane in a gas cylinder. We know its total mass, but what physical quantities are significant? Many properties of material samples, such as gases, depend upon the number of atoms or molecules present and their individual masses. The mass, m, of an atom or molecule can be expressed in kg. This will often be the case in questions you encounter. However, a typical atomic mass is 10^{-26} kg, which is very small for everyday use and scientists have developed several different measures of mass and quantity to cope with this. Some of them have similar names and it is important to distinguish them. Notice the important role of carbon-12 in these quantities.

Top tip

You may come across the unit Da (dalton). This is equal to u and is replacing it in scientific work. It will not be used in WJEC exam questions.

(a) The atomic mass unit, u

The **atomic mass unit**, u, is defined as one twelfth of the mass of a carbon-12 atom. From this definition and accurate determinations of the mass of carbon-12 atoms, the value of u is given by:

$$1\ \text{u} = 1.660\,539\,066\,60 \times 10^{-27}\ \text{kg}$$

The WJEC Data Booklet gives this value to 3 sf (1.66×10^{-27} kg) for use in exams. Atomic and molecular masses are often expressed in terms of u. See Top tip.

3.3.2 Knowledge check

The mass of an atom is 7.02 u. Express this mass in kg.

Example

According to a science data book the mass of a uranium-235 atom is 235.04 u. Express this mass in kg, to an appropriate number of sf [$1\ \text{u} = 1.66 \times 10^{-27}$ kg]

Answer

$m_{U235} = 235.04 \times 1.66 \times 10^{-27}\ \text{kg} = 3.90 \times 10^{-25}\ \text{kg}$ (3 sf)

(b) The relative molecular (or atomic) mass, M_r

M_r is the ratio of the mass, m, of a molecule (or atom) to the unified atomic mass unit,

i.e. $$M_r = \frac{m}{1\ \text{u}}$$

Notice that **M_r is numerically equal to the molecular mass in u**. It has no unit as it is the ratio of two masses. [For chemists: the relative molecular mass is equal to the sum of the relative atomic masses of the atoms in a molecule.]

Example

The relative molecular mass of an organic compound is 94.0. Determine the molecular mass in (a) u and (b) kg.

Answer

(a) $m = 94.0\ \text{u}$

(b) $m = 94.0 \times 1.66 \times 10^{-27}\ \text{kg}$

$= 1.56 \times 10^{-25}\ \text{kg}$

(c) The Avogadro constant and the mole

The mole and the **Avogadro constant** deal with the number of atoms or molecules in a sample. The number of atoms or molecules in a macroscopic sample is inconveniently large for everyday use. For example, the number of water molecules in a cup of tea is about 5×10^{24}. Hence the **amount of substance**, which is the measure of the number of atoms or molecules in it, is expressed more conveniently in moles. The abbreviation for **mole** is mol.

Note that the mass of carbon specified is 12 grams not kilograms (probably because it is in line with what chemists can get into a test tube). This gives rise to a factor of 1000 in several equations which follow. The actual number of particles in a mole (e.g. 12 g of carbon-12) is called the Avogadro constant, N_A.

The value of N_A given in the WJEC Data Booklet is 6.02×10^{23} mol^{-1}. Note that the numerical value of N_A is the reciprocal of the value of u expressed in grams (i.e. approximately 1.66×10^{-24}).

Key terms

The **Avogadro constant**, N_A, is the number of particles per mole.
UNIT: mol^{-1}

The **mole** (mol) is the SI unit of *amount of substance*. It is the amount containing as many particles (e.g. molecules) as there are atoms in 12 g (exactly) of carbon-12.

Knowledge check 3.3.3

Calculate the length of side of a graphite cube consisting of 1.0 mol of carbon atoms.

Density of graphite = 2 300 kg m^{-3}.
Mass of a carbon atom = 12 u (exactly).

(d) Molar mass, *M*

The **molar mass** of an element or compound is just the mass per mole. The basic SI unit of the molar mass is kg mol^{-1}. Expressed in this unit, the relationship between M and M_r is

$$M/\text{kg mol}^{-1} = \frac{M_r}{1000}$$

So the element helium, which has a relative molecular mass of 4.00 has a molar mass of 0.004 00 kg mol^{-1}. If M is expressed in g mol^{-1} (grams per mole), then it is numerically equal to M_r, e.g. helium has a molecular mass of 4.00 g mol^{-1}.

When working with gases, we shall often need to calculate the *amount* of a gas in a sample, i.e. the number of moles, given the mass, m, and the molar mass, M. We use the symbol n for the number of moles. As the molar mass is the mass per mole, we can write the following equation:

$$n = \frac{\text{total mass}}{\text{molar mass}} \quad \text{or, in symbols,} \quad n = \frac{m}{M}$$

Top tip

This equation is given in the WJEC Data Booklet.

Knowledge check 3.3.4

The relative molecular mass of a hydrocarbon is 42. Calculate its molecular mass in kg.

[For Chemists: identify the hydrocarbon]

Top tip

This equation is given in the WJEC Data Booklet.

Molar mass and nucleon number

For the purposes of mole calculations, it is a very good approximation to write that the molar mass in g mol^{-1} is equal to the nucleon number (the total number of protons and neutrons present). In fact, for all stable nuclides making this approximation involves an error of much less than 1%, e.g. iron-56 has a molar mass of 55.94 g mol^{-1} which can be taken as 56.0 with a percentage error of 0.11%. The only place where you cannot make this approximation is in calculations of nuclear binding energy in Section 3.6.

Knowledge check 3.3.5

The nucleon number of oxygen (O) is 16. The formula of an oxygen molecule is O_2.

(a) State the molar mass of (i) atomic oxygen, (ii) molecular oxygen.
(b) State the relative molecular mass of O_2.
(c) State the molecular mass of O_2 in u.
(d) Calculate the molecular mass of O_2 in kg. [$1\text{ u} = 1.66 \times 10^{-27}$ kg]

New definition of the mole and Avogadro constant

On 20 May 2019 a new definition of the Avogadro constant was adopted by the scientific community. The SI unit of amount of substance, i.e. the mole is now *defined* as being exactly $6.022\,140\,76 \times 10^{23}$ elementary entities (which could be atoms, molecules, ions, electrons...). The Avogadro constant therefore now has the exact value of $6.022\,140\,76 \times 10^{23}$ mol^{-1}. There is no effect on everyday scientific calculations as the value chosen is deliberately identical to previous measurements.

Top tip

The WJEC exams will continue to use the previous definitions for the lifetime of the current specification.

3.3.3 The ideal gas equation

Low pressure gases which are well above their boiling points behave in a very simple way, as we shall see. Such gases are referred to as **ideal gases** and in Section 3.3.3 we will consider their properties.

If we investigate the relationships between the pressure, p, volume, V, and kelvin temperature, T, of a sample of an ideal gas (such as dry air) we find that:

- at constant temperature: $p \propto \frac{1}{V}$ or $pV = \text{constant}$
- at constant pressure: $V \propto T$ or $\frac{V}{T} = \text{constant}$
- at constant volume: $p \propto T$ or $\frac{p}{T} = \text{constant}$

In fact these are not independent relationships as any one of them can be derived by combining the other two (see Stretch & challenge). In the same way we can combine them to give the relationship (for a fixed sample of gas):

$$pV \propto T$$

Writing this as $pV = kT$, the constant k will be proportional to the amount of gas present (double the amount at constant pressure will give double the value of pV because V is doubled). Combining this with Avogadro's law (see Study point) leads to:

$$pV = nRT$$

where n is the number of moles (i.e. the amount) of gas, and

R is a constant called the *molar gas constant*, $8.31\ \text{J mol}^{-1}\ \text{K}^{-1}$

This equation is called the *equation of state for an ideal gas* or (slightly more catchily) the ideal gas equation. We are now in a position to define an **ideal gas**. No gas exactly obeys the ideal gas equation but some common gases (oxygen, nitrogen, helium, methane) do so to a very good approximation under everyday conditions and as do other gases (carbon dioxide, propane, butane) at very low pressure.

This example illustrates how we can use the equation and a few pitfalls to avoid (see Study point).

Example

Calculate the mass of gas in a medical oxygen (O_2) cylinder of height 80 cm and internal diameter 15 cm. At 20 °C the gas pressure is 20 MPa.

Answer

Volume of gas, $V = \pi\left(\frac{d}{2}\right)^2 h = \pi \times (0.075\ \text{m})^2 \times 0.80\ \text{m} = 0.0141\ \text{m}^3$

$pV = nRT,\ \therefore n = \frac{pV}{RT} = \frac{20 \times 10^6\ \text{Pa} \times 0.0141\ \text{m}^3}{8.31\ \text{J mol}^{-1}\text{K}^{-1}(20 + 273.15)\text{K}} = 115.8\ \text{mol}$

$\therefore$ Mass of oxygen $= nM = 115.8\ \text{mol} \times 32\ \text{g mol}^{-1} = 3700\ \text{g} = 3.7\ \text{kg}$.

Stretch & challenge

Consider a sample of gas with pressure, volume and temperature values of p_0, V_0 and T_0 respectively.

(a) Allow it to change to p_1, V_0 and T_1 via the intermediate state p_1, V_1 and T_0.

Derive the relationship $p \propto T$ from $p \propto \frac{1}{V}$ and $V \propto T$.

(b) Now allow it to change from p_0, V_0, T_0 to p_1, V_1, T_1 via an intermediate state of your choice to derive $pV \propto T$.

Key terms

An **ideal gas** is one which strictly obeys the equation $pV = nRT$.

Study point

The law of multiple proportions suggests strongly that, 'Under the same conditions of temperature and pressure, equal volumes of all gases contain the same number of molecules,' which is *Avogadro's law*. The electrolysis of water to give two volumes of hydrogen to one of oxygen is a nice example of this because

$2H_2O \rightarrow 2H_2 + O_2$.

3.3.6 Knowledge check

Show that the unit, of R, $\text{J mol}^{-1}\ \text{K}^{-1}$ makes the ideal gas equation homogeneous.

Study point

In the working of the example notice:

1. Using radius not diameter when applying $\pi r^2 h$.
2. Converting the °C to K.
3. The molar mass of oxygen is $32\ \text{g mol}^{-1}$ because it is O_2.

3.3.4 The behaviour of molecules in gases

Kinetic theory, seeks to explain the properties of materials (gases, liquids and solids) in terms of the motion and interactions of molecules. We'll concentrate on gases but refer to solids and liquids where appropriate.

(a) Molecular basics

We start from the assumption that gases (as well as liquids and solids) consist of large numbers of molecules (see Study point) in permanent motion.

Molecules of a gas have plenty of space to move around in (Fig. 3.3.4). This explains why we can compress gases easily, unlike liquids and solids. At atmospheric pressure a typical separation ~ 3 nm is about 10 times the size of a molecule. It also explains why gases diffuse into each other much more quickly than do solids and liquids.

Molecular speeds are typically a few hundred $\mathrm{m\ s^{-1}}$ (we'll calculate them later), so even with all this space, they must collide frequently with one another. They also collide frequently with the container wall.

Fig. 3.3.4 Molecular spacing

(b) Molecular collisions

Making the simple assumption that the kinetic energy of the molecules of a gas depends upon the temperature of the gas and with the observation that gas in a cylinder doesn't spontaneously cool down below room temperature, we must conclude that intermolecular collisions must be elastic – at least on the average – meaning that there is no loss in kinetic energy.

Let's examine this in more detail. What does 'on the average' mean? The diagram in the Stretch & challenge was of two monatomic molecules, e.g. helium. Could some kinetic energy be lost in the collision? It is possible for kinetic energy to be lost and transferred to one of the atoms, putting it into an excited state. On a subsequent collision this energy can be transferred back to kinetic energy (see AS textbook Fig. 2.8.3).

The scenario depicted in Fig. 3.3.5 gives a way of losing translational kinetic energy for diatomic molecules (e.g. O_2 and N_2). Rotational kinetic energy is picked up in the collision, slowing the molecules down. Again, subsequent collisions can convert this back to translational KE.[2]

So some translational kinetic energy can be lost in a collision; in other collisions it can be gained. On average the translational KE is constant. In any case, the effect of the frequent collisions is to randomise the velocities and hence the kinetic energies of the gas molecules. Fig. 3.3.6 shows the probability function, P, for molecular speeds. Just as with the intermolecular forces graph (Fig. 3.3.3) you do not need to know this graph for the examination but you should be aware that a small number of molecules have values of kinetic energy much greater than the mean.

2 Because of their very low moments of inertia (see Option C) and for quantum mechanical reasons, monatomic molecules cannot possess rotational kinetic energy.

Study point

Rather than have two different names for the particles in gases, they are always referred to as *molecules*, even in monatomic gases such as helium and argon. This saves us from saying, 'Air consists of *molecules* of nitrogen, oxygen, water and carbon dioxide and *atoms* of argon'.

Study point

The *bromine diffusion* experiment illustrates both the high speed of molecules (rapid motion of bromine into a vacuum) and the frequent collisions (the much slower rate of diffusion into air). There is a nice clip of this on a popular video-sharing website.

Stretch & challenge

The diagram shows two molecules on a collision course.

Using Fig. 3.3.5, discuss what happens to their velocities just before, during and just after the collision.

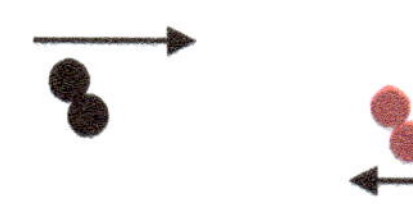

(a) Before collision

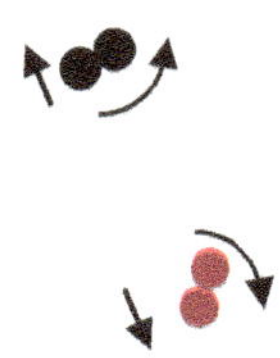

(b) After collision

Fig. 3.3.5 Collision between diatomic molecules

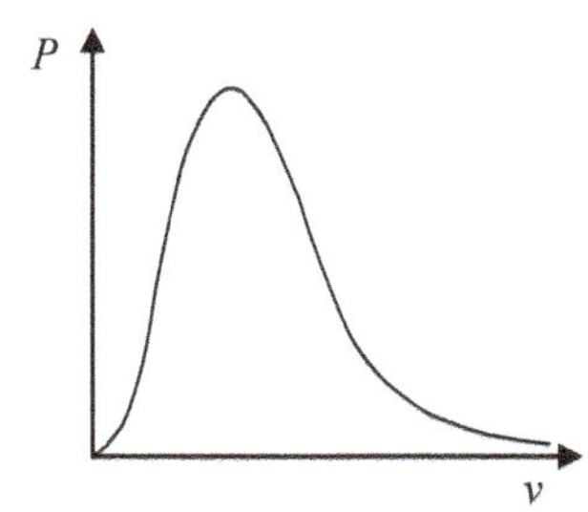

Fig. 3.3.6 Molecular speed distribution

3.3.5 Molecular motion and gas pressure

The aim of this section is to use our knowledge of molecules and Newtonian mechanics to refine the qualitative pre-A level explanation of gas pressure to make quantitative statements.

(a) Kinetic theory assumptions

We have already made three assumptions about gases:

1. Gases consist of a large number of molecules in constant motion in (otherwise) empty space.
2. Collisions between gas molecules are elastic (on average).
3. The mean kinetic energy of the gas molecules depends only on the temperature.

We'll make a few more assumptions for ideal gases:

4. The volume of the molecules themselves is a negligible fraction of the volume occupied by the gas.

   We can justify this by reference to KC 3.3.7. The helium molecules occupy less than 0.1% of the total space.
5. The molecules exert negligible forces on one another except during collisions.

   If the molecules occupy less than 0.1% of the space then their mean separation must be at least ten times their diameter. Looking back at Fig. 3.3.3, this means that most of the time they are well out of the range of intermolecular forces.
6. The effect of gravity is negligible.

   At first sight this is a bit strange but the molecular speeds are so high that, in the short distances travelled between collisions, the effect of gravity is tiny. See KC 3.3.8.

   Taken together, assumptions 5 and 6 mean that molecules travel at a constant velocity between collisions.

(b) Pressure and molecular motion

Imagine a single molecule of mass m in a box, as in Fig. 3.3.7. We ask the question, 'What pressure does it exert on face **A**?'

At the instant of the diagram let the velocity c have components c_x, c_y and c_z. Eventually, the molecule will hit face **A**. Because we are assuming elastic collisions, it will bounce off with velocity $(-c_x, c_y, c_z)$

$\therefore$ Change in momentum of molecule at **A** $= -2mc_x$

After bouncing off a few more walls it will reach **A** again after a time, Δt given by

$$\Delta t = \frac{2l_x}{c_x}$$

at which point its momentum will change by $-2mc_x$ again. So its rate of change of momentum at **A** is $-2mc_x \times \dfrac{c_x}{2l_x} = -\dfrac{mc_x^2}{l_x}$.

But, from N2, rate of change of momentum = force

$\therefore$ The mean force exerted by **A** on the molecule $= -\dfrac{mc_x^2}{l_x}$

3.3.7 Knowledge check

The density of liquid water is 1.0 g cm^{-3}.

The molar mass of water of 18 g mol^{-1}.

(a) Show that the 'diameter' of a water molecule ~400 pm.

(b) Estimate the separation of water molecules in steam at 100 °C and 1 atm. (Density of steam = 0.6 kg m^{-3})

3.3.8 Knowledge check

A molecule travels horizontally at 300 m s^{-1}. How far does it drop:

(a) if it could cross the 8 m wide lab without colliding?

(b) in the mean distance between collisions of ~0.1 μm?

Study point

The SI unit of pressure is the **pascal** (Pa) which is equivalent to N m^{-2}.

Engineers often express pressure in **bar**: 1 bar = 10^5 Pa

An everyday unit is the **atmosphere** (atm):

1 atm = 1.01325 bar
= 1.01325×10^5 Pa

Top tip

The derivation of the kinetic theory equation is quite long and you will not be examined on it. However, you are advised to make sure you understand its principles. Equation [3] is given in the Data booklet.

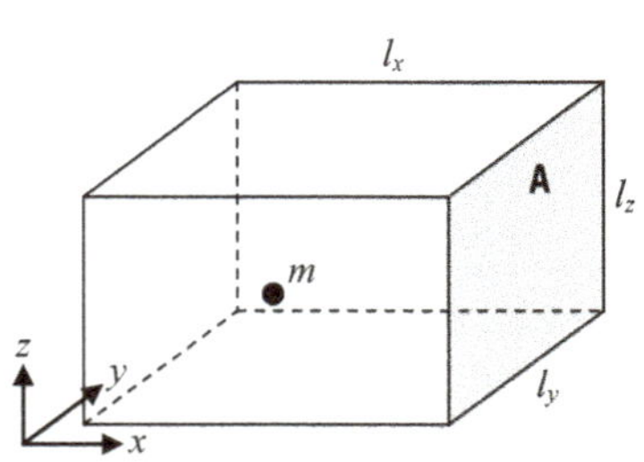

Fig. 3.3.7 Molecule in a box

$\therefore$ By N3, the mean force exerted by the molecule on **A** = $\dfrac{mc_x^2}{l_x}$

And the mean pressure, p, exerted on **A** by the molecule is given by:

$$p = \frac{\text{force}}{\text{area}} = \frac{mc_x^2}{l_x(l_y l_z)} = \frac{mc_x^2}{V}$$

where V is the volume of the box.

We now boldly introduce a lot more molecules so there are N molecules in total and we'll call their x-velocity components c_{1x}, c_{2x}, ...c_{Nx}. Their pressures on face **A** add to give:

$$p = \frac{mc_{1x}^2}{V} + \frac{mc_{2x}^2}{V} + \ldots \frac{mc_{Nx}^2}{V} = \frac{Nm\overline{c_x^2}}{V} \qquad [1]$$

where $\overline{c_x^2}$ is the mean of the c_x^2 values. See Study point on the opposite page.

We'll write this pressure in terms of c rather than c_x. Applying Pythagoras' theorem (in 3 dimensions) for each molecule: $c^2 = c_x^2 + c_y^2 + c_z^2$

$$\therefore \quad \overline{c^2} = \overline{c_x^2} + \overline{c_y^2} + \overline{c_z^2} \qquad [2]$$

Given assumption 3 above, there is no special direction and after multiple collisions the distribution of the components of velocity in all directions would be the same, even if they started out different.

This must mean that $\overline{c_x^2} = \overline{c_y^2} = \overline{c_z^2}$

which together with equation [2] means that $\overline{c_x^2} = \frac{1}{3}\overline{c^2}$

Substituting in equation [1] ⟶ $\quad \therefore p = \dfrac{1}{3}\dfrac{Nm\overline{c^2}}{V}$

which is more conveniently written $\quad pV = \frac{1}{3}Nm\overline{c^2} \qquad [3]$

(c) What does $\overline{c^2}$ mean?

Imagine a group of five nitrogen molecules which are travelling along the x axis. Their velocities, c, are:

$100 \text{ m s}^{-1} \qquad -200 \text{ m s}^{-1} \qquad -300 \text{ m s}^{-1} \qquad 400 \text{ m s}^{-1} \qquad 500 \text{ m s}^{-1}$

We'll work out various means:

1. Mean velocity, $\bar{c} = \dfrac{100 - 200 - 300 + 400 + 500}{5} \text{ m s}^{-1} = 100 \text{ m s}^{-1}$
2. Mean speed, $\overline{|c|} = \dfrac{100 + 200 + 300 + 400 + 500}{5} \text{ m s}^{-1} = 300 \text{ m s}^{-1}$
3. **Mean square speed**, $\overline{c^2} = \dfrac{100^2 + 200^2 + 300^2 + 400^2 + 500^2}{5} \text{ m}^2 \text{ s}^{-2}$

$$= 110\,000 \text{ m}^2 \text{ s}^{-2}$$

Notice the difference between the mean speed and the mean velocity. All the speeds are positive – direction is not significant.

We now introduce the **rms speed**, c_{rms} which is the square root of the mean square speed. For the above molecules:

4. Rms speed, $c_{\text{rms}} = \sqrt{\overline{c^2}} = \sqrt{110\,000 \text{ m}^2 \text{ s}^{-2}} = 331 \text{ m s}^{-1}$.

This value is the most often used for molecules because the mean kinetic energy, $\overline{E_k}$, can be calculated from

$$\overline{E_k} = \frac{1}{2}mc_{\text{rms}}^2$$

Study point

Can we really just add up the pressure contributions from the N molecules? Won't they collide which will change the values of c_x?

Yes, but when they collide, momentum and energy will be conserved and so the total change of momentum each second at face **A** is just the same as if they hadn't collided. See Stretch & challenge.

Stretch & challenge

Consider a 1 m long box with two 'molecules' as shown.

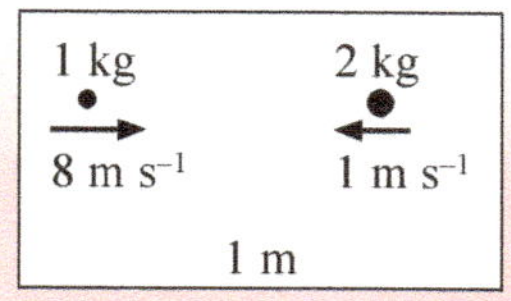

(a) Assuming no collisions, calculate the mean force exerted by the two molecules on the end faces.

(b) Let the molecules collide head on as shown. Apply the principle of conservation of momentum and energy to show that the velocities after the collision are -4 m s^{-1} and 5 m s^{-1} (right is positive).

(c) Show that the mean force exerted by the two molecules on the ends following this one-off collision is the same as in part (a).

Key terms

The quantity $\overline{c^2}$ in equation [3] is called the **mean square speed**. It is the mean value of the squares of the molecular speeds [squaring it makes it a scalar quantity].

Taking the square root, gives $\sqrt{\overline{c^2}}$ which is called the **root mean square speed** (or **rms speed**) which is written c_{rms}.

Link

If you study AC electricity (Option A in Unit 4) you will meet the rms current and pd too.

3.3.9 **Knowledge check**

Notice that the rms speed for the five molecules in Section 3.3.5(c) is greater than the mean speed.

(a) Explain why this must always be the case (unless the molecules have identical speeds).

(b) Copy Fig. 3.3.6 and label suggested positions of the mean and rms speeds.

Study point

Note that, from the definitions:

$\overline{c^2} = c_{\text{rms}}^2$

Example

For the five nitrogen molecules, calculate:

(a) their mean momentum
(b) their mean kinetic energy.

Answer

(a) Mass of N_2 molecules = 28 u = 28 × 1.66 × 10^{-27} kg = 4.6 × 10^{-26} kg

Mean momentum = mass × mean velocity

= 4.6 × 10^{-26} kg × 100 m s^{-1}

= 4.6 × 10^{-24} N s

(b) Mean KE $= \frac{1}{2}mc_{\text{rms}}^2 = \frac{1}{2} \times 4.6 \times 10^{-26}\ \text{kg} \times 110\,000\ \text{m}^2\ \text{s}^{-2}$

$= 2.53 \times 10^{-21}$ J

(d) Comparison with the ideal gas equation

The equation of state for an ideal gas, $pV = nRT$, under the conditions of a constant temperature and a fixed amount of gas, reduces to

$$pV = \text{constant}$$

Is equation [3] above consistent with this? The left-hand sides are the same. The right-hand side of equation [3], $\frac{1}{3}Nm\overline{c^2}$, is proportional to the kinetic energy of the molecules, which, according to our assumption is constant at constant temperature.

How have our other assumptions figured in the derivation of [3]? We mentioned the lack of gravitational effect. The others are:

1. Negligible volume of molecules: No two molecules can be in the same place, i.e. the space occupied by a molecule is not available for any other. This means that the effective volume for the molecules would be less than $l_x l_y l_z$ if the molecules did not have negligible volume.
2. Negligible intermolecular forces: If the molecules exerted significant forces on one another, the speed of impact with the container walls would not be the same as just before impact (it could be more or it could be less because of interactions with the other gas molecules and the wall molecules).

Without being explicit during the derivation, we have assumed that both these effects are negligible.

Conclusion: The molecular model appears to reproduce the pV aspect of the ideal gas equation.

Stretch & challenge

The radius of an argon atom is 71 pm. The centre of a second argon atom cannot come within 2 radii of the centre of the first argon atom. 1 mole of argon atoms occupies a volume of 0.024 m^3. Show that the fraction of the volume which the atoms effectively occupy is only 0.03% of the total.

3.3.6 Molecular energy

Consider again the equation $pV = \frac{1}{3}Nm\overline{c^2}$ [1]

Equation [1] tells us that we can calculate the translational kinetic energy of the molecules if we know just the pressure and volume of a gas: we don't need to know the temperature, the number of molecules or the mass of the molecules!

The translational KE of the molecules, $E_{\text{k trans}} = \frac{1}{2}Nm\overline{c^2}$

$= \frac{1}{2} \times 3pV$, from equation [1]

so $E_{\text{k trans}} = \frac{3}{2}pV$ [2]

Example

Calculate the kinetic energy of the molecules in a helium cylinder of volume $0.020\ \text{m}^3$ at a pressure of $10\ \text{MPa}$.

Answer

$E_{\text{k trans}} = \frac{3}{2}pV = 1.5 \times 10\ \text{MPa} \times 0.020\ \text{m}^3 = 300\ \text{kJ}$

Note: Helium is monatomic so this energy is the total kinetic energy of the molecules. See footnote to Section 3.3.4(b).

(a) Molecular energy and temperature

We can also write $E_{\text{k trans}}$ in terms of temperature by substituting from equation [2] into $pV = nRT$, to give

$$E_{\text{k trans}} = \frac{3}{2}nRT \quad [3]$$

So the molecules of one mole of gas have a translational kinetic energy of $\frac{3}{2}RT$. We can go further and calculate the mean translational kinetic energy per molecule:

$$\text{Mean KE per molecule} = \frac{1}{2}m\overline{c^2} = \frac{3}{2}\frac{R}{N_A}T = \frac{3}{2}kT \quad [4]$$

The constant k is called the Boltzmann constant and has a value of $1.38 \times 10^{-23}\ \text{J K}^{-1}$. Note that this is the mean KE per molecule. Looking back at the molecular speed distribution in Fig. 3.3.6 it should be obvious that a small fraction of molecules would have a significantly greater energy.

We now have a new way of calculating the total translational kinetic energy of the molecules of a gas sample – by using the Boltzmann constant, k.

If we have a sample of gas, consisting of N molecules, at a temperature T, then equation [4] tells us that the total translational kinetic energy, is given by

$$E_{\text{k trans}} = \frac{3}{2}NkT \quad [5]$$

So, we have a choice of ways of calculating the energy of the molecules of a sample of gas. This is illustrated in the following example.

Study point

How can the molecules of argon in a $1\ \text{m}^3$ container of argon at 1 atm and 600 K have the same kinetic energy as those in the same container at 1 atm and 300 K?

Answer: Because the mean molecular KE is double, but there are only half as many molecules!

Knowledge check 3.3.10

If the cylinder of helium in the Example is at 25 °C, calculate:

(a) the number of moles and

(b) the number of molecules of helium in the cylinder.

Knowledge check 3.3.11

Use

$R = 8.31\ \text{J mol}^{-1}\ \text{K}^{-1}$

and

$N_A = 6.02 \times 10^{23}\ \text{mol}^{-1}$

to confirm the value and unit of the Boltzmann constant given in the text.

Knowledge check 3.3.12

(a) An ideal gas has a temperature of 300 K. Calculate the mean translational kinetic energy of its molecules.

(b) Calculate the ratio of the rms speeds of neon molecules ($M_r = 20.0$) and helium molecules ($M_r = 4.0$) at the same temperature.

Example

Calculate the total energy of $1.00\ \text{kg}$ of helium at a temperature of $290\ \text{K}$.
Molar mass of helium = $4.0\ \text{g mol}^{-1}$

Answer

Method 1, using R [$8.31\ \text{J mol}^{-1}\ \text{K}^{-1}$]

Amount of helium = $\frac{1.00\ \text{kg}}{0.0040\ \text{kg mol}^{-1}} = 250\ \text{mol}$

$\therefore E_{\text{k trans}}$ = total energy = $\frac{3}{2}nRT = \frac{3}{2} \times 250 \times 8.31 \times 290 = 9.0 \times 10^5\ \text{J}$ (to 2 sf)

Method 2, using k [$1.38 \times 10^{-23}\ \text{J K}^{-1}$]

Molecular mass of helium = $4.0\ \text{u}$

$\therefore$ Number of helium atoms = $\frac{1.00\ \text{kg}}{4.0 \times 1.38 \times 10^{-27}\ \text{kg}} = 1.506 \times 10^{26}$

$\therefore E_{\text{k trans}} = \frac{3}{2}NkT = \frac{3}{2} \times 1.506 \times 10^{26} \times 1.38 \times 10^{-23} \times 290 = 9.0 \times 10^5\ \text{J}$ (to 2 sf)

Remember that the necessary constant values, R and k, for these methods, as well as N_A and u are given in the Data booklet.

See Knowledge check 3.3.13 for yet another way of answering this question.

3.3.13 Knowledge check

For the sample of helium in the example:

(a) Calculate the number of moles of helium.

(b) Hence, calculate the number of molecules using the Avogadro constant ($6.02 \times 10^{23}\ \text{mol}^{-1}$).

(c) Hence, calculate the total energy.

Top tip

You will not be examined on the contents of sections (b) and (c). These sections are included to give you a greater insight into the nature of molecular interactions.

(b) Molar heat capacity: monatomic gases

The relationship between temperature and molecular energy provides us with another test of the kinetic theory because the heat needed to raise the temperature of a gas can be measured experimentally and compared with the theoretical prediction. Equation [3] in 3.3.6(a) tells us that the energy of the molecules in a monatomic gas is $\frac{3}{2}nRT$, so the energy needed to raise the temperature of $1\ \text{mol}$ of gas per kelvin is $\frac{3}{2}R$, i.e. $12.5\ \text{J mol}^{-1}\ \text{K}^{-1}$. This quantity is called the **molar heat capacity at constant volume**[3], C_V. Table 3.3.1 gives a few values of C_V.

Gas	C_V / J mol^{-1} K^{-1}	C_V / R
Helium (He)	12.5	1.50
Argon (Ar)	12.5	1.50
Nitrogen (N_2)	20.6	2.49
Oxygen (O_2)	21.1	2.54
Carbon monoxide (CO)	20.7	2.49

Table 3.3.1 Molar heat capacities of some common gases

The theoretical prediction for the two monatomic gases is spot on. We'll now look at the diatomic gases.

(c) Molar heat capacity: diatomic gases

The values of the molar heat capacity for the diatomic gases, nitrogen and oxygen and carbon monoxide, are consistently greater than those for the monatomic gases – very close to $\frac{5}{2}R$ rather than $\frac{3}{2}R$. Why should this be? The answer is that the pressure of a gas is determined by the translational KE of the molecules $\frac{1}{2}Nm\overline{c^2}$, but diatomic gases can possess energy of rotation as well – see Fig. 3.3.8. As we saw earlier, Fig. 3.3.5, energy can be interchanged between the translational and rotational motion. According to equipartition theory each independent way of possessing energy (known as degrees of freedom) of the gas molecules should have the same energy. The three independent translational degrees of freedom (x, y and z motion) have $\frac{3}{2}nRT$ between them, i.e. $\frac{1}{2}nRT$ each. There are two rotational degrees of freedom too making the total molecular energy $5 \times \frac{1}{2}nRT = \frac{5}{2}nRT$. Hence the molar heat capacity at constant volume should be $\frac{5}{2}R$ – very close to the observed values.

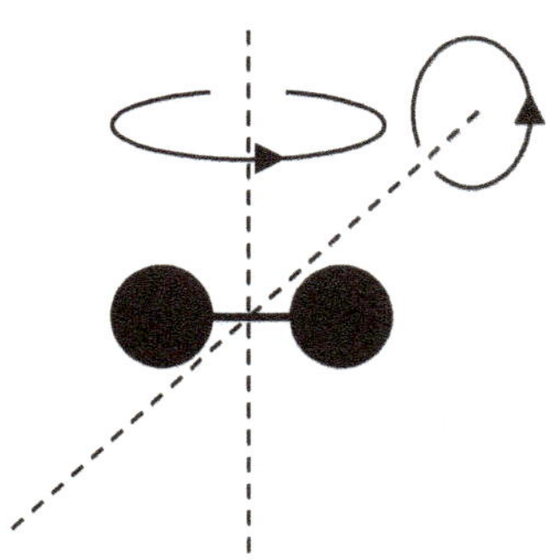

Fig. 3.3.8 Rotational degrees of freedom

Stretch & challenge

There is no kinetic energy of rotation about the axis along the bond in Fig. 3.3.8 for the same reason that monatomic molecules have no rotational kinetic energy.

3 The significance of the phrase 'at constant volume' will become clear in Section 3.4 Thermal physics.

Test yourself 3.3

1. The equation of state for an ideal gas can be written in the following ways:

   $pV = nRT$ and $pV = NkT$.

   State the quantity that each of the symbols in these equations represents and give the appropriate SI unit.

2. State what is meant by an ideal gas.

3. Nigel was asked to identify the quantities in the equation $pV = \frac{1}{3}Nm\overline{c^2}$. He wrote, '$N$ is the number of moles of the gas and m is the molar mass.' **These were both incorrect**.

   (a) State what he should have written for N and m.
   (b) State what the product Nm represents in this equation.
   (c) When he used the equation to calculate the pressure in a gas, his answer was correct in spite of his misidentification of N and m. Explain this.

4. The molar mass of a gas is $18\ \mathrm{g\ mol^{-1}}$.

   (a) State the relative molecular mass.
   (b) State the molecular mass (i) in u, and (ii) in kg.
   (d) [For chemists] Identify the gas.

5. The molar mass of gold is $196.97\ \mathrm{g\ mol^{-1}}$. Calculate the number of gold atoms in a $1.000\ \mathrm{kg}$ gold bar.

6. A cylinder contains $47\ \mathrm{kg}$ of propane gas (C_3H_8). Calculate: (a) The number of moles and (b) the number of molecules in the cylinder.

7. The density of pure salt (NaCl) is $2.163 \times 10^3\ \mathrm{kg\ m^{-3}}$. Calculate the number of sodium ions in a $1.000\ \mathrm{mm^3}$ salt crystal.

   $m(\mathrm{Na}) = 22.990\,\mathrm{u}$ $m(\mathrm{Cl}) = 35.453\,\mathrm{u}$ $1\,\mathrm{u} = 1.660 \times 10^{-27}\ \mathrm{kg}$

8. (a) Use the data below to calculate the relative molecular mass of uranium hexafluoride, UF_6.
   (b) State the molar mass of UF_6 (i) in $\mathrm{g\ mol^{-1}}$, (ii) in $\mathrm{kg\ mol^{-1}}$.
   (c) 0.7% of natural uranium is $^{235}\mathrm{U}$. Calculate the number of moles of $^{235}\mathrm{U}$ in 1 tonne (i.e. $1000\ \mathrm{kg}$) of uranium hexafluoride.

   $m(\mathrm{U}) = 238.02\ \mathrm{u}$ $m(\mathrm{F}) = 19.00\ \mathrm{u}$

9. The density of dry air at atmospheric pressure of $1.013 \times 10^5\ \mathrm{Pa}$ and $15\ ^\circ\mathrm{C}$ is $1.225\ \mathrm{kg\ m^{-3}}$. Use these data to calculate:

   (a) The rms speed of air molecules at $15\ ^\circ\mathrm{C}$.
   (b) The number of moles of air molecules in $1.0\ \mathrm{m^3}$ of air at $15\ ^\circ\mathrm{C}$.
   (c) The mean molar mass of air molecules, comparing your answer with that expected from the composition of air [$78.09\%\ N_2$ (rmm = 28), $20.95\%\ O_2$ (rmm = 32), 0.93% Ar (rmm = 40)].

10. In an experiment to determine the molar gas constant, R, a student weighs an oxygen cylinder, of volume $0.0100\ \mathrm{m^3}$ and reads the pressure gauge. She releases some oxygen, weighs the cylinder again and reads the final pressure (after allowing the cylinder to re-equilibrate with the room). She measures room temperature. Use her results to calculate a value for R. [1 atm = 101.3 kPa]

    Results:
    Temperature of room = $18.5\ ^\circ\mathrm{C}$; Initial pressure = 70.0 atm; Final pressure = 37.5 atm
    Initial mass of cylinder + oxygen = 2.832 kg; Final mass of cylinder + oxygen = 2.397 kg

11. Five molecules of oxygen have speeds in m s^{-1} of 150, 250, 300, 350 and 400.

(a) Calculate (i) the mean speed, (ii) the mean square speed and (iii) the rms speed of the molecules.
(b) Calculate the temperature of oxygen with the same rms speed as that calculated in part (a).

12. An interstellar gas cloud consists mainly of atomic hydrogen, molecular hydrogen and helium, with small quantities of atomic and molecular oxygen. [m(H) = 1.0 u; m(He) = 4.0 u ; m(O) = 16.0 u]

(a) Compare the rms speeds of these particles with that of atomic hydrogen.
(b) Calculate the rms speeds of the molecules at 50 K.

13. The atmosphere exerts a pressure of 100 kPa at ground level. To a good approximation its composition may be taken as $\frac{4}{5}N_2$ and $\frac{1}{5}O_2$.

(a) Estimate the number of molecules in the atmosphere. [Hint: consider the gravitational force on the atmosphere.] [Radius of Earth = 6370 km.]
(b) It has been suggested that every breath you take contains one or two molecules from the dying breath of Julius Caesar. Investigate whether this is plausible.

14. A sample of a monatomic ideal gas is contained in a cylinder of volume 0.679 m^3 has a pressure of 1.19 MPa.

(a) Calculate the total kinetic energy of the molecules in the gas.
(b) The temperature of the gas is 17 °C. Calculate:
(i) the mean molecular kinetic energy of the gas and (ii) the number of moles of the gas.

15. A university student used a high-pressure compressor to study the p–V isotherms [graphs of pressure against volume at constant temperature] of carbon dioxide at 40 °C and 50 °C and obtained the results shown in the graphs.

Use data from the graphs to discuss whether carbon dioxide behaves as an ideal gas at each of these temperatures.

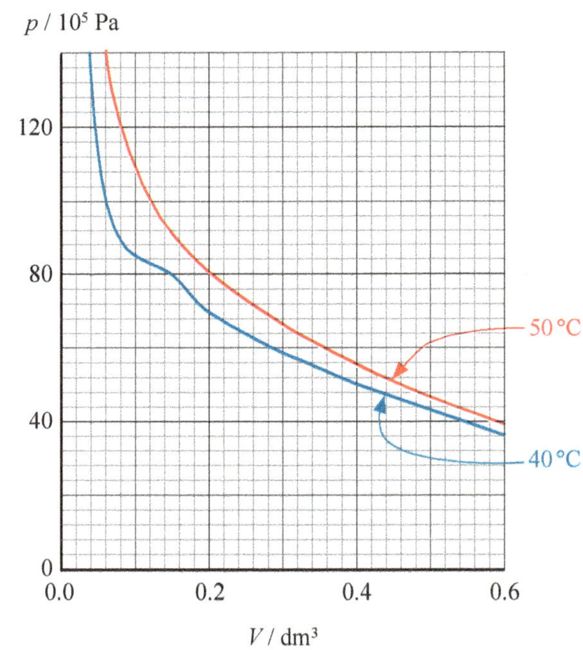

3.4 Thermal physics

The impulse for developing this central area of physics was the industrial revolution. The high grade chemical energy store of coal was used to pump water out of mines, transport people and goods, drive mill machinery and, eventually, generate electricity. Steam engines and internal combustion engines were developed to fulfil these needs and the questions of how to improve their power output and maximise their efficiency became central. The science of *thermodynamics* was developed in response to these requirements.

Fig. 3.4.2 Stott Park bobbin-mill engine

Fig. 3.4.3 Steam turbine under construction

Fig. 3.4.1 LNER's 60103, The Flying Scotsman

Study point

80% of the world's electrical power is currently (!) generated using steam turbines. This is down from a high of 90% because of the introduction of combined cycle gas turbines, photovoltaic panels and wind turbines.

3.4.1 Thermodynamic systems

Thermodynamics starts off with the concept of the **system**. This is just the collection of particles that we are considering. A system has a *boundary*, i.e. a surface which limits its extent; the universe outside is referred to as its *surroundings*.

In the context of heat engines, such as those illustrated in Figs. 3.4.1–3, the system is a collection of gas molecules and the boundary is the walls of a cylinder and piston. The engine is designed to extract energy from the gas molecules as they collide with the piston and make it move, i.e. they do **work**.

We'll now look at how we express the energy interchanges between a system and its surroundings.

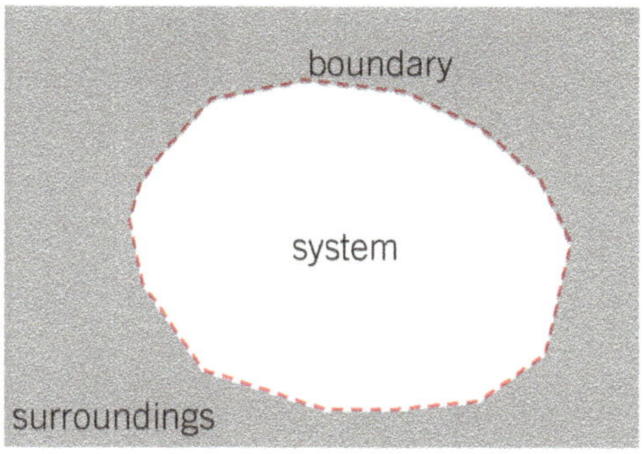

Fig. 3.4.4 System, boundary and surroundings

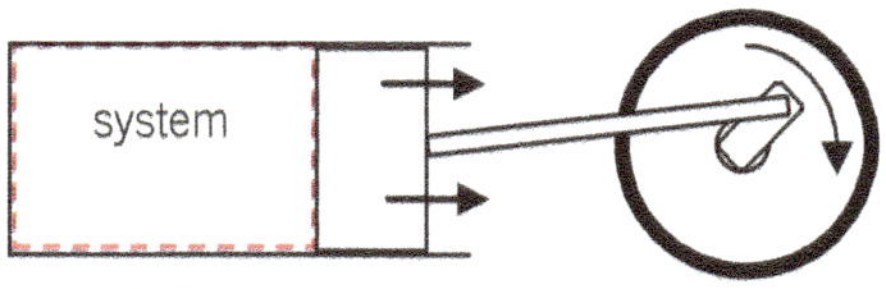

Fig. 3.4.5 Gas in a cylinder as a system

Study point

A thermodynamic **system** is the collection of particles, e.g. atoms and molecules, within a volume of space.

A system which does not allow energy or matter to enter or leave is called an **isolated system**.

A **closed system** is one which matter cannot enter or leave (so an isolated system is also closed).

A **heat engine** is a system which converts thermal energy into mechanical work in a cyclic process.

3.4.2 Internal energy, U

The first of the key concepts is that of the **internal energy** of a system. Put simply this is the sum of the energy possessed by all the particles. How can the particles possess energy? They can possess:

- Kinetic energy (translational and rotational)
- Potential energy.

Key terms

The **internal energy** of a system is the sum of the potential and kinetic energies of its particles.

Study point

Changes in the bond energies of molecules are only relevant when chemical reactions occur such as combustion.

We shall not consider changes in nuclear energy states in this context.

We have already looked at excited energy states in the context of lasers.

3.4.1 Knowledge check

Calculate the internal energy of:

(a) 3 moles of a monatomic gas at a temperature of 300 K.

(b) 4 kg of helium at 0 °C. (Molar mass = 4.0 g/mol)

Study point

The equation $U = \frac{3}{2}nRT$ tells us that the internal energy of a sample of ideal gas is a function only of temperature. This is not true of all systems, e.g. 1 kg of steam at 100 °C has a much higher internal energy than 1 kg of liquid water at 100 °C.

3.4.2 Knowledge check

Calculate the pressure of the gas in the model system if the mass of the piston is 1 kg and the diameter of the cylinder is 10 cm. Atmospheric pressure = 1.013×10^5 Pa.

We know from earlier work that the term *potential energy* covers a variety of different types, such as gravitational and elastic. In terms of thermodynamic systems we also have to consider the interactions between the particles, so there is also the potential energy due to:

- Intermolecular forces
- Chemical bonds
- Excited energy states in atoms
- Nuclear energy states.

Luckily, we are only going to be concerned with the *change* in internal energy, ΔU, so any forms of potential energy that stay constant need not be considered (see Study point). If the system is a sample of an ideal gas, then intermolecular forces can be ignored and we can consider U to consist entirely of the molecular kinetic energy.

i.e for a monatomic gas: $U = \frac{3}{2}nRT$ [1]

which we can also write as $U = \frac{3}{2}pV$ (see Section 3.3.6).

Example

Compare the internal energy of 2 mol of argon at 0 °C with that of 3 mol of helium at 200 °C.

Answer

$$U = \frac{3}{2}nRT \therefore \frac{U_{Ar}}{U_{He}} = \frac{n_{Ar}T_{Ar}}{n_{He}T_{He}} = \frac{2\text{ mol} \times 273.15\text{ K}}{3\text{ mol} \times 473.15\text{ K}} = 0.38$$

The $\frac{3}{2}pV$ form of equation [1] gives us a way of monitoring the internal energy of a simple system. We'll also meet a practical version of Fig. 3.4.6 in our specified practical work in Section 3.4.8 – the capillary tube containing dry air in Fig. 3.4.19.

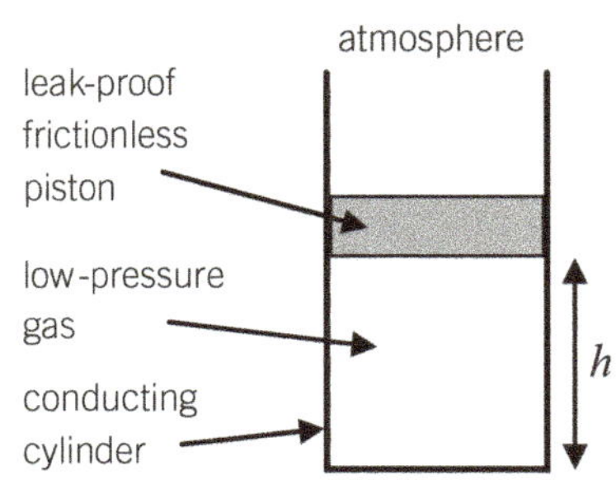

Fig. 3.4.6 Model system

3.4.3 Energy transfer by heat

(a) Temperature and thermal equilibrium

You'll use the variation of volume with temperature, in your specified practical work, to estimate a value for absolute zero. Here we'll use the same idea, in a 'thought experiment' to think about heat.

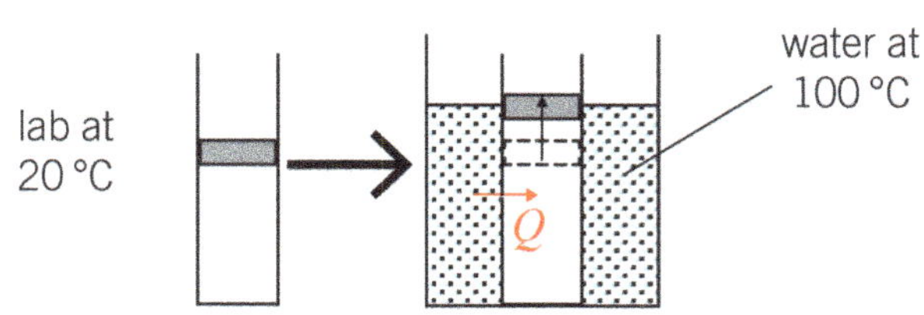

Fig. 3.4.7 Heat flow

If we take our cylinder of gas from the lab bench, at 20 °C, and plunge it into water at 100 °C (see Fig. 3.4.7) we notice that the gas expands. You can use the ideal gas equation (Section 3.3.3) to calculate by how much it expands. Because the volume of the gas has increased (at constant pressure) we know that the internal energy of the gas has increased. Where has this extra energy come from? It has moved into the gas by conduction through the cylinder walls, driven by the temperature difference between the water and the gas.

The name we give to this spontaneous flow of energy due to a temperature difference is **heat**, which we give the symbol Q. Notice from the definition of heat that we cannot talk about the 'heat content' of an object – heat is a movement of energy.

Before we moved the cylinder, the volume of the gas was constant, showing that its internal energy was constant. No heat was flowing because the temperature of the gas was the same as the temperature of the lab (20 °C). We say that the gas and the lab are in thermal equilibrium.

The change in pressure can also reveal heat flow as shown in the following example.

Example

A flask, with a volume of 250×10^{-6} m^3, contains of argon (a monatomic ideal gas) at 17 °C and a pressure of 1.0 kPa. Calculate the heat flow into the gas when the flask is plunged into water at 100 °C.

Answer

$pV = nRT$, so the number of moles, $n = \frac{pV}{RT} = \frac{1.0 \times 10^5 \times 250 \times 10^{-6}}{8.31 \times 290} = 0.0104$ mol

$\therefore \Delta U = \frac{3}{2}nRT = \frac{3}{2} \times 0.0104 \times 8.31 \times (100 - 17) = 10.8$ J

$\therefore$ Heat flow $Q = 10.8$ J

Note: This is not the only way of answering the example. We could use $U = \frac{3}{2}pV$, leading to $\Delta U = \frac{3}{2}(p_{100} - p_{17})V$. This requires the use of the ideal gas law to calculate p_{100}, the pressure at 100 °C (373 K), so it is no quicker. See Knowledge check 3.4.3.

(b) Temperature and internal energy

$U \propto T$ for an ideal gas. Indeed, we can be more specific: for a monatomic ideal gas $U = \frac{3}{2}nRT$. And generally the higher the temperature the greater the value of U for any system. However, consider what happens if we heat 1 kg of ice up from −100 °C until it has all boiled away as steam. An approximate graph of temperature, θ, against heat added, Q, is shown in Fig. 3.4.8.

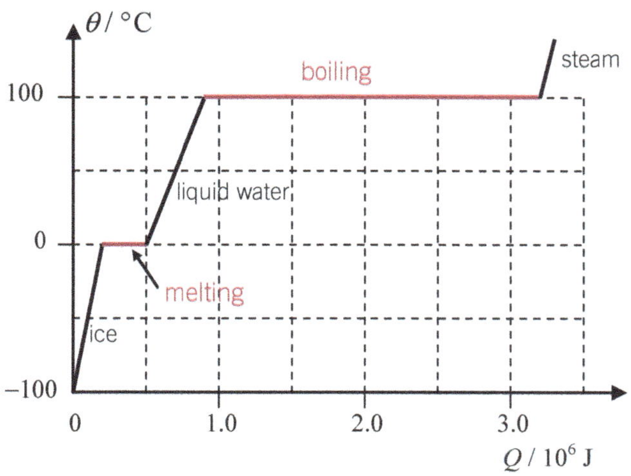

Fig. 3.4.8 The effect of heat on the temperature of 1 kg of water

Key terms

Heat: Q, the energy flow from a region at higher temperature to a region at lower temperature, due to the temperature difference.

Knowledge check 3.4.3

A sample of a monatomic gas has a volume of 0.50 m^3 at 300 K. It is heated to 450 K at constant pressure of 2.0×10^5 Pa.

Calculate (a) the increase in volume, (b) the increase in internal energy and (c) the heat flow into the gas.

Study point

Two systems in thermal contact are in **thermal equilibrium** if there is no flow of heat between them. They are said to be at the same temperature.

Knowledge check 3.4.4

The diagram shows 1 mol of helium and 1 mol of argon in thermal equilibrium. They are insulated from the surroundings.

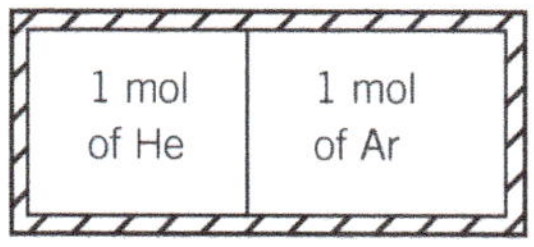

Compare the internal energies of the two gas samples. Explain your answer.

Study point

This book doesn't cover the practicalities of quantifying temperature. To do so, we need a system which has a measurable change with temperature (e.g. the height, h, of our model system in Fig. 3.4.6) and some standardised temperatures, e.g. absolute zero (0 K) and the triple point of water (273.16 K). This will not be examined.

Stretch & challenge

Compare the internal energy of:

(a) 1 mol of helium in thermal equilibrium with 1 mol of oxygen.

(b) 1 mol of nitrogen at 500 °C with 3 mol of argon at 500 K.

[Hint: See Section 3.4.6]

If we examine Fig 3.4.8, and taking the heat added as an indication of the increase in internal energy (see Study point), we see that the temperature does not always rise with internal energy. When water is changing state, there is a significant increase in internal energy without a rise in temperature. So, although we can say that two monatomic ideal gases at the same temperature have the same energy per mole, we cannot say it for other materials.

Study point

The rise in internal energy, ΔU is actually slightly less than Q, as we'll see later. The difference is less than 5% and occurs almost entirely during the change of state from liquid to gas.

Why is there a large change in internal energy when we melt or vaporise a material, even if the temperature remains the same?

When we change state, e.g. liquid to gas, we need to add energy to break a lot of inter-molecular bonds in separating the molecules, so even if the temperature of the substance stays the same, the total internal energy of the gas molecules is much greater than that of the liquid.

3.4.5 Knowledge check

Compare the internal energy of 0.50 m^3 of He at a pressure of 500 kPa with that of 2 mol of Ar at 400 K.

Example – using unfamiliar data

Use the graph on page 51 to estimate the effect on 10 kg of ice at 0 °C of adding 5.0 MJ of heat.

Answer

The effect will be the same as heating 1.0 kg by 0.5 MJ. From the graph, the first 0.3 MJ will just melt the ice. The remaining 0.2 MJ raises the temperature to about 50 °C.

Absolute zero

For all systems, as the temperature falls, the internal energy decreases. As the temperature drops towards absolute zero, 0 K, the internal energy falls to a minimum value. For real substances (as opposed to ideal gases), this minimum value of internal energy is, strangely, not equal to zero because of a quantum mechanical effect called zero point energy. One of the results of the motion associated with the zero point energy is that helium at atmospheric pressure does not freeze.

3.4.4 Energy transfer by work

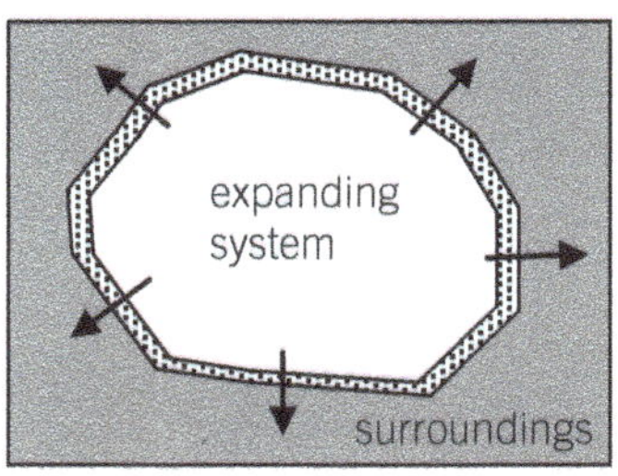

Fig. 3.4.9 A system doing work on the surroundings

We started this section by noting that the stimulus for developing the science of thermodynamics was to understand and improve the operation of heat engines. These engines do **work** on their surroundings, e.g. by driving vehicles, turning generators and pumping out mines. As we saw in the AS course:

- Work is the product of force and distance moved in the direction of the force.
- The energy transfer is equal to the work done.

Fig. 3.4.9 shows a system expanding into its surroundings. Assuming that the surroundings contain molecules (i.e. the expansion isn't into a vacuum) the system will need to exert a force on the surroundings and therefore does work as it expands. Consequently, energy is transferred to the surroundings and, unless energy is supplied to the system (e.g. by heat), its internal energy must decrease.

Key terms

The **work** done by a system is $p\Delta V$ where p is the pressure exerted and ΔV is the increase in volume.

(a) The work done by an expanding gas at constant pressure

The gas in the cylinder in Fig. 3.4.10 is doing work by expanding and pushing back the piston. If it exerts a force, F, and expands a distance, Δx, then:

Work done, $W = F\Delta x$

If p = the gas pressure and A = cylinder cross-sectional area

then $F = pA$ and $W = pA\Delta x$

But $A\Delta x = \Delta V$, the increase in volume of the gas

$\therefore$ $W = p\Delta V$ as long as the pressure remains constant.

The expression $p\Delta V$ simply gives the energy transfer to the piston and whatever it is connected to (crank, dynamo, etc.); it doesn't indicate what form the energy takes, whether kinetic, gravitational potential, elastic…

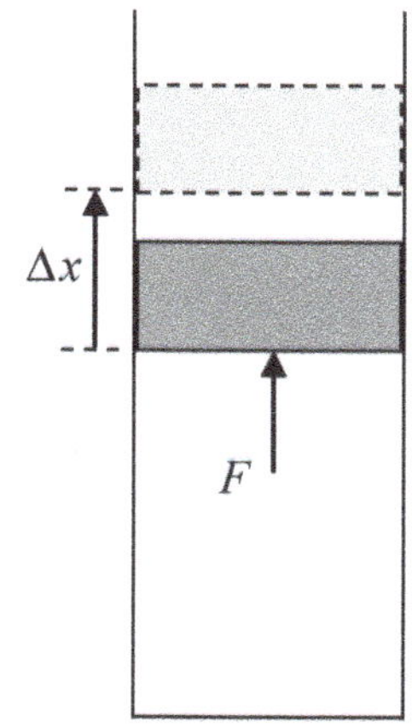

Fig. 3.4.10 Gas doing work against a piston

Study point

Surely the idea of a constant pressure expansion is not possible? In principle it is. The gas expansion in Fig. 3.4.7 was of this type. There was a source of heat which kept the pressure up as the gas expanded. In a gas turbine the pressure of the gas feed is constant.

(b) The work done by an expanding gas with varying pressure

The equation $W = p\Delta V$ for a constant pressure expansion tells us that, if the pressure varies during the expansion, the total work done for an expansion from volume V_1 to V_2 is given by:

$$W = \int_{V_1}^{V_2} p(V)\mathrm{d}V$$

i.e. the area between the p–V graph and the V-axis.

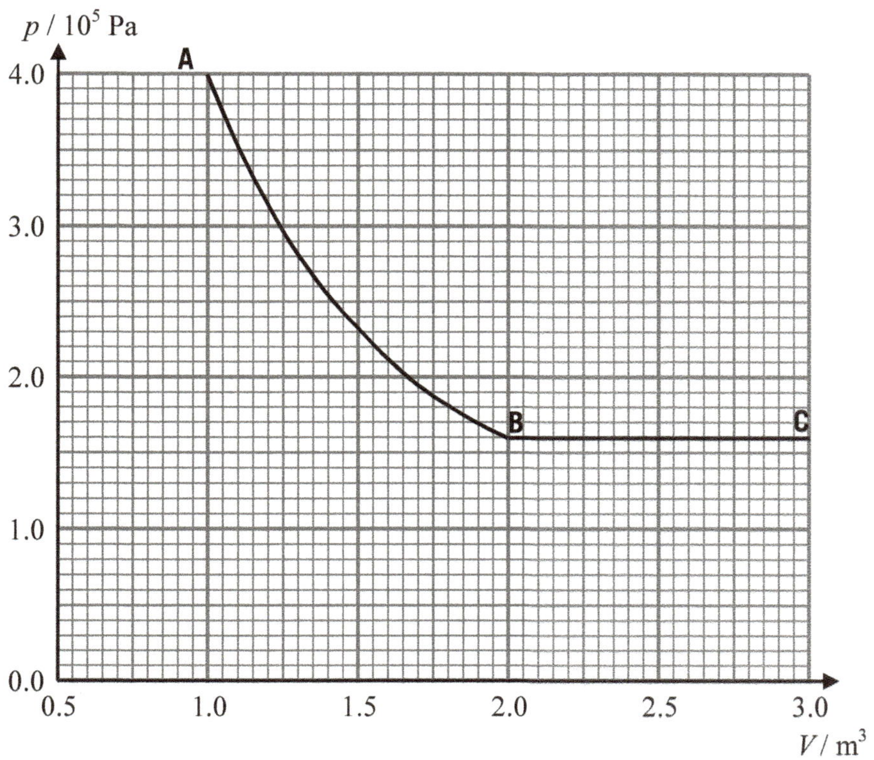

Fig. 3.4.11 The expansion of methane

Knowledge check 3.4.6

For the expansion of the methane sample in Fig. 3.4.11

(a) show that $T_C > T_A > T_B$, and

(b) find the ratio $\frac{T_B}{T_A}$,

where T_A is the temperature at **A**, etc.

Stretch & challenge

(a) For an ideal gas

$pV = nRT$

If an ideal gas expands at constant temperature from V_1 to V_2 show that, W, the work done is given by

$W = nRT \ln \frac{V_1}{V_2}$

(b) Calculate the work done by the methane sample in Fig. 3.4.11 if it expanded from A to 4 m^3 at constant temperature of 600 K.

Example

A sample of methane (CH_4) expands as shown in Fig. 3.4.11. Calculate the work it does in doing so.

Answer

Work done = Area under graph

= Area under **AB** + Area under **BC**

Area under AB = 493 small squares [see AS textbook Section 4.5.3]

$= 493 \times (0.1 \times 10^5 \text{ Pa} \times 0.05 \text{ m}^3)$

$= 246\,500 \text{ J}$

Area under BC $= 1.00 \text{ m}^3 \times 1.6 \times 10^5 \text{ Pa} = 160\,000 \text{ J}$

$\therefore$ Total work done $= 246\,500 \text{ J} + 160\,000 \text{ J} = 410 \text{ kJ}$ (2 sf)

3.4.7 Knowledge check

Calculate the mass of the methane sample in Fig. 3.4.11 given that the temperature at **A** is 600 K.

M_r (methane) = 18

Fig. 3.4.12 Working to make a fire

(c) Electrical work

The main focus of this chapter is on the work done by gases, e.g. in heat engines, but this is not the only form of work. If an energy transfer is directly caused by a temperature difference, we call it heat and give it the symbol, Q. Any other form of energy transfer is called work, W, even if it involves a rise in temperature, e.g. in the traditional way of making a fire illustrated by the survivalist in Fig. 3.4.12.

The hallmark of work is that it involves an ordered transfer of energy. In order to turn the round wooden stick pressing on the wooden block, the survivalist pushes a bow backwards and forwards; all the molecules of the bow string are moving in the same direction. Heat involves a disordered transfer, whether it be by conduction (random movement of molecules / electrons colliding) or radiation (random absorption of individual photons).

Now we'll consider an electrical circuit, containing a cell and a lamp, in terms of heat and work.

For the moment we'll consider the system consisting of the lamp only. There is energy being transferred in and out:

3.4.8 Knowledge check

Consider the cell (including the internal resistance) as a thermodynamic system. What are the energy inputs and outputs in time t if the cell's temperature is constant? Give your answer in algebraic form.

Energy in: transferred from the cell by means of an electric current. This is an ordered flow of charge, so it is work:

Electrical work input, $W_{in} = VIt$

Energy out: transferred by radiation (mainly infrared). This is disordered so it is categorised as heat, Q. Once the lamp has reached a steady temperature, $Q_{out} = W_{in} = VIt$.

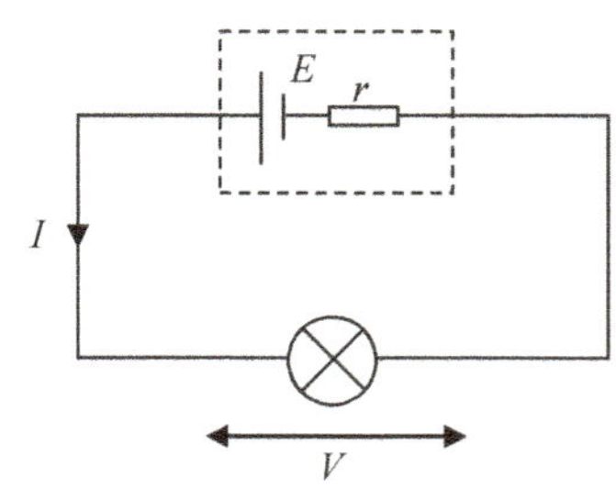

Fig. 3.4.13 Electrical circuit

We should now go on to consider what is happening to the internal energy of the lamp (and the cell) but we'll leave that until after the next section.

3.4.5 The first law of thermodynamics

The law of conservation of energy applied to thermodynamic systems is referred to as the **first law of thermodynamics**.

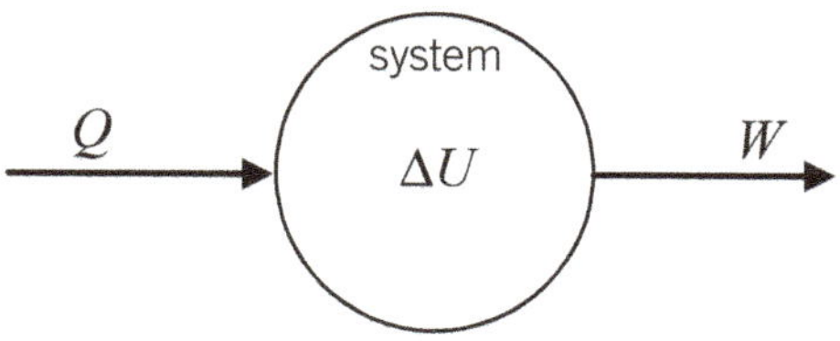

Fig. 3.4.14 $\Delta U = Q - W$

The relationship between U, Q and W is:

$$\Delta U = Q - W$$

where Q = heat input

and W = work output

Note carefully the direction of W and Q.

Question: Why has U got a Δ sign but Q and W haven't?

Answer: U, the internal energy, is a *state variable* of the system. At any one moment, U has a definite value. ΔU is the change in the value.

W is the work which is done by the system. The system does not 'have' a quantity of work, so its work cannot change.

Similarly, the system does not possess heat, so its heat cannot change. Heat is the flow of energy driven by a temperature gradient. It is worth looking at several examples of the application of this law.

Example 1

A 3 mol sample of monatomic gas is heated (e.g. by an electrical heater). Its temperature changes from 300 K to 400 K. Calculate the values of ΔU, W and Q if the change is at constant pressure.

Answer

$\Delta U = \frac{3}{2}nR\Delta T = 1.5 \times 3 \text{ mol} \times 8.31 \text{ J mol}^{-1} \text{ K}^{-1} \times 100 \text{ K} = 2490 \text{ J}$

$pV = nRT$, $\therefore$ at constant pressure $p\Delta V = nR\Delta T$ (see foot note[1])

$\therefore W = nR\Delta T = 3 \text{ mol} \times 8.31 \text{ J mol}^{-1} \text{ K}^{-1} \times 100 \text{ K} = 1660 \text{ J}$

$\therefore Q = \Delta U + W = 2490 \text{ J} + 1660 \text{ J} = 4150 \text{ J}$

Example 2

A cell of emf 9 V and internal resistance 1 Ω delivers a current of 1 A. Analyse the energy changes in a period of 1 minute in terms of the first law of thermodynamics. Assume that the cell has reached a constant temperature.

Answer

Terminal pd $V = E - Ir = 9 \text{ V} - 1 \text{ A} \times 1\ \Omega = 8 \text{ V}$

Work output $W_{out} = VIt = 8 \text{ V} \times 1 \text{ A} \times 60 \text{ s} = 480 \text{ J}$

Heat output $Q_{out} = I^2rt = (1 \text{ A})^2 \times 1\ \Omega \times 60 \text{ s} = 60 \text{ J}$

$\therefore$ Heat input, $Q_{in} = -60 \text{ J}$

$\therefore \Delta U = Q - W = -60 \text{ J} - 480 \text{ J} = -540 \text{ J}$

i.e. the cell has done 480 J of electrical work on the circuit and has emitted 60 J of heat, so its internal energy has decreased by 540 J.

1 The value of W doesn't depend on the pressure. If the p were doubled, the V values and therefore ΔV would be halved, so $p\Delta V$ stays the same.

Key terms

The **first law of thermodynamics**: The increase, ΔU, in the internal energy of a system is $\Delta U = Q - W$, in which Q is the heat **entering** the system and W is the work done **by** the system.

Study point

Each of Q, W and ΔU can be positive, negative or zero, e.g.

1. If you compress a gas, work is done **on** (not by) the gas, so $W < 0$.
2. When a cup of tea cools down, it **gives out** heat, so $Q < 0$.

Knowledge check 3.4.9

For the equation

$\Delta U = Q - W$

state the sign of:

(a) ΔU if $Q > 0$ and $W = 0$

(b) Q if $\Delta U = 0$ and $W < 0$

(c) W if $Q = 0$ and $\Delta U > 0$

(d) ΔU if $W > Q > 0$

(e) ΔU if $Q > 0$ and $W < 0$.

Knowledge check 3.4.10

State (no calculation needed) the answers to Example 1 if the volume of the gas and not the pressure were constant.

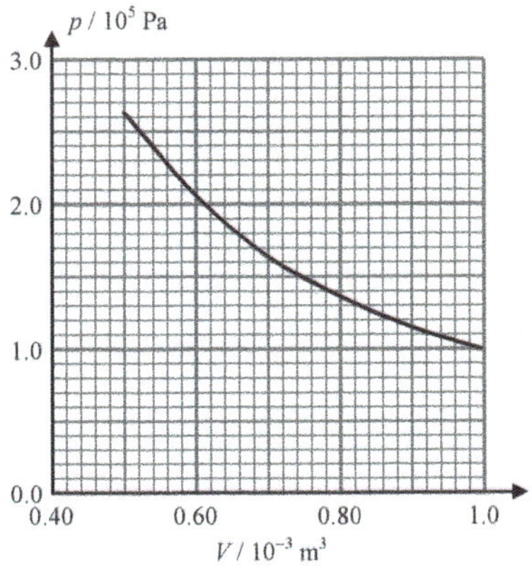

Fig. 3.4.15 Compressing a gas

Example 3

A well-insulated sample of an ideal gas at $300\ \text{K}$ is compressed from a volume of $1.0 \times 10^{-3}\ \text{m}^3$ to half this volume; the pressure increases by a factor of 2.63 as shown in Fig. 3.4.15.

(a) Calculate the work done on the gas.
(b) State the heat supplied to the gas and the change in internal energy.

Answer

(a) Work done = area under graph
Using the trapezoidal rule (see AS Textbook Section 4.5.3) with $\Delta V = 0.1 \times 10^{-3}\ \text{m}^2$, the author obtained
Work done = $80\ \text{J}$

(b) $Q = 0$ because the gas is thermally insulated.
$W = -80\ \text{J}$ because the work is done **on** the gas
$\Delta U = Q - W = 0 - (-80) = 80\ \text{J}$

Stretch & challenge

The internal energy of methane, $U = 3nRT$

(a) Use equipartition theory (Section 3.4.6) and the shape of the molecule to explain why.
(b) Calculate the values of Q, W and ΔU for sections AB and BC of Fig. 3.4.11.

3.4.6 Heat and temperature change

When we heat a system, its temperature usually increases (see Study point). The first law of thermodynamics applies to this process, so we'll examine a typical case and see what the values of ΔU, Q and W are.

Study point

Heating ice at $0\ ^\circ\text{C}$, or water at $100\ ^\circ\text{C}$, results in a change of state rather than temperature.

See Fig. 3.4.8.

(a) Applying the first law of thermodynamics

Consider a $1\ \text{kg}$ sample of water at $20\ ^\circ\text{C}$. If we heat it up, e.g. using an electric kettle, its temperature rises. Experimentally, if we supply $200\ \text{kJ}$ of heat its temperature increases to almost $70\ ^\circ\text{C}$ (an increase of $50\ ^\circ\text{C}$).

$\therefore\ Q = 200\ \text{kJ}.$

$\therefore\ \Delta U + W = 200\ \text{kJ}.$

W is the work done against the atmosphere. $\therefore\ W = p_A \Delta V$, where p_A is atmospheric pressure $\sim 1.0 \times 10^5\ \text{Pa}$.

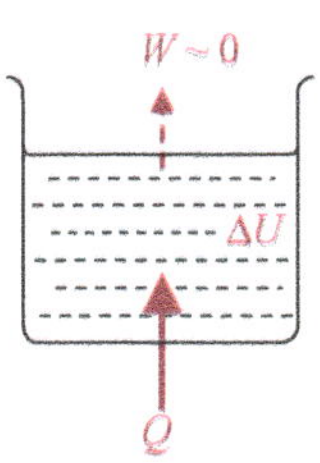

Fig. 3.4.16 Heating water

Experimentally, the volume of the water increases from $1000\ \text{cm}^3$ to about $1020\ \text{cm}^3$, i.e. $\Delta V = 20\ \text{cm}^3\ [2.0 \times 10^{-5}\ \text{m}^3]$.

$\therefore\ W = p\Delta V \sim 1.0 \times 10^5\ \text{Pa} \times 2 \times 10^{-5}\ \text{m}^3 = 2\ \text{J}$

So, because water only expands by a small fraction, W is negligible and therefore we can write, $W \approx 0$ and $\Delta U \approx Q = 200\ \text{kJ}$. Actually, water expands more than most engineering materials; for example, $1000\ \text{cm}^3$ steel only expands by $2\ \text{cm}^3$ over the same temperature range, so the approximations $W \approx 0$ and $\Delta U \approx Q$ are even better for other solids and liquids.

Stretch & challenge

A website gives the coefficient of linear expansivity of cobalt (i.e. the fractional increase in *length* per degree) as $12 \times 10^{-6}\ \text{K}^{-1}$. Find the increase in *volume* of a $1\ \text{m}^3$ cube when its temperature is raised by $10\ ^\circ\text{C}$ (i.e. $10\ \text{K}$).

(b) Specific heat capacity

The rise in temperature, $\Delta\theta$, of an object, provided no change of state occurs is found experimentally to be closely proportional to the heat input, Q. For an element, compound or mixture of fixed composition, $\Delta\theta$ is inversely proportional to the mass of material. So, for a particular material, we can write

$$Q = mc\Delta\theta$$

where c is a constant called the **specific heat capacity** (shc) of the material. The word *specific* in this context means *per unit mass*. A verbal definition of shc is 'the heat per unit mass needed to raise the temperature of a material per degree'. It is easier to define it using the equation (but you must remember to define the symbols).

Example

A 0.50 kg block of iron, initially at 20 °C, is heated for 5 minutes with a 25 W heater. Calculate the final temperature. [$c_{\text{iron}} = 450 \text{ J kg}^{-1} \text{ K}^{-1}$]

Answer

Heat input $Q = Pt = 25 \text{ W} \times 300 \text{ s} = 7500 \text{ J}$

$$Q = mc\Delta\theta, \therefore \Delta\theta = \frac{Q}{mc} = \frac{7500 \text{ J}}{0.50 \text{ kg} \times 450 \text{ J kg}^{-1} \text{K}^{-1}} = 33.3 \text{ K} = 33.3 \text{ °C}$$

$\therefore$ Final temperature = 20 °C + 33.3 °C = 53 °C (2 sf)

Note: The *difference* between two temperatures is the same regardless of whether they are expressed in degrees celsius or kelvin:

$$80 \text{ °C} - 20 \text{ °C} = 60 \text{ °C} \quad \text{and} \quad 353 \text{ K} - 293 \text{ K} = 60 \text{ K}$$

Hence the unit of c can equally be written J kg^{-1} °C^{-1}.

Key terms

The **specific heat capacity**, c, of a substance is defined by
$Q = mc\Delta\theta$

where Q is the heat input, m the mass and $\Delta\theta$ the temperature change.

UNIT: J kg^{-1} K^{-1} (or J kg^{-1} °C^{-1})

Study point

The term **specific heat capacity** is an unfortunate and misleading one. It suggests:

(a) That heat is something that can be contained, which does not agree with the definition of heat.

(b) And that a substance can only contain a certain amount of this.

The term dates back to the long-discarded 18th-century theory of heat as a fluid (called caloric) which could be added or taken away from objects.

(c) Heat transfer

We often need to consider the transfer of heat between systems. If two systems are in thermal contact, then heat will flow from the system at the higher temperature into the one at lower temperature. This will continue until the systems are in thermal equilibrium, i.e. their temperatures are equal.

Assuming that there is no heat exchange with the surroundings we can say that the heat flow from one system is equal to the heat flow into the other, which will allow us to calculate the equilibrium temperature.

Consider the two systems in Fig. 3.4.17. θ_1 and θ_2 are the initial temperatures. Let the final temperature be θ. Then:

heat loss by system 1 = heat gain by system 2

$$\therefore m_1c_1(\theta_1 - \theta) = m_2c_2(\theta - \theta_2)$$

We don't need to learn this equation but we need to be able to apply the principle of heat gain = heat loss together with $Q = mc\Delta\theta$.

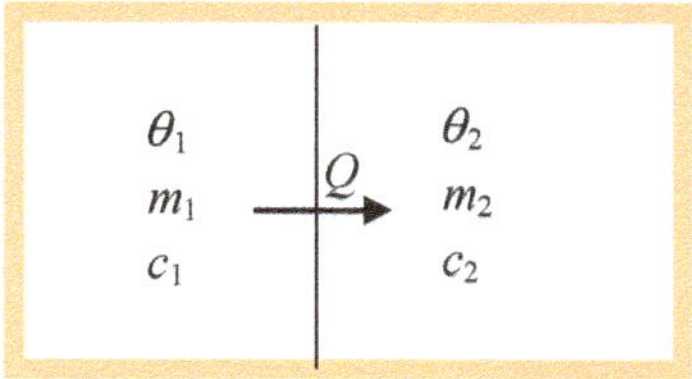

Fig. 3.4.17 Heat transfer

Study point

If the temperature of system 1 changes from θ_1 to θ then:
temperature fall = $\theta_1 - \theta$

If the temperature of system 2 changes from θ_2 to θ
temperature rise = $\theta - \theta_2$

Study point

In the example, it doesn't matter whether we express the masses in g or kg (or tonnes!) as long as both are in the same units. Similarly the specific heat capacities can be in kJ kg^{-1} K^{-1} or J kg^{-1} K^{-1} (or even calories per ounce per °C) as long as the units of the two sides are the same.

Example

1.0 kg of water at 100 °C is poured into a 0.50 kg porcelain teapot at 20 °C. Calculate the equilibrium temperature (assuming no heat loss to the surroundings).

$c_{water} = 4.18$ kJ kg^{-1} K^{-1}; $c_{porcelain} = 1.07$ kJ kg^{-1} K^{-1}

Answer

Let equilibrium temperature be θ.

heat lost by water = heat gained by teapot

$\therefore$ Using $Q = mc\Delta\theta$:

$$1.0 \times 4.18\,(100 - \theta) = 0.50 \times 1.07\,(\theta - 20)$$

A bit of algebra leads to: $\theta = \frac{428.7}{4.715} = 91$ °C (2 sf)[2]

3.4.7 Engines for doing work and pumping heat

The push for the development of the science of thermodynamics was the need to understand and develop better heat engines – machines for using the energy released by combustion to do work. Heat engines themselves will not be examined but you do need to understand hcw to use p–V diagrams [a.k.a. indicator diagrams] to investigate the transfer of heat and work. We took a first look at this in Example 3 in Section 3.4.5. We'll now consider a machine in which a 12.0 mol sample of a monatomic ideal gas is taken around a cycle of expansion and compression, **ABCDA**.

3.4.11 Knowledge check

Calculate W_{out} and W_{in} by finding the area under **AB** and **CD** respectively and hence show that the net work output is 42 kJ.

Study point

The net work done in one cycle [i.e. $W_{out} - W_{in}$ in Fig. 3.4.18] is the area enclosed by the loop. With curved graphs it is often easier to calculate W_{out} and W_{in} separately

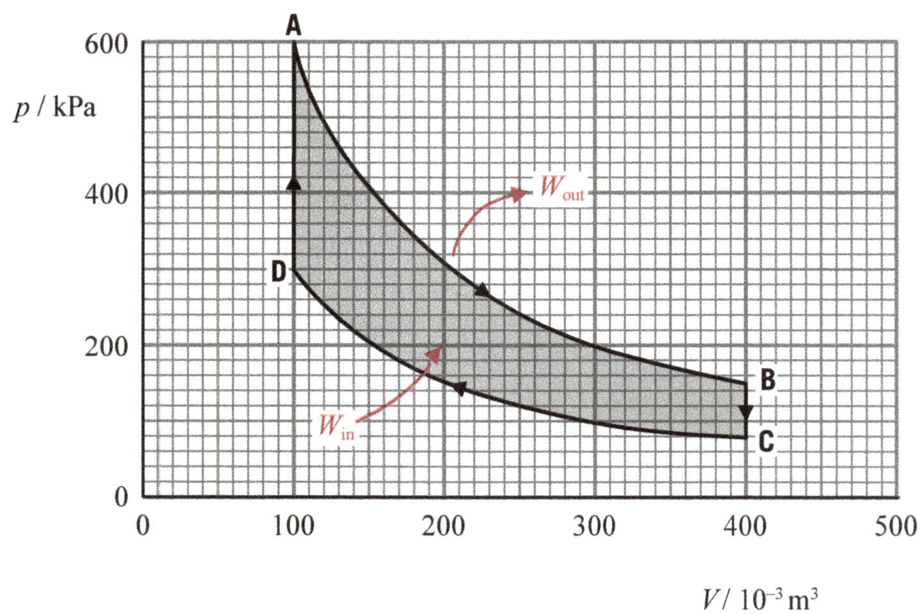

Fig. 3.4.18 Estimating absolute zero

3.4.12 Knowledge check

Calculate the temperature at each of **A**, **B**, **C** and **D**.

We'll look at the principles of how this cycle can be made to do work – leaving the task of the calculations to you. The work done by a gas in expanding is the area under the p–V graph, so it is clear that the gas does more work in **AB** (W_{out}) than is done on it in **CD** (W_{in}).

2 Tea brews ('mashes' in Yorkshire) best as close to 100 °C as possible which is why you should always pre-warm the pot.

Task 1: (KC 3.4.11) Calculate W_{out} and W_{in} and hence show that the net work done by the gas is 42 kJ.

Task 2: (KC 3.4.12) Calculate the temperature of the gas at **A**, **B**, **C** and **D**.

Task 3: (KC 3.4.13) Internal energy: Show that $U_A = U_B$ and $U_C = U_D$ and calculate these quantities.

Clearly, heat must be supplied between **D** and **A** ($Q_{DA} > 0$) and extracted between **B** and **C** ($Q_{AC} < 0$). What about **AB** and **CD**? In both these parts work is done and $\Delta U = 0$ so heat must be exchanged:

Task 4: (KC 3.4.14) Show that the total heat input **D** → **B** is 128 kJ.

If we want to calculate the heat output from **B** to **D**, we could now do the same calculation as in Task 4. But we can also note that the net work output around the cycle is 42 kJ so the net heat input must also be 42 kJ (conservation of energy) and so the heat output from **B** to **D** is 128 kJ − 42 kJ = 86 kJ.

Power output

Consider an engine with 4 cylinders, each with an indicator diagram as in Fig. 3.4.18. If the engine was running at 1500 cycles per minute, the power output, P, would be given by:

$$P = 4 \times 42 \text{ kJ} \times \frac{1500}{60 \text{ s}} = 4.2 \text{ MW}$$

Efficiency of a heat engine

In most heat engines, the heat lost (e.g. in **BC** and **CD** in Fig. 3.4.18) is simply thrown away. This is what happens in a car engine or a thermal power station (but see Study point). Hence the efficiency of the heat engine is given by:

$$\text{Efficiency} = \frac{42 \text{ kJ}}{128 \text{ kJ}} \times 100\% = 33\%$$

Not impressive! Can it be made better? Yes, in principle. If the temperature of AB were increased, this would help. See KC 3.4.15.

Heat pumps

Heat doesn't flow spontaneously from a low temperature to a high temperature. But we can give it a little nudge by running a heat engine in reverse. If the cycle in Fig. 3.4.18 were reversed, for a net work input of 42 kJ, there would be a heat flow of 128 kJ. This is the principle of the heat pump – though usually the fluid used isn't an ideal gas but one which can be liquefied by compressing it. Heating your home with heat pumps is much more efficient with a lower level of carbon emissions, assuming it is powered by electricity which is generated by renewable resources. The use of ground source heat pumps is currently (2021) being seriously investigated as part of the contribution towards a net zero carbon economy.

Knowledge check 3.4.13

Use $U = \frac{3}{2}pV$, $\left[\text{or } U = \frac{3}{2}nRT\right]$ to calculate the internal energy at each of **A**, **B**, **C** and **D** and hence show that $U_A = U_B$ and $U_C = U_D$.

Knowledge check 3.4.14

Use $\Delta U = Q - W$ to show that the total heat input **D** → **B** is 128 kJ.

Study point

In some European countries, Combined Heat and Power (CHP) stations use the waste heat for domestic heating for the local community.

Knowledge check 3.4.15

The cycle in Fig 3.4.18 is changed so that **A** and **B** are at a temperature of 900 K. Calculate the efficiency of a heat engine based upon this.

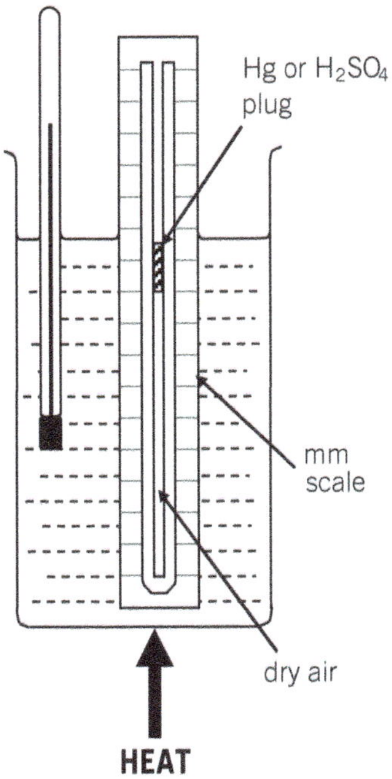

Fig. 3.4.19 Variation of volume with temperature

3.4.8 Specified practical work

(a) Estimation of absolute zero using the gas laws

The variation of volume of a gas with temperature at constant pressure can be investigated using the apparatus in Fig. 3.4.19. The water in the beaker is heated slowly in stages using a bunsen burner. The temperature and the length of the air column in the capillary tube are measured at each stage. The volume of the air is proportional to the length of the air column.

The p-T variation is undertaken similarly using the apparatus in Fig. 3.4.20.

For both investigations, the dependent variable (length or pressure) is plotted against the temperature, a best-fit straight line drawn and the intercept on the temperature axis noted, as shown in Fig. 3.4.21.

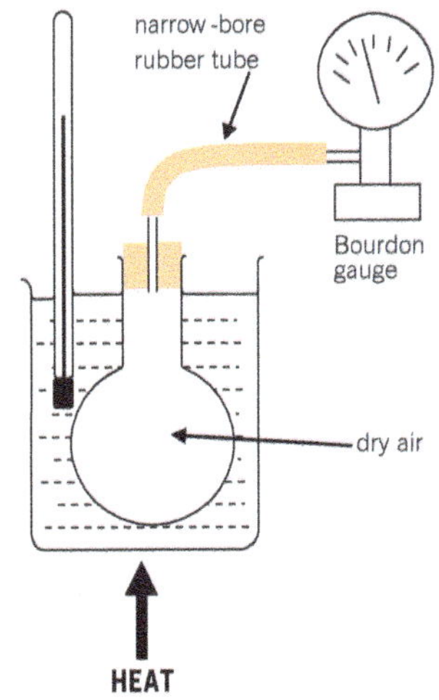

Fig. 3.4.20 Variation of pressure with temperature

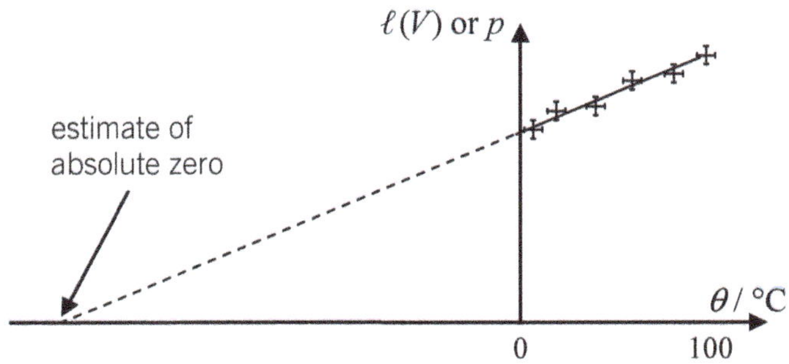

Fig. 3.4.21 Estimating absolute zero

3.4.16 Knowledge check

The equation of the best-fit graph of p (in atm) against θ (in °C) is

$p = (0.0038 \pm 0.0002)\theta + 1.00$

Estimate the value of absolute zero in °C.

(b) Measurement of the specific heat capacity of a solid

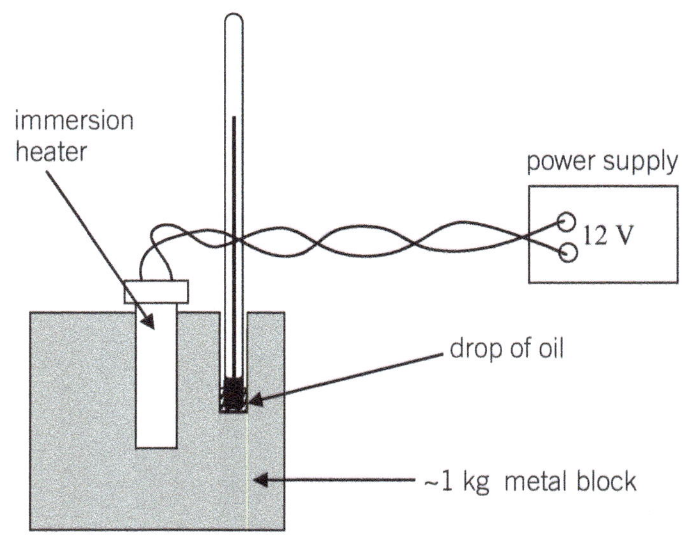

Fig. 3.4.22 Measuring the specific heat capacity of a metal

Study point

Experiments to determine specific heat capacity of solids at A level are generally restricted to metallic solids because their high thermal conductivity allows them to reach thermal equilibrium quickly.

The drop of oil is to provide a good thermal contact between the thermometer and the metal block.

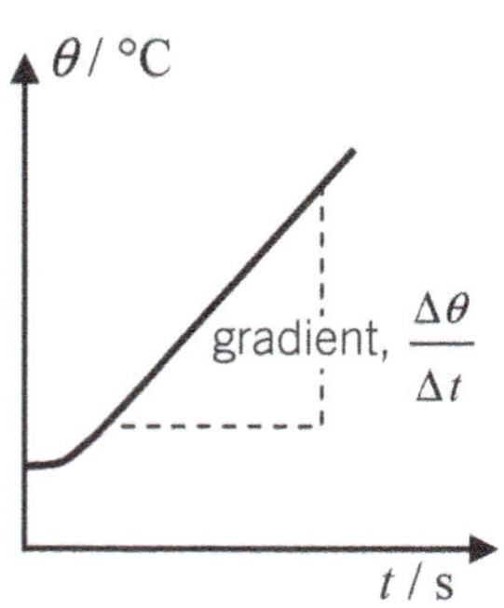

Fig. 3.4.23 Temperature–time graph for determining c

If necessary, an initial experiment is undertaken to determine the power of the immersion heater, by measuring the current and pd and applying $P = IV$.

The immersion heater is switched on and the temperature displayed on the thermometer noted at regular intervals, e.g. every 30 s. A graph of temperature against time is plotted and the gradient of the linear portion determined (see Fig. 3.4.23).

The initial horizontal portion of the graph is a delay until the heat from the immersion heater reaches the position of the thermometer. After this has occurred

$$P\Delta t = mc\Delta\theta$$

$$\therefore c = \frac{P}{m \times \text{gradient}}$$

Test yourself 3.4

1. A cylinder of volume $0.060\ \text{m}^3$ contains helium at a pressure of $8.0\ \text{MPa}$.

   (a) Calculate the internal energy of the helium.
   (b) Explain why your answer to (a) doesn't depend upon the temperature of the helium.
   (c) The mass of helium in the cylinder is $800\ \text{g}$. Calculate:
   (i) the temperature (ii) the rms speed of the helium molecules.

2. An oxygen cylinder contains $3.2\ \text{kg}$ of oxygen at $27\ °\text{C}$.

   (a) Calculate the translational kinetic energy of the oxygen molecules.
   (b) Calculate the rms speed of the oxygen molecules.
   (c)

   Stretch & challenge

   The internal energy of the oxygen molecules is $620\ \text{kJ}$ (2 sf). Account for the difference between this figure and your answer to part (a).

3. A rigid, insulating container has two compartments as shown, each containing $500\ \text{mol}$ of a monatomic ideal gas. [Take 1 atm as $100\ \text{kPa}$]

A	B
$2\ \text{m}^3$	$4\ \text{m}^3$
12 atm	4 atm

   (a) Calculate the internal energy in each compartment.
   (b) Calculate the temperature in each compartment.
   (c) The insulation between the two halves is now removed but the rigid dividing wall remains in place. Calculate:
   (i) The equilibrium temperature.
   (ii) The pressure in each compartment at equilibrium.
   iii) The net heat flow between the two compartments.
   (d) The dividing wall is now removed so the gas molecules can mix. Calculate the final temperature and pressure.

4. Consider the following situations and for the given system, state whether the quantities Q, ΔU and W from the first law of thermodynamics, written in the form $\Delta U = Q - W$, are positive, negative or zero.

   (a) A sample of ideal gas is compressed rapidly, so that no heat has time to escape.
   (b) The temperature of the sample of air in Fig. 3.4.20 is raised.
   (c) A block of steel at 20 °C is placed into a furnace at 500 °C.
   (d) A sample of gas at 2 atm in a cylinder is allowed to expand slowly against a piston, so that its temperature remains constant.
   (e) A medical cylinder of oxygen is placed in a warm room after being taken out of a cold room.
   (f) The gas in a cylinder is allowed to escape into a vacuum.
   (g) A mixture of methane and oxygen, in a sealed, rigid thermally insulated container, spontaneously burns to produce carbon dioxide and water. [The system is all the particles initially in the methane and oxygen.]

5. It takes 2.26 MJ of heat to vaporise 1 kg water at 100 °C. The density of steam at 100 °C is 0.60 kg m^{-3}.

(a) Estimate the work which the water does in expanding into steam against the atmosphere.
(b) Calculate the fraction of 2.26 MJ that your answer to part (a) represents.
(c) State the values of Q, ΔU and W in the equation $\Delta U = Q - W$.

6. In a coffee shop, steam is bubbled through 250 g of milk to raise its temperature from 20 °C to 100 °C. The specific heat capacity of milk is $3800 \text{ J kg}^{-1} \text{ K}^{-1}$.

(a) Calculate the heat input into the milk.
(b) State the values of Q, ΔU and W for the milk in the equation $\Delta U = Q - W$.

7. A sample of an ideal gas is taken around the sequence of pressures and volumes, **ABCDA**. Calculate:

(a) The ratios $\frac{U_B}{U_A}$, $\frac{U_C}{U_A}$ and $\frac{U_D}{U_A}$.
(b) The ratios $\frac{T_B}{T_A}$, $\frac{T_C}{T_A}$ and $\frac{T_D}{T_A}$.
(c) The values of ΔU, Q and W for each of the four stages **A→B**, **B→C**, **C→D** and **D→A**.
(d) The total values of ΔU, Q and W for the whole cycle.
(e) The number of moles of gas given that $T_A = 350$ K.

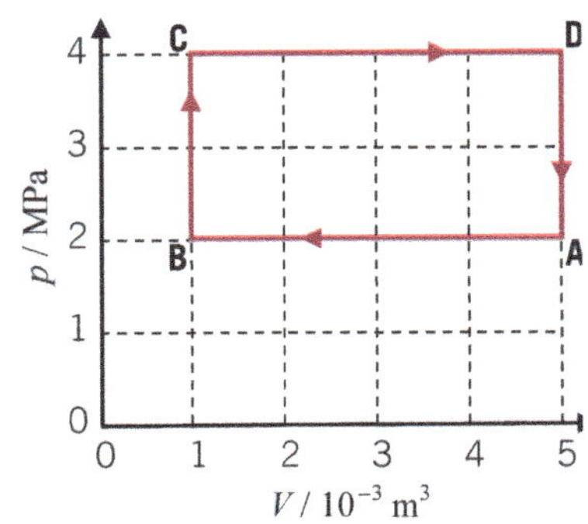

8. A hypothetical engine consists of six cylinders each of which contains a sample of the gas as in Q7. The gas in each cylinder is taken around the sequence **ABCDA** 1500 times per minute. Calculate the power output of the engine.

9. A 250 cm^3 flask contains dry air. The flask is connected to a pressure gauge, which is calibrated in bar [1 bar = 1.00×10^5 Pa]. The flask is placed in a large beaker containing an ice/water mixture at 0 °C. Temperature and pressure readings are taken as the beaker is gradually heated up to 100 °C. The results are as follows.

Temperature / °C	0	15	34	48	64	85	100
Pressure / bar	0.95	1.02	1.06	1.11	1.18	1.22	1.30

By plotting a suitable graph, use the results, including ±0.02 bar error bars in pressure to calculate a value for absolute zero with its absolute uncertainty. Show your working.

10. Travel immersion heaters can be used to for making hot drinks in hotel rooms. Nia uses a travel immersion heater, which is advertised as 230 V, 0.50 kW, to measure the specific heat capacity of water.

She heats up 0.700 kg of water in a 0.250 kg glass beaker and measures the temperature every 30 seconds.

Results:

Time / minutes	0	1	2	3	4	5
Temperature / °C	19.5	29.0	39.5	48.5	58.5	68.0

Use these results to calculate a value for the specific heat capacity of water.

[Specific heat capacity of glass = $840 \text{ J kg}^{-1} \text{ °C}^{-1}$.]

3.5 Nuclear decay

The story of Henri Becquerel's serendipitous discovery of radioactivity is quite well known. He discovered that a sample of uranium salt had fogged a sealed photographic plate through its wrapper. This discovery was shocking because it appeared to violate the principle of conservation of energy – the uranium salt continued to affect film emulsions without any obvious energy source.[1] Even though this was a fortuitous initial observation, his scientific rigour in not just rejecting it but in investigating it thoroughly made the 1903 Nobel Prize for physics, which he shared with Marie and Pierre Curie, highly deserved. The extract from his lab notebook in Fig. 3.5.1 includes the shadow of a metal Maltese cross which had been placed between the salt and the photographic plate.

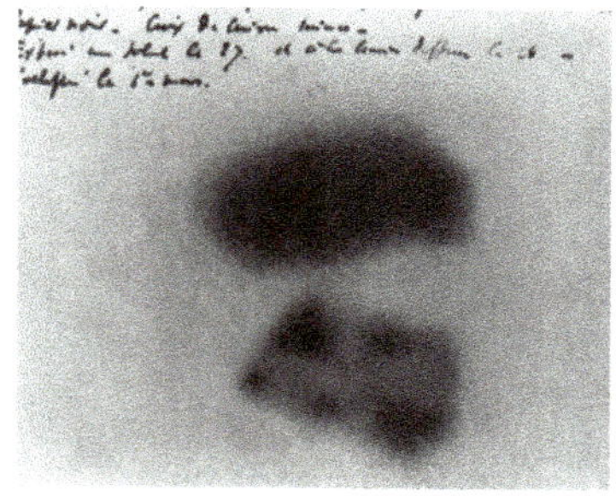

Fig. 3.5.1 Extract from Becquerel's laboratory notebook

The discovery of radioactivity paved the way for research into the structure of the atom and the later development of quantum theory, so its importance cannot be overstated.

Key terms

Nuclear radiation is the emission of energy in the form of waves or particles from the atomic nucleus. The term is also used for the emitted waves or particles themselves.

Nuclear decay (or radioactive decay) is the process by which the nucleus of an atom loses energy by emitting (nuclear) radiation.

3.5.1 Nuclear decay

Following the discovery of **nuclear radiation**, by Henri Becquerel, it was quickly established that there were three kinds with different penetrating powers, which he called α, β and γ radiation. The range of each type of radiation depends upon their energy and the medium but the following are typical:

- α range: 3 cm of air; 0.02 mm of glass
- β range: 1 m of air; 3 mm aluminium
- γ range: 500 m of air; 20 cm aluminium; 5 cm lead

The alpha particles given out by a radio-nuclide have a well-defined energy, so they penetrate a set distance in a particular substance. The α particles, which produced the tracks in Fig. 3.5.2, have two energies, so two ranges. β particles are produced with a continuous spectrum of energies (see AS textbook, p 87) so have no clearly-defined range. We shall discuss the range of nuclear radiation in Section 3.5.2.

Fig. 3.5.2 Alpha particle tracks

We now know that these types of radiation are caused by a change in the nucleus of the atom, allowing it to assume a lower energy configuration. This process is called **nuclear decay**. The atomic nucleus is composed of protons and neutrons, themselves composite particles consisting of combinations of up and down quarks. In the same way that atoms have energy levels, so do nuclei. A nucleus with excess energy can drop down to lower energy levels in various ways, which we shall now explore.

Recap on nuclear terms and symbols

Symbols like $^{14}_{6}C$ and $^{235}_{92}U$ will be used a lot when discussing nuclear decay and nuclear fission. What do these symbols tell us? The following diagram should make it clear:

Top tip

The WJEC terms and definitions booklet includes an explanation of the $^{A}_{Z}X$ notation, in terms of the mass number and atomic number.

The atomic number (or proton number), Z, is the number of protons in the nucleus and the mass number (or nucleon number), A, is the total number of nucleons (i.e. protons and neutrons) in the nucleus. The symbol can be used for the nucleus as well as the neutral atom. As this chapter deals with nuclear decay, we shall usually be using the symbol to indicate the nucleus.

1 Naturally, physicists proceeded to include nuclear energy in the recognised forms of energy, thus preserving the inviolability of the principle!

>>> **Study point**

Alpha radiation (α): Fast-moving particles consisting of two protons and two neutrons (i.e. helium nuclei) ejected from the nucleus.

Beta radiation (β): High-speed electrons (β^-) or positrons (β^+) ejected from the nucleus.

Gamma radiation (γ): High energy photons ejected from the nuclei of radioactive atoms.

>>> **Study point**

Note that the symbols ${}^{A}_{Z}X$ in a decay equation represent nuclei rather than atoms. The sum of the values of A and Z on the two sides of the equation must be equal, i.e. the equation must balance.

3.5.1 Knowledge check

The atomic number of uranium is 92. Uranium-235 undergoes α decay to an isotope of thorium.

Write the decay equation for this process.

3.5.2 Knowledge check

Write the decay equation for ${}^{106}Te$.

3.5.3 Knowledge check

(a) Which conservation laws are illustrated in the neutron decay equation?

(b) What indications are there that it is a weak decay?

>>> **Study point**

In nuclear equations, the top line of numbers represents nucleon conservation; the bottom line, charge conservation. Also when writing β-decay equations it is quite common to omit the neutrino.

One thing to notice is that the chemical symbol (X) gives the same information as the atomic number (Z) because, for example, carbon atoms always have 6 protons in their nuclei (otherwise they wouldn't be carbon!). Hence it is quite common to write, ${}^{14}C$, ${}^{235}U$, etc., leaving out the atomic number.

(a) Alpha (α) decay

The combination of two protons and two neutrons is particularly stable. It is the composition of the nucleus of a ${}^{4}_{2}He$ atom. The identity of the α particle with ionised helium was demonstrated in an elegant experiment[2] in 1907.

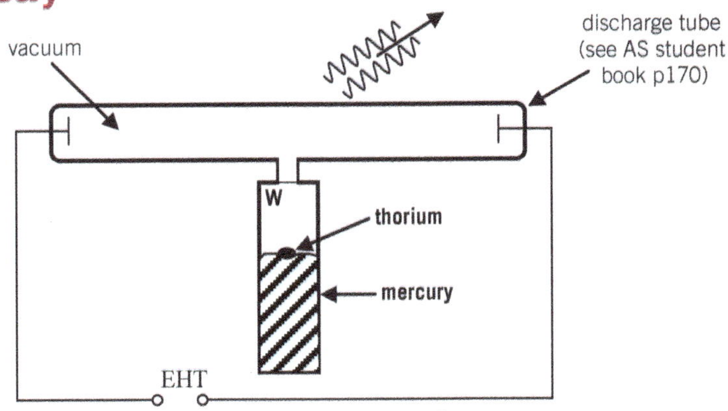

Fig. 3.5.3 The nature of α radiation

Thorium, an α-emitter, was floated on mercury. The α particles could pass through the very thin window, **W**, into the gas discharge tube. After a few days, the characteristic lines of helium were seen in the spectrum of the radiation.

Almost 100% of naturally occurring thorium is the isotope ${}^{232}_{90}Th$. The decay equation is:

$$ {}^{232}_{90}Th \rightarrow {}^{228}_{88}Ra + {}^{4}_{2}He, $$

where the symbol ${}^{4}_{2}He$ is used for the α particle. Generalising from this example, we see that in an alpha decay, from a nucleus ${}^{A}_{Z}X$ the decay products are:

- a daughter nucleus, ${}^{A-4}_{Z-2}Y$ and
- an alpha particle ${}^{4}_{2}He$

α decay only occurs in heavy elements. The lightest known α-decay nuclide is telliurium-106 (${}^{106}_{52}Te$) which decays to an isotope of tin (Sn).

(b) Beta (β) decay

The more common form is β^- decay, which takes place in neutron-rich nuclei. A neutron decays, via the weak interaction, into a proton with the emission of an electron and an electron anti-neutrino. Using the symbolism of particle physics:

$$ n \rightarrow p + e^- + \overline{\nu}_e $$

In radioactive decay equations it is usual to use ${}^{0}_{-1}\beta$ to represent the beta particle; ${}^{0}_{-1}e$ can also be used.

The equation for the decay of potassium-42 is

$$ {}^{42}_{19}K \rightarrow {}^{42}_{20}Ca + {}^{0}_{-1}\beta + {}^{0}_{0}\overline{\nu}_e $$

So a β^- decay from ${}^{A}_{Z}X$ produces:

- a daughter nucleus ${}^{A}_{Z-1}Y$
- a β^- particle, i.e. an electron, e^-
- an electron anti-neutrino.

2 By Ernest Rutherford and Thomas Royds.

Some nuclei, created in fusion processes, can decay by the emission of a positron, rather than an electron. In the context of radioactivity this is called a β^+ decay and the particle is called a β^+ particle, e.g.

$$^{13}_{7}\text{N} \rightarrow {}^{13}_{6}\text{C} + {}^{0}_{1}\beta\ (+\ {}^{0}_{0}\nu_e)$$

Note the difference in the nuclear symbols for the β^+ and β^- particles. In order for this to happen, a proton within the ^{13}N nucleus must transform into a neutron (with the emission of a positron and a neutrino), i.e.

$$\text{p} \rightarrow \text{n} + \text{e}^+ + \nu_e$$

But how is this possible without violating the conservation of mass/energy? The proton has a mass/energy of 938.3 MeV; the neutron is more massive, weighing in at 939.6 MeV. And then there is the positron with a mass/energy of 0.5 MeV, to say nothing of the kinetic energy of the neutrino. The answer lies in the nuclear energy levels (see Stretch & challenge).

Maths tip

Physicists and chemists often use the unified mass unit, u, to express the masses of atoms and molecules.

Nuclear and particle physicists often use the energy units and the formula $E = mc^2$ to express the mass/energy of particles in the units MeV or GeV.

Stretch & challenge

Just like electrons in atoms, the neutrons and protons in a nucleus each have their own sets of energy levels. This concept is illustrated in Fig. 3.5.4 for a proton-rich nucleus. The proton in the top energy level can transform and drop down into the vacant neutron energy level – but only if ΔE is enough. Suggest a minimum value of ΔE for β^+ decay to occur.

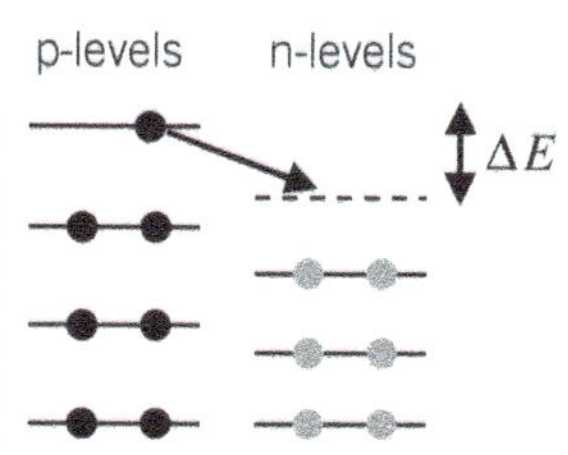

Fig. 3.5.4 Nuclear energy levels

Knowledge check 3.5.4

A nucleus, X, has atomic number Z and mass number A. It undergoes β^+ decay. State the products of the decay.

(c) Gamma (γ) decay

The daughter nucleus from an α or a β decay is often formed in an excited state. That means that either a proton or a neutron has a vacant lower energy state to drop down into. When it does this, the energy is carried away by a high energy photon which is indicated by γ. The proton and nucleon numbers (Z and A) of the nucleus are unchanged in the process.

Knowledge check 3.5.5

In a γ decay the mass of the nucleus decreases because of the $E = mc^2$ relationship. Calculate the decrease in mass caused by the emission of a 1 MeV photon. Express your answer in (a) MeV, (b) kg and (c) u.

(d) Some less common decay modes

The decay modes you need to know for the exam are α, β and γ but there are a few others that you might come across:

Electron capture (a.k.a. K capture[3])

This occurs in some proton-rich nuclei. An electron in the 1s orbital, the lowest atomic energy level, spends a tiny fraction of its time actually inside the nucleus. The shading in Fig. 3.5.5 represents qualitatively the probability density function of the 1s orbital – this is the probability of finding the electron in any particular small region. It is actually a maximum at the centre of the atom, but because the nucleus is so tiny (~10^{-15} m) compared to the orbital (~10^{-10}) the electron spends only a tiny fraction of its time there. Nevertheless, when it does, the following reaction can occur:

$$\text{p} + \text{e}^- \rightarrow \text{n} + \nu_e$$

Of course, mass/energy must still be conserved.

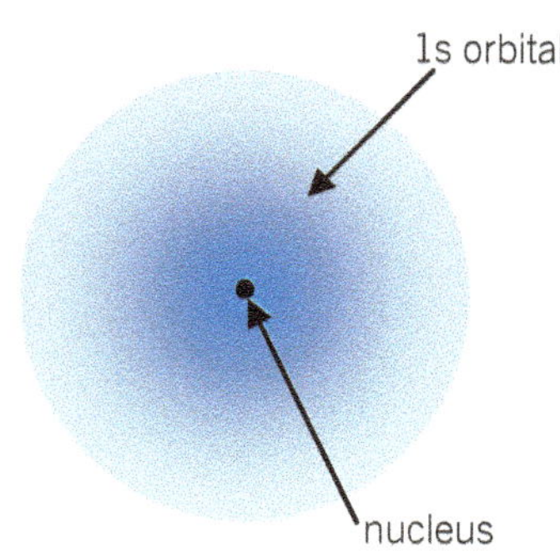

Fig. 3.5.5 Nucleus and 1s orbital

3 Chemists historically refer to the 1s orbital as the K-shell.

Spontaneous fission

Most ^{235}U nuclei decay by α emission. However, in about 2×10^{-7} of the decays, the nucleus splits spontaneously into two with the emission of several (on average, 1.86) neutrons. These neutrons can induce other ^{235}U nuclei to split in the process of induced fission. Spontaneous fission also occurs in ^{238}U, ^{239}Pu and ^{240}Pu. This form of decay is different from **neutron emission** which can occur in some artificial neutron-rich nuclides, e.g. ^{10}Li and ^{17}N – with a very short half-life (typically < 1 ns). **Proton emission** has also been observed in artificial proton-rich nuclides, e.g. ^{151}Lu, again with a very short half-life (81 ms).

3.5.6 Knowledge check

State the effect on the nucleus $^{A}_{Z}X$ of (a) K capture (b) proton emission.

3.5.2 The range (penetrating power) of nuclear radiation

The different penetrating powers of α, β and γ radiation are well known and form the basis of pre-A level experiments on radioactivity. You should remember the following:

- The penetrating power of all forms depends upon their energy: the greater the energy the greater the range.
- The denser the material the better it is at absorbing radiation of any type, i.e. the lower the range of the radiation.
- Gamma rays are the most penetrating (tens or hundreds of metres of air, several cm of dense metals, e.g. lead), then beta particles (about 1 m of air, a couple of mm of light metals), then alpha (stopped by thin paper, 3–5 cm of air, the dead epidermis of the human skin).

In order to understand these differences we need to understand the absorption mechanisms operating.

Study point

Neutrinos, whether from β emission or from nuclear fission, have a very long range – estimated as 5 light-years in lead! This is because they only interact via the weak interaction. They are never detected in school laboratory radioactivity experiments.

3.5.7 Knowledge check

Write the decay equation for N-11 by proton emission.

(a) The absorption of charged particles

When a charged particle, e.g. an alpha particle, penetrates a material, it is able to pass through the constituent atoms – though neither the particle nor the atoms come through unscathed.

Study point

From Fig. 3.5.6 you will see why α particles are referred to as ionising radiation. This is also true of β and γ (and UV and X-rays).

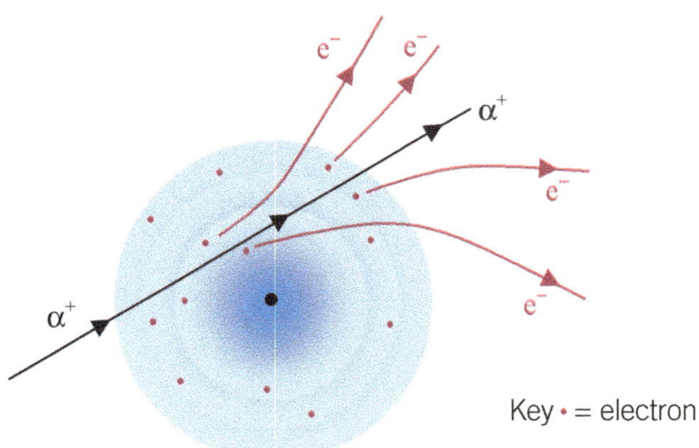

Fig. 3.5.6 Effect of an alpha particle on an atom

If an alpha particle with a typical energy of several MeV passes through an atom, it will attract electrons (because of the opposite charges). It doesn't need to collide directly with an electron – the electric attraction will cause an energy transfer and pull several electrons from the atom. The energy transfer is typically a few tens of eV. The effect on the α particle is to slow it down slightly, so it passes through the next atom a bit more slowly, increasing the ionising effect (see Study point). Eventually it runs out of energy, picks up a couple of stray electrons and becomes a helium atom.

So what about β particles? The effect is the same – it doesn't matter whether they are β^+ or β^- they will still kick out electrons. But there is one difference. Alpha and beta particles have the same energy range (~100 keV – 5 MeV) but electrons and positrons are much less massive, so they travel more quickly (very close to the speed of light in fact). You should be able to explain why they are more weakly ionising than alpha particles and therefore why they have a bigger range.

(b) The absorption of gamma rays

These are ionising, too, but their interaction with the electrons of atoms is different in two important ways:

1. They are uncharged so they can only affect an atomic electron if they pass very close. Therefore they pass through a large number of atoms before interacting.
2. They lose all their energy in one interaction (remember the photoelectric effect).

Suppose we have a beam of γ-ray photons passing through a uniform material. A certain fraction, say 0.9 (i.e. 90%), gets through the first 10 cm. What about the next 10 cm? The 'surviving' photons have not lost any energy, so they have just the same chance, 0.9, of passing though the next 10 cm. So after 20 cm, the fraction 'surviving' is $(0.9)^2 = 0.81$. After 30 cm, we have $(0.9)^3 = 0.73$, etc. Fig. 3.5.7 is a graph of intensity against distance, from an initial 100 units. This relationship is investigated in the specified practical work in Section 3.5.8. It is an exponential decay curve, and has the same form as that of activity against time for a radioactive nuclide – see Section 3.5.6.

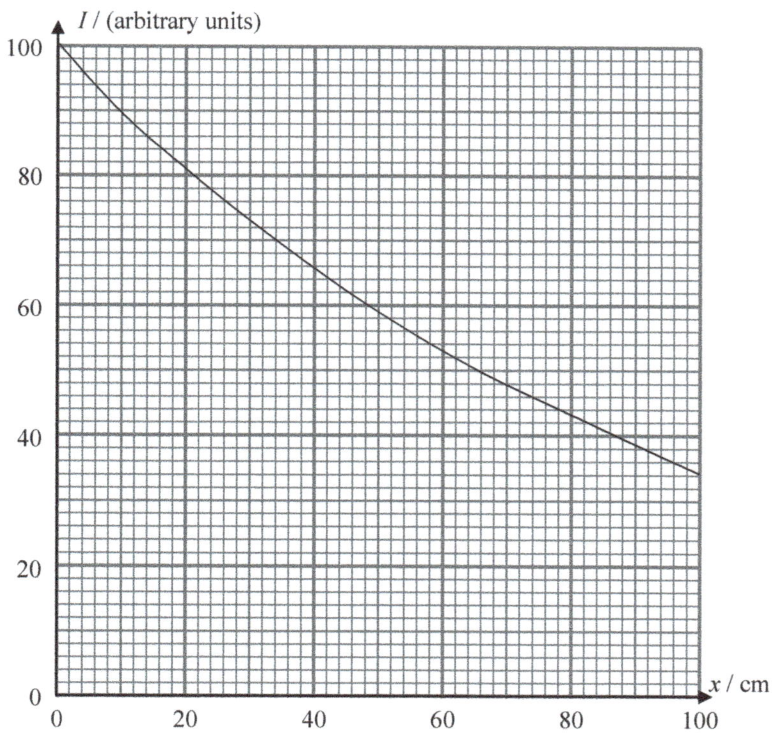

Fig. 3.5.7 Drop in intensity of a γ beam with thickness of absorber

Study point

If the alpha particle exerts a force F on an electron for a time Δt it transfers a momentum $F\Delta t$. So the longer it spends in an atom, the more disruptive it is. If you look back at Fig. 3.5.2 you'll see that the tracks become thicker towards the end because of the increased ionising effect.

Knowledge check 3.5.8

A 5 MeV alpha particle, loses an average of 0.1 keV of kinetic energy each time is passes through at atom.

(a) Estimate the number of atoms it passes through before coming to a halt.

(b) Explain why it has a greater range in gases than in solids.

Knowledge check 3.5.9

For the intensity distance relationship in Fig. 3.5.7,

(a) State the fraction of the photons which penetrate 1.0 m.

(b) Use your answer to (a) to estimate the fraction which penetrate 5 m.

Stretch & challenge

See Section M3.3 for the discussion on exponential decay. The intensity, I, is given by

$I = I_0 e^{-\mu x}$

where μ is the fraction penetrating per unit distance. For the data given, the value of μ is $0.0105\ \text{cm}^{-1}$. and $I_0 = 100$ a.u. Use these values to find x for which $I = 10$.

Knowledge check 3.5.10

Explain why strongly ionising radiation has a shorter range than weakly ionising radiation.

3.5.11 Knowledge check

The ionic radius of metallic aluminium ($Z = 13$) is 67.5 pm.

For lead ($Z = 82$) it is 91.5 pm. Estimate the ratio of the electron concentrations in aluminium and lead.

(c) Why are denser materials more effective in absorbing radiation?

Most of the mass of a material is in the nuclei, so the denser the material, the more protons and neutrons there are. The numbers of protons and electrons are the same. Comparing a gas to a solid, there are a lot fewer atoms (in a gas) per unit volume, so there are a lot fewer electrons and it is these which absorb the radiation.

Comparing solids of different elements: atoms are all more or less the same size – with an atomic radius of something like 0.1 nm. As we increase the value of Z, the more the inner electron shells are pulled towards the nucleus (by its greater charge) but the atomic radius is hardly changed. This means that the concentration of electrons increases with the density. Hence denser means more absorption.

Top tip

This very brief review of radiation detectors is for completeness, to give you some idea of how they work. **You will not be examined on Section 3.5.3.**

3.5.3 Detecting and measuring nuclear radiation

Nuclear radiation is generally detected by the ionisation that it produces. Because it is only weakly ionising, γ is therefore more difficult to detect than α and β. A detector for α radiation must also be constructed with a very thin window otherwise the radiation will not be able to enter the detector.

Study point

The GM tube is able to detect α, β and γ radiation. The α detection is directional because the α particles can only enter the tube via the mica window. Even the weakly ionising γ rays produce sufficient electrons to be registered.

The role of the bromine is to mop up excess electrons and so reduce the duration of the pulse and hence shorten the 'dead time' during which no events can be registered.

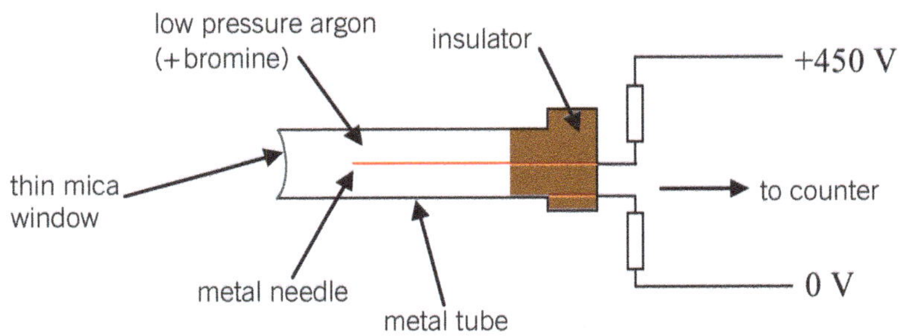

Fig. 3.5.8 The Geiger-Müller tube

The nuclear radiation ionises argon atoms. The liberated electrons, from the argon atoms, are attracted to the positively charged metal needle (typical potential ~400 V) and pick up enough kinetic energy to cause further ionisation on the way. The resulting pulse of current produces a voltage pulse across the resistor which is registered on the counter.

Solid state detectors, which are essentially reverse-biased silicon diodes, are commonly used. The advantages of solid state detectors are that they operate on low voltages (typically 3–5 V) and their dead time is very small, so they can detect high levels of radiation. Their performance does degrade with time as they accumulate radiation damage.

3.5.4 The random nature of decay

(a) The variability of results

If you listen to the output speaker of a Geiger counter (i.e. a GM tube connected to a counter), watch the rising count display or watch the needle of an analogue rate meter, you'll soon notice that the counts arrive at random times. This needs to be borne in mind when you are measuring the **background radiation** or observing a source.

The following are 20 readings, each over 1 minute, of the background radiation in Aberystwyth:
24, 20, 26, 29, 27, 24, 25, 22, 25, 21, 21, 28, 27, 25, 18, 24, 19, 27, 32, 28

The readings all fluctuate around ~25 (the mean value is 24.6). The range appears to be ± 7 counts per minute (cpm). What happens if we time for longer? These are 10 minute readings:
268, 227, 247, 258, 230, 240, 232, 226, 266, 236, 248, 242, 246, 269, 264, 230, 260, 237, 238, 267

The mean value agrees nicely with the mean of the 1-minute readings (see KC 3.5.12). The results have a larger spread than the 1-minute readings but when we convert them to cpm we find a *smaller* spread: extreme values 22.7–26.9 i.e. ± 2. The difference is highlighted in Fig. 3.5.9.

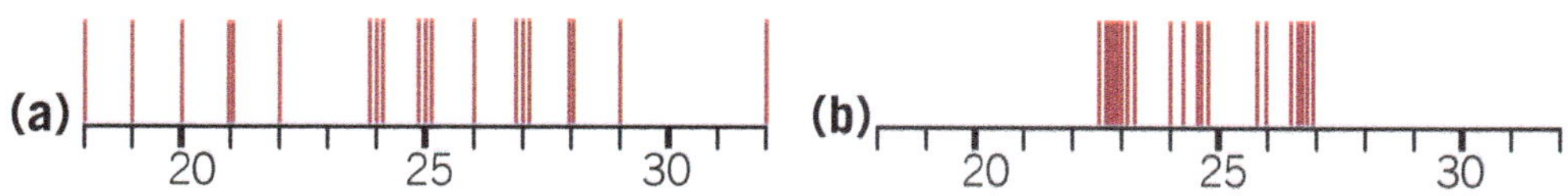

Fig. 3.5.9 Comparison of cpm variation with (a) a 1-minute and (b) a 10-minute counting period

(b) Long counting period or repeat readings?

As long as the activity of the radioactive sample is constant over the time-scale of the experiment (e.g. much shorter than the half-life) it doesn't matter whether one long reading is taken or repeat readings adding up to the same time, e.g. measure the background radiation over 10 minutes or do it 10 times over one minute (or 5 times over 2 minutes...). In all cases, the mean rate (i.e. cpm or cps) will be the total count divided by the total time. For example, the total of the first 10 of the 1-minute readings is 243 counts, giving a mean rate of 24.3 cpm.

(c) Correcting for background radiation

This is done by subtracting the background count rate from the count rate with the source present or the background count from the count with the source, (over the same period of time).

Study point

In the context of laboratory investigations into radioactive sources, **background radiation** is the nuclear radiation which is detected in the absence of the source under investigation.

Knowledge check 3.5.12

Calculate the mean counts per minute of the 10-minute background readings.

Stretch & challenge

Detailed statistical analysis predicts that, if the long-term mean count over a period is N, the distribution of a large number of individual readings will have a standard deviation of $\sqrt{N}$: ~70% of readings will be within $N \pm \sqrt{N}$ and ~95% will be within $N \pm 2\sqrt{N}$. Check the 1-min and 10-min values – and comment.

Knowledge check 3.5.13

Calculate the mean count rates of the 1-minute background readings in batches of 10: i.e. 1st – 10th (done); 2nd – 11th; 3rd – 12th, etc. Compare the results with the 10-minute variation.

Stretch & challenge

A GM tube placed near a radioactive source detects 1659 counts in 10 seconds. The counting time is controlled by a manual switch. Does the random decay process or the uncertainty in handling the switch have a bigger effect on the reading?

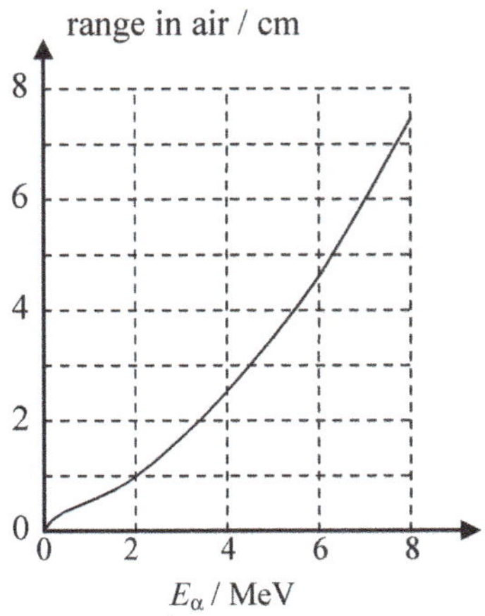

Fig. 3.5.10 Range of α particles in air as a function of energy

3.5.5 Distinguishing between α, β and γ radiation

(a) Using penetrating power

You will recall from your pre-A level knowledge that we can use penetrating power to distinguish between α, β and γ radiation. The rule of thumb is:

- If it can't penetrate a sheet of paper or more than ~3–5 cm air it is α.
- If it can penetrate the above but is stopped by thin metal, e.g. 1–2 mm steel; 2–3 mm aluminium, it is β (either β^- or β^+).
- If it gets through all of the above it is γ.

The range of any of the radiations actually depends on their energy. Fig. 3.5.10 shows how the range of α particles varies with their kinetic energy.

Here is a slightly tough GCSE example (it would usually come with a diagram) but the numbers are easy:

Study point

To answer the example we need to assume that:

1. The aluminium is thick enough to absorb all the β and none of the γ.
2. The paper absorbs 100% of the α and none of the β and γ.

These are both approximations but the first one is less true than the second. The aluminium will absorb some of the γ. It will also give out some X-rays produced by the decelerating electrons but these are unlikely to be detected.

Example

A GM tube is placed 2 cm from a source of nuclear radiation. It detects 525 counts per minute (cpm). With a paper absorber, the count rate falls to 400 cpm; with a 2 mm thick aluminium absorber it is 225 cpm. What are the fractions of α, β and γ radiation in the emissions from the source?

The background reading was 250 counts in 10 minutes.

Answer

(We need to make certain assumptions – see Study point.)

The background count rate $= \dfrac{250 \text{ counts}}{10 \text{ min}} = 25 \text{ cpm}$

Thus the total radiation detected from the source = 525 – 25 = 500 cpm

Using the obvious symbols:

$$\alpha + \beta + \gamma + 25 = 525 \qquad [1]$$

$$\beta + \gamma + 25 = 400 \qquad [2]$$

$$\gamma + 25 = 225 \qquad [3]$$

From [3] the γ is thus 200 out of 500 = 0.40 of the total, i.e. 40%

Subtracting [3] from [2] → β = 175

So the β is 175 out of 500 = 0.35 of the total, i.e. 35%.

Then either by requiring that the fractions all add up to 1.00 (100%) or by subtracting [2] from [1], α is 0.25 of the total, i.e 25%.

Note: alternatively subtract the background at the beginning to give:

$\alpha + \beta + \gamma = 500$; $\beta + \gamma = 400$; $\gamma = 200$ and then proceed as above.

Stretch & challenge

Explain, in terms of the mechanism of absorption, why the graph in Fig. 3.5.10 curves upwards, i.e. the range of a 4 MeV α particle is more than double than one of 2 MeV.

(b) Deflection of radiation in a magnetic field

Beams of charged particles constitute electric currents and so are deflected in magnetic fields. See Section 4.4. The force is always at right angles to both the magnetic field and the velocity of the particles. Thus, if the particles are travelling at right angles to the magnetic field, their path is circular, with the deflection in a direction given by Fleming's left hand rule. The radius, r, of the circular path is given by:

$$r = \frac{p}{qB}$$

where p is the momentum of the particles, q their charge and B the flux density of the magnetic field (see Section 4.4.3).

You will often come across a diagram similar to Fig. 3.5.11 to illustrate the deflections (or lack of them) for α, β and γ radiation. In one respect, the diagram gives the correct idea: α and β, having opposite charges, deflect in opposite directions and γ (being uncharged) is undeflected.

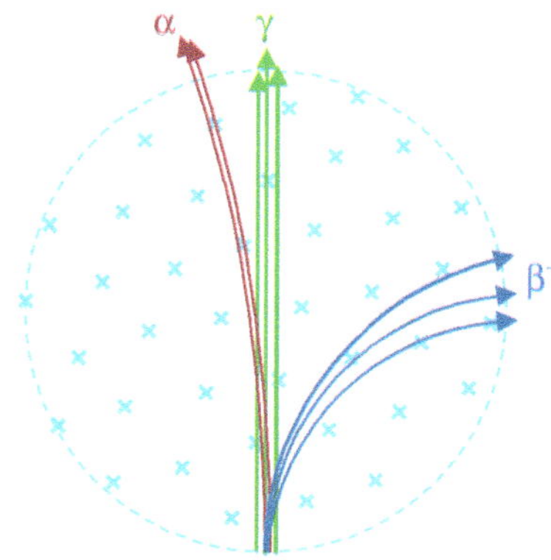

Fig. 3.5.11 Unrealistic magnetic deflection diagram

In another respect, Fig. 3.5.11 is very unrealistic. The masses of α and β are very different (a helium nucleus is about $8000\times$ as massive as an electron). If the α and β have the same energy, E, their momenta are not the same. Ignoring, for the moment, the need for a relativistic treatment:

$$E = \frac{p^2}{2m} \therefore p = \sqrt{2Em}$$

$\therefore$ The radius of the α particles' circle is hugely greater than that of the β particles. Allowing for the double charge on α, the ratio is $\sqrt{2000}$, i.e. about 45 (see footnote).[4]

Stretch & challenge

One 'realistic' aspect of Fig. 3.5.11 is that, whereas the tracks of the α particles are shown keeping together, those of the β particles are shown spreading out. Why should this be?

Example

Calculate the radius of curvature of 1 MeV α particle tracks in a 0.2 T magnetic field. [$1\text{ MeV} = 1.60 \times 10^{-13}\text{ J}$; $m_\alpha = 6.64 \times 10^{-27}\text{ kg}$]

Answer

$E = \frac{p^2}{2m}$, so the momentum is given by:

$$p = \sqrt{2mE} = \sqrt{2 \times 6.64 \times 10^{-27} \times 1.60 \times 10^{-13}} = 4.61 \times 10^{-20}\text{ N s}$$

$\therefore$ The radius of curvature, $r = \frac{p}{qB} = \frac{4.61 \times 10^{-20}}{2 \times 1.60 \times 10^{-19} \times 0.2} = 0.72\text{ m}$

Knowledge check 3.5.14

The 'unrealistic' Fig. 3.5.11 shows the tracks of β^- particles. How would the diagram be different if β^+ tracks were shown?

3.5.6 The exponential decay of radioactive substances

In this section we shall be looking at the **activity**, A, of a radioactive material and how it changes over time. This is number of decays per second. A typical school source has an activity of ~100 kBq, i.e. ~10^5 decays per second.

Key terms

The **activity**, A, of a radioactive source is the number of radioactive decays per unit time.

UNIT: becquerel (Bq) $\equiv$ s^{-1}

4 For α and β particles with a kinetic energy 1 MeV, a relativistic treatment gives the radius of the α particles to be approximately $20\times$ that of the β particles.

3.5.15 Knowledge check

The activity of radioactive sources used to be expressed in *curies*, a non-SI unit. The curie used to be defined as the activity of 1 g of ^{226}Ra, which has a decay constant of $1.37 \times 10^{-11}\ s^{-1}$. Express the curie in becquerel.

Stretch & challenge

A short-lived radioactive nuclide has $\lambda = 1.0 \times 10^{-6}$ s. Why can we calculate the fraction of the atoms decaying in one minute by multiplying by 60 s but not the fraction decaying in one year by multiplying by 3.16×10^7 s [the number of seconds in a year]?

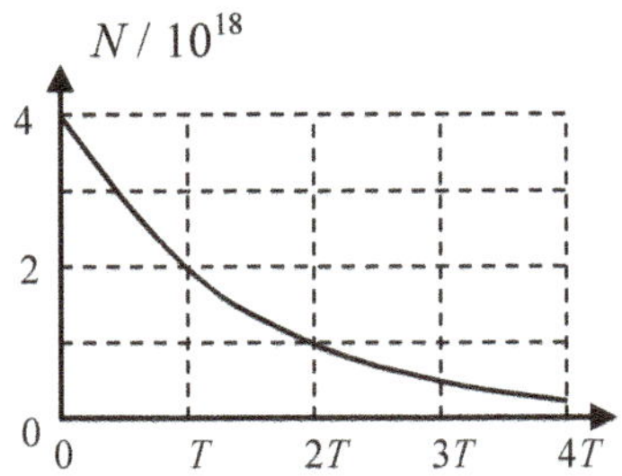

Fig. 3.5.12 Decay curve (1)

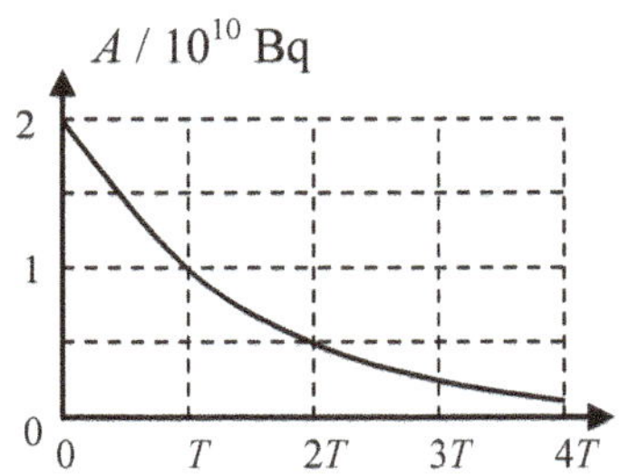

Fig. 3.5.13 Decay curve (2)

3.5.16 Knowledge check

Calculate the value of λ in the radioactive sample in Section 3.5.6(b).

Key terms

The **decay constant**, λ, is the fraction of the nuclei of a radioactive nuclide which decay per unit time.

UNIT: s^{-1}

The **half-life** is the time taken for the number of nuclei of a radioactive nuclide (or the activity) to halve.

(a) Activity and decay constant

The decay of an unstable atomic nucleus is a random event. It is entirely unaffected by temperature, pressure or any other physical condition. We cannot predict when an individual nucleus will decay but we can determine the probability that a nucleus will do so within any time period. For example, a $^{14}_{6}C$ nucleus has a probability of 3.94×10^{-12} of decaying each second. We say that the **decay constant**, λ, of $^{14}_{6}C$ is $3.94 \times 10^{-12}\ s^{-1}$. This means that, in a sample of N carbon-14 atoms, $3.94 \times 10^{-12}N$ will decay each second.

The relationship between N, A and λ is: $A = \lambda N$.

Example

The decay constant of ^{235}U is $3.14 \times 10^{-17}\ s^{-1}$. Calculate the activity of 1 kg of ^{235}U. [$N_A = 6.02 \times 10^{23}\ mol^{-1}$]

Answer

The molar mass of $^{235}U = 235\ g\ mol^{-1} = 0.235\ kg\ mol^{-1}$.

$\therefore$ The number of atoms in 1 kg of ^{235}U, $N = \frac{1.0\ kg}{0.235\ kg\ mol^{-1}} N_A = 2.56 \times 10^{24}$

$\therefore$ Activity, $A = \lambda N = 3.14 \times 10^{-17}\ s^{-1} \times 2.56 \times 10^{24} = 80\ MBq$

(b) Half-life

The unpredictable, random decay leads to the **half-life** law. The rate of decay of a sample of single radioactive nuclide is proportional to the number of atoms; if the number remaining halves, the rate of decay halves. Hence, the time taken for the number remaining to fall to any given fraction (e.g. a half, 0.1, e^{-1}) is constant.

Suppose we have a sample of 4×10^{18} atoms of a radioactive nuclide with an initial activity of 2×10^{10} Bq. Fig. 3.5.12 shows how the number remaining, N, varies with time, where T is the (as yet unknown) half-life.

This type of graph is called an exponential decay graph (see Section M3.3 p262). Because $A = \lambda N$, the graphs of N against t and A against t show the same characteristic half-life. Fig. 3.5.13 is the $A(t)$ graph corresponding to the $N(t)$ graph in Fig. 3.5.12.

(c) The formulae of radioactive decay curves

We normally write the initial values of N and A as N_0 and A_0. We know that after x half-lives these values have been reduced by a factor 2^x so we can write:

$$N = \frac{N_0}{2^x} \text{ and } A = \frac{A_0}{2^x},$$

or more conveniently

$$N = N_0 2^{-x} \text{ and } A = A_0 2^{-x}.$$

It is important to note that x can take any value, not just positive whole numbers. So for example, after 3.7 half-lives the activity of a sample will have fallen to 0.077 times the original (because $2^{-3.7} = 0.077$ to 2 sf). You should make sure you can use your calculator to obtain this!

In Section M3.2 we derive the following alternative formulae in terms of time rather than the number of half-lives:

$$N = N_0 e^{-\lambda t} \text{ and } A = A_0 e^{-\lambda t}.$$

The Stretch & challenge in the margin asks you to show that these formulae are equivalent to the 2^{-x} ones, but first we need to derive the relationship between λ and the half-life, $T_{\frac{1}{2}}$.

Starting from $A = A_0 e^{-\lambda t}$

By definition, after time $t = T_{\frac{1}{2}}$, $A = \frac{1}{2}A_0$,

∴ Rearranging $e^{-\lambda T_{\frac{1}{2}}} = \frac{A_0}{2A_0}$, i.e. $e^{-\lambda T_{\frac{1}{2}}} = \frac{1}{2}$

Taking logs $\ln\left(e^{-\lambda T_{\frac{1}{2}}}\right) = -\ln 2$

∴ $-\lambda T_{\frac{1}{2}} = -\ln 2$, i.e $\lambda = \frac{\ln 2}{T_{\frac{1}{2}}}$

Exam questions involving λ and the half-life require care with units because half-life is commonly expressed in hours, days or years as appropriate, whereas the activity A is in Bq (i.e. s^{-1}).

Example

Calculate the activity of 1 ng of ^{14}C, which has a half-life of 5570 years.

Answer

Decay constant, $\lambda = \frac{\ln 2}{T_{\frac{1}{2}}} = \frac{\ln 2}{5570 \times 3600 \times 24 \times 365.25} = 3.94 \times 10^{-12}\ s^{-1}$

No. of atoms, $N = \frac{m}{M_r} N_A = \frac{1.0 \times 10^{-9}\ g}{14\ g\ mol^{-1}} \times 6.02 \times 10^{23}\ mol^{-1} = 4.3 \times 10^{13}$.

The activity, $A = \lambda N$ ∴ $A = 3.94 \times 10^{-12}\ s^{-1} \times 4.3 \times 10^{13} = 170$ Bq.

(d) Using log-linear (semi-log) graphs

We have seen that the activity of a radio-nuclide decays according to the equation $A = A_0 e^{-\lambda t}$, resulting in the familiar exponential decay graph. We can turn this relationship into a linear graph by taking logs:

$$\ln A = \ln A_0 - \lambda t \qquad \text{[See Maths tip]}$$

So a graph of $\ln A$ against t should be straight line of gradient $-\lambda$ and intercept $\ln A_0$ on the $\ln A$-axis. The advantage of a straight-line graph over a non-linear one is that the eye is much better at comparing straight lines to slightly random data points than curved lines. So, in an experiment to determine the half-life of a radioactive material, a more accurate result is likely from a log-linear graph.

Sample data

A teacher monitored the decay of a sample of radon-220 over a period of 2 minutes by noting down the count in successive 10 s intervals. The results and decay graph were as follows:

Time, t / s	0	10	20	30	40	50	60	70	80	90	100	110	120
Count, C	540	495	432	361	310	290	245	220	174	165	157	125	112

Link

See Section 3.3 for the way to use the molar mass and the Avogadro constant to calculate the number of particles of a substance of a given mass.

Remember that the units of mass and molar mass need to be consistent, i.e. g and g mol^{-1} or kg and kg mol^{-1} respectively.

Top tip

You need to learn the derivation of $\lambda = \frac{\ln 2}{T_{\frac{1}{2}}}$. You could be asked for it in the exam.

Stretch & challenge

Derive the equation

$$A = A_0 2^{-x}$$

from $A = A_0 e^{-\lambda t}$

Hint: Write the time, t as $t = xT_{\frac{1}{2}}$, use $\lambda = \frac{\ln 2}{T_{\frac{1}{2}}}$ and use the relationship between natural logs and powers of e.

Knowledge check 3.5.17

For the radioactive sample in Section 3.5.6(b), calculate:

(a) The half-life.

(b) The activity after 10 years.

Maths tip

From $A = A_0 e^{-\lambda t}$

∴ $\ln A = \ln(A_0 e^{-\lambda t})$

∴ $\ln A = \ln(A_0 e^{-\lambda t})$

But $\ln ab = \ln a + \ln b$

∴ $\ln A = \ln A_0 + \ln(e^{-\lambda t})$

And $\ln(e^x) = x$

∴ $\ln A = \ln A_0 - \lambda t$

Study point

It doesn't matter that the counts are over 10 s periods. There's no need to convert these to s^{-1} or cpm.

Study point

Ways of finding the half-life from Fig. 3.5.14:

First draw a best-fit decay curve then:

1. Find two times where the counts are a factor of 2^n apart, e.g. 400 and 200, 500 and 125, or
2. Choose any two widely separated points with counts C_0 and C_1. Find x from $C_1 = C_0 2^{-x}$ and calculate the half-life from the time difference, or
3. As 2 but use $C_1 = C_0 e^{-\lambda \Delta t}$ to find λ and use it to calculate the half-life.

3.5.18 Knowledge check

Find the half-life of ^{220}Rn using all three methods given in the Study point.

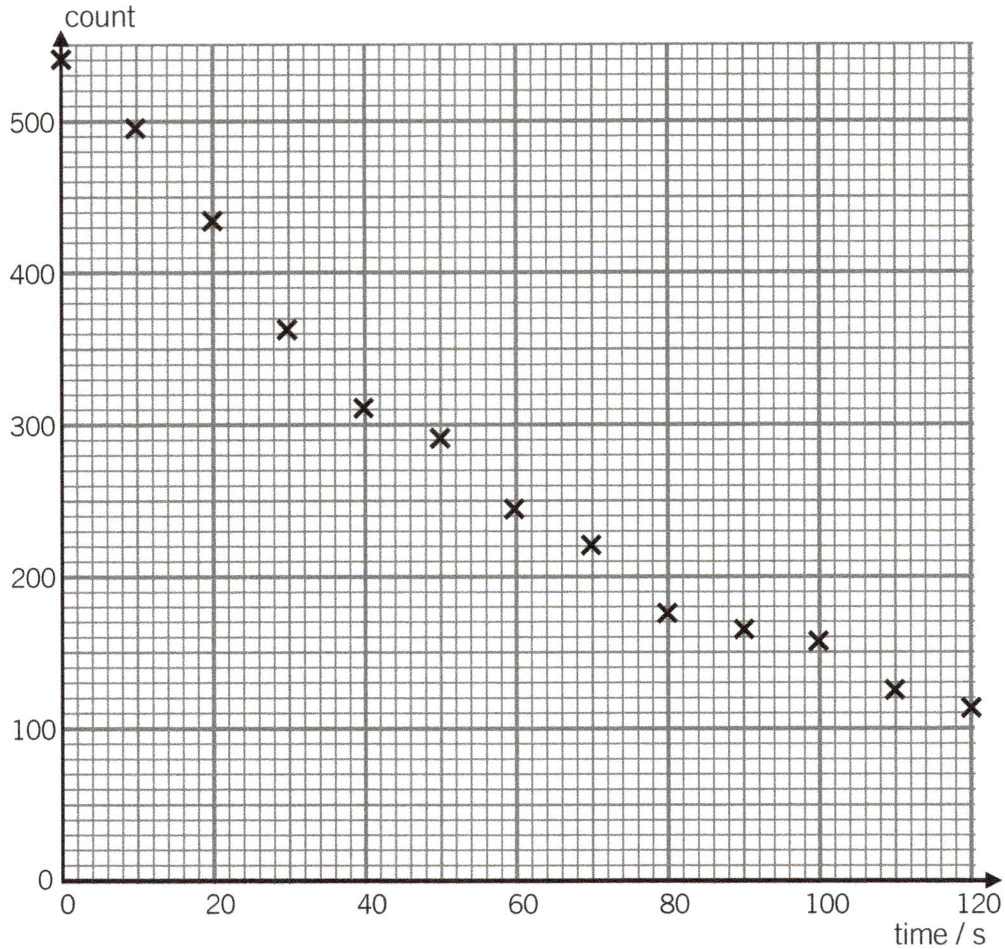

Fig. 3.5.14 Decay of radon-220

We can use this graph to determine the half-life of ^{220}Rn (see Study point). A better way is to plot $\ln C$ against t, draw a best-fit straight line, determine λ from the gradient and calculate $T_{\frac{1}{2}}$. See Test yourself 3.5 Q1.

3.5.7 Calculations using the decay equations

The simplest calculations involve determining the activity, number of atoms or mass remaining after a given time, given the half-life or decay constant. This just requires inserting the data into the relevant equation:

e.g. $N = N_0 2^{-x}$ or $A = A_0 e^{-\lambda t}$

Example

Iodine-131 has a half-life of 8.1 days. A sample of ^{131}I has an activity of 3.0×10^6 Bq. Calculate the activity after 21 days.

Answer

We'll do this using half-lives. See KC 3.5.19 to do the same question in a different way.

21 days $= \frac{21}{8.1} = 2.59$ half-lives.

$\therefore$ Using $A = A_0 2^{-x}$, activity after 21 days $= 3.0 \times 10^6 \times 2^{-2.59}$ Bq

$= 500$ kBq

Calculations which require you to calculate λ or $T_{\frac{1}{2}}$ are more complicated because we need to use the relationship between logs and exponents (see Maths tip).

Example

Iodine-128 has a half-life of 25 minutes. Calculate the time taken for a 5.0 MBq sample to decay to 10 kBq.

Answer

This time we'll work using the decay constant, λ (in s^{-1}).

$\lambda = \frac{\ln 2}{25 \times 60} = 4.62 \times 10^{-4}\ s^{-1}$

$\therefore$ Using $A = A_0 e^{-\lambda t}$, $e^{-4.62 \times 10^{-4} t} = \frac{10 \times 10^3}{5 \times 10^6} = 0.002$

$\therefore$ Taking logs: $-4.62 \times 10^{-4} t = \ln(0.002) = -6.21$

$\therefore t = \frac{-6.21}{-4.62 \times 10^{-4}} = 13\,500$ s (3 hr 44 min)

3.5.8 Specified practical work

(a) Investigating random decay using a dice analogy

Many centres do not have the facilities to collect and observe short-lived radio-isotopes but the random nature of decay can be illustrated using the fact that cubical dice have a 1 in 6 chance of landing on any specific side.

A simple way to use dice to illustrate radioactive decay is to consider them to be radioactive nuclei and any that land with (say) 1 upwards have 'decayed'. So the procedure is:

1. Roll a large known number (N_0) of dice.
2. Remove any that display a 1 and count the remainder (N_1).
3. Roll the remaining dice and remove the 1s.
4. Count the remainder (N_2) and repeat to obtain N_3, N_4 … .

Study point

In answering Knowledge check 3.5.19 it doesn't matter whether we use seconds, minutes, hours or days as the unit of time as long as λ and t are in consistent units; e.g. λ in s^{-1} and t in s or λ in day^{-1} and t in days. The value of λt will always be the same so $e^{-\lambda t}$ will always have the same value. Repeat Knowledge check 3.5.19 using a different unit of time.

Knowledge check 3.5.19

(a) Calculate the decay constant, λ, of ^{131}I.

(b) Show, using $A = A_0 e^{-\lambda t}$, that the activity after 21 days is 500 kBq.

Maths tip

Make sure you can use these relationships:

$\ln(e^x) = x$; $e^{\ln x} = x$

$\ln(2^x) = x\ln 2$; $e^{x \ln 2} = 2^x$

Knowledge check 3.5.20

Answer the ^{128}I example using half-life and $A = A_0 2^{-x}$.

Hint: First solve $A = A_0 2^{-x}$ for x using the given A and A_0.

Study point

As with the 'repeats' issue in Section 3.5.4, it doesn't matter whether all the dice are rolled at once. It can also be done with a smaller number of dice which are repeatedly rolled and counted.

Knowledge check 3.5.21

Using the method of Section 3.5.8(a) with $N_0 = 1000$, calculate N_1, N_2 and N_{10}.

Theory

Each die has a 1 in 6 chance of decaying, so on the average[5] the fraction remaining 'live' after any roll is $\frac{5}{6}$. After two rolls it is $\frac{5}{6} \times \frac{5}{6} = \left(\frac{5}{6}\right)^2$ and after n rolls it is $\left(\frac{5}{6}\right)^n$.

We can find the 'half-life', i.e the number of rolls needed to halve the number of undecayed dice, by solving the equation:

$$\left(\frac{5}{6}\right)^n = \frac{1}{2}$$

Taking logs and a bit of manipulation gives: $n = \frac{\ln 2}{\ln 1.2} = 3.8$ (2 sf)

This model can be varied, e.g. by letting 1 or 2 indicate a decayed die or by use of 12-sided or 20-sided dice.

(b) Investigating the variation of intensity of gamma rays with distance

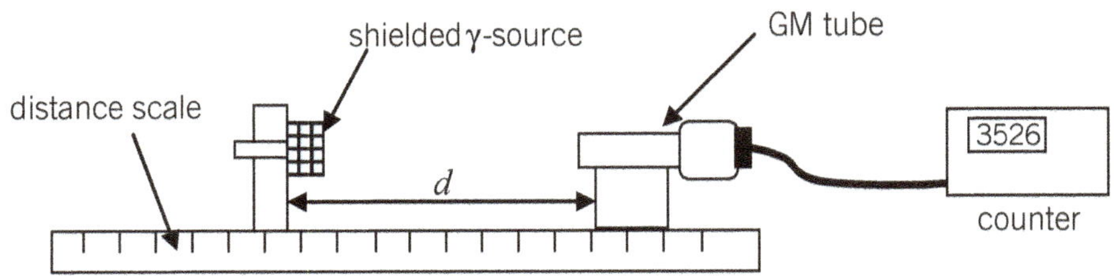

Fig. 3.5.15 Inverse square law for γ rays

If the separation of the source and detector is large compared with the size of the source, we expect the intensity of the gamma rays to be proportional to the inverse square of the separation. Unfortunately we know neither the exact position of the source within its shielding nor the effective position of the detector in the GM tube. With the set-up above we therefore expect the count rate C due to the source (i.e. corrected for background) to depend upon d according to:

$$C = \frac{k}{(d - \varepsilon)^2}$$

where k and ε are constants.

If we take the square root of the equation and invert it we get:

$$\frac{1}{\sqrt{C}} = \frac{d}{\sqrt{k}} - \frac{\varepsilon}{\sqrt{k}}.$$

Hence, if the inverse square law holds, a graph of $\frac{1}{\sqrt{C}}$ against d will be a straight line.

The procedure is quite straightforward and is the subject of Test yourself 3.5 Q4.

3.5.22 Knowledge check

A student used a large number of dodecahedral (i.e 12-sided) dice as a model decay system using the method in Section 3.5.8(a). Show that the expected 'half-life' is almost exactly 8 rolls.

3.5.23 Knowledge check

If the units of C and d are s^{-1} and cm respectively, state the units of k and ε.

Stretch & challenge

In a S Wales school, the mean background count is 21 per minute. The count with a γ source present is 760 over a 5-minute period. Estimate:

(a) C and $\frac{1}{\sqrt{C}}$ in counts per second, and

(b) the uncertainties in C and $\frac{1}{\sqrt{C}}$.

3.5.24 Knowledge check

Explain how a value for ε can be found from the graph of $\frac{1}{\sqrt{C}}$ against d.

5 The 'on the average' is important here. If N_1 is expected to be about 900, a variation of ±30 is also to be expected. See Stretch & challenge in Section 3.5.4.

Test yourself 3.5

1. From a graph of $\ln C$ against t for the results of the ^{220}Rn decay in Fig. 3.5.14 find:

   (a) The gradient and the intercept on the $\ln C$ axis.
   (b) The decay constant, λ and half-life of the decay.

2. Consult a periodic table to help you write decay equations for the following decay processes:

   (a) The α decay of $^{192}_{78}Pt$
   (b) The β^- decay of $^{181}_{72}Hf$
   (c) The β^+ decay of $^{48}_{23}V$
   (d) The K-capture decay of $^{59}_{28}Ni$
   (e) The decay of $^{77}_{32}Ge$ into $^{77}_{33}As$
   (f) The decay of $^{65}_{30}Zn$ by the emission of a positive particle
   (g) The decay of $^{65}_{30}Zn$ by the absorption of a negative particle
   (h) The transformation of $^{239}_{92}U$ into a plutonium isotope by two successive β^- decays
   (i) The decay of $^{17}_{7}N$ by the emission of an electron and a neutron.

3. The table gives the properties of some isotopes of neon:

Isotope	$^{18}_{10}Ne$	$^{19}_{10}Ne$	$^{20}_{10}Ne$	$^{21}_{10}Ne$	$^{22}_{10}Ne$	$^{23}_{10}Ne$	$^{24}_{10}Ne$
decay mode	β^+	β^+	stable	stable	stable	β^-	β^-

   Suggest reasons for the pattern in the table.

4. A group of students investigated the inverse square law for gamma radiation. They obtained the following measurements:

   Background count = 130 in 5 minutes

Distance, d / cm	10	15	20	25	30	50	70
Count	755	282	155	230	275	240	365
Count time / min	1	1	1	2	3	5	10

   (a) Suggest why they measured the counts over different times.
   (b) Draw up another table of counts per minute, C, corrected for background radiation.
   (c) Draw a graph of $\frac{1}{\sqrt{C}}$ against d.
   (d) Use your graph to confirm the validity of the inverse square law and calculate the values of k and ε (See Section 3.5.8(b)).

5. The isotopes of natural uranium have the following relative abundances and half-lives:

Isotope	$^{234}_{92}U$	$^{235}_{92}U$	$^{238}_{92}U$
Relative abundance	0.01%	0.72%	99.27%
Half-life / years	2.5×10^5	7.1×10^8	4.5×10^9

   Calculate:
   (a) The decay constant for each isotope.
   (b) The fraction of the activity of a sample of natural uranium which is due to each of the isotopes.
   (c) The activity of 1 kg of natural uranium.

6. All three uranium isotopes in Q5 decay by α emission. Explain why the total detectable emissions from 1 kg of uranium in the form of a compact lump would be much less than the answer to 5(c) and why such a lump might feel slightly warm to the touch.

7. The short half-life of ^{234}U compared to the age of the Earth (2.56×10^9 years) means that any of this isotope that was in the Earth originally has long since decayed to homeopathic amounts (i.e. zero).

(a) Show that this statement is correct. [Hint: $2^{-10} \sim 10^{-3}$]
(b) That ^{234}U exists currently, albeit in small amounts, is because it is constantly being produced (indirectly) by the decay of ^{238}U. The daughter products of the decay of ^{238}U include α and β^- emitters only. Give the shortest possible decay sequence for the production of ^{234}U from ^{238}U. [Consult a periodic table if necessary.]

8. (a) Explain why ^{235}U cannot be produced as a result of the decay of ^{238}U.
(b) Use information from questions 5 and 7 to calculate the isotopic abundances of ^{235}U and ^{238}U at the time that the Earth was formed.
(c) Assuming that ^{235}U and ^{238}U were formed in approximately equal amounts in the supernova explosion which provided the heavy elements for the nebula out of which the Solar System condensed, estimate when this supernova occurred.

9. Geologists determine the age of igneous rocks using potassium-argon dating, involving measuring the ratio of $^{40}_{19}$K to $^{40}_{18}$Ar in the rocks. The basis of the method is that ^{40}K is radioactive with a half-life of 1.3×10^{10} years, with 10% of the decays being to ^{40}Ar and the remainder to $^{40}_{20}$Ca (calcium-40). A main assumption is that igneous rocks contain no argon initially.

(a) [For chemists] Explain why the assumption is likely to be valid.
(b) There are two modes of decay of $^{40}_{19}$K to $^{40}_{18}$Ar. Write decay equations for both of these and for the decay of $^{40}_{19}$K to $^{40}_{20}$Ca.
(c) Explain, qualitatively how the argon to potassium ratio depends upon the age of the rock and state a second assumption that must be made.
(d) Calculate the decay constant for the decay of ^{40}K.
(e) Construct a table to show the abundance of the ^{40}K in 100 million year steps up to 600 million years from an initial value of 100. Include in your table the abundance of ^{40}Ar and the ratio Ar:K.
(f) Use the table to estimate the age of an igneous rock with an Ar:K ratio of 0.020.
(g)

Stretch & challenge

(i) Show that, R, the ratio $\frac{\text{Ar}}{\text{K}}$ is given by $R = \frac{1 - e^{-\lambda t}}{10e^{-\lambda t}}$

(ii) Hence show that $t = \frac{1}{\lambda}\ln(1 + 10R)$

(iii) Use the result to verify the answer to (f)

10. A GM tube is used to investigate a radioactive source. The background radiation is measured over a period of 10 minutes. The source is then placed 2 cm from the GM tube and the counts taken over 10 minutes, with and without a sheet of paper in between the source and detector.

Results:

No source present:	250 counts
No absorber:	570 counts
Paper absorber:	525 counts

Discuss whether the results show that the source is an α-emitter.

11. The photo-plates that Becquerel used were wrapped in black paper; the Maltese cross was made of thin steel. What can you conclude about the nature of the radiation that fogged the plate?

12. A GM tube is set up close to a γ-emitter. The corrected count rate with no absorber is $24.0\ \text{s}^{-1}$. When a 5 mm-thick lead absorber is placed between source and detector the corrected count rate drops to $18.0\ \text{s}^{-1}$. What count rate is expected with (a) with a 10 mm lead absorber and (b) with a 15 mm absorber?

3.6 Nuclear energy

Studies in radioactivity, developments in the knowledge of atomic and nuclear structure, careful measurements of nuclear masses and Einstein's Special Theory of Relativity in the first half of the 20th century led to the ability to harness the most powerful sources of energy on the Earth: the energy in very low and very high mass nuclei. Nuclear fusion is the energy source of stars; nuclear fission is made possible by the existence of fissile nuclei which are among the debris of exploding stars. Knowledge of this section will add to your understanding of particle physics (in the AS textbook) and radioactivity. We start with a largely misunderstood concept: $E = mc^2$ and mass-energy.

Fig. 3.6.1 The Crab nebula – seeding space with new elements

3.6.1 Mass-energy

Through careful experiments in the 18th and 19th centuries, chemists developed the law of **conservation of mass**, according to which the mass of the products of a reaction is equal to that of the reactants. For example:

	C	+	O_2	→	CO_2
Mass/u	12.01	+	32.00	=	44.01

At the same time physicists and chemists were developing the law of conservation of energy: if a falling ball loses 100 J of gravitational potential energy and only gains 20 J of kinetic energy, the balance (80 J) must appear in some other form, e.g an increase of the internal energy of the atmosphere and/or the ball.

Study point

In the reaction between carbon and oxygen, the masses of the individual atoms and molecules are given, expressed in u. (See Section 3.3).

Knowledge check 3.6.1

State the masses of the reactants and product when 1.000 mol of carbon reacts with oxygen to form carbon dioxide.

(a) $E = mc^2$

In developing his Special Theory of Relativity Einstein concluded that mass and energy are two aspects of the same quantity, which we can call mass-energy and that this quantity is conserved. We can express mass-energy in units of mass (kg) or units of energy (J), the exchange rate between them being c^2 , where c is the speed of light *in vacuo*,[1] i.e.

$$E = mc^2$$

Key terms

Conservation of mass/energy: Energy cannot be lost or gained, only transferred from one form to another. We can measure the energy in a body by multiplying its mass by c^2.

Example

To see why chemists didn't spot this earlier, consider again the C + O_2 reaction. The energy released by burning 1 mol of carbon is 394 kJ (see Study point).

(a) Calculate the mass of this energy, and
(b) express this as a fraction of the reacting masses.

Answer

(a) $m = \frac{E}{c^2} = \frac{394 \times 10^3 \text{ J}}{(3.00 \times 10^8 \text{ m s}^{-1})^2} = 4.38 \times 10^{-12} \text{ kg}$

(b) Total reacting mass = 12.01 g + 32.00 g = 44.01 g = 0.004401 kg.

$\therefore$ Fraction of reacting masses = $\frac{4.38 \times 10^{-12} \text{ kg}}{44.01 \times 10^{-3} \text{ kg}} = 1.0 \times 10^{-10}$

Study point

In chemists' language, the *enthalpy* change, ΔH, in the reaction

$$C + O_2 \rightarrow CO_2$$

is –394 kJ mol^{-1}.

Knowledge check 3.6.2

Calculate the increase in mass of a 1 kg cube of iron which is heated up (e.g. using a bunsen burner) from 20 °C to 520 °C. [Specific heat capacity of iron = 450 J kg^{-1} K^{-1}].

1 *In vacuo* is just the Latin for 'in a vacuum'. Scientists like to show off sometimes!

Discussion

1. The first thing to note is that chemists would need to know the molar masses of carbon, oxygen and carbon dioxide to 10 significant figures in order to spot this mass-energy effect. Notice that the effect in KC 3.6.2 (a typical 'physics' process) is even smaller.
2. Where is this mass-energy? It all (i.e. 44.01 g) starts off in the reactants. It is all there in the 'hot' product (the CO_2) but the 4.38 ng is the mass of the energy that dissipates in the environment. That means that the atmosphere gets 4.38 ng heavier due to its extra internal energy; the carbon dioxide is 4.38 ng lighter because its atoms are more tightly bound together (i.e. they have a lower potential energy) than the reactants.

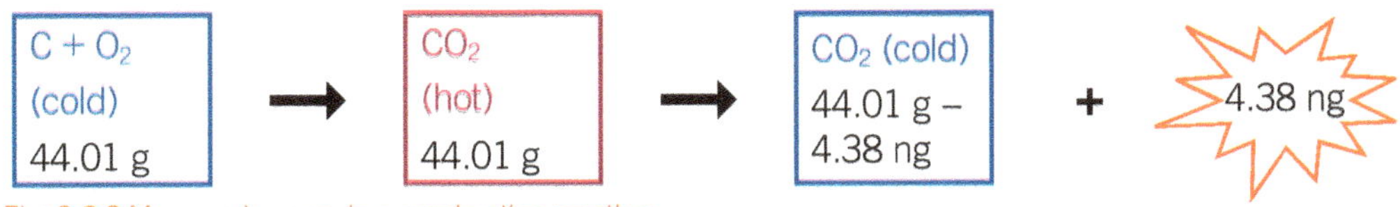

Fig. 3.6.2 Mass and energy in a combustion reaction

(b) The energy equivalence of u

As nuclear physicists generally express atomic and nuclear masses in u and energy in eV, it will help to be able to convert between the two. The conversion is:

The energy equivalent of 1 u = 931 MeV (to 3 sf)

You should be able to derive this relationship using the following data:

$1\ \text{u} = 1.660\,538\,922 \times 10^{-27}\ \text{kg}$ $\quad c = 299\,792\,458\ \text{m s}^{-1}$ $\quad e = 1.602\,176\,62 \times 10^{-19}\ \text{C}$

(c) $E = mc^2$ and nuclear energy

We've seen that $E = mc^2$ has general applicability; it does not apply only to nuclear energy. Yet $E = mc^2$ is usually seen as paving the way for the development of nuclear energy and specifically nuclear weapons. Why should this be? The reason is the energy involved in nuclear changes is a much greater fraction of the mass-energy of the particles. We'll illustrate this by looking at the nuclear fusion process at the heart of the Sun: the fusion of 4 hydrogen-1 **nuclei** (i.e. protons) to a helium-4 nucleus. There are several stages (and several routes) to this process but it can be summarised as follows.

$$4\,{}^1_1\text{H} + 2\,{}^{\ 0}_{-1}\text{e} \rightarrow {}^4_2\text{He} + 2\nu_e$$

The example shows how to calculate the mass change among the particles in the reaction.

Example

Use the following data to calculate the percentage loss in mass of the particles involved in the fusion of 4 protons to a helium nucleus as in the above reaction.

$m({}^1_1\text{H}) = 1.007\,276\ \text{u}$; $\quad m({}^4_2\text{H}) = 4.001\,508\ \text{u}$; $\quad m_e = 0.000\,548\ \text{u}$

Answer

Initial mass $= 4m\,({}^1_1\text{H}) + 2m_e = 4 \times 1.007\,276\ \text{u} + 2 \times 0.000\,548 = 4.030\,200\ \text{u}$

$$\therefore\ \%\ \text{Mass loss} = \frac{4.030\,200\ \text{u} - 4.001\,508\ \text{u}}{4.030\,200\ \text{u}} \times 100\% = 0.71\%$$

The mass loss in this reaction $= 4.030\,200\ \text{u} - 4.001\,276\ \text{u} = 0.015\,12\ \text{u}$

Where has this mass gone? It is the mass of the energy which is released, which we can calculate using $E = mc^2$ or using the equivalence, 1 u = 931 MeV. (See KC 3.6.3)

3.6.3 Knowledge check

Express the energy release in the $4\text{H} \rightarrow \text{He}$ reaction:

(a) in J (b) in MeV

Compare this with the 13.6 eV energy release in the formation of H-1 from a proton and an electron.

3.6.4 Knowledge check

Consider the fusion of 1.0 kg of hydrogen to helium.

(a) Use the result of the example to show that the energy released is 6.3×10^{14} J.

(b) The typical annual household electrical energy use is 3.0 MW h [Wales 2020 data]. How many annual household consumptions does the energy in part (a) represent?

3.6.5 Knowledge check

Use the values of u, c and e to derive the equivalence

1 u ≡ 931 MeV (to 3 sf)

Study point

If we add 2 electrons to each side of the equation, the equation becomes:

$$4\,{}^1_1\text{H} \rightarrow {}^4_2\text{He} + 2\nu_e,$$

where, this time, the symbols represent the atoms of hydrogen and helium. So, we can calculate the energy yield using *atomic masses*.

3.6.2 Nuclear binding energy

A hydrogen-1 atom consists of a proton and an electron. If the hydrogen atom is in its ground state, it takes 13.6 eV to separate these particles. We say that the **binding energy** of a hydrogen atom is 13.6 eV. It is another way of saying that the sum of the potential and kinetic energies of the atomic particles is −13.6 eV (with respect to a proton and an electron at rest at infinite separation).

It also takes energy to disassemble the nuclei of heavier atoms (i.e. ones that consist of more than just a single proton). This energy is called the **nuclear binding energy**. We can use the measured masses of the particles and $E = mc^2$ to calculate it for any nucleus, as follows:

1. We calculate the **mass deficit**, Δm of the nucleus. The nuclear mass is less than the sum of the masses of the individual protons and neutrons. The difference is the mass deficit.
2. We calculate the **binding energy** using $E = \Delta mc^2$, or $1\ \text{u} \equiv 931\ \text{MeV}$.

Example

Calculate the binding energy of a ${}^{12}_{6}\text{C}$ nucleus from the following data:

$m_p = 1.007\,276\ \text{u},\ m_n = 1.008\,665\ \text{u},\ m_e = 0.000\,549\ \text{u}$

Answer [See Knowledge check 3.6.6 for an alternative way of calculating this]

By definition, the mass of a ${}^{12}\text{C}$ atom is 12 u exactly.

There are 6 electrons in a ${}^{12}\text{C}$ atom

∴ Mass of a ${}^{12}\text{C}$ nucleus, $m_{^{12}\text{C}} = (12 - 6 \times 0.000\,549)\ \text{u}$

$$= 11.996\,706\ \text{u}$$

There are 6 protons and 6 neutrons in a ${}^{12}\text{C}$ atom.

$$6m_p + 6m_n = 6 \times 1.007\,276\ \text{u} + 6 \times 1.008\,665\ \text{u}$$

$$= 12.095\,646\ \text{u}$$

∴ Mass deficit, $\Delta m = 12.095\,646\ \text{u} - 11.996\,706\ \text{u}$

$$= 0.098\,940\ \text{u}$$

Leading to: binding energy of ${}^{12}\text{C}$ nucleus = 92.1 MeV

Key terms

The **binding energy** of a nucleus is the energy required to disassemble the nucleus into its constituent protons and neutrons.

UNIT: J or MeV

Stretch & challenge

The data in the ${}^{12}_{6}\text{C}$ binding energy example are given to 7 sf. By looking at the calculation, discuss the number of figures actually required to calculate the binding energy to 3 sf.

Knowledge check 3.6.6

The mass of a ${}^{12}_{6}\text{C}$ atom is 12 u (exactly). Use the following data to calculate the binding energy of a ${}^{12}_{6}\text{C}$ nucleus:

$m({}^{1}_{1}\text{H})$ atom = 1.007 825 u

$m_{neutron}$ = 1.008 665 u

[Hint: Consider the ${}^{12}\text{C}$ atom as 6 H1 atoms plus 6 neutrons]

Knowledge check 3.6.7

Repeat the ${}^{12}_{6}\text{C}$ binding energy example by converting the mass deficit to kg and using $E = \Delta mc^2$. Give your answer in (a) J and (b) MeV.

3.6.3 Binding energy per nucleon

(a) Calculating the binding energy per nucleon

Imagine building up stable nuclei by adding more and more nucleons (i.e. protons and neutrons). As each extra nucleon is added, the binding energy must increase. If it didn't, that extra nucleon would not be bound to the others. Hence we'd expect the binding energy of stable nuclei to increase with nucleon number, which is generally what we find. However, to determine how tightly the *individual* nucleons are bound we need to calculate the **binding energy per nucleon**:

$$\text{BE / nucleon} = \frac{\text{binding energy}}{\text{nucleon number}}$$

The binding energy per nucleon of $^{12}_{6}\text{C} = \frac{92.1\text{ MeV}}{12} = 7.68\text{ MeV nuc}^{-1}$

Your attention is drawn to the table of atomic masses (Table 3.6.3) of selected nuclides at the end of Section 3.6. It contains all stable nuclides up to $Z = 20$ (calcium) and others which are used as illustrations in Knowledge check questions and problems in Test yourself 3.6.

(b) The stability of low-mass nuclei

The **nuclear binding energy curve** is one of the most important diagrams in nuclear physics. Section 3.6.5 will consider it in relationship to nuclear fission and fusion. For the moment we will just examine the lower slopes of the relationship, illustrated in Fig. 3.6.3, on which are plotted the binding energies per nucleon of the 20 lightest stable atoms.

We notice that there is a positive correlation of BE/nuc with nucleon number (A): as we progress up the along the Periodic table the nuclei become more and more stable in this region (see Study point).

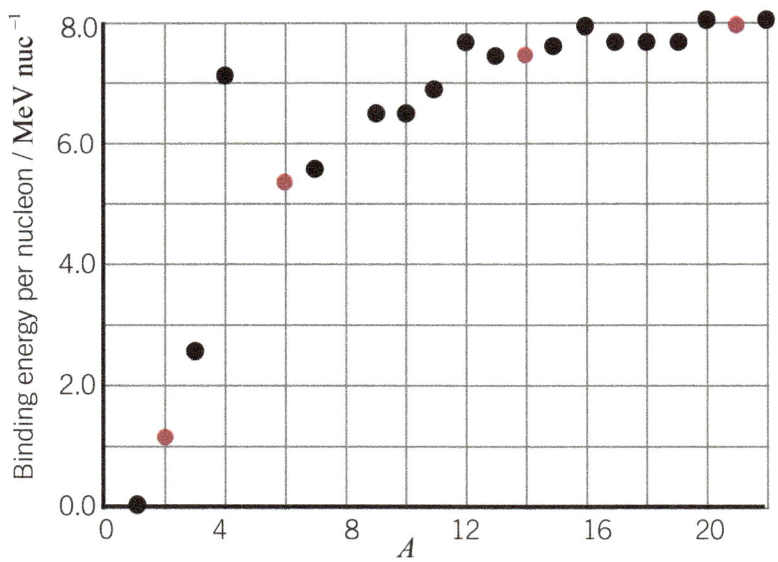

Fig. 3.6.3 Plot of stable nuclides up to $A = 22$

The next obvious feature of the plot is the spike at $A = 4$, which represents $^{4}_{2}\text{He}$. This nuclide is particularly tightly bound. Notice also that there are smaller spikes at $A = 12$, 16 and 20: the nuclides $^{12}_{6}\text{C}$, $^{16}_{8}\text{O}$ and $^{20}_{10}\text{Ne}$ which are multiples of the $^{4}_{2}\text{He}$ structure. This, together with the α decay mode of nuclei, strongly suggests that the 2p+2n combination exists within nuclei.

Study point

A **nuclide** is a distinct variety of atom or nucleus, indicated by a specific symbol, $^{A}_{Z}\text{X}$.

3.6.8 Knowledge check

Use the data in the table of atomic masses (Table 3.6.3, p89) to calculate (a) the nuclear binding energy and (b) the binding energy per nucleon of $^{31}_{15}\text{P}$, giving your answer in MeV and MeV nuc^{-1} respectively.

Study point

The **nuclear binding energy curve** is a plot of the binding energy per nucleon against nucleon number for known nuclides.

Study point

We'll see later that for nuclides above $^{62}_{28}\text{Ni}$ the stability decreases with A.

3.6.9 Knowledge check

In Fig. 3.6.3:

(a) Use the table of atomic masses to identify the nuclides marked in red.

(b) Explain why the binding energy per nucleon is zero for $A = 1$

Study point

Notice that there are no stable nuclides with 5 or 8 nucleons. These stability gaps (the half-life of $^{8}_{4}\text{Be}$ is only 10^{-16} s; $^{5}_{2}\text{He}$ is ~10^{-20} s and $^{5}_{3}\text{Li}$ ~ 10^{-24} s) exercise an important brake on the development of stars beyond the hydrogen burning phase.

(c) The energetics of β⁻ decay

The nuclide $^{27}_{12}\mathrm{Mg}$ decays by β⁻ decay to $^{27}_{13}\mathrm{Al}$.

$$^{27}_{12}\mathrm{Mg} \rightarrow {}^{27}_{13}\mathrm{Al} + {}^{0}_{-1}\mathrm{e} + {}^{0}_{-1}\overline{\nu_e}$$

Why should this be and what is the energy release?

The clue is in the masses of the particles involved. All we have to do is work out the change in mass, Δm and convert this to energy units.

Example

Calculate the energy release in the β⁻ decay of $^{27}_{12}\mathrm{Mg}$.

Atomic masses: $^{27}_{12}\mathrm{Mg} = 26.984\ 35\ \mathrm{u}$; $^{27}_{13}\mathrm{Al} = 26.981\ 535\ \mathrm{u}$

Answer

Mass deficit $= 26.984\ 35\ \mathrm{u} - 26.981\ 535\ \mathrm{u}$

$= 0.002\ 815\ \mathrm{u}$

∴ Energy released $= 0.002\ 815 \times 931 = 2.62\ \mathrm{MeV}$ (to 3 sf)

Note: We can use the atomic masses of $^{27}_{12}\mathrm{Mg}$ and $^{27}_{13}\mathrm{Al}$ because the $^{27}_{13}\mathrm{Al}$ requires another electron to make up its full complement of 13 – this is (essentially) provided by the decay of the $^{27}_{12}\mathrm{Mg}$. We mustn't make the mistake of adding in the electron mass on the right.

(d) The energetics of β⁺ decay

The nuclide $^{23}_{12}\mathrm{Mg}$ (atomic mass $22.994\ 14\ \mathrm{u}$) decays by β⁺ emission to give $^{23}_{11}\mathrm{Na}$ (atomic mass $22.989\ 773$). How much energy is released in the decay? We can do this in the same way but it is not quite as neat!

$\Delta m = (22.989\ 773\ \mathrm{u} - 11m_e + m_e) - (22.994\ 14\ \mathrm{u} - 12\ m_e)$

$= -0.004\ 367\ \mathrm{u} + 2m_e$ with $m_e = 0.000\ 549\ \mathrm{u}$

$= -0.003\ 269\ \mathrm{u}$ leading to Energy release $= 3.05\ \mathrm{MeV}$

However, this is not the total energy released in the decay, because the β⁺ particle is a positron, the anti-particle of the electron. This will collide with an electron and these will annihilate releasing another $2m_e$ of energy (as two photons). So the total energy release has a mass of $0.004\ 367\ \mathrm{u}$, i.e. $4.06\ \mathrm{MeV}$.

(e) The energetics of α decay

The same principles can be applied to α decay. Because only two particles are produced (the daughter nucleus and the α particle) we can use conservation of momentum to determine the energy and hence the speed of the α particles.

Example

Using the answer of Knowledge check 3.6.12(b), calculate the energy of the α particles emitted in the decay of $^{180}_{74}\mathrm{W}$.

Study point

In working out the energy release in β⁻ decays, the masses of the electrons can be ignored. Follow the working to see why. They just cancel out. We can consider the emitted electron in the decay as providing the extra electron needed to make up the 13 in the aluminium atom.

Knowledge check 3.6.10

Repeat the calculation of the energy released in the β⁻ decay of $^{27}_{12}\mathrm{Mg}$ by finding the mass loss in kg and using $E = mc^2$. Express your answer in J and then convert to MeV.

Study point

The method of calculation in the Example uses the same idea as Knowledge check 3.6.6.

Knowledge check 3.6.11

The half-life of $^{27}\mathrm{Mg}$ is 9.5 minutes. If it were possible to produce a 1 μg sample of $^{27}\mathrm{Mg}$

(a) show that its initial power output would be ~10 W, and

(b) calculate its power output after 3 hours.

Study point

The energy released in β decay is shared between 3 particles: the β particle, the neutrino and a small fraction to the daughter nucleus. For this reason, the β particles have a spread of energies. The energy spectrum of the β particles was the evidence for the existence of the neutrino (see AS textbook, Section 1.7.4).

Knowledge check 3.6.12

The nuclide $^{180}_{74}\mathrm{W}$ ($179.946\ 70\ \mathrm{u}$) decays by α emission into a nuclide of mass $175.941\ 51\ \mathrm{u}$.

(a) Write the decay equation.

(b) Show that the energy released is $3.86 \times 10^{-13}\ \mathrm{J}$.

Study point

For this part of the calculation, 3 sf accuracy is enough for the masses. Hence we can use the mass number as equal to the atomic mass in u.

Stretch & challenge

Show that the energy of the α particle is given by:

$$E_\alpha = \frac{M_{nuc}}{M_{nuc} + m_\alpha} E_{tot}$$

Use this to check the answer to the example.

Answer

Let v_α and v_{nuc} be the speeds of the α particles and nucleus.

Then total $\text{KE} = 3.86 \times 10^{-13} = \left(\frac{1}{2}4.00 v_\alpha^2 + \frac{1}{2}176 v_{nuc}^2\right) \times 1.66 \times 10^{-27}$

(See Study point)

Simplifying: $4.00 v_\alpha^2 + 176 v_{nuc}^2 = 4.65 \times 10^{14}$ [1]

Conservation of momentum: $4.00 v_\alpha = 176 v_{nuc}$ [2]

Eliminating v_{nuc} in [1] by substituting from [2] leads to

$$v_\alpha^2 = \frac{4.65 \times 10^{14}}{4.09} \text{ (you should check this).}$$

$$\therefore E_\alpha = \frac{1}{2} m_a v_\alpha^2 = \frac{1}{2} 4.00 \times 1.66 \times 10^{-27} \times \frac{4.65 \times 10^{14}}{4.09} = 3.77 \times 10^{-13} \text{ J}$$

Link

To follow the electrostatics working in this section you will need to consult Section 4.2.

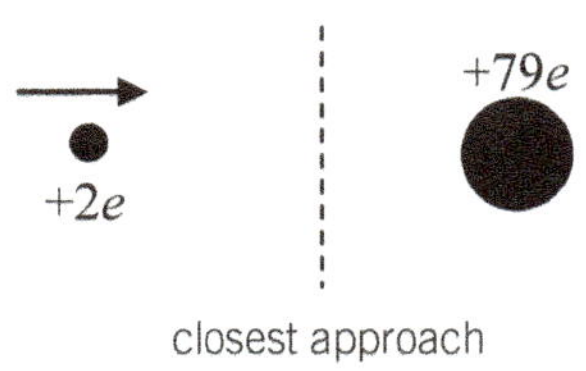

Fig. 3.6.4 Head-on α collision with a gold nucleus

Nuc	r/fm	Nuc	r/fm
^{11}B	2.41	^{120}Sn	4.65
^{16}O	2.70	^{140}Ce	4.88
^{24}Mg	3.06	^{184}W	5.37
^{40}Ca	3.48	^{197}Au	5.43
^{56}Fe	3.74	^{208}Pb	5.50
^{80}Se	4.14	^{238}U	5.86

Table 3.6.1 Selected nuclear radii

Study point

The data for atomic nuclei obtained from high energy electron beam experiments are easier to interpret than α particle data because they (the electrons) only feel the electromagnetic force; α particles are also influenced by the strong force when they come within a few fm of the nucleons.

3.6.13 Knowledge check

Taking $k = 1.1$ fm, show that the density of a nucleus is approximately 3×10^{17} kg m^{-3}.

3.6.4 Physical properties of nuclei

This section is to give an insight into some further aspects of nuclear physics. It is fascinating but not for examination.

(a) Nuclear radius

The discovery of the atomic nucleus by Rutherford, Geiger and Marsden was by the scattering of α particles by gold atoms. The distribution of the scattered α particles was as would be expected for a point charge, rather than for a distributed positive charge as was expected if the + charge on the atom was distributed throughout. We can calculate the closest distance of approach that the 5 MeV α particles could make to a nucleus of charge +79e (a gold nucleus). We'll need to use ideas from Unit 4 (see Section 4.2.3). From this, we can calculate an upper limit for the size of a gold nucleus by considering the head-on collision of an alpha particle ($Z = 2$) with a gold nucleus ($Z = 79$) (see Study point).

For a 5 MeV alpha particle, the electrostatic potential energy at the closest approach is 5 MeV, i.e. 8.0×10^{-13} J.

Applying the potential energy formula[2]

$$8.0 \times 10^{-13} = \frac{2 \times 79 \times (1.6 \times 10^{-19})^2}{4\pi\varepsilon_0 d}$$

gives us 4.6×10^{-14} m. So we conclude that the radius of the gold nucleus must be less than this. The most accurate measurements are made using the diffraction of high-energy electrons from nuclei and a few values are given in Table 3.6.1. Notice that the figure for gold is about a tenth of that mentioned above (see Study point).

(b) Nuclear density

Analysis of the data in Table 3.6.1 (see Test yourself 3.6 Question 11) gives us the relationship r / fm $= kA^{1/3}$ for the heavier nuclides, with a value of $k \sim 1.1$ fm. This suggests that the density of nuclei is approximately constant, so we can picture the protons and neutrons in nuclei as rather like marbles in spherical bags. If the radius of the bag is multiplied by 1.5, the number of marbles is multiplied by $1.5^3 \sim 3.5$ (e.g. ^{56}Fe and ^{197}Au).

2 Note that we are ignoring the recoil of the gold nucleus here. It makes very little difference to the outcome. You might like to check this – it'll take a page of algebra!

From this we can obtain a figure for the density of the nuclear material of about 3×10^{17} kg m^{-3}. This is left as an exercise (KC 3.6.13). A pea-sized object of nuclear material would have a mass of over 30 million tonnes.

(c) Nuclear composition and stability

We cannot give a full picture of this huge topic but we can make some introductory remarks. The *N-Z* stability plot, Fig. 3.6.5, contains a wealth of detail which can only be revealed at a larger scale. It is worth using a search engine to find a large-scale version. The zig-zag black line contains all the stable nuclides that exist (note a couple of gaps). Here are a few of the features:

- There are no stable nuclides with more than 83 protons.
- The ratio of neutrons to protons in stable light nuclei ($Z < 20$) is very close to 1, e.g. ${}^{4}_{2}$He, ${}^{16}_{8}$O, ${}^{24}_{12}$Mg. For heavier stable nuclei, the ratio increases with Z. ${}^{209}_{83}$Bi has 126 neutrons and 83 protons: a ratio of 1.5.
- Nuclides above the line of stability (i.e. an excess of neutrons) are β^- unstable; below the line is the β^+ instability region.

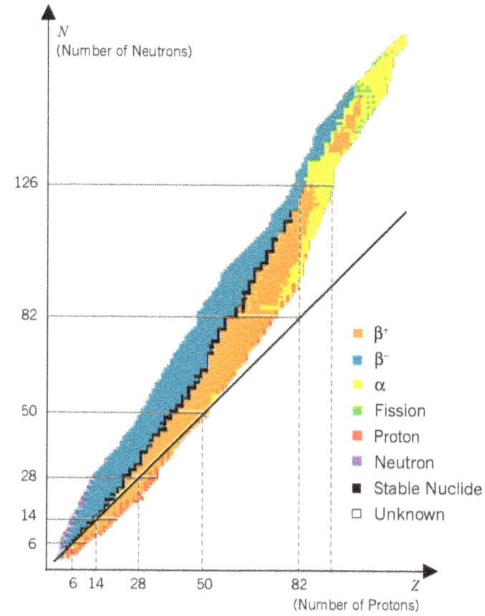

Fig. 3.6.5 *N-Z* stability plot

Apart from ${}^{1}_{1}$H all stable nuclei contain neutrons; combinations of more than one proton (with no neutrons) are never stable. Why not? Protons are positively charged so repel one another according to Coulomb's law. All hadrons attract at short distances (less than 4–5 fm) with a force called the **nuclear force**. The strong force can only stabilise groups of nucleons if there are uncharged neutrons as well as protons.

Study point

Quarks are held together in hadrons by the strong force. Outside a hadron there is a residual short-range interaction called the **nuclear force** which is equivalent to the van der Waal's force between neutral atoms.

As we saw in the Stretch & challenge in Section 3.5.1(b), β^+ decay can occur when there is an unoccupied neutron energy level at least 1.8 MeV below the highest occupied proton energy level (electron capture decay is also possible). Similarly β^- decay arises if there is an unoccupied proton energy level of low enough energy.

Because of the Coulomb repulsion of the protons, their energy levels are further apart than those of the neutrons. For stability against β decay, the highest occupied neutron and proton energy levels must be approximately the same, so a stable nucleus needs to have more neutrons than protons.

As an example, consider the isotopes of niobium (Table 3.6.2). We conclude that the energy levels of the 41st proton and 52nd neutron are approximately the same, but the 53rd neutron is at a higher energy level and is able to decay into a proton occupying the 42nd position. Similarly, if there are only 51 neutrons, there must be a vacant neutron energy level at least 1.8 MeV below the top proton energy level allowing β^+ decay.

For nuclei with more than 83 protons, the nuclear force cannot overcome the coulomb repulsion, whatever the number of neutrons. Hence there are no stable nuclei with $Z > 83$.

Nuclide	decay	½ life
${}^{97}_{41}$Nb	β^-	72 min
${}^{96}_{41}$Nb	β^-	23 h
${}^{95}_{41}$Nb	β^-	35 d
${}^{94}_{41}$Nb	β^-	2.0×10^4 y
${}^{93}_{41}$Nb	stable	–
${}^{92}_{41}$Nb	β^+	3.5×10^6 y
${}^{91}_{41}$Nb	β^+	680 y
${}^{90}_{41}$Nb	β^+	14.6 h
${}^{89}_{41}$Nb	β^+	2.0 h

Table 3.6.2 Decay of isotopes of niobium

3.6.5 Nuclear fission and fusion

(a) Nuclear binding energy curve

The extraction of energy by splitting or merging nuclei (i.e. fission or fusion) is made possible by the form of the binding energy curve: the plot of binding energy per nucleon against nucleon number, Fig. 3.6.6. The red arrows indicate the direction of the nuclear changes in the two processes.

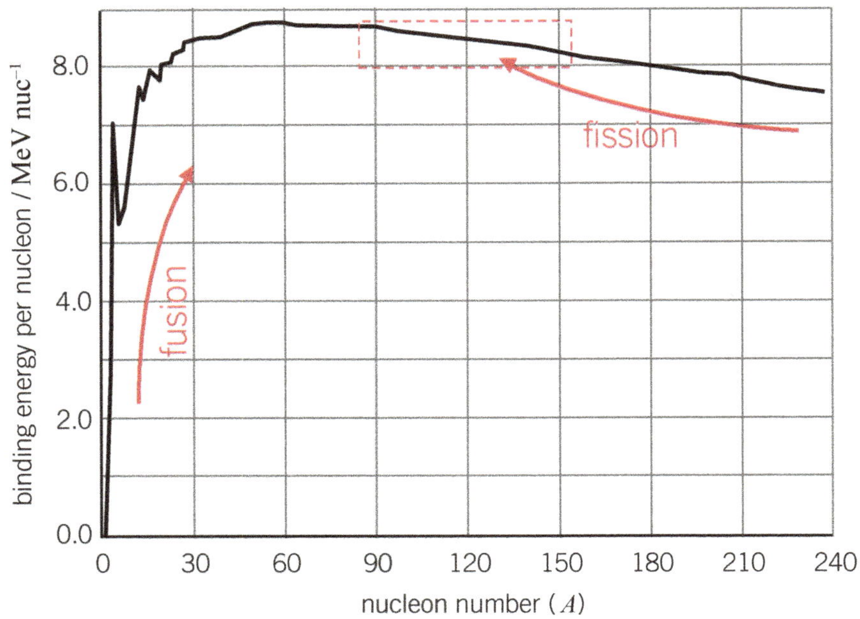

Fig. 3.6.6 Nuclear binding energy curve

We examined the low mass end of the curve in Section 3.6.3(b) and you should compare the two plots for consistency. There are many important features of the curve with which you should become familiar: the Study point gives the main ones.

Study point

Note that binding energy per nucleon is not a continuous function of A. The nucleon number can only take integer values and so the 'curve' is actually a set of closely spaced points. There are also several different nuclides for each value of A, not all of them unstable.

Study point

Features of the binding energy curve:

- When $A = 1$, BE/nuc = 0.
- Large spike of 7 MeV at ^4_2He.
- Spikes at multiples of ^4_2He.
- Maximum of 8.8 MeV nuc^{-1} for A between 56 and 62.
- Uranium isotopes have a BE of 7.6 MeV nuc^{-1}.
- Fission products of uranium have a BE of ~8.5 MeV nuc^{-1}.

Example

$^{62}_{28}\text{Ni}$ has an atomic mass of 61.928 345 u. Calculate its binding energy per nucleon.

Answer

Nuclear mass of $^{62}_{28}\text{Ni}$ = atomic mass – 28 × electronic mass

$$= 61.928\,345\text{ u} - 28 \times 0.000\,549\text{ u}$$

$$= 61.912\,973\text{ u}$$

∴ Mass deficit $= 28m_\text{p} + 34m_\text{n} - 61.912\,973 = 0.585\,365\text{ u}$

∴ BE per nucleon $= \dfrac{0.585\,365\text{ u} \times 931.5\text{ MeV u}^{-1}}{62\text{ nuc}} = 8.79\text{ MeV nuc}^{-1}$

3.6.14 Knowledge check

Show that the binding energy per nucleon of $^{235}_{92}\text{U}$ is 7.6 MeV nuc^{-1} (to 2 sf)

Atomic mass = 235.043 930 u

(b) Nuclear fission

This section provides a much more detailed and practical picture of nuclear fission than is required by the specification, which concentrates on the energy released in the process. Hence students can skip over everything apart from the calculations if they wish.

Fig. 3.6.5 shows a sprinkling of very high mass nuclides which can undergo **spontaneous fission**. The only **primordial nuclide**, which can decay in this way[3] is $^{235}_{92}\text{U}$. This nuclide decays almost 100% by α emission but in 2 out of 10^9 of the decays of ^{235}U, the nucleus splits into two large fission fragments (the *decay products* or *daughter nuclei*) and several neutrons, e.g.

$$^{235}_{92}\text{U} \rightarrow {}^{125}_{50}\text{Sn} + {}^{107}_{42}\text{Mo} + 3^1_0\text{n}$$

The two daughter nuclei, ^{125}Sn and ^{107}Mo, are both β^- radioactive. This is because they have an excess of neutrons: ^{235}U has a neutron to proton ratio of 1.55 against 1.3–1.4 which is the range for stable nuclides, such as those for tin and molybdenum.

If a ^{235}U nucleus absorbs a neutron, e.g. from a fission event, the nucleus is likely to split in a process called **induced fission**. This is much more likely to happen with **thermal neutrons**. Fast neutrons are much less likely to induce fission. A chain reaction is possible, in which the neutrons from fission events go on to induce further fissions. This can be:

- A *controlled chain reaction* in a nuclear reactor, in which excess neutrons are absorbed by control rods so that only an average of one emitted neutron goes on to induce further fission and the neutrons are slowed down (*thermalised*) by a moderator and the energy extracted as heat.
- An *uncontrolled chain reaction* in a nuclear weapon.

The energy released in nuclear fission can be calculated by the difference in the binding energy of the ^{235}U (7.6 MeV nuc^{-1}) and the fission products (8.5 MeV nuc^{-1}).

Example

Estimate the energy release in the fission of a nucleus of $^{235}_{92}\text{U}$.

Answer

The difference in the binding energy per nucleon = 8.5 – 7.6 MeV nuc^{-1}

= 0.9 MeV nuc^{-1} (1 sf)

There are 235 nucleons in $^{235}_{92}\text{U}$

∴ The energy release = 0.9 MeV nuc^{-1} × 235 = 200 MeV (1 sf)

The immediate energy release is divided up as:

- ~83% as kinetic energy of the daughter nuclei (see Test yourself 3.6 Q8).
- ~2.5% as kinetic energy of fast neutrons
- ~3.5% as gamma rays.

This energy (~89% of the total) becomes internal energy of the surrounding material (see Study point). The remaining 11% is emitted over time in the beta decay of the neutron rich daughter nuclides, of which about half is in the beta particles and half in the electron anti-neutrinos. There is a huge range of half-lives in the fission products from milliseconds to several 10^5 years. The spent fuel rods need keeping in cooling ponds until the short lived fission products have decayed.

3 This may not be quite true. $^{232}_{90}\text{Th}$ minerals have some evidence of spontaneous fission products.

Study point

Spontaneous fission is the splitting of a nucleus without stimulation by an incident neutron.

Induced fission is the splitting of a nucleus caused by the absorption of a neutron.

Thermal neutrons are relatively slow-moving. They have a kinetic energy of the order of $\frac{1}{40}$eV, equivalent to kT at room temperature.

Primordial nuclides are ones which have existed since the formation of the Earth.

Knowledge check 3.6.15

Estimate the energy released from 1 kg of nuclear fuel in a fission reactor. Assume the nuclear binding energy of the fuel and fission products are 7.6 MeV nuc^{-1} and 8.5 MeV nuc^{-1} respectively. Show your working.

Hint:

(a) Estimate the number of nucleons in 1 kg of material.

(b) Calculate the difference in the binding energy of 1 kg or fuel and 1 kg of fission products.

Study point

The figure of 200 MeV for the energy release per fission event is a useful figure to remember. It is also worth estimating the energy release per mol and per kg.

Study point

From a thermodynamics point of view, if we include the nuclear energy in the internal energy of the fuel, then all the fission has done is to change one form of internal energy into another.

Study point

Tritium, ${}^3_1\text{H}$, is made by exposing ${}^6_3\text{Li}$ to neutrons, e.g. in a fission reactor:

$${}^6_3\text{Li} + {}^0_1\text{n} \rightarrow {}^4_2\text{He} + {}^3_1\text{H}$$

Because ${}^4_2\text{He}$ is produced, this is an exothermic reaction, adding to the yield of the fission reactor.

3.6.16 Knowledge check

Calculate the energy yield of the tritium production reaction in the Study point.

Atomic mass of ${}^3_1\text{H} = 3.016\ 049\ \text{u}$

3.6.17 Knowledge check

The figure of 3 fm was obtained from the sum of the radii of the deuterium and tritium nuclei. Show that this figure agrees with the kA^3 formula.

Study point

The Sun gets by with a core temperature of about 13 million K. However, there are rather a large number of hydrogen nuclei there, so even with a weak reaction the power output is 6×10^{24} W!

(c) Nuclear fusion

As with nuclear fission, the details of the process of nuclear fusion are not required in the core specification. Option D, Energy and the environment in Unit 4 requires a deeper understanding. The Unit 3 specification concentrates on the calculation of the energy released in the process. The temperature needed for two nuclei to approach closely enough for fusion to occur is a good example of synoptic material, drawing on ideas from Unit 4 as well as Kinetic theory in Unit 3.

The binding energy curve – Fig. 3.6.6 – shows us that, if we combine low mass nuclei the resulting nucleus is more tightly bound, so we obtain an energy output. Actually, the local peak of ${}^4_2\text{H}$ allows for energy to be extracted from above this point too (see the ${}^6\text{Li}$ reaction in the Study point). The simplest possible reaction is

$${}^1_1\text{H} + {}^1_1\text{H} \rightarrow {}^2_1\text{H} + {}^0_1\text{e}^+ + \nu_e.$$

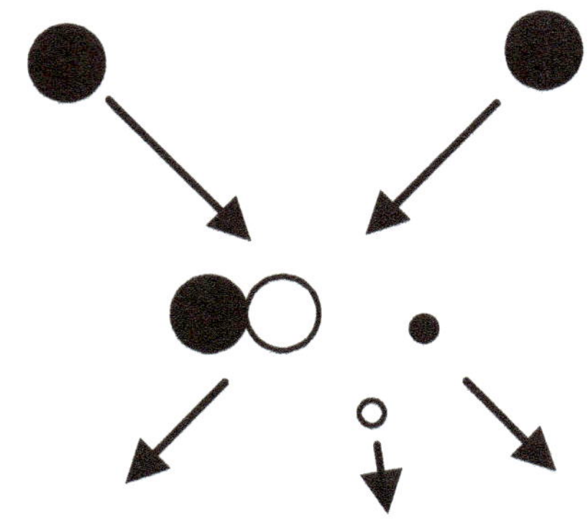

Fig. 3.6.7 H-H fusion

This reaction has several drawbacks: it is relatively low yield (2 MeV including the subsequent e^+e^- annihilation), energy is lost in the neutrino and it is a weak interaction (and therefore unlikely to happen). This last point is the most serious. This reaction is the base of the proton-proton chain in the Sun, where an individual proton in the core has a lifetime of about 10^9 years before making a successful collision!

A more promising reaction is:

$${}^2_1\text{H} + {}^3_1\text{H} \rightarrow {}^4_2\text{He} + {}^0_1\text{n},$$

where ${}^3_1\text{H}$ is tritium, consisting of a proton and two neutrons. Approximately 1.5 atoms in 10^4 of hydrogen is deuterium (a primordial nuclide, dating from the Big Bang), so it is present in large amounts in sea-water. Tritium needs to be made artificially, which is easily done in a fission reactor (see Study point).

The reaction is controlled by the strong interaction and so, as long as we can bring the reactants close enough together (~3 fm), they should fuse. The problem is that they are both positively charged so we have to overcome the Coulomb repulsion. This means raising the reactants to a high temperature. We can estimate the temperature needed as follows, using ideas from Section 4.2:

Assume the nuclei need to be brought to 3 fm. The potential energy is:

$$E_p = \frac{1}{4\pi\varepsilon_0}\frac{Q_1Q_2}{r} = 9 \times 10^9 \frac{(1.6 \times 10^{-19})^2}{3 \times 10^{-15}} = 8 \times 10^{-14}\ \text{J}\ [0.5\ \text{MeV}].$$

In order to achieve this, the thermal energy of the reactants each needs to be half this. So, using $E_k = \frac{3}{2}kT$, for the mean kinetic energy to be this high needs:

$$T = \frac{2}{3}\frac{E_k}{k} = \frac{2}{3}\frac{4 \times 10^{-14}\ \text{J}}{1.38 \times 10^{-23}\ \text{J K}^{-1}} \sim 2 \times 10^9\ \text{K}$$

Actually we don't need 2000 million K because we don't need the mean kinetic to be this high: we just need a small but significant fraction to have this. Engineers have achieved temperatures of 10^8 K and observed fusion taking place; however, they have not yet maintained the required density for long enough for the fusion to be self sustaining.

Nuclide	mass / u
$^{1}_{1}H$	1.007825
$^{2}_{1}H$	2.014102
$^{3}_{2}He$	3.01603
$^{4}_{2}He$	4.002604
$^{6}_{3}Li$	6.015126
$^{7}_{3}Li$	7.016005
$^{9}_{4}Be$	9.012186
$^{10}_{5}B$	10.012939
$^{11}_{5}B$	11.009305
$^{12}_{6}C$	12.000000
$^{13}_{6}C$	13.003354
$^{14}_{7}N$	14.003074
$^{15}_{7}N$	15.000108
$^{16}_{8}O$	15.994915
$^{17}_{8}O$	16.999133
$^{18}_{8}O$	17.99916
$^{19}_{9}F$	18.9984
$^{20}_{10}Ne$	19.99244
$^{21}_{10}Ne$	20.993849
$^{22}_{10}Ne$	21.991384
$^{23}_{11}Na$	22.989773
$^{24}_{12}Mg$	23.985045
$^{25}_{12}Mg$	24.98584
$^{26}_{12}Mg$	25.982591
$^{27}_{13}Al$	26.981535
$^{28}_{14}Si$	27.97693
$^{29}_{14}Si$	28.97649
$^{30}_{14}Si$	29.97376
$^{31}_{15}P$	30.973763
$^{32}_{16}S$	31.972074
$^{33}_{16}S$	32.97146
$^{34}_{16}S$	33.96786
$^{36}_{16}S$	35.96709
$^{35}_{17}Cl$	34.96885
$^{37}_{17}Cl$	36.9659
$^{36}_{18}Ar$	35.96755
$^{38}_{18}Ar$	37.96572
$^{40}_{18}Ar$	39.962384
$^{39}_{19}K$	38.96371
$^{41}_{19}K$	40.96183
$^{40}_{20}Ca$	39.96259
$^{42}_{20}Ca$	41.95863
$^{43}_{20}Ca$	42.95878
$^{44}_{20}Ca$	43.95549
$^{46}_{20}Ca$	45.95369
$^{48}_{20}Ca$	47.95236

Additional data: $m_{\text{proton}} = 1.007\ 276$ $m_{\text{neutron}} = 1.008\ 665$ u $m_{\text{electron}} = 0.000\ 549$ u

Table 3.6.3 The atomic masses of the stable nuclides for $Z = 1–20$.

Study point

As well as calculating the binding energy you can use this table to search for interesting patterns, e.g. the number of stable nuclei with odd and even values of Z, the number of nuclei with odd and even values of Z and N. Notice that the least nucleon number with two different stable nuclides is 36 – $^{36}_{16}S$ and $^{36}_{18}Ar$.

Test yourself 3.6

1. A coal-fired power station, running continuously during the winter months, generates electricity with an efficiency of 35%. Its power output is 2.0 GW. Calculate:

   (a) The thermal energy by burning the fuel in a two-month (60-day) period.
   (b) The mass difference between the reactants (coal + oxygen) and the products of the burning (carbon dioxide and water vapour).

2. The Sun loses 4.0 million tonnes (1 tonne = 10^3 kg) per second in its radiation. Calculate:

   (a) The power output of the Sun.
   (b) The power incident on the Earth.

   [Radius of Earth's orbit = 150 million km; radius of Earth = 6400 km.]

3. Some nuclei decay by electron capture in which an inner electron (1s orbital) is absorbed. ${}^{7}_{4}\text{Be}$ (atomic mass 7.016 930 u) decays by electron capture into a nuclide of mass 7.016 004 u.

   (a) Write the equation for the decay.
   (b) Explain which of the nuclear interactions (strong, e-m, weak) is responsible for the decay.
   (c) Calculate the energy released in the decay and explain which particle carries this energy.

4. The nuclide ${}^{241}_{95}\text{Am}$ (atomic mass 241.056 69 u) decays into an isotope of neptunium (Np) of mass 237.048 173 u, with a half-life of 460 years.

   (a) Write the decay equation.
   (b) Calculate the energy, in MeV, released in the decay.
   (c) Use the principle of conservation of momentum to calculate the recoil speed of the nucleus.
   (d) A smoke detector contains 1 mg of ${}^{241}_{95}\text{Am}$. Calculate its activity.

5. The primordial nuclide ${}^{40}_{19}\text{K}$ (atomic mass 39.963 998 u) has three modes of decay: β^- (89.28%), electron capture (10.72%) and β^+ decay (0.001%), with a half-life of 1.3×10^{-9} years.

   (a) Identify the daughter nuclides in the three modes of the decay and write the decay equations.
   (b) Calculate the energy released in each decay mode.

6. The nuclear fuel in uranium fission reactors contains a large proportion of the non-fissile ${}^{238}\text{U}$. A ${}^{238}\text{U}$ nucleus absorbs a neutron and the resulting nucleus decays in two stages into the fissile isotope of plutonium, ${}^{239}_{94}\text{Pu}$.

   $${}^{238}_{92}\text{U} + {}^{1}_{0}\text{n} \rightarrow {}^{239}_{92}\text{U} \xrightarrow{\beta^-} {}^{239}_{93}\text{Np} \xrightarrow{\beta^-} {}^{239}_{94}\text{Pu}$$

   (a) Write decay equations for the decay of ${}^{239}\text{U}$ and ${}^{239}\text{Np}$.
   (b) Use the atomic masses of the nuclides to calculate the energy released in the two β^- reactions.
   (c) Explain briefly why only a fraction of this energy release is absorbed within the reactor.

   Masses: ${}^{239}_{92}\text{U} = 239.054\ 293$ u; ${}^{239}_{93}\text{Np} = 239.052\ 939$ u; ${}^{239}_{94}\text{Pu} = 239.052\ 163$ u

7. After about 1 billion years as a Red Giant, the helium-rich core in a solar-mass star achieves a high enough temperature and pressure to start 'burning' helium in the triple-alpha reaction:

   ${}^{4}_{2}\text{He} + {}^{4}_{2}\text{He} + {}^{4}_{2}\text{He} \rightarrow {}^{12}_{6}\text{C}$ Calculate the energy released in this reaction.

Nuclide	atomic mass / u
$^{92}_{37}Rb$	91.919 730
^{92}Zr	91.905 041
$^{140}_{55}Cs$	139.917 282
^{140}Ce	139.905 439
^{235}U	235.043 9

8. One of the possible fission reactions in a ^{235}U reactor is:

$$^{235}_{92}U + ^{1}_{0}n \rightarrow ^{140}_{55}Cs + ^{92}_{37}Rb + 4^{1}_{0}n$$

^{92}Rb decays by successive β^- emissions: $Rb \rightarrow Sr \rightarrow Y \rightarrow Zr$ (stable)

^{140}Cs decays by successive β^- emission: $Cs \rightarrow Ba \rightarrow La \rightarrow Ce$ (stable)

Use the table of atomic masses to calculate:

(a) The energy given off in the initial fission.
(b) The energy given off in the subsequent β^- decays.
(c) The energy given off by the fission of $1\ kg$ of ^{235}U.

9. The proton-proton fusion chain in the Sun starts with the following two steps:

$$^{1}_{1}H + ^{1}_{1}H \rightarrow ^{2}_{1}H + ^{0}_{1}e^+ + \nu_e \qquad [1]$$

$$^{1}_{1}H + ^{2}_{1}H \rightarrow ^{3}_{2}He + \gamma \qquad [2]$$

69% of the subsequent energy release is from the following reaction:

$$^{3}_{2}He + ^{3}_{2}He \rightarrow ^{4}_{2}He + 2^{1}_{1}H \qquad [3]$$

Calculate:

(a) The energy in MeV release in each step [include the e^+e^- annihilation].
(b) The total energy released in the fusion of $1\ kg$ of hydrogen.

10. Estimate the energy released in the fission fragments of a nuclear fission by considering the change in their electric potential energy as they separate. This is a synoptic question which uses the equation for the electrical potential energy of two charges in Section 4.2.

Consider the tin and molybdenum nuclei as they are about to separate in the fission event in Section 3.6.5 (b):

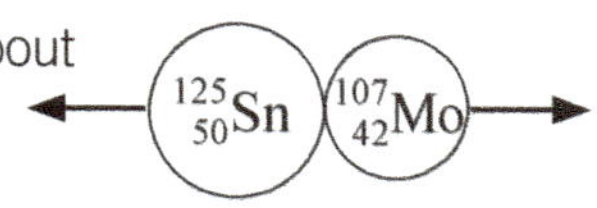

Calculate:

(a) The charge on each nucleus.
(b) The radius of each nucleus (Hint: see Section 3.6.4).
(c) The potential energy of the two charges and hence the energy released as they separate. [They are spherically symmetric so you can treat them as point charges, separated by the sum of their radii.]
You should obtain an answer which is (at least to a fair approximation) in agreement with the calculation in 3.6.5(b).
(d)

Stretch & challenge

The final speed that each of the fragments achieves.

11. Use Table 3.6.2 to test the relationship $r = kA^{1/3}$ and determine a value for k. Is it advantageous to draw a $\ln r - \ln A$ graph or one of r against $A^{1/3}$?

Unit 3 equations

Because of the requirement to set synoptic questions, the WJEC Data Booklet for Unit 3 includes the equations for AS Units 1 and 2 as well as the core content of Unit 4. The equations below are the ones specific to Unit 3. The descriptions given below are not given in the Data Booklet – just the symbol equations.

Equation	Description
$\omega = \frac{\theta}{t}$	ω = angular velocity, θ = angle, t = time
$v = \omega r$	v = speed around circumference; r = radius
$a = \omega^2 r = \frac{v^2}{r}$	centripetal acceleration
$F = m\omega^2 r = \frac{mv^2}{r}$	centripetal force
$a = -\omega^2 r$	simple harmonic motion (SHM) definition
$x = A\cos(\omega t + \varepsilon)$	displacement in SHM
$v = -A\sin(\omega t + \varepsilon)$	velocity in SHM
$T = \frac{2\pi}{\omega}$	T = period; ω = angular frequency
$T = 2\pi\sqrt{\frac{m}{k}}$; $T = 2\pi\sqrt{\frac{l}{g}}$	period of mass on spring; period of simple pendulum
$pV = nRT$; $pV = NkT$	ideal gas equation in terms of moles / in terms of molecules
$p = \frac{1}{3}\rho\overline{c^2} = \frac{Nm}{V}\overline{c^2}$	kinetic theory equation
$M/\text{kg} = \frac{M_r}{1000}$	M = molar mass; M_r = relative molecular mass
$n = \frac{\text{total mass}}{\text{molar mass}}$	n = amount of material expressed in moles
$k = \frac{R}{N_A}$	k = Boltzmann constant; R = ideal gas constant; N_A = Avogadro constant
$U = \frac{3}{2}nRT = \frac{3}{2}NkT$	U = internal energy of a monatomic ideal gas
$W = p\Delta V$	W = work done by a system on expansion
$\Delta U = Q - W$	First law of thermodynamics; Q = heat entering a system
$Q = mc\Delta\theta$	c = specific heat capacity; $\Delta\theta$ = change in temperature
$A = \lambda N$	A = activity of a radioactive sample; λ = decay constant
$N = N_0 e^{-\lambda t} = \frac{N_0}{2^x}$	N_0 = initial number of radioactive atoms; x = number of half-lives
$A = A_0 e^{-\lambda t} = \frac{A_0}{2^x}$	A_0 = initial activity
$\lambda = \frac{\ln 2}{T_{1/2}}$	$T_{1/2}$ = half life
$E = mc^2$	Einstein's mass/energy equation; c = speed of light in a vacuum

Unit 3

1 A fairground ride consists of a large empty cylinder of radius $2.80\ \text{m}$ with its axis of rotation vertical. A person of mass $66.2\ \text{kg}$ walks into the cylinder and stands against its inner wall. The cylinder is rotated around the axis. When the rotation rate is 36.0 revolutions per minute the floor is lowered, but the person remains fixed in place against the wall.

This is a student's diagram:

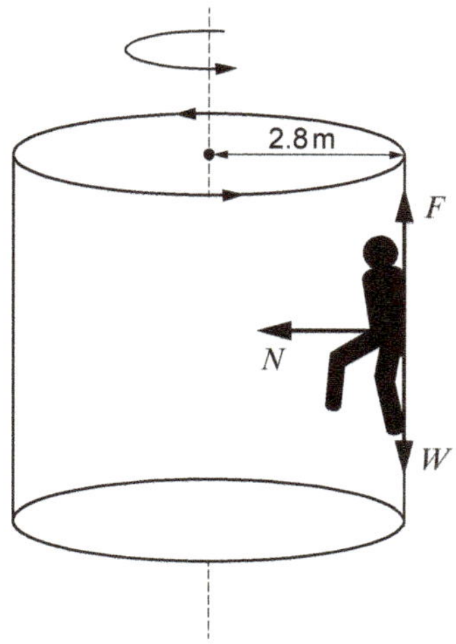

(a) Define the *radian*. **[1]**

(b) Calculate:

(i) the period of rotation, T; **[2]**

(ii) the speed, v, at which the person is moving in the circle. **[2]**

(c) Three forces act on the person: the normal contact force from the wall, N, the weight, W, of the person, and the frictional force, F, from the wall. These are shown in the student's diagram.

Explain why the values of N and F are approximately $2\,600\ \text{N}$ and $650\ \text{N}$ respectively. **[4]**

(d) The maximum value that a frictional force can take is μN where μ is a dimensionless constant.

(i) Show that the person does not slide down the wall provided that μ is larger than approximately 0.25. **[2]**

(ii) A student says that as the cylinder slows down, the floor will need to be raised to support the person when the angular velocity reduces below a certain value. Justify this and determine this angular velocity if μ has a value of 0.45. **[4]**

(Total 15 marks)

[*WJEC A level Physics 2019 Unit 3 Q2*]

2. Simple harmonic motion (SHM) may be represented by:

$a = -\omega^2 x$

(a) State what quantities are represented by: **[1]**

a

ω

x

(b) If the amplitude of the oscillation of an object moving with SHM is 0.012 m and the period is 0.40 s:

(i) determine the maximum acceleration; **[3]**

(ii) sketch a graph showing the variation of a with x. Insert values on the a-axis. Values are given on the x-axis. **[3]**

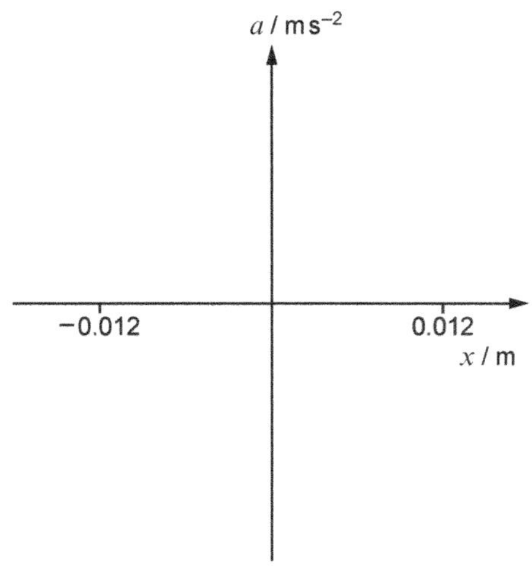

(c) The oscillating object in part (b) is at its centre position when $t = 0$ and moving in the positive x direction. Insert the missing values in the expression for x in metres. **[3]**

$x = \square \cos(\square t + \square)$

(d) An oscillating system may be driven by an external force. Describe and explain:

(i) an application of forced oscillations that is useful; **[2]**

(ii) an example of forced oscillations that should be avoided. **[3]**

(Total 15 marks)

[*WJEC A level Physics 2019 Unit 3 Q3*]

3. Describe an experiment that uses a pendulum to determine the acceleration due to gravity with graphical analysis. **[6]**

(Total 6 marks)

[*WJEC A level Physics 2019 Unit 3 Q4*]

4. A sealed container with a leakproof piston of cross-sectional area $0.040\ \text{m}^2$ at one end has an initial volume of $8.5 \times 10^{-3}\ \text{m}^3$. It contains oxygen gas at a pressure of $5.0 \times 10^5\ \text{Pa}$ and a temperature of $285\ \text{K}$.

(Relative molecular mass of oxygen = 32)

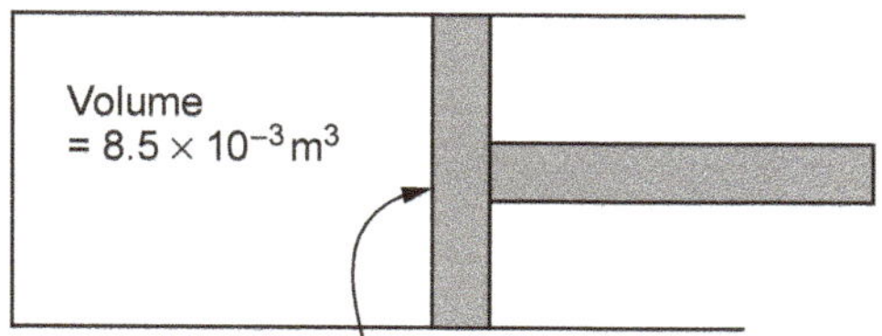

(a) Calculate:

(i) the number of moles of oxygen gas in the container; **[2]**

(ii) the number of molecules in the container; **[1]**

(iii) the rms speed of the molecules; **[3]**

(iv) the force exerted by the gas on the piston. **[1]**

(b) The gas is allowed to expand at a constant temperature of $285\ \text{K}$ to a volume of $10.2 \times 10^{-3}\ \text{m}^3$.

(i) Calculate the final pressure of the gas. **[2]**

(ii) During the process, the gas does $773\ \text{J}$ of work. Determine the heat flowing into the gas, explaining your reasoning. **[2]**

(iii) Following the expansion, the gas is returned to its initial state by a two-stage process:

– a decrease in volume at constant pressure;

– an increase in pressure at constant volume.

A student says that approximately $60\ \text{J}$ of heat flows into the gas during the overall cycle (the expansion and return of the gas to its original state). Justify this. Include a sketch of the paths on a p–V diagram with values on both axes. **[4]**

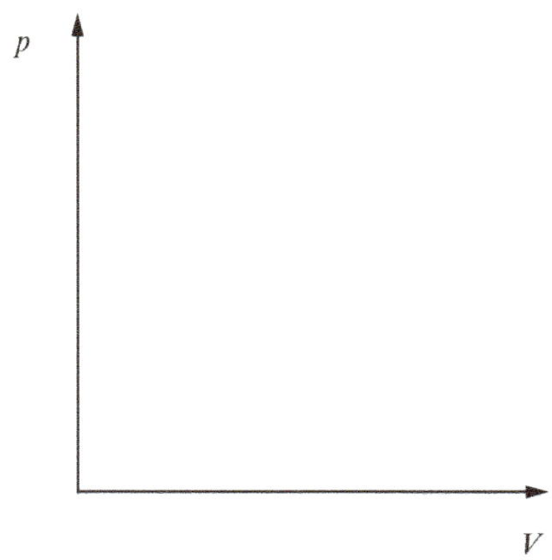

(Total 15 marks)

[*WJEC A level Physics 2019 Unit 3 Q5*]

5. (a) Complete the following decay equations:

(i) ${}^{228}_{90}\text{Th} \rightarrow {}^{\cdots\cdots}_{\cdots\cdots}\text{Ra} + {}^{\cdots\cdots}_{\cdots\cdots}\alpha$ **[1]**

(ii) ${}^{90}_{38}\text{Sr} \rightarrow {}^{\cdots\cdots}_{\cdots\cdots}\text{Y} + {}^{\cdots\cdots}_{\cdots\cdots}\beta$ **[1]**

(b) Calculate the binding energy per nucleon for ${}^{90}_{38}\text{Sr}$. **[4]**

$m_{\text{proton}} = 1.007\ 276\ \text{u}$

$m_{\text{neutron}} = 1.008\ 664\ \text{u}$

$m_{\text{electron}} = 0.000\ 549\ \text{u}$

atomic mass of Sr = 89.907 738 u

1 u = 931 MeV

(c) Radioactive decay may be investigated in the laboratory using a dice analogy.

A student has 564 identical wooden dice. These dice are tetrahedrons, with one face painted black. A tetrahedron is a solid with four identical faces, each face being an equilateral triangle.

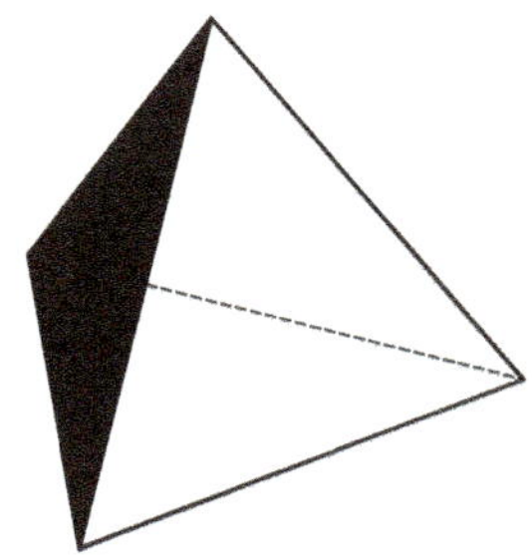

(i) If the student throws one of the dice, what is the probability of the tetrahedron landing on the black face? **[1]**

(ii) The student then throws all dice and counts and discards those that landed on the black face. The remaining dice are thrown again and the process is repeated multiple times. The number of dice discarded after each throw is recorded in the table, and the number remaining after each throw is calculated. The results are shown in the table, which also gives the fraction of dice remaining, where:

$$\text{Fraction of dice remaining after a throw} = \frac{\text{number of dice remaining after throw, } R}{\text{number of dice thrown in that throw, } T}$$

Throw number, n	Number of dice thrown, T	Number of dice discarded after throw	Number of dice remaining after throw, R	Fraction remaining after throw, R/T
1	564	138	426	0.76
2	426	116	310	0.73
3	310	87	223	0.72
4	223	52	171	0.77
5	171	39	132	0.77
6	132	34	98	0.74
7	98	10	88	0.90
8	88	27	61	0.69
9	61	18	43	0.70
10	43	8	35	0.81

I. Explain why the expected fraction remaining after n throws is $(0.75)^n$. **[2]**

II. Hence predict the expected number remaining after 10 throws. **[1]**

III. Discuss to what extent the observations agree with those expected from theory. **[3]**

(Total 13 marks)

[*WJEC A level Physics 2019 Unit 3 Q6*]

Unit 4

Fields and options

This unit has a single main theme of fields and a set of four options, of which you should study one.

The study of fields of force is one of the basic pillars of physics. It is used to describe influences which extend through space, so that the effect takes place remotely from the cause. Three types of field are studied:

- **Gravitational field**. All material particles exert forces on one another which are proportional to their masses. These interactions are described in terms of a gravitational field which surrounds each particle. Historically this was the first field to be described in Newton's theory of gravity.
- **Electrostatic field**. All charged particles also exert forces on one another, in fields which are very similar to gravitational fields – they both decrease as the inverse square of the distance. The electric field is the origin of the adhesive forces between atoms and molecules.
- **Magnetic field**. This is a more complicated field than the other two, being caused by charged particles in motion. Its properties are useful in electrical motors and generators.

In addition to these basic areas of study, there are sections which focus on applications of the fields:

- **Capacitors** are devices for storing electrical energy and operate on the basis of the electrostatic fields between metal plates.
- **Orbits and the wider universe** describes how the motion of astronomical objects in combination with the knowledge of gravitational fields allows us to gain insights into the structure of the universe

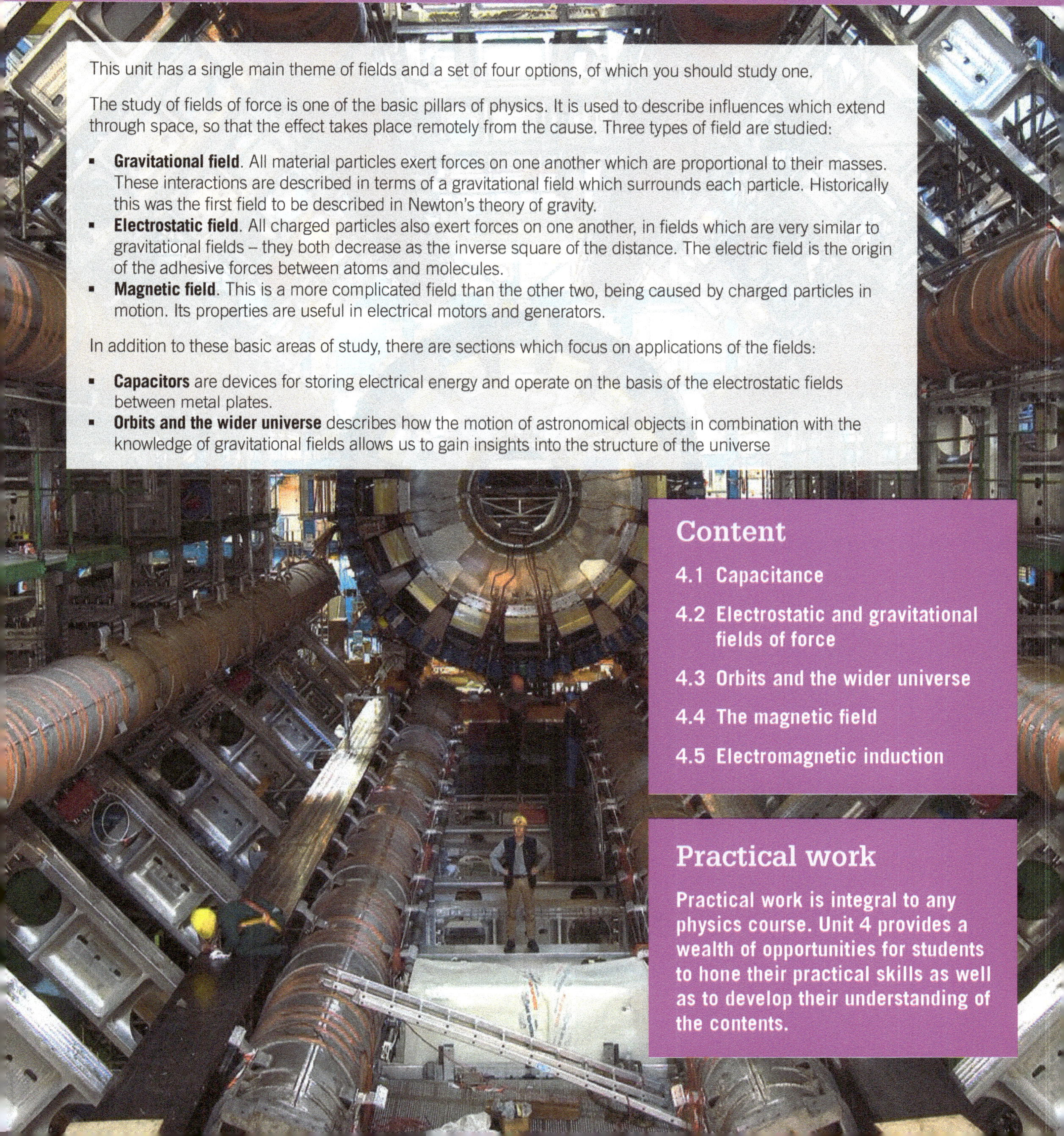

Content

Practical work

Practical work is integral to any physics course. Unit 4 provides a wealth of opportunities for students to hone their practical skills as well as to develop their understanding of the contents.

4.1 Capacitance

Key terms

A **capacitor** consists of two conductors (so-called 'plates') separated by an insulator called a *dielectric*.

Capacitors are simple devices used widely in electronics. We learn about their behaviour. Then we introduce the fundamental idea of an electric field, and explain how a uniform electric field can be produced in a capacitor. We finish by studying the motion of a charged particle in a uniform electric field.

4.1.1 Capacitors and capacitance

What is a capacitor? This is easily stated – see Key terms. What does a capacitor do? It stores equal and opposite charges on its plates when there is a pd between the plates.

In practice almost all capacitors are of the parallel plate type, with an insulating material of constant thickness (often less than $0.1\ \text{mm}$) between the plates. The circuit symbol for a capacitor is as shown in Fig. 4.1.3 (without '+' and '–' signs).

Study point

All the cylindrical capacitors in Fig. 4.1.1 are electrolytic except the silvery one.

Supercapacitors (usually of capacitance more than 1 F) have dielectrics that are only about two atoms thick. There's more than one type and the technology involved is as much chemical as physical.

Capacitors come in all shapes and sizes (Fig. 4.1.1). A cylindrical case is likely to contain two strips of metal foil (the plates) and two strips of thin plastic (the dielectric) rolled up like a Swiss roll (Fig. 4.1.2). See also the Study point. Capacitors are usually marked with the maximum pd that can be applied safely.

The capacitors with exposed plates, to the right in Fig. 4.1.1, are 'variable'. One set of plates can be moved to change the area of overlap between them and another set. The dielectric is air.

Capacitors with a capacitance of more than about $1\ \mu\text{F}$ (see Section 4.1.2) and less than about $0.1\ \text{F}$ are usually *electrolytic*. The dielectric is a thin film of oxide on one plate. The capacitor must be connected the right way round in a circuit, so that the oxide-coated plate is at a positive potential relative to the other plate, or the film is destroyed.

4.1.1 Knowledge check

Why are *two* insulating strips needed in the 'rolled up' capacitor of Fig. 4.1.2?

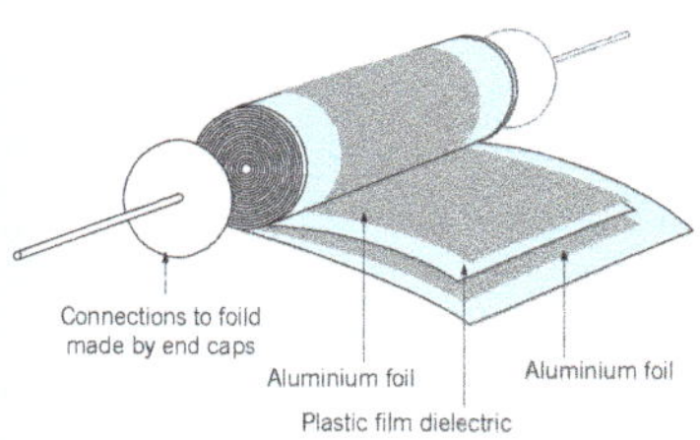

Fig. 4.1.2 One type of capacitor

Fig. 4.1.1 Capacitor smörgåsbord

Capacitance

If a battery or DC power supply is connected across a capacitor, as in Fig. 4.1.3(a), some free electrons are removed from one plate and transferred via the battery to the other. The plates gain equal and opposite charges (on facing surfaces). This 'charging' stops when the pd between the plates is equal to the emf of the battery (Fig. 4.1.3(b)). It usually takes a small fraction of a second, unless a resistor is included in series or the battery has a high internal resistance.

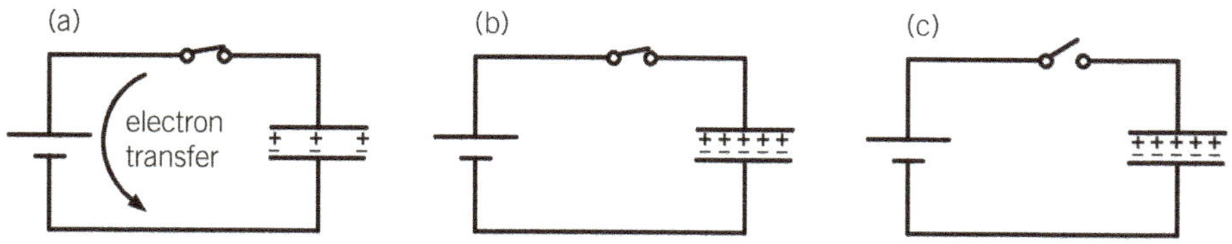

Fig. 4.1.3 Charging and isolating a capacitor

If we then disconnect the battery, as in Fig. 4.1.3(c), the charges are isolated on the plates, maintaining the pd between the plates.

We can investigate how the charge ($\pm Q$) on the capacitor plates is related to the pd, V, between them using the apparatus in Fig. 4.1.4(a). We charge the capacitor by switching the two-way switch to the left briefly, and then to the right, so the capacitor discharges into the coulomb meter. We repeat for different pds, selected using the variable DC power supply. Fig. 4.1.4(b) is a typical graph of the results.

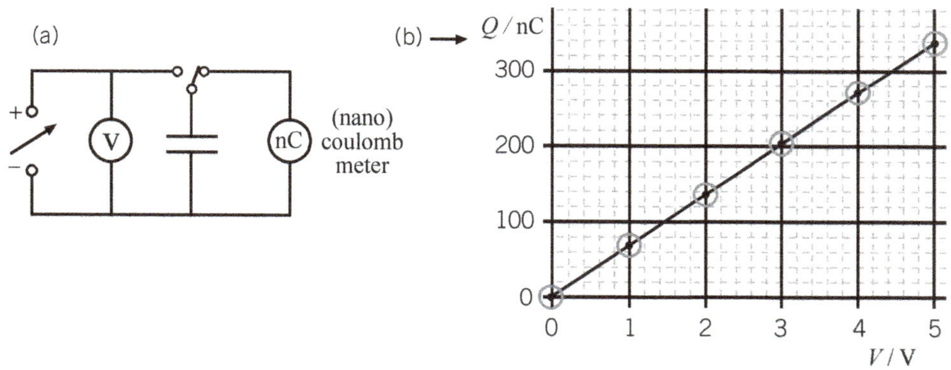

Fig. 4.1.4 Q against V for a capacitor

The charge stored on either plate is proportional to the pd, V. This is the case for any normal capacitor. So we can write

$$Q = CV$$

where C is a constant for the capacitor, called its **capacitance**, and is the gradient of the graph of Q against V.

Capacitors are made with a huge range of values, from a few picofarads (pF), used mainly in circuits involving high frequency alternating currents, up to hundreds of farads, used for backing up computer memories' power supplies or helping batteries to cope with spikes of demand for current. Time to brush up your prefixes for powers of ten!

Study point

Don't confuse C (italic type) standing for the quantity, capacitance, with C (upright type) standing for the unit of charge, coulomb.

Knowledge check 4.1.2

Express the farad in terms of the base SI units, m, kg, s and A.

Hint: remember that $\text{V} = \text{J C}^{-1}$

Knowledge check 4.1.3

Calculate the capacitance of the capacitor to which Fig. 4.1.4 applies. Give your answer in both nF and μF.

Knowledge check 4.1.4

Determine the pd needed to store a charge of 4.8 mC on a 160 μF capacitor. What should be checked before applying the pd?

Key terms

The **capacitance**, C, of a capacitor is the constant ratio

$$\frac{\text{charge on either plate}}{\text{pd between plates}}$$

That is $C = \frac{Q}{V}$

UNIT: C V^{-1} = farad (F)

4.1.2 The capacitance of a parallel plate capacitor

How do we design a capacitor to have a particular capacitance? There is a neat formula, testable by experiment, for the capacitance of a parallel plate capacitor (Fig. 4.1.5) with a vacuum or air between its plates (as long as their separation, d, is much less than their side length, $\sqrt{A}$). It is

$$C = \frac{\varepsilon_0 A}{d},$$

where ε_0 is a constant with a rather scary 19th-century name: *the permittivity of free space*. Most of us call it 'epsilon nought'. By experiment,

$$\varepsilon_0 = 8.85 \times 10^{-12}\ \text{F m}^{-1}$$

Note that A is the area of a single surface of either plate.

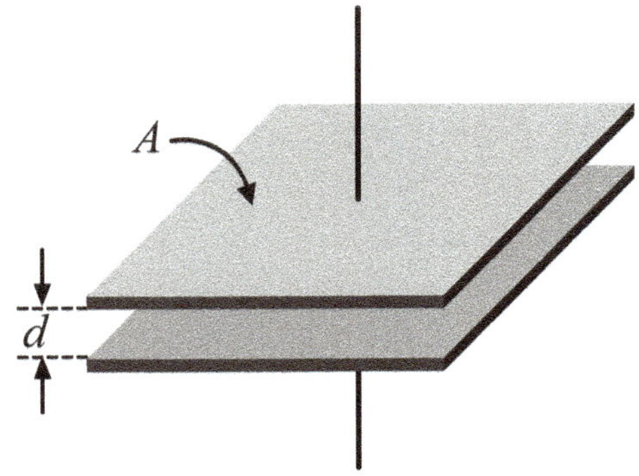

Fig. 4.1.5 Parallel plate capacitor

The dielectric

Most capacitors used in electronics have (solid) dielectrics, such as films of tantalum (V) oxide, thin sheets of polypropene or thin slices of ceramic material. A dielectric does more than maintain the insulating gap between the plates: it makes the capacitance more than it would be if the gap had nothing (or just air) inside it. In other words it makes the capacitance larger than $\frac{\varepsilon_0 A}{d}$; often by many times. See Study points.

4.1.3 Capacitor combinations

Combinations of capacitors behave electrically like single capacitors. This can sometimes be useful when a certain value of capacitance is needed, but no single capacitor of this value is available.

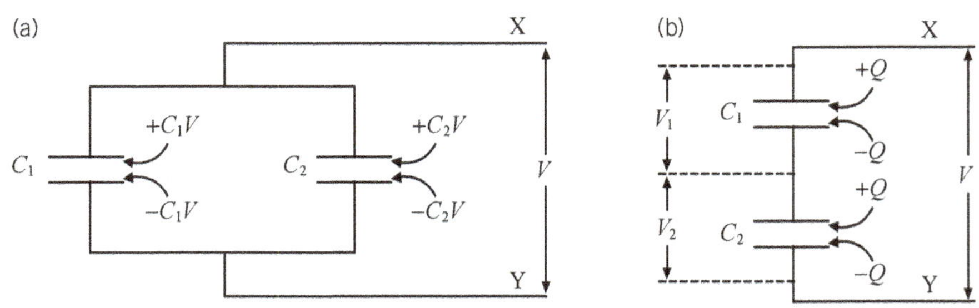

Fig. 4.1.6 (a) Capacitors in parallel (b) Capacitors in series

We'll connect capacitors together in parallel and series, as in Fig. 4.1.6(a) and (b) and find a formula for the equivalent capacitance in each case, just as we did for resistors in series and parallel in the AS/Year 12 textbook.

4.1.5 Knowledge check

Calculate the plate area needed for a 150 pF air-spaced capacitor with a gap of 1.5 mm between the plates.

Study point

It's not surprising to find that the capacitance is proportional to the plate area, A, since it's the plates that have the charge. The Study point in Section 4.1.6(d) explains the inverse relationship between C and d.

Study point

A dielectric increases capacitance because of an effect called 'dielectric polarisation': a temporary shift in the positions of nuclei and electrons within the dielectric's molecules, due to the *electric field* (Section 4.1.5) set up by charges on the plates. Certain ceramic dielectrics will increase the capacitance by a factor of more than 1000.

(a) Capacitors in parallel

From Fig. 4.1.6(a), each capacitor has the same pd, V, across it. The charges on the plates of C_1 and C_2 are therefore $\pm C_1V$ and $\pm C_2V$. So the charge that flowed through wires X and Y is

$$Q = C_1V + C_2V \qquad \text{i.e. } Q = (C_1 + C_2)V$$

So if we wanted to replace the parallel combination with a single equivalent capacitance, C, we'd choose:

$$C = \frac{Q}{V} = \frac{(C_1 + C_2)V}{V}, \qquad \text{i.e. } C = C_1 + C_2$$

Even if there are more than two capacitors in parallel we simply add individual capacitances. See Study point.

Notice that the formula for **capacitors in parallel** is of the same form as the formula for **resistors in series**.

A useful consistency check is to consider two parallel-plate air-spaced capacitors, with the same value of d, connected in parallel.

Then $\quad C = C_1 + C_2$

So $\quad C = \frac{\varepsilon_0 A_1}{d} + \frac{\varepsilon_0 A_2}{d} = \frac{\varepsilon_0 (A_1 + A_2)}{d}$

So the capacitors behave like a single capacitor with plate area $(A_1 + A_2)$, as one would expect.

Study point

Remember that for capacitors **in parallel**

$C = C_1 + C_2 + \ldots$

For capacitors **in series**

$\frac{1}{C} = \frac{1}{C_1} + \frac{1}{C_2} + \ldots$

(b) Capacitors in series

When a pd, V, is applied as shown in Fig. 4.1.6(b), each capacitor's plates gain equal and opposite charges. These charges are the same for both capacitors, because the 'island' plates (the plates connected together) are insulated from everything else and must have a *total* charge of zero – assuming that neither capacitor had any charge before connection.

Because pds in series add,

$$V = V_1 + V_2 = \frac{Q}{C_1} + \frac{Q}{C_2} \text{ so } \qquad \frac{V}{Q} = \frac{1}{C_1} + \frac{1}{C_2}$$

Since the charge that flowed through the wires X and Y is Q, the equivalent capacitance of the combination is $C = \frac{Q}{V}$.

Therefore $\quad \frac{1}{C} = \frac{1}{C_1} + \frac{1}{C_2}$

If there are more than two capacitors we simply go on adding reciprocals.

We notice that the formula for **capacitors in series** is of the same form as the formula for **resistors in parallel**.

Stretch & challenge

Perform a consistency check, modelled on the one for capacitors in parallel, by considering two parallel plate air-spaced capacitors, with the same plate area, A, in series.

Knowledge check 4.1.6

Calculate the equivalent capacitance of a 22 μF and a 33 μF capacitor connected (a) in parallel and (b) in series.

Study point

For two capacitors in series a little algebra is a worthwhile investment…

$$\frac{1}{C} = \frac{1}{C_1} + \frac{1}{C_2} = \frac{C_2}{C_1C_2} + \frac{C_1}{C_2C_1}$$

that is $\frac{1}{C} = \frac{C_1 + C_2}{C_1C_2}$

We now have a single fraction on each side of the equation. The reciprocals of these fractions are also equal, so

$$C = \frac{C_1C_2}{C_1 + C_2} = \frac{\text{product}}{\text{sum}}$$

WARNING! This doesn't work for more than two capacitors at a time.

Top tip

The useful equation $C = \frac{C_1C_2}{C_1 + C_2}$ isn't in the WJEC Data booklet.

4.1.7 Knowledge check

Suppose that, in Fig. 4.1.7(c), 10.0 V is applied between X and Y.

Calculate the charge and pd for (a) the 22 nF and (b) the 47 nF capacitor.

Stretch & challenge

Draw a circuit containing two 100 μF capacitors and one 47 μF capacitor whose combined capacitance is 60 μF (to 2 sf).

Show that your combination *does* have this capacitance.

Study point

There's no simple, wholly satisfactory way of treating significant figures in a case like this, so we retained 3 and rounded to 2 at the end, as data was only to 2 sf Spot the apparent inconsistencies in the final answers!

(c) Capacitor combination calculations

As examples we find the equivalent capacitance for (a) and (b) in Fig. 4.1.7.

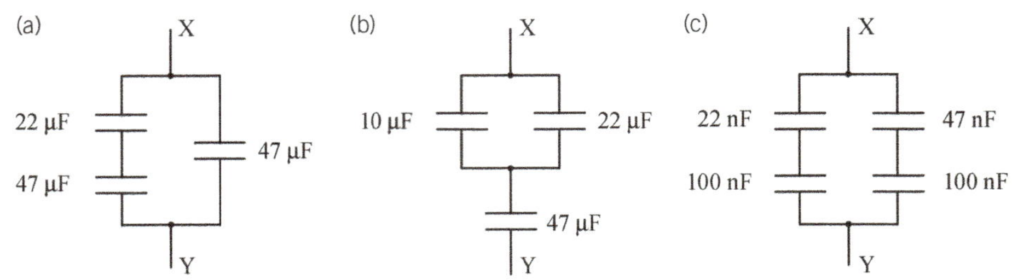

Fig. 4.1.7 Three combinations of capacitors

The equivalent capacitance in Fig. 4.1.7(a)

We first deal with the left-hand side: 22 μF and 47 μF in series. As for two resistors in parallel, so also for two capacitors in series, there's a simpler and less accident-prone version of the 'reciprocal of sum of reciprocals' formula. It's derived in the Study point. We'll use it now…

$$C = \frac{\text{product}}{\text{sum}} = \frac{22\ \mu\text{F} \times 47\ \mu\text{F}}{69\ \mu\text{F}} = 15\ \mu\text{F}$$

[Make sure you understand how the units work out. Write 22 μF as 22×10^{-6} F and so on if this helps.]

The full combination is therefore 15 μF in parallel with 47 μF.

So, finally, C = 15 μF + 47 μF = 62 μF.

The equivalent capacitance in Fig. 4.1.7(b)

For the capacitors in parallel, the combined capacitance is 10 μF + 22 μF = 32 μF. But this is in series with 47 μF.

So overall capacitance , $C = \frac{\text{product}}{\text{sum}} = \frac{32\ \mu\text{F} \times 47\ \mu\text{F}}{79\ \mu\text{F}} = 19\ \mu\text{F}$

Charges and pds for each capacitor in Fig. 4.1.7(b)

Suppose a pd of 12.0 V is applied between X and Y. If required, we can work out charges and pds for all the capacitors, as we now show…

For the combination, $Q = CV$ = 19 μF × 12.0 V = 228 μC.

Since the combination behaves as a 32 μF capacitor in series with a 47 μF capacitor, the charges on these are both ±228 μC.

The pd across the 47 μF is therefore:

$$V = \frac{Q}{C} = \frac{228\ \mu\text{C}}{47\ \mu\text{F}} = 4.85\ \text{V}.$$

This leaves 12.0 V – 4.85 V = 7.15 V across the 32 μF combination.

So charge on 10 μF capacitor = CV = 10 μF × 7.15 V = 71.5 μC,
and charge on 22 μF capacitor = CV = 22 μF × 7.15 V = 157 μC.

Conclusion (see Study point):
10 μF: 7.2V, 72 μC;   22 μF: 7.2V, 160 μC;   47 μF: 4.9 V, 230 μC.

4.1.4 Energy stored in a charged capacitor

One way of demonstrating that a charged capacitor stores energy is to use the energy to produce light (and infra-red!), as in Fig. 4.1.8. After charging the capacitor (switch to the left), we can discharge it through the lamp (switch to the right).

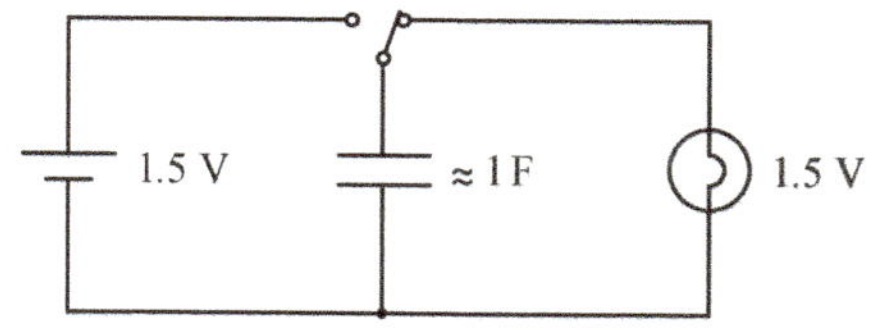

Fig. 4.1.8 Demonstrating energy stored in a charged capacitor

It's easy to work out *how much* energy is stored. We consider a capacitor discharging through any conducting device (Fig. 4.1.9(a)). We've shown a resistor, but what follows doesn't require Ohm's law to apply.

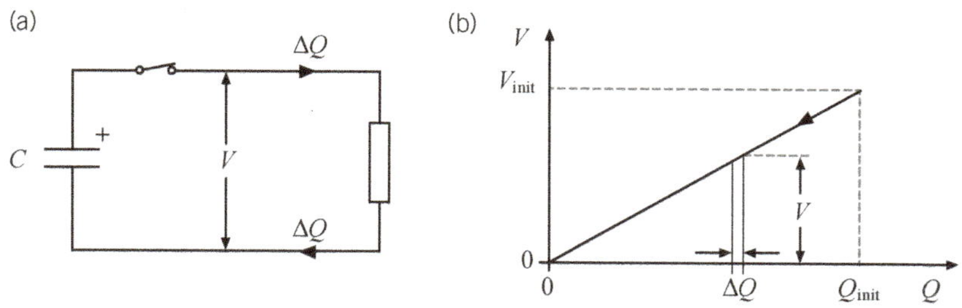

Fig. 4.1.9 Capacitor discharging through a resistor

Electrons flow off the negative plate through the resistor to the positive plate. As the charge, Q, on the plates decreases, the pd, V, between them falls, as in Fig. 4.1.9(b). This is simply a consequence of $V = Q/C$.

The energy, ΔU, transferred from the capacitor and dissipated in the resistor when a small portion, ΔQ, of charge, flows is

$$\Delta U = V\Delta Q$$

in which V is the (mean) pd between the capacitor plates *while that small portion of charge is flowing*. [Remember that pd, V, means energy converted from electrical PE per unit charge passing.]

$V\Delta Q$ is the 'area' of the narrow vertical strip drawn under the graph, and the total energy released by the capacitor is the sum of the areas of these strips, that is the area under the whole graph line, which is $\frac{1}{2} Q_{\text{init}} V_{\text{init}}$ (that is $\frac{1}{2}$ base × height for the triangle, but with units of C × V = J).

Now that we've taken account of the changing pd and charge, we can drop the 'init' subscripts, and write the energy stored simply as

$$U = \tfrac{1}{2}QV.$$

Using the defining equation for capacitance ($C = \frac{Q}{V}$) we can express U in terms of a ingle variable, either Q or V. See Study point.

Knowledge check 4.1.8

The graph shows how the energy stored in a capacitor depends on the pd.

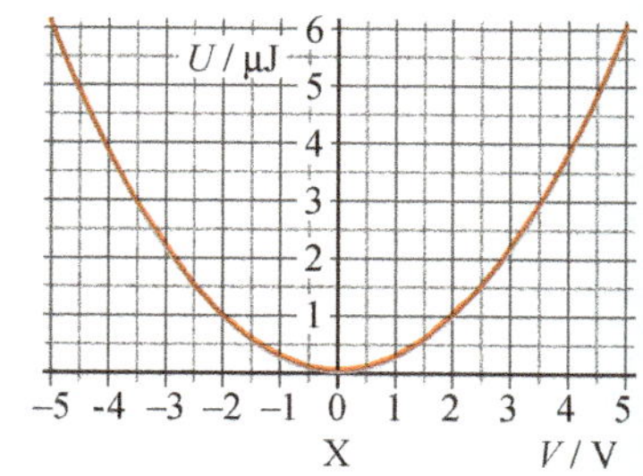

Fig. 4.1.10 Energy in a capacitor

(a) Describe the shape of the curve.

(b) Determine the capacitance.

(c) Explain briefly whether or not the capacitor is likely to be an electrolytic capacitor.

Study point

The energy stored in a capacitor of capacitance C when there is a pd V between its plates and a charge $\pm Q$ on them is

$$U = \tfrac{1}{2}QV = \tfrac{1}{2}CV^2 = \tfrac{1}{2}\frac{Q^2}{C}$$

These are all equivalent. Use whichever suits the case in hand.

Top tip

The useful equations $U = \frac{1}{2}CV^2$ and $U = \frac{1}{2}\frac{Q^2}{C}$ are not in the WJEC Data booklet.

Knowledge check 4.1.9

In a prototype torch, a 'supercapacitor', of 60 F, charged to a pd of 2.4 V, is discharged through an LED (with a current-limiting resistor in series) until the pd has fallen to 1.8 V, when the light is deemed to be too dim. If the mean power supplied is 0.30 W, calculate the length of time the LED was on.

4.1.5 Charging and discharging a capacitor through a resistor

We shall show that capacitors charge and discharge according to an exponential decay law, i.e. the mathematical form is the same as with radioactive decay and damped oscillatory motion. We'll study discharging first. This might seem back-to-front, but the discharging case is simpler.

(a) Discharging

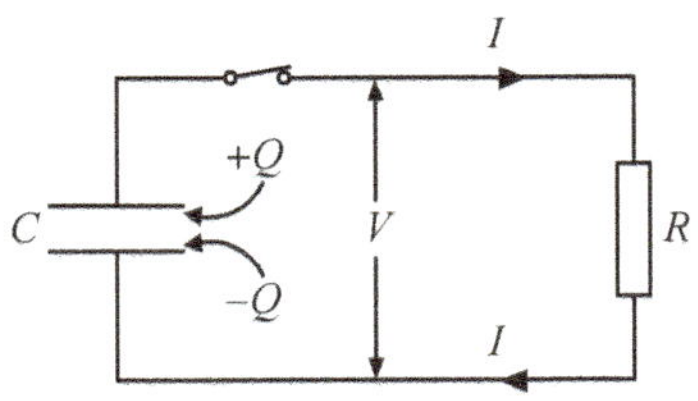

Fig. 4.1.11 *RC* discharge

The capacitor in Fig. 4.1.11 is initially charged and is discharged through the resistor by closing the switch at $t = 0$. At any time after that, the pds across the capacitor and resistor are equal, so

$$IR = \frac{Q}{C}.$$

Now for the key step. The current, I, is the rate of flow of charge, which in this case means the rate of loss of the charges, $\pm Q$, held on the plates of the capacitor.

In mathematical language, $I = -\frac{\Delta Q}{\Delta t}$.

Substituting for I into the previous equation and rearranging gives us:

$$\frac{\Delta Q}{\Delta t} = -\frac{Q}{RC}.$$

Notice two things about this equation:
The top lines have the same unit (C), so the bottom lines have the same unit as each other, meaning that the unit of RC is s. We can check this:

Unit of $RC = \Omega\ \text{F} = \text{V A}^{-1} \times \text{C V}^{-1} = \text{A}^{-1}\ \text{C} = \text{A}^{-1} \times \text{A s} = \text{s}$.

This product RC is called the **time constant** with the symbol τ (see below).

- Both R and C are constant, so the equation says that the rate of change of Q is proportional to Q. From section M3.3, we see that the discharge equation is:

$$Q = Q_0 e^{-t/RC}.$$

where Q_0 is the initial charge on the capacitor ($+Q_0$ on one plate, $-Q_0$ on the other).

Because $V \propto Q$ and $I \propto V$, the variations of V and I with time are just the same,

i.e. $V = V_0 e^{-t/RC}$ and $I = I_0 e^{-t/RC}$

This is called an **exponential decay** equation.

- In the equation $Q = Q_0 e^{-t/RC}$, the index, $-\frac{t}{RC}$, is a negative number without units, because the unit of RC, like the unit of t, is s.
- The numerical value of $e^{-t/RC}$ is found using the e^x (inverse $\ln x$) facility on a scientific calculator.
  For example, if $R = 680\ \text{k}\Omega$, $C = 4.7\ \mu\text{F}$, $t = 9.6\ s$, then
  $$-\frac{t}{RC} = -\frac{9.6\ \text{s}}{4.7 \times 10^{-6}\ \text{F} \times 6.8 \times 10^5\ \Omega} = -3.00 \qquad \text{giving } e^{-t/RC} = 0.0496$$
- Q_0 is the initial value of Q. We can confirm this by remembering that anything raised to the power of zero is 1, so when $t = 0$,
  $Q = Q_0 e^{-t/RC} = Q_0 \times 1 = Q_0$.

Key terms

The **time constant**, τ, of an RC system undergoing exponential decay is the time taken for the charge to fall to $\frac{1}{e}$ (= 0.37) of its original charge. $\tau = RC$

The significance of the time constant, τ

As RC has the units of time, it makes sense to ask: what value does Q have when $t = RC$, that is when the capacitor has been discharging for a time RC?

Putting $t = RC$, we have $Q = Q_0e^{-t/RC} = Q_0e^{-1} = 0.368\ Q_0$.

In other words the capacitor has 37% of its initial charge.

The time constant, then, is a measure of how quickly (or slowly) the charge decays.

Determining t knowing Q, Q_0, R and C

Taking logs of the equation $Q = Q_0e^{-t/RC}$ gives us:

$$-\frac{t}{RC} = \ln\left(\frac{Q}{Q_0}\right).$$

Example

Calculate the time it takes for Q to halve:

Answer

Putting $Q = \frac{1}{2}Q_0$, $-\frac{t}{RC} = \ln\left(\frac{1}{2}\right) = -0.639$

$\therefore$ Time to halve $= 0.639\ RC$

The exponential decay curve

Fig. 4.1.12 is a typical graph of Q against t for a capacitor discharging through a resistor. (Ignore the tangent line for the time being.)

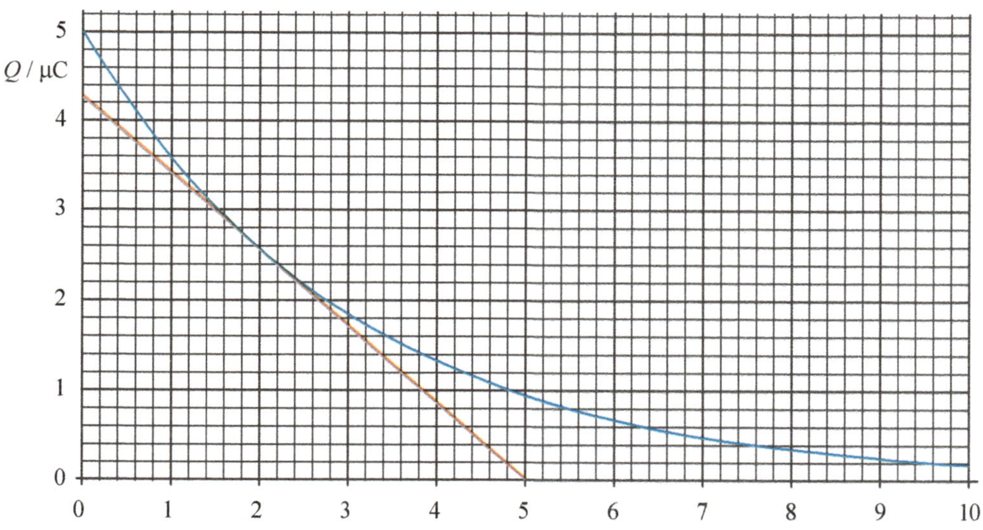

Fig. 4.1.12 Exponential decay curve for a CR circuit

The feature of the curve that confirms the decay as **exponential** is that, in equal time intervals, Q always decays to the same fraction of its value at the start of the interval. An important example is a time interval of τ, for which, as we've seen, the fraction is 0.368…

At $t = 0$, $Q = 5.00\ \mu C$, so, at time $t = \tau$, $Q = 0.368 \times 5.00\ \mu C = 1.84\ \mu C$

The graph shows this to be at a time of 3.0 s. So $\tau = 3.0$ s.

Over an interval of 3.0 s, *starting at any time*, we find that the charge at the end of the interval is 37% of the charge at the start. Do check this.

Study point

$\ln\left(\frac{1}{2}\right) = -\ln 2$ follows from the definition of a logarithm as a power. You should convince yourself that for any positive value of x, $\ln\left(\frac{1}{x}\right) = -\ln x$.

[This is a special case ($n = -1$) of the rule: $\ln(x^n) = n \ln x$.]

Knowledge check 4.1.10

A 15 μF capacitor is charged to 18 μC. At time $t = 0$, a 2.2 MΩ resistor is connected across it. Calculate:

(a) The time constant.

(b) The charge remaining after 5.0 s.

(c) The time taken for the charge to fall to 4.5 μC.

Knowledge check 4.1.11

The graph of Fig. 4.1.12 was plotted using a 0.50 μF capacitor.

(a) State the initial pd.

(b) Calculate the resistance (knowing τ).

Knowledge check 4.1.12

(a) In Fig. 4.1.12 a tangent has been drawn to the graph at $t = 2.0$ s. Calculate its gradient.

(b) State what electrical quantity this gradient represents.

(c) Determine this quantity by another method… Read the value of Q at $t = 2.0$ s and use the value of τ already found (see main text.)

Knowledge check 4.1.13

A resistor of 150 kΩ is connected across a charged capacitor of 6.8 μF.

(a) Calculate the time for its charge to fall to half of its original value.

(b) Calculate the time for the energy to fall to half of its original value.

(c) What is the relationship between these times?

Variation of pd and current with time

As the charge, Q, decays exponentially with time, so also do the pd, V, and the current, I, (see Fig. 4.1.11). In detail…

Putting $Q = CV$ and $Q_0 = CV_0$ into $Q = Q_0e^{-t/RC}$ and dividing by C,
$V = V_0e^{-t/RC}$ in which V_0 is the pd at time $t = 0$.
Putting $V = IR$ and $V_0 = I_0R$ into $V = V_0e^{-t/RC}$ and dividing by R,
$I = I_0e^{-t/RC}$ in which I_0 is the current at time $t = 0$.

(b) Charging

This is more complicated mathematically than discharging the capacitor, so we'll use the discharging result to help us. The capacitor will be charged using the circuit in Fig. 4.1.13. At time $t = 0$, the capacitor has zero charge. At any time after $t = 0$, the pds across R and C add up to V_0.

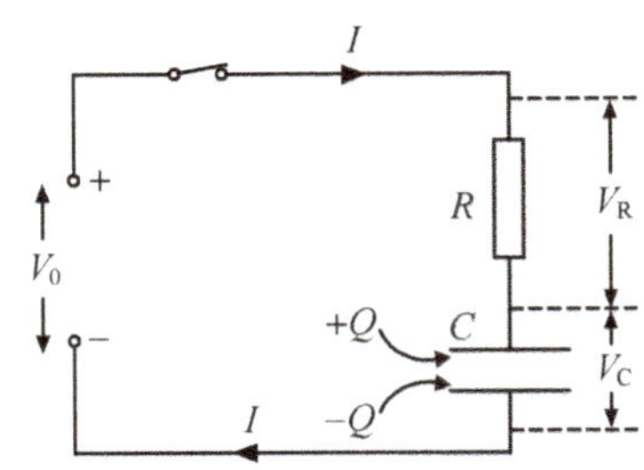

Fig. 4.1.13 Capacitor charging through a resistor

In symbols, $V_C + V_R = V_0$ so $\frac{Q}{C} + IR = V_0$ that is $\frac{Q}{C} + R\frac{\Delta Q}{\Delta t} = V_0$.

Note that, this time, $I = +\frac{\Delta Q}{\Delta t}$, as the capacitor is charging.

We can work out the general shape of the graph of Q against t by some very simple reasoning…

- When $t = 0$, $Q = 0$. So $V_C = 0$. So $V_R = V_0 - V_C = V_0$. The resistor has the full supply pd across it, so $I = +\frac{V_0}{R}$. This is the initial rate at which Q is rising.
- As Q rises, V_C rises, so V_R falls. Therefore I falls, so Q's *rate* of rise decreases, approaching zero as V_C approaches V_0. The maximum possible charge, Q_0, is simply $Q_0 = CV_0$.

4.1.14 Knowledge check

When the capacitor in Fig. 4.1.13 is fully charged it stores energy $\frac{1}{2}Q_0V_0$. Charge Q_0 has passed through the source of V_0, which must therefore have supplied energy Q_0V_0. What has happened to the other $\frac{1}{2}Q_0V_0$?

So the charging curve is just an upside down version of the discharging curve, with the equation:

$$Q = Q_0\left(1 - e^{-t/RC}\right)$$

where, this time, Q_0 represents the **final** value of charge, when it has been charging for much longer than RC. The charging graph is as follows, where we have used the same values of R, C and V_0 as in Fig. 4.1.12.

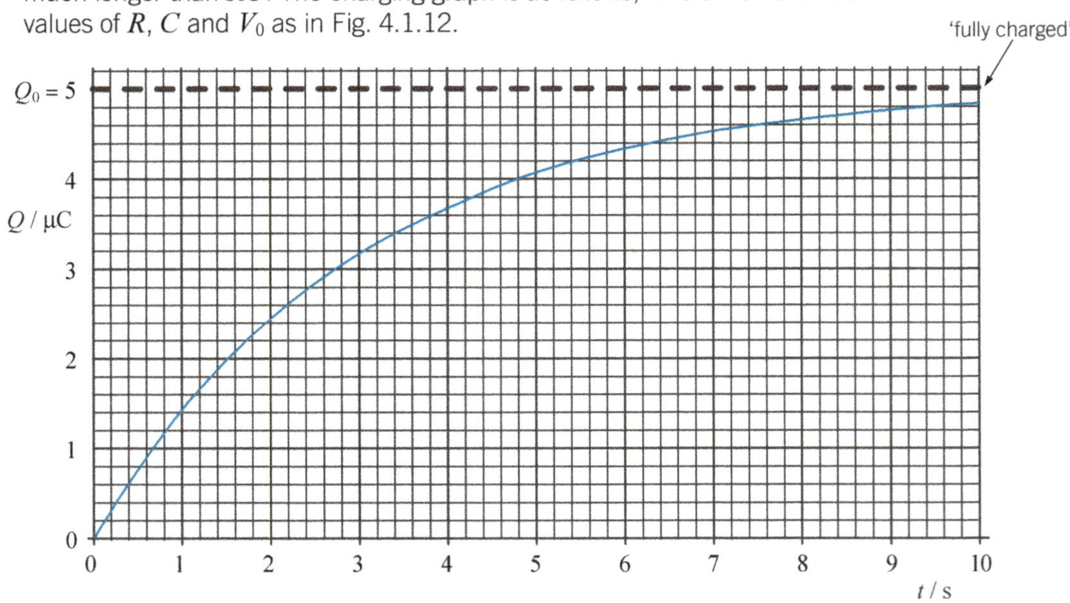

Fig. 4.1.14 Q against t for a capacitor charging through a resistor

4.1.15 Knowledge check

A steady pd of 9.0 V is applied across a 2.2 μF capacitor in series with a 1.5 MΩ resistor. Calculate:

(a) The time taken for the pd across the capacitor to reach 5.7 V.

(b) The pd an equal time later.

Now the **time constant** represents the time taken for the charge to **rise** to $(1 - 37\%) = 63\%$ of its eventual value.

Variation of pd and current and with time

It's easy to show (in the order given) that, if $V_0 = \frac{Q_0}{C}$, $I_0 = \frac{V_0}{R}$, then

$$V_C = V_0\left(1 - e^{-t/RC}\right) \qquad V_R = V_0 e^{-t/RC} \qquad I = I_0 e^{-t/RC}.$$

4.1.6 Electric fields

(a) The idea of an electric field

There is an electric field between the plates of a charged capacitor, around a plastic comb which has been rubbed, around the top sphere of a Van de Graaff generator, in the air between clouds and the ground (and between the clouds themselves) in a thunder-storm, in the water around certain fish when they send out electrical pulses ...

We say that *there is an electric field at any point, P, where a charged body (at rest) experiences a force proportional to its charge.*

We shall call a charged body used to test for the presence of an electric field a *test charge*. Ideally it would have only a tiny charge (so as not to displace the very charges that are giving rise to the field!), and would occupy very little space, so as to respond to the field 'at a point'.

In Fig. 4.1.16 the test charge (hardly ideal) is a small, light sphere, coated with graphite and given a positive charge by touching the positive plate (to which it loses some electrons). It then experiences a force to the right when dangled in the gap – until it touches the negative plate. There is clearly an electric field between the plates!

Fig. 4.1.15 An 'electric eel'

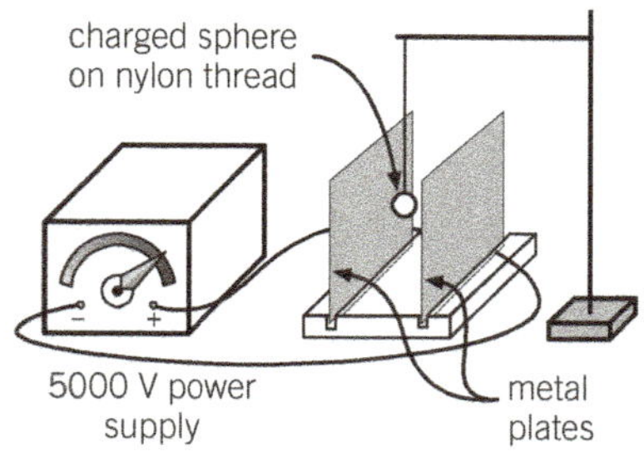

Fig. 4.1.16 Demonstrating an electric field

(b) Electric field strength

Having defined what we *mean* by an electric field in terms of a test charge experiencing a force, we extend the idea in a very natural way to define the **electric field strength** vector – see Key terms.

Note that if we double the test charge, the force will double, so we get the same value for E. If we change from a positive test charge to a negative one the force direction will reverse, but dividing F by a negative quantity (q) reverses the direction again! In other words, E, defined as F/q , tells us about the field in which we place q, independently of the magnitude or sign of q itself. That's just what we want.

Key terms

The **electric field strength**, E, at a point, P, is defined as

$$E = \frac{\text{force on test charge, } q\text{, at P}}{q} = \frac{F}{q}$$

It is a vector whose direction is that of the force on a positive test charge.

UNIT: $\text{N C}^{-1} = \text{V m}^{-1}$
(see Study point).

Study point

We'll soon see how V m^{-1} arises naturally as a unit for E. In the meantime note that...

$\text{V m}^{-1} = (\text{J C}^{-1})m^{-1} = (\text{N m C}^{-1})\text{m}^{-1} = \text{N C}^{-1}$.

(c) Setting up a uniform electric field

Not only do electric charges *respond* to electric fields, they also *create* electric fields around them. We study the field due to a 'point' charge in Section 4.2. Now, though, we consider the field due to a special configuration of charges – those on the metal plates of a parallel plate capacitor, when a battery has been connected to them (Fig. 4.1.17).

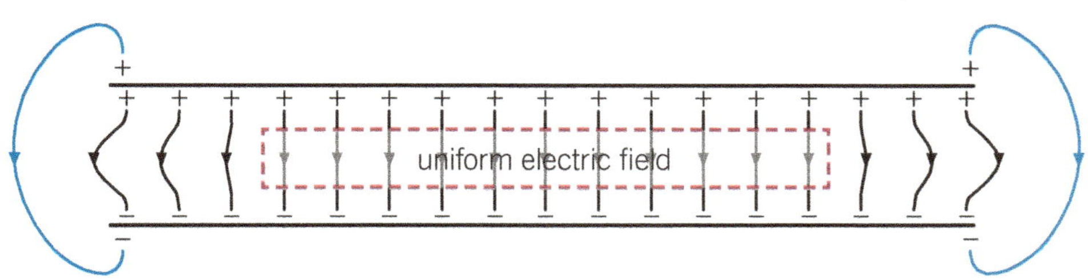

Fig. 4.1.17 Electric field between charged parallel plates

Knowledge check 4.1.16

A small sphere with a charge of −1.2 nC is placed at a point P above a Van de Graaff generator. The sphere experiences a force of 1.8×10^{-3} N directly upwards. Determine the electric field strength at P.

Key terms

There is a **uniform electric field** in a region where the field strength is the same in magnitude and direction at all points.

The **electrical potential energy** of a charge is the work it can do because of its position in an electric field.

The **potential difference**, ΔV, between two points, A and B, in an electric field is the work done per unit charge (W/q) by the field on a test charge, q, going from A to B. If W/q is positive, A is at the higher potential.

UNITS: J C^{-1} = volt (V)

In a uniform field $\Delta V = -E\,\Delta x$, that is $E = -\frac{\Delta V}{\Delta x}$.

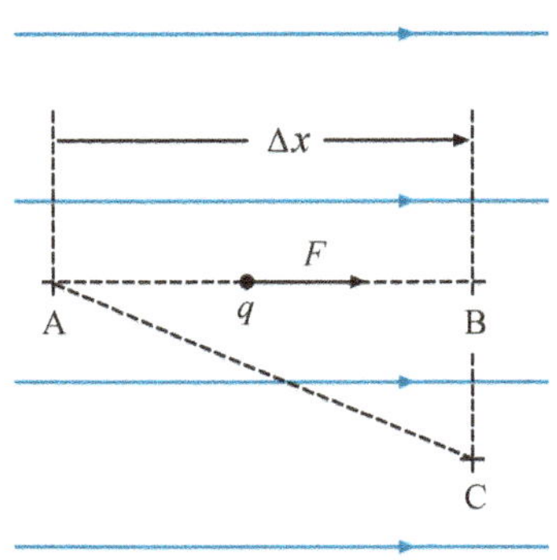

Fig. 4.1.18 Work done on q

Study point

Between the plates of a parallel plate capacitor the electric field strength can be shown to be proportional to the 'charge density'. That is: $E \propto \frac{Q}{A}$.

But $E = \frac{V}{d}$, so $\frac{Q}{A} \propto \frac{V}{d}$ so $\frac{Q}{V} \propto \frac{A}{d}$.

... in agreement with $C = \frac{\varepsilon_0 A}{d}$.

4.1.17 Knowledge check

A pd of 750 V is applied between flat metal plates separated by a gap of 5.00 mm. A dust particle in the gap carries a charge of 18 pC. Calculate the force on it.

If the gap between the plates is small compared with the length of side of the plates (if they are square) or their diameter (if they are discs), there is a **uniform electric field** in the gap, except near the edges. This uniformity is represented in Fig. 4.1.17 by parallel and evenly spaced *electric field lines*. More about these in Section 4.2.

(d) Potential difference

The charged sphere in Fig. 4.1.16 will pick up speed as it goes across the gap. As it gains kinetic energy it must lose some other sort of energy. This is **electrical potential energy**.

Suppose that a test charge, q, is displaced by Δx from A to B in the direction of a uniform electric field (Fig. 4.1.18). It loses electrical potential energy (EPE) equal to the work done on it by the field.

So Change in EPE of q = –(force on q) × Δx

$$\therefore \quad \frac{\text{change in EPE of } q}{q} = \frac{\text{force on } q}{q} \times \Delta x$$

That is: $\Delta V = -E\,\Delta x$

The last step follows from the definitions of **potential difference**, ΔV, and **electric field strength**, E (see previous page).

There are several points to note:

- ΔV, like E, is defined so as not to depend on the magnitude or sign of the test charge, q, (as long as q isn't too large).
- We would have reached the same value for ΔV by taking q on any route between A and B. For example, the work done on q, divided by q, in going from A to C would also be $E\Delta x$, because only the displacement component, Δx, in the direction of E counts. No further work is done in going from C to B. We say that the pd between two points in an electric field due to static charges is *path-independent*.
- The equation $\Delta V = -E\,\Delta x$ can be re-arranged as

  $$E = -\frac{\Delta V}{\Delta x} = \textit{potential gradient}$$

  In particular if we apply a pd V between parallel plates distance d apart, the magnitude, $|E|$ of the field strength in the gap is

  $$|E| = \frac{V}{d}$$

  For example, for the plates in Fig. 4.1.17 if $V = 1500$ V, $d = 7.5$ mm:

  $$|E| = \frac{V}{d} = \frac{1500\text{ V}}{7.5 \times 10^{-3}\text{m}} = 2.0 \times 10^5\text{ V m}^{-1}.$$ [The field is downwards.]

  Note how the units, V m^{-1}, emerge naturally, as promised earlier.

(e) Electric fields in conductors

In the last section we went back to basics to develop the idea of *potential difference* (pd) in the context of an electric field. But this is the very same pd that features in the study of electric circuits, implying that there's an electric field inside any conductor with a pd across it. Hence the forces experienced by the mobile charge carriers in it, making them drift through the conductor (despite collisions with ions).

If – as in the plates of Fig. 4.1.17 – there is *no* movement of charge in a conductor, there can't be any macroscopic electric field in it (that is a field extending over regions which are large compared with an atomic diameter) – or the free charges *would* be moving! Since $\Delta V = -E\Delta x$, and E is zero throughout the conductor, there is no pd between any points in the conductor, even if it carries a static charge. 'Static' is the key word.

4.1.7 Motion of a charged particle in a uniform electric field

We'll assume that forces on the particle, other than from the electric field, are negligible. This generally requires the particle to be in a vacuum, so there are no collision forces (resistive forces) on it. There's usually no need to worry about gravitational forces: see Study point.

The force, qE, on a charged particle is the same in magnitude and direction wherever the particle is in a uniform E field. This implies that:

- Its component of velocity at right angles to the field doesn't change.
- Its component of velocity in the direction of the field changes at a constant rate, given by:

$$\text{acceleration} = \frac{\text{resultant force}}{\text{mass}} = \frac{qE}{m}.$$

So the particle moves like a resistance-free projectile, except that the acceleration is unlikely to be $9.8\ \text{m s}^{-2}$, and may well not be downwards. The particle's path is either straight or parabolic.

(a) Straight line motion

As an example, suppose a particle of mass m and (positive) charge q is released from rest near the positive plate in the set-up of Fig. 4.1.19. How long will it take to reach the other plate, and at what speed will it arrive?

To find the time taken we use $x = ut + \frac{1}{2}at^2$

in which $x = d$, $u = 0$, $a = \frac{qE}{m} = \frac{qV}{md}$

Making these substitutions, and re-arranging, we find $t = \sqrt{2d \times \frac{md}{qV}} = \sqrt{\frac{2m}{qV}} \times d$.

For the final speed we'll first use $v^2 = u^2 + 2ax$

Substituting as before, we get $v = \sqrt{2\frac{qV}{md}d}$, that is $v = \sqrt{\frac{2qV}{m}}$

For a given particle, we note that the final speed depends only on V, not on d. In fact it doesn't even depend on the field being uniform, as we can show by deriving the equation again by a *very useful argument…*

(b) Speed change related to potential difference

If the particle experiences no forces other than from the electric field, then, however it moves in the field (in a straight line or curve),

KE gained by particle = EPE lost by particle

So, if particle starts from rest, $\frac{1}{2}mv^2 = qV$, so $v = \sqrt{\frac{2qV}{m}}$

in which v is the speed acquired due to moving through a potential difference (fall) of V. We can, of course, apply the same principle to a particle that gains EPE and loses KE. See Study point for a more formal general approach.

Study point

Consider a proton in a (weak) electric field of $10\ \text{V m}^{-1}$ near the Earth's surface…

Force from electric field = qE
$= 1.60 \times 10^{-19}\ \text{C} \times 10\ \text{N C}^{-1}$
$= 1.6 \times 10^{-18}\ \text{N}$

Force from gravitational field = mg
$= 1.67 \times 10^{-27}\ \text{kg} \times 9.81\ \text{N kg}^{-1}$
$= 1.6 \times 10^{-26}\ \text{N}\ (qE \times 10^{-8})$

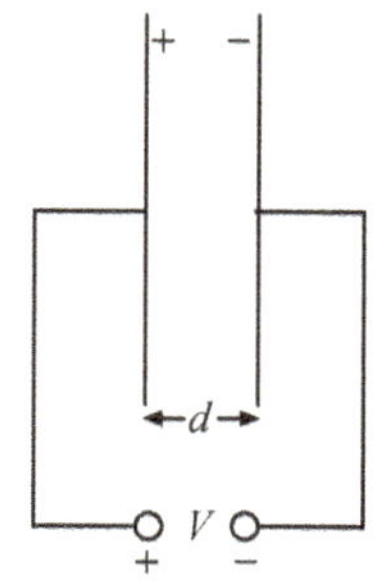

Fig. 4.1.19 Gap to be crossed

Knowledge check 4.1.18

Explain why multiplying the final speed, v, by the journey time, t, for the example of straight line motion in an E field, gives $2d$ rather than d.

Knowledge check 4.1.19

An electron enters a uniform electric field of $750\ \text{V m}^{-1}$, travelling at a velocity of $2.20 \times 10^7\ \text{m s}^{-1}$ in the direction of the field. Calculate how long it takes to return to its point of entry, and the total distance it travels.

Study point

In general, for a particle going from A to B in an electric field

$$\frac{1}{2}mv_B^2 - \frac{1}{2}mv_A^2 = -q\Delta V$$

in which ΔV is the potential difference (rise) in going from A to B. The value of q (as well as of ΔV) needs to be entered with the correct sign.

Study point

The electric field between the plates in Fig 4.1.20 is only very approximately uniform and the field extends significantly beyond the edges of the plates. This is because the distance between the plates isn't really small compared with their length and breadth.

4.1.20 Knowledge check

This question is about the electron beam shown in Fig. 4.1.20. Assume a uniform E field between the plates.

The electrons leave the gun at a speed of 3.20×10^7 m s^{-1}.

Calculate:

(a) V_{acc}.

(b) The time taken for an electron to traverse the screen (left to right).

(c) The acceleration in the y-direction responsible for the observed y-displacement (20mm).

(d) The deflecting pd, V_y, given that the separation, b, of the deflecting plates is 45 mm.

(e) The y velocity component of an electron when it reaches the right-hand edge of the screen.

(f) The magnitude and direction of the velocity of an electron when it reaches the right-hand edge of the screen.

[$e = 1.60 \times 10^{-19}$ C

$m_e = 9.11 \times 10^{-31}$ kg]

(c) Motion in two dimensions

One way of demonstrating this uses the apparatus of Fig. 4.1.20, and we'll take it as our example.

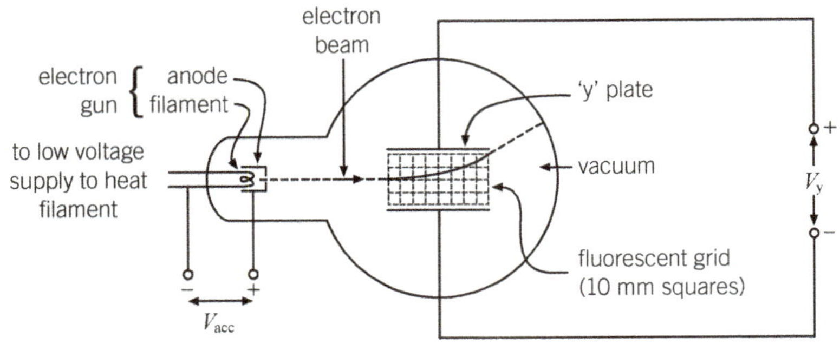

Fig. 4.1.20 Deflection of electrons in a transverse electric field

The electron gun

This produces a fine beam of fast electrons. The electrons are emitted from a metal filament heated to a high temperature by passing a current through it. They are accelerated towards a cylindrical metal 'anode' by means of a pd, V_{acc}, applied between filament and anode as shown. Many electrons are captured by the anode and returned straight to the filament via the V_{acc} power supply, but some escape through a small hole in the anode's end-cap, to form the beam. The speed, v, of these escapees is easily found, using $\frac{1}{2}mv^2 = qV_{acc}$. See (b) above. [We neglect the relatively small KE with which the electrons are emitted from the filament.]

Parabolic path

The electron beam enters a transverse (crosswise) electric field between parallel plates with a pd, V_y, applied between them. We shall assume that the field is uniform and doesn't extend beyond the edges of the plates – but see Study point.

In this region the electrons' x-wise velocity component, v_x (= v), remains unchanged, and equal to the speed v at which they entered the field. So their x-wise displacement component time t after entering will be

$$x = v_x t \qquad \text{so} \qquad x = vt$$

The (y-direction) acceleration is $\frac{(-e)}{m}E_y = \frac{(-e)}{m}\left(-\frac{V_y}{b}\right) = \frac{eV_y}{mb}$

in which b is the distance apart of the plates.

Applying $x = ut + \frac{1}{2}at^2$ to the y component of motion, $y = \frac{eV_x}{2mb}t^2$

The key thing to understand here is that, for the y motion, the initial velocity is zero.

The electron beam (Fig. 4.1.20) brushes against a fluorescent screen between the deflecting plates, revealing the path of the electrons. We know that, according to our assumptions, this is part of a parabola because we can eliminate t between the x and y equations to give

$$y = \frac{eV_y}{2mb}\left(\frac{x}{v}\right)^2 \qquad \text{that is} \qquad y \propto x^2.$$

4.1.8 Specified practical work

We outline two 'classic' investigations of capacitor behaviour.

(a) Investigating the discharging of a capacitor through a resistor

It's possible to confirm simply and very convincingly that the decay of charge is exponential. Having charged the capacitor (switch to the left in Fig. 4.1.21), we let it discharge through the resistor (switch to the right), recording the pd, V, (see Study point) at roughly equally spaced times, t.

Planning

Here are some points to consider:

- How does one choose the values of C and R? The time constant, RC, needs to be long enough for ten or so pairs of (V, t) readings to be taken comfortably while the pd drops to (say) a tenth of its original value. If using a data-logger its capabilities need to be known.
- The voltmeter display, especially if digital, may not be readable if it's changing rapidly. It's worth considering doing the charge and discharge procedure afresh for each measurement, stopping the discharge at a different time (switch central) before reading the pd. The voltmeter's position would have to be different from that shown in Fig. 4.1.21!

Using the results

If we plot V (or CV) against t, we expect a graph like Fig. 4.1.12. The time constant, τ, can be determined as explained in Section 4.1.5, starting from various points on the graph, to check that the decay is exponential.

A less fiddly and more accurate way of checking the relationship and determining τ is to plot $\ln (V/\text{volt})$ against t. (We have to divide V by its unit before taking the logarithm, because only numbers have logarithms!) This gives a straight-line graph, because, as shown earlier,

$$V = V_0 e^{-t/RC}.$$

In inverse form $\ln\left(\frac{V}{V_0}\right) = -\frac{t}{RC}$ that is $\ln\left(\frac{V/\text{volt}}{V_0/\text{volt}}\right) = -\frac{t}{RC}$.

Because $\ln \frac{a}{b} = \ln a - \ln b$ we have $\ln (V/\text{volt}) = -\frac{t}{RC} + \ln (V_0/\text{volt})$

Comparing this with $y = mx + c$, we'd expect that plotting $\ln (V/\text{volt})$ as y and t as x would give a straight line of gradient $m = -\frac{1}{RC}$.

Determining C from the graph gradient

R is the resistance of the parallel combination of resistor and voltmeter. The resistor's resistance can be measured with a digital multimeter. A typical resistance for a digital multimeter on it 'volts' range is $10\ \text{M}\Omega$. Knowing the value of R, we can determine C, as

$$C = -\frac{1}{R \times \text{gradient}}$$

Charging a capacitor through a resistor

A possible circuit is given Fig. 4.1.22. In this case $V_R = V_0 e^{-t/RC}$. Plotting $\ln (V_R/\text{volt})$ against t should therefore give a straight-line graph.

Study point

Why measure the pd, V, when it's the charge, Q, we're interested in? It's trivial to calculate Q from $Q = CV$, and voltmeters are more convenient than coulombmeters.

Equally sensible would be to use a (micro)ammeter in series with the resistor. Then $Q = CV = CIR$.

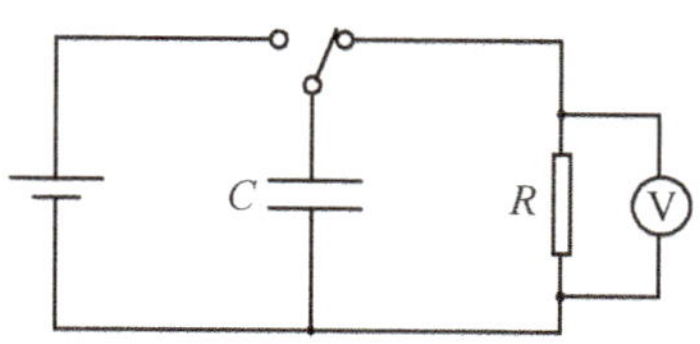

Fig. 4.1.21 Circuit for investigating discharging

Study point

Here's a reminder of the rules that apply to logarithms to any base.

- $\log 1 = 0$
- $\log(ab) = \log a + \log b$
- $\log(\frac{a}{b}) = \log a - \log b$
- $\log(x^n) = n\log x$

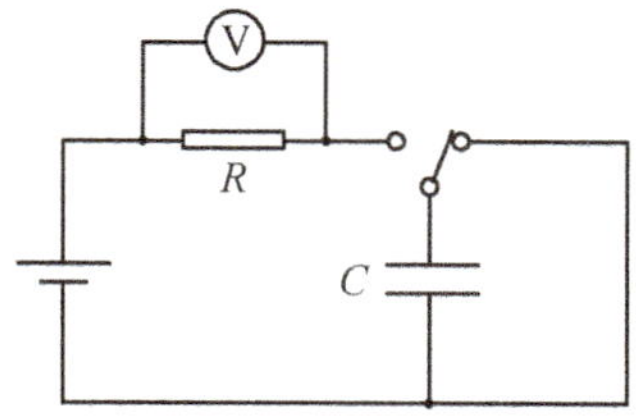

Fig. 4.1.22 Circuit for investigating charging

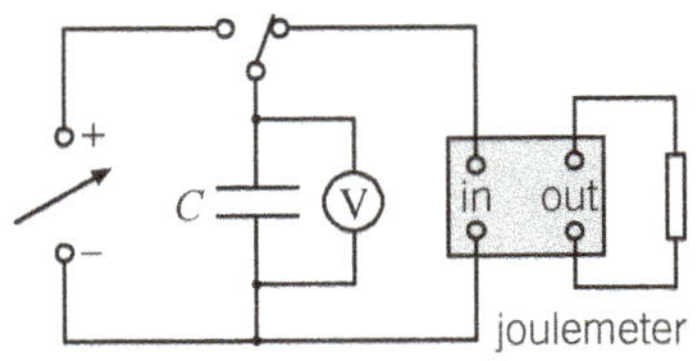

Fig. 4.1.23 Measuring energy stored in a capacitor

(b) Investigating the energy stored in a capacitor

Method 1: Using a joulemeter

The capacitor is charged (switch to the left in Fig. 4.1.23) to a measured pd and discharged through a resistor (switch to the right). The energy, ΔU, supplied to the resistor is measured by the joulemeter, interposed between capacitor and resistor as shown.

The capacitor value needs to be chosen, together with the range of pds, so that the energy stored is within the joulemeter's capabilities. For convenience the resistor value needs to be such that the time constant, RC, is no more than a few seconds.

The capacitor will not discharge completely in a finite time. One way to deal with this is deliberately to stop the discharge when the pd has fallen to the same arbitrary small value, V_{min}, each time. Then

$$\Delta U = \tfrac{1}{2}CV^2 - \tfrac{1}{2}CV_{\text{min}}^2$$

A graph of ΔU against V^2 should therefore be a straight line of gradient $\tfrac{1}{2}C$ and intercept $-\tfrac{1}{2}CV_{\text{min}}^2$.

Method 2: By measuring a temperature rise

A joulemeter will deliver reliable readings with little effort, but how it does so may be, to some extent, a mystery. A less 'black box' approach is to discharge the capacitor through a resistor in the form of a coil of wire and to measure the temperature rise.

4.1.21 Knowledge check

For method 2

(a) Show that, if heat loss is neglected, then $mc(\theta_2 - \theta_1) = 5 \times \tfrac{1}{2}C[V^2 - (1.00\text{ V})^2]$ in which m is the mass of the copper and c is the specific heat capacity of copper.

(b) Use the results in the table given in the Example to determine a value for C. You will also need these:

$m = 0.077$ kg

$c = 385$ J kg^{-1} °C^{-1}.

Study point

Method 2 is an alternative to the method in the WJEC experimental guide.

Study point

Informal analysis suggests an uncertainty in the ratio in the last column, for $V = 3.00$ V, of about 10% due to random errors. Although the figure may be better for $V = 4.00$ V and 5.00 V, systematic errors due to heat loss are probably greater for higher pds.

Example

The coil shown in Fig. 4.1.24 was wound from copper wire of total mass 0.077 kg and resistance 2.35 Ω. It was wrapped in thermal insulation. A digital thermometer was used to measure the temperature at its centre.

Fig. 4.1.24 Coil (thermal insulation removed) and thermometer

An initial temperature reading, θ_1, was taken. Then the coil was connected across a supercapacitor of nominal value 5.0 F (and maximum allowed pd 5.4 V), pre-charged to a pd of 3.00 V, until the pd fell to 1.00 V. This was done 5 times in quick succession and the maximum temperature, θ_2, reached at the end was recorded. The procedure was repeated for starting pds of 4.00 V and 5.00 V. Here are the results.

V/V	$(V/\text{V})^2 - 1.00^2$	θ_1/°C	θ_2/°C	$(\theta_2 - \theta_1)$/°C	$\dfrac{(\theta_2 - \theta_1)}{V^2 - (1.00\text{ V})^2}$ /°C V^{-2}
3.00	8.0	15.3	17.6	2.3	0.28
4.00	15.0	16.1	20.4	4.3	0.29
5.00	24.0	14.4	21.9	7.5	0.31

If the energy stored in the capacitor is proportional to the pd squared, the ratio in the last column should be constant. See Study point.

Test yourself 4.1

1. A parallel plate capacitor is made from two rectangular metal plates of size $1.0\ \text{m} \times 2.0\ \text{m}$, held slightly apart by small pieces of plastic of thickness $0.1\ \text{mm}$. The two plates are connected to a $500\ \text{V}$ DC supply. Ignoring the dielectric effect of the plastic pieces, calculate:

   (a) the capacitance; (b) the charge on each sheet of metal
   (c) the energy stored (d) the intensity of the electric field between the plates.

2. Two pieces of aluminium foil of the same area as the plates in Q1 are rolled together to make a capacitor, separated by two sheets of plastic wrap (as in Fig. 4.1.2) of thickness $12.5\ \mu\text{m}$. Rolling the plates effectively doubles the surface area because charge is stored on both faces of each. Estimate the capacitance of the arrangement ignoring the dielectric effect of the plastic (which increases the capacitance by a further factor).

3. Calculate the effective capacitance of two capacitors of value C and $2C$ connected (a) in series and (b) in parallel.

4. Two initially uncharged capacitors of value C and $2C$ and a resistor R are connected in the circuit shown. The switch is then closed. The power supply has negligible internal resistance.

   (a) State the pd across each component, and the current, immediately after the switch is closed.
   (b) State in terms of V_s the values of the following after the charging current has dropped to half its initial value:
      (i) the current
      (ii) the pd across the resistor
      (iii) the charge in each of the four capacitor plates
      (iv) the pd across each capacitor
      (v) the energy stored in each capacitor
      (vi) the power dissipated in the resistor.
   (c) State the values of quantities in part (b) after charge has stopped flowing.

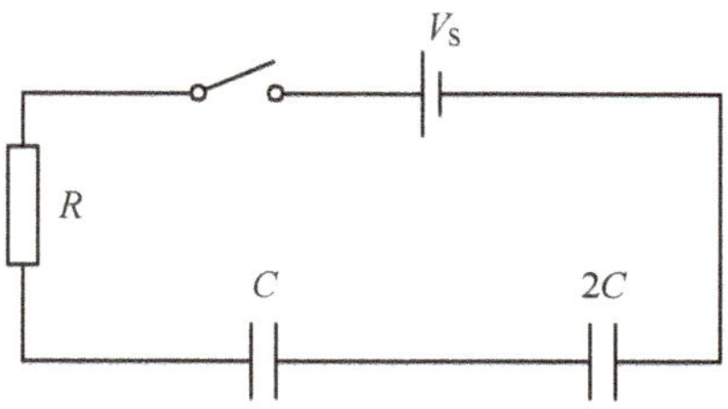

5. What values of capacitance can you obtain using three capacitors of value $10\ \mu\text{F}$, $22\ \mu\text{F}$ and $33\ \mu\text{F}$?

6. A capacitor is allowed to discharge through a $1.0\ \text{k}\Omega$ resistor. It takes $10\ \text{s}$ for the pd across the capacitor to fall from $15\ \text{V}$ to $6\ \text{V}$. Calculate the capacitance.

7. A capacitor is charged to $25\ \text{V}$ and allowed to discharge through a resistor. The energy transfer is measured by an energy meter, which indicates an energy of $35\ \text{mJ}$ when the capacitor is fully discharged. Calculate the capacitance.

8. A capacitor of value $220\ \mu\text{F}$ is charged to $10\ \text{V}$ and the power supply disconnected. The capacitor is then connected across a $100\ \mu\text{F}$ capacitor. Calculate:

   (a) The initial charge on the $220\ \mu\text{F}$ capacitor and the energy stored.
   (b) The final charge on each capacitor. [Hint: the final pds across the capacitors are the same.]
   (c) The final energy stored.

9. A capacitor is made from two metal plates held on insulating handles, a distance d apart as shown. The capacitor is charged by momentarily connecting it to a battery. The capacitance, pd, charge and energy of the capacitor are C, V, Q and U respectively. The plates are pulled apart to a separation of $2d$. State and account for the new values of capacitance, pd, charge and energy.

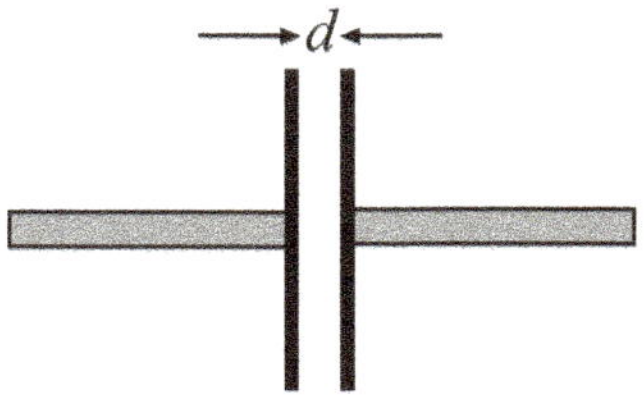

   Assume that the capacitor equation $C = \frac{\varepsilon_0 A}{d}$ is valid.

10. The procedure in Q9 is repeated but this time the capacitor remains connected to the battery. State and account for the values of capacitance, pd, charge and energy after the plates are pulled apart.

11. A 4700 μF capacitor is charged to 12 V. It is allowed to discharge through a 100 Ω resistor.

(a) Calculate the initial charge stored on the capacitor plates.
(b) Write an equation for the discharge current, I as a function of time.
(c) Sketch a graph of I against t and show how this graph relates to the answer to part (a).
(d) Calculate the initial energy stored in the capacitor.
(e) Write an equation for the energy stored as a function of time, $U(t)$. [Hint: First write $V(t)$.]
(f) Write an equation for the power dissipated in the resistor as a function of time, $P(t)$.
(g) Draw a graph of P against t and state how this graph relates to the answer to part (d).

12. (a) A negative ion of mass 4.65×10^{-26} kg and charge 3.20×10^{-19} C is at a point P in an electric field. Its acceleration is 3.10×10^{9} m s^{-2} upwards. Determine the magnitude and direction of the field at point P.
(b) Point P in part (a) is in the gap between two plane horizontal metal plates that are 15.0 mm apart. Calculate the potential difference that has been applied between the plates to produce the electric field.

13. A water droplet in a thunder-cloud has a mass of 3.32×10^{-13} kg and has acquired an electric charge (by losing electrons in collisions with air molecules). An upward electric field of strength 1.70×10^{6} N C^{-1} is just great enough to prevent the drop from falling. Determine how many electrons the drop has lost.

14. A potential difference of 1500 V is placed between two metal plates separated by a vacuum. The negative plate is heated and electrons are emitted from it (at negligible speed).

(a) Use the concept of energy to calculate the electrons' speed when, after acceleration by the electric field, they reach the positive plate.
(b) A student argues that the answer to (a) must depend on the distance between the plates, because increasing the distance will increase the work done by the accelerating force. Explain briefly why this argument is invalid.

15. A beam of electrons with an initial velocity of 3.0×10^{7} m s^{-1} in the x direction enters, at point E, a region where there is a uniform electric field of strength 12 000 V m^{-1} in the y direction. At a point P in the beam the electrons' velocity is at 45° to the x and y directions.

(a) Show that the time an electron takes to go from E to P is approximately 15 ns.
(b) Determine the x displacement component of P from E.
(c) Determine the y displacement component of P from E.

4.2 Electrostatic and gravitational fields of force

The electrostatic forces between charged bodies and the gravitational forces between all bodies with mass are, we now believe, very different in origin. One pointer is that we don't have positive and negative mass; gravitational forces are always attractive. Conveniently, though, for bodies that can be treated as point-like, the strength of the force over a wide range of distances is govered by an equation which is essentially the same in both cases – an *inverse square law*. That's why we've included both fields in this section. We start with forces between stationary charged bodies, because we can make good use of the idea of *electric field strength*, introduced in the last section.

Fig. 4.2.1 Charles-Augustin de Coulomb

4.2.1 Forces between stationary point charges: Coulomb's law

In the early 1780s, Charles-Augustin de Coulomb (Fig. 4.2.1) measured the forces between small, charged, conducting spheres. He used a 'torsion balance' (Fig. 4.2.2) in which the moment of the repulsive force on one of the spheres was balanced by the moment due to the twisting of a fibre. He found that:

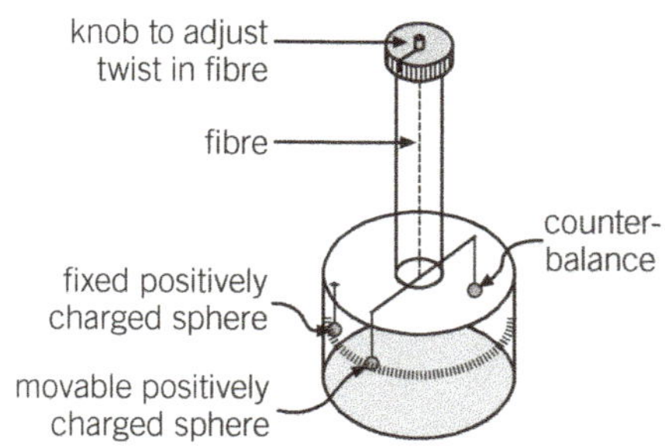

Fig. 4.2.2 Coulomb's experiment

- When the separation between the centres of the spheres was doubled, the force went down by a factor close to 4. When the separation was tripled, the force went down by a factor close to 9.
- When the charge on either of the spheres was halved, the force halved, and so on. [Coulomb halved the charge on a sphere by touching it on to an identical, initially uncharged, sphere.]

Generalising from his results, Coulomb suggested that an exact 'inverse square' law applied for *point charges*, that is charged bodies that are very small in size compared with their separation. **Coulomb's law** is stated in Key terms using SI units (not around in Coulomb's day!). The law has been amply confirmed by more accurate, but less direct, means.

Key terms

Coulomb's law

The force between point charges, Q_1 and Q_2, in a vacuum and separated by distance r is given by

$$F = \frac{Q_1 Q_2}{4\pi\varepsilon_0 r^2}$$

in which ε_0 is the *permittivity of free space* $= 8.85 \times 10^{-12}$ F m^{-1}

[NB $\frac{1}{4\pi\varepsilon_0} = 9.0 \times 10^9$ F^{-1} m (2 sf)]

Using the Coulomb's law equation

- If Q_1 and Q_2 are both positive or both negative, F clearly comes out as positive, so we can interpret a positive F as a repulsive force. If the charges are of opposite signs, F will be negative: attractive!
- The equation holds almost exactly if the charges are separated by air at ordinary densities. Only when the separating medium is a solid or liquid insulator is the force substantially reduced (because the charges will be surrounded by 'haloes' of opposite charge due to displacement of charge in the molecules of the medium). This is beyond A-level.
- We can write Coulomb's law as $F = \frac{1}{4\pi\varepsilon_0} \times \frac{Q_1 Q_2}{r^2}$. Clearly $\frac{1}{4\pi\varepsilon_0}$ is itself a constant.

  Using values of ε_0 and π to four significant figures, and rounding to three, we find that

$$\frac{1}{4\pi\varepsilon_0} = 8.99 \times 10^9 \text{ F}^{-1}\text{ m}$$

Knowledge check 4.2.1

In Section 4.1.2, ε_0 was shown to have the units F m^{-1}. Yet the Coulomb's law equation demands that its units be C^2 N^{-1} m^{-2}. Show, step by step, that F m^{-1} = C^2 N^{-1} m^{-2}.

Knowledge check 4.2.2

(a) In a simple theoretical 'model' of the hydrogen atom, the electron orbits the nucleus at a distance of $5.29 \times 10–11$ m. Calculate the force on the electron.

(b) Using a similar model for the He+ helium ion (proton number = 2) the force on the one electron orbiting the nucleus is 8.00 times the force of part (a). Calculate the distance of the electron from the nucleus.

For most purposes it's fine to use the very memorable $9 \times 10^9\ \mathrm{F^{-1}\ m}$, which is actually correct to 2 sf. So why on earth do we state Coulomb's law as we do, rather than writing $\frac{1}{4\pi\varepsilon_0}$ as a single constant? Recall that we first met ε_0 in the parallel plate capacitor equation – which has no 4π in it (Section 4.1.2). By including 4π in the Coulomb's law equation we avoid having to have it in the parallel plate equation. The 4π is more appropriate in the Coulomb's law equation as this deals with spherically symmetric force fields around point charges. A deeper question is why the two equations are connected at all. We'll hint at the answer in Section 4.2.6.

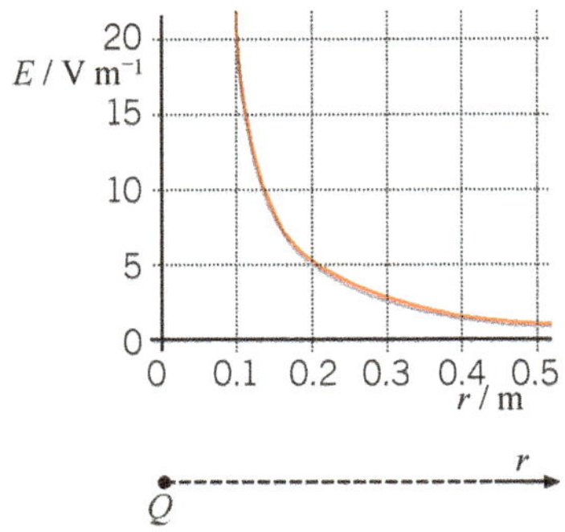

Fig. 4.2.3 E against r for a point charge

4.2.3 Knowledge check

Calculate the charge, Q, whose field strength is plotted in Fig. 4.2.3.

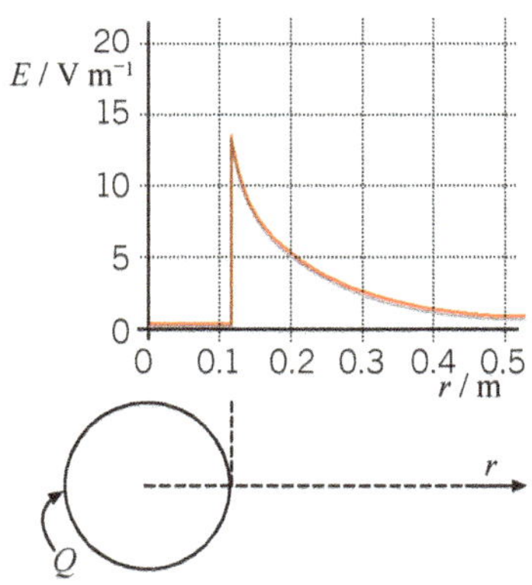

Fig. 4.2.4 E against r for a charged spherical surface

Study point

A test charge for measuring electric field strength should be a very small charge. Why? [Think of it being used to measure the field due to a charged conducting sphere.]

4.2.2 The electric field strength due to a point charge or equivalent

Electric field strength, E was defined (as force per unit charge on a small testing charge) in Section 4.1.6.

In the Coulomb's law equation we can regard one of the charges as the source of an electric field and the other as a testing charge. Calling these charges Q and q respectively, the equation becomes

$$F = \frac{Qq}{4\pi\varepsilon_0 r^2} \quad \text{that is} \quad \frac{F}{q} = \frac{Q}{4\pi\varepsilon_0 r^2}$$

But $\frac{F}{q}$ is, by definition, the magnitude of the electric field strength, E, at distance r from Q.

Hence the magnitude of the electric field strength at distance r from a point charge, Q, or from the centre of a spherically symmetric charge distribution of total charge Q, and outside the charge distribution, is

$$E = \frac{Q}{4\pi\varepsilon_0 r^2}.$$

The direction of the field is radially outwards from Q if Q is positive, and radially inwards if Q is negative.

Fig. 4.2.3 shows how E varies with distance r, according to this equation. The value of Q was chosen to make the graph easy to plot – see KC 4.2.3. Two points are worth mentioning:

- A doubling of r results in a quartering of E and so on. Even a sketch-graph of an inverse square law relationship should use this as a basis.
- E goes to infinity as r goes to zero. Physicists call a point where something of interest can't be calculated a 'singularity'. (It is unlikely that there exist any true point charges for which the inverse square law holds right down to $r = 0$.)

Outside a spherically symmetric charge distribution, such as a conducting sphere with charge spread evenly on its surface, the field strength is the same as if all the charge were concentrated in the centre. (In the space *enclosed by* a charged spherical surface, the field strength due to the surface charge is zero!) Compare Figs 4.2.3 and 4.2.4.

In fact, if a spherical conductor is given charge, free charge carriers will move in such a way that the charge *will* spread evenly over the surface by mutual repulsion. However, any nearby external charges will disturb the symmetry. For example, in Coulomb's experiment, it's not, in the first instance, the *spheres* that repel, but the charges on them, and being effectively mobile these will move so that the far sides of the spheres have more charge than the facing sides. So the spheres don't quite behave as point charges at their centres! See Study point.

4.2.3 Electrical potential, V

(a) Definition

How close can an α particle with 7.7 MeV of kinetic energy approach a compact body of charge 1.26×10^{-17} C? This was a key calculation in Rutherford's estimate of the size of an atomic nucleus. Like other similar calculations it is made much easier by using the concept of potential, which we now develop...

We start by defining the **potential energy** (PE) of a testing charge, q, at a point P in an electric field.

How do we set about calculating this when the field is due to a point charge, Q? First recall that *Work* means force × distance in direction of force. The force on q is qE, radially outwards from Q. But E keeps getting less as we go from P to infinity (Fig. 4.2.3), because $E = \dfrac{Q}{4\pi\varepsilon_0 r^2}$.

We must therefore calculate the work done on q in small steps at a time (distance Δr), *so* small that E hardly changes over any one step. Then ...

Work done on q as q moves distance Δr further from $Q = qE\,\Delta r$.

As shown in Fig. 4.2.5, $E\Delta r$ is represented on a graph of E against r by the area of a narrow strip (shaded) 'under' the curve at the appropriate r. The *total* work done on q is q multiplied by the sum of the areas of all the narrow strips from P to infinity. So

PE of q at P (due to Q) $= q \times$ (area under E vs r graph from P to infinity)

There is a mathematical technique (see Maths tip) for calculating the area. It comes out to be $\dfrac{Q}{4\pi\varepsilon_0 r_P}$.

The subscript 'P' has now served its purpose, so we can drop it and say:

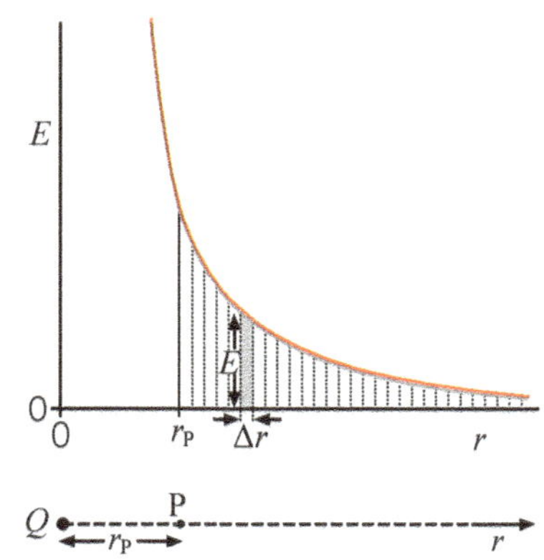

Fig. 4.2.5 Area under E vs r graph

For q at a distance r from Q:

$$\text{PE} = q \times \frac{Q}{4\pi\varepsilon_0 r}$$

The quantity $V = \dfrac{Q}{4\pi\varepsilon_0 r}$ and is called the electrical **potential** at distance r from Q.

In fact, for a field due to any distribution of stationary charges, the potential, V, at a point is defined by the equation

i.e. $V = \dfrac{\text{PE of test charge } q}{q}$

That is, PE of test charge $= qV$.

Key terms

The **potential energy** of a test charge, q, at a point P in an electric field is the work done by the field on q, as q goes from P to infinity.

The **potential**, V, at a point P in an electric field is the work done by the field *per unit charge* on a test charge as it goes from P to infinity.

UNIT: J C^{-1} = volt (V)

Study point

'Infinity' means so far from the charge(s) giving rise to the field that the field strength is negligible.

Study point

Some people define potential energy and potential at P in terms of work done by some external force (equal and opposite to the force of the electric field on q) as q goes *from* infinity *to* P. These definitions are entirely equivalent to ours.

Maths tip

If you've met integral calculus this is for you...

Shaded area in Fig. 4.2.5 is

$$\int_{r_P}^{\infty} E\,dr = \frac{Q}{4\pi\varepsilon_0}\int_{r_P}^{\infty}\frac{1}{r^2}dr$$

$$= \frac{Q}{4\pi\varepsilon_0}\left[-\frac{1}{r}\right]_{r_P}^{\infty} = \frac{Q}{4\pi\varepsilon_0}\left[0 - \left(-\frac{1}{r_P}\right)\right]$$

which tidies up to give the result quoted in the main text.

4.2.4 Knowledge check

In Fig. 4.2.6 a tangent has been drawn to the V against r graph at $r = 0.20$ m.

(a) Determine the gradient of this tangent.

(b) Read off the value of E at $r = 0.20$ m from Fig. 4.2.3, and comment on whether it is in agreement with $E = -\frac{\Delta V}{\Delta x}$.

(b) Graph of V against r for a point charge

Fig. 4.2.6 is just such a graph. Note the inverse proportionality between V and r: when r is doubled, V halves and so on, in accordance with the equation $V = \frac{Q}{4\pi\varepsilon_0 r}$. Fig. 4.2.6 should be compared with Fig. 4.2.3, in which E is plotted against r for the same value of Q and an *inverse square* law applies.

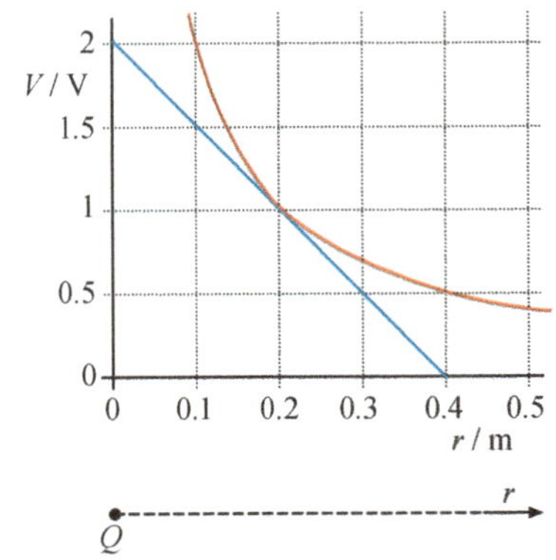

Fig. 4.2.6 V against r for a point charge

(c) Potential gradient

In Section 4.1.6(d) we derived the equation

$$E = -\frac{\Delta V}{\Delta x} = -\text{ potential gradient}$$

In the form $\Delta V = -E\Delta x$ it is based on the now familiar reasoning that

$$\frac{\text{work done on } q}{q} = \frac{\text{force on } q \times \text{(small) distance gone by } q \text{ in direction of force}}{q}.$$

In Section 4.1.6, we were concerned with uniform fields, where Δx can have any value. However, we can apply the potential gradient equation to non-uniform fields, provided that $\frac{\Delta V}{\Delta x}$ is taken as the value that this fraction approaches as Δx becomes smaller and smaller. This makes it equal to the gradient of a tangent drawn to a graph of V against distance in direction of E (that is the r direction for our point charge, Q).

Knowledge check 4.2.4 recommends a consistency check you should make on Figs 4.2.6 and 4.2.3 using $E = -\frac{\Delta V}{\Delta x}$. A more complete check, on the underlying equations for V and E, is suggested in Stretch & challenge.

Stretch & challenge

Show that the *difference* in potential between points B and A at distances $(r + \Delta r)$ and r from a charge Q is

$$\Delta V = V_B - V_A = \frac{Q}{4\pi\varepsilon_0}\left\{\frac{-\Delta r}{r(r+\Delta r)}\right\}$$

Show further that, if Δr is extremely small compared to r, this equation collapses to $\Delta V = -E\Delta r$, in which E is the field strength at distance r from Q.

(d) The zero of potential

According to $V = \frac{Q}{4\pi\varepsilon_0 r}$, V approaches zero as r becomes very large. This is a direct consequence of our defining potential in terms of work done on a testing charge going from a point *to infinity*, rather than to some other point at a finite distance (r_0, let's say) from Q. Had we chosen the r_0 option, the equation for potential would be

$$V = \frac{Q}{4\pi\varepsilon_0 r} - \frac{Q}{4\pi\varepsilon_0 r_0}$$

in which the second term is a constant. In fact only *differences* in potential (pds) have any physical significance, and either formula for potential would give the same value for the pd between any two particular points. We choose the option that gives us the simpler equation.

A reason sometimes given for taking the potential to be zero at infinity is that at infinity the (zero strength) field can do no more work on a positive testing charge. The argument fails, though, if Q is negative so its field is directed inwards *towards* Q!

Note that if Q *is* negative, then, according to $V = \frac{Q}{4\pi\varepsilon_0 r}$, V is negative everywhere and approaches zero for very large r.

4.2.5 Knowledge check

The top sphere (radius 0.12 m) of a Van de Graaff generator has a potential of −80 kV. Calculate:

(a) the charge on the sphere, stating your assumption(s).

(b) the electric field strength just outside the sphere.

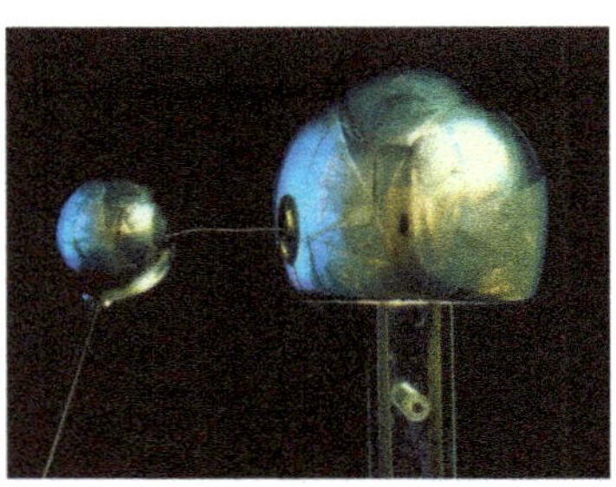
Fig. 4.2.7 Van de Graaff generator

(e) Using the idea of potential: an example

How close can an α particle with 7.7 MeV of kinetic energy approach a compact body (nucleus) of charge 1.26×10^{-17} C? This was the question posed at the beginning of Section 4.2.3. We can now answer it.

Fig. 4.2.8 α particle approaching a nucleus (not to scale!)

The closest approach will be when the α is heading straight at the nucleus and momentarily stops (at point P, say), having been continuously retarded by the repulsive force from the nucleus. We shall assume that the α is infinitely far from the nucleus initially, and that the nucleus is stationary throughout. Since energy is conserved,

Initial electrical PE + initial KE = electrical PE at P + KE at P

So $$0 + 7.7 \times 10^6 \text{ eV} = 2e \times \frac{1.26 \times 10^{-17}\text{ C}}{4\pi\varepsilon_0 r_P} + 0$$

Here we have remembered that the charge on the α is $+2e$. A neat trick is to divide through by e, as $7.7 \times 10^6 \text{ eV} = 7.7 \times 10^6 \text{ V} \times e$. Dividing and rearranging,

$$r_P = \frac{2}{7.7 \times 10^6 \text{ V}} \times \frac{1}{4\pi\varepsilon_0} \times 1.26 \times 10^{-17}\text{ C}$$

leading to $r_P = 2.9 \times 10^{-14}$ m

The significance of this result is explained briefly in the top Study point.

4.2.4 *E* and *V* due to more than one point charge

In principle we can calculate the electric field strength and potential at a point P due to any spatial distribution of charges. Here's the procedure:

- View the charge distribution as a collection of point charges.
- Calculate E and V at P due to each point charge.
- Add the field strengths at P as *vectors* and the potentials at P as *scalars*.

Actually *carrying out* the procedure and establishing the results for even something as simple as a spherical charge (see Study point) is a little beyond A-Level. We shall be content with finding field strengths and potentials due to just two or three point charges.

Example

Calculate the field strength and potential at point **B** in Fig. 4.2.9 (a) due to the charges shown at **A** and **C**.

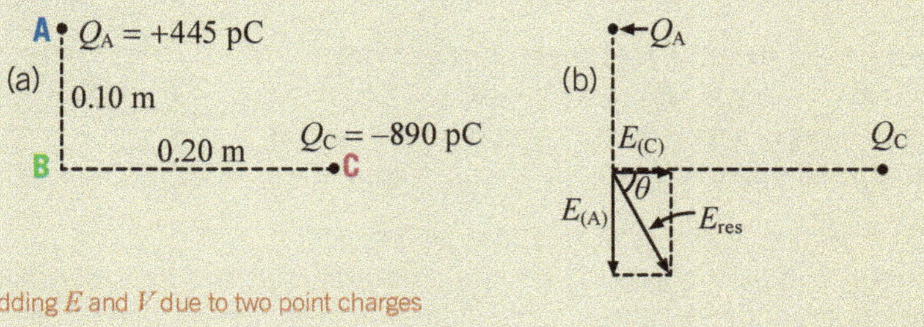

Fig. 4.2.9 Adding E and V due to two point charges

Knowledge check 4.2.6

The surface of a sphere of radius 5.0 mm carries a uniformly spread charge of +1.1 pC. An electron is emitted from it at a speed of 7.3×10^5 m s^{-1}.

Use energy conservation to determine the furthest distance from the sphere's centre that the electron can travel.

Study point

Rutherford concluded that a gold nucleus (charge = 1.26×10^{-17} C) must, if spherical, have a radius of less than r_p, because the paths of α particles through gold foil showed no signs of the particles experiencing any forces but electrostatic repulsive ones.

Study point

If we carry out the procedure in Section 4.2.4 for a spherically symmetric charge distribution we find the following:

1. E and V outside the charge distribution are the same as if all the charge were concentrated at the centre.
2. Inside a uniformly charged spherical surface, the field strength is 0 and the potential is everywhere the same as just outside the surface.

Knowledge check 4.2.7

In Fig 4.2.9 how much work is done against the electric field by an external force in moving a small charge q to B from a very long way away?

4.2.8 Knowledge check

For the electric field strength at C in Fig. 4.2.10, calculate

(a) the x-component, [Add the x-components of the field strengths due to Q_A and Q_B.]
(b) the y-component,
(c) the magnitude and direction of the resultant.

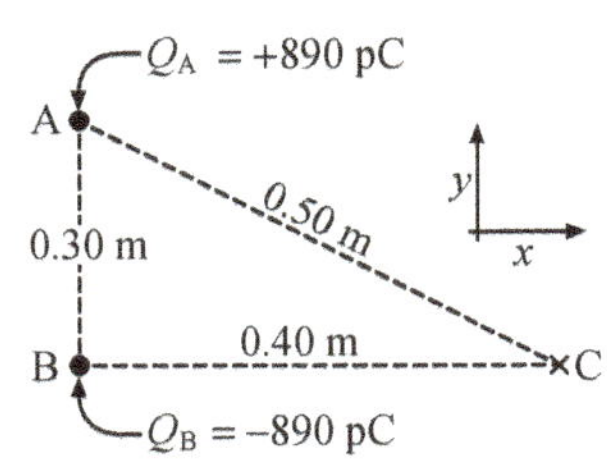

Fig. 4.2.10 Two point charges

Answer

The field strength at **B** due to Q_A is $E_{(A)} = \dfrac{445 \times 10^{-12}\ \text{C}}{4\pi\varepsilon_0 (0.10\ \text{m})^2} = 400\ \text{V m}^{-1} \downarrow$

Because $Q_C = -2\ Q_A$, but Q_C is twice as far from **B** as Q_A, the field strength at **B** due to Q_C is: $E_{(C)} = 200\ \text{V m}^{-1} \rightarrow$

The resultant field at **B** is obtained by vector addition (Fig. 4.2.9 (b)). Applying Pythagoras' theorem and using trigonometry we obtain $E_{res} = 447\ \text{V m}^{-1}$; $\theta = 63°$.

$E_{res} = \sqrt{400^2 + 200^2}\ \text{V m}^{-1} = 447\ \text{V m}^{-1}$ and $\theta = \tan^{-1}\dfrac{400}{200} = 63°$.

The potential at **B** if Q_A were the only charge would be

$V_{(A)} = \dfrac{445 \times 10^{-12}\ \text{C}}{4\pi\varepsilon_0 \times 0.10\ \text{m}} = 40\ \text{V}$

Because $Q_C = -2\ Q_A$, but Q_C is twice as far from **B** as Q_A, $V_{(C)} = -40\ \text{V}$.

The potential at **B** is the scalar sum of the potentials due to Q_A and Q_B.

Thus $V = V_{(A)} + V_{(C)} = 40\ \text{V} + (-40\ \text{V}) = 0$

Speculating from the example

A positively charged particle (charge q, mass m) released from rest at B in Fig. 4.2.9, would initially gain velocity, at a rate $\dfrac{qE}{m}$, in the direction of the resultant field. This implies that it gains KE and loses electrical PE – at first.

As the particle changes its position, though, the field strength changes, and it is no trivial business to work out the path it will follow. Will it, for example, avoid capture by Q_C? To find out, we could use a computer-executed step-by-step numerical method, calculating acceleration, velocity, displacement over and over again for very short time intervals.

If the particle's trajectory took it far from the charges at A and C, its net change in electrical PE since it left B would approach zero, since $V = 0$ at B and, by definition, at infinity. Therefore the particle's kinetic energy would approach its value at B – zero! What would happen then?

4.2.9 Knowledge check

Referring to Fig. 4.2.10, calculate the change in electrical potential energy of a charge of +3.0 pC if it is taken from C to infinity. Explain whether this PE change will be positive or negative.

4.2.5 Electric field lines and equipotential surfaces

Key terms

An **electric field line** is a line whose direction at each point along it is the direction of the electric field at that point.

Electric field lines are huge aids to visualising how electric field strength and potential vary from point to point in an electric field.

Study point

It follows from the definition above that electric field lines can't cross or meet. When the field arises from more than one charge, there is a single *resultant* field at each point.

(a) Electric field lines

In principle we could show the field direction at as many points as we please using small arrows (Fig. 4.2.11). It is usually clearer to select a few 'chains' of such arrows, running head of one to tail of the next. These are electric field lines.

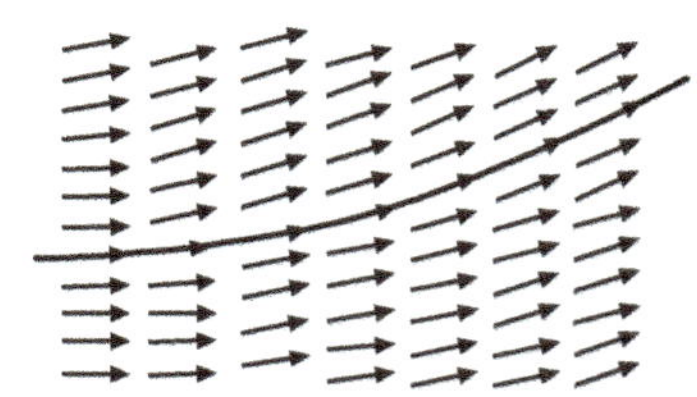

Fig. 4.2.11 Electric field line

Figs 4.2.12 (a) and (b) show some field lines for 'isolated' positive and negative point charges. The lines are radially outward for the positive charge and radially inward for the negative. Ignore the broken lines for the time being.

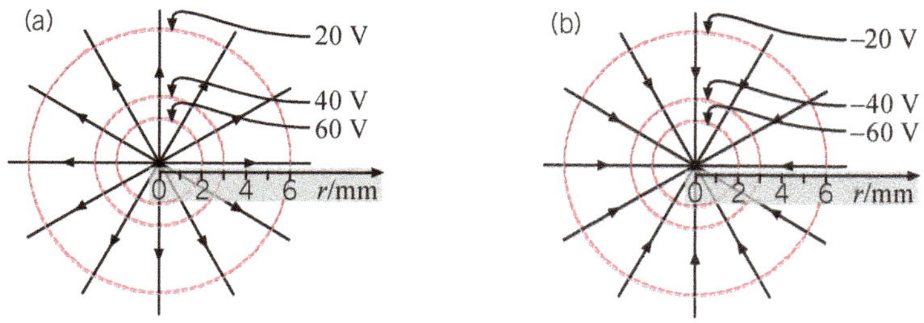

Fig. 4.2.12 Electric field lines and equipotential surfaces due to
(a) an isolated positive charge (b) an isolated negative charge

Although, in the first instance, electric field lines tell us the direction of the field, they have other properties.

Where the field lines come closer together, the field is stronger. This is clearly the case for the isolated point charges, but holds true for the field lines from any configuration of charges, such as the equal and opposite charges in Fig. 4.2.13.

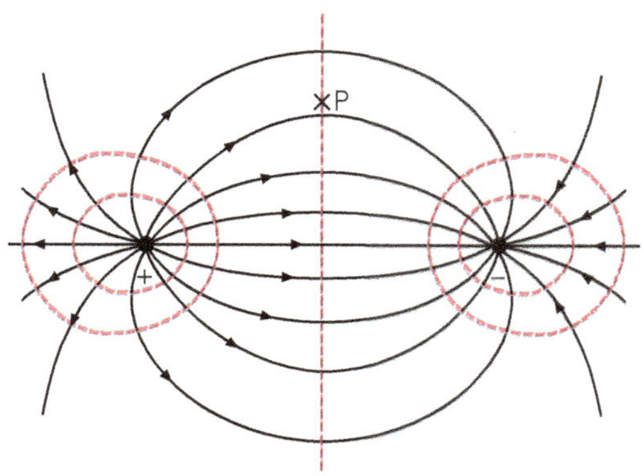

Fig. 4.2.13 Electric field lines due to equal and opposite charges (an electric *dipole*)

When the only charges present are stationary, electric field lines start on positive charges and end on negative charges as in Fig. 4.2.13. This is even true for 'isolated' charges. In Fig. 4.2.12 (a), field lines starting on the positive charge will end on negative charges 'induced', by electron displacement, on surrounding objects, specifically on the surfaces that face the charge. If these induced charges are too distant to distort the radial field near the 'original' charge, we call it 'isolated'. The negative charge in Fig. 4.2.12 (b) is also isolated in this sense, but the field lines ending on it do originate on induced positive charges elsewhere!

(b) Equipotential surfaces

The name says it all: see Key terms. For an isolated point charge the **equipotential surfaces** are spherical surfaces centred on the charge. The surfaces (shown in section as broken lines in Figs 4.2.12 and 4.2.13) are by convention drawn at equal intervals of potential. Note these additional properties:

- The surfaces are closer together where the field is stronger. This follows from the equation $\Delta V = -E\Delta x$. Rearranging, we have $\Delta x = \frac{\Delta V}{E}$, (*accurate* only when ΔV and Δx are very small).
- The electric field lines are at right angles to the equipotential surfaces as they pass through them. See Figs 4.2.12 and 4.2.13. This is a little like the lines of greatest slope on a hillside being at right angles to the contours (lines of constant height).

Knowledge check 4.2.10

Calculate the electric field strength at P in Fig. 4.2.13. P is 50 mm from each charge. The charges are of ± 0.15 nC, and are 80 mm apart.

Study point

An electric field surrounds an area through which there is a changing *magnetic* field, such as the core of a working transformer. In this case the electric field lines are closed loops – not starting and finishing. This is one type of electromagnetic induction; see Section 4.4. Note that *moving* charges are involved (in setting up the changing magnetic field).

Knowledge check 4.2.11

Calculate the potential difference between the equipotential surfaces at 1.0 mm and 10.0 mm from the charge in Fig. 4.2.12 (a).

Key terms

An **equipotential surface** in an electric field is an imaginary surface on which all points are at the same potential.

Knowledge check 4.2.12

Why are the oval equipotential surfaces around either charge in Fig. 4.2.13 closer on the 'facing' sides of the charges than on the far sides?

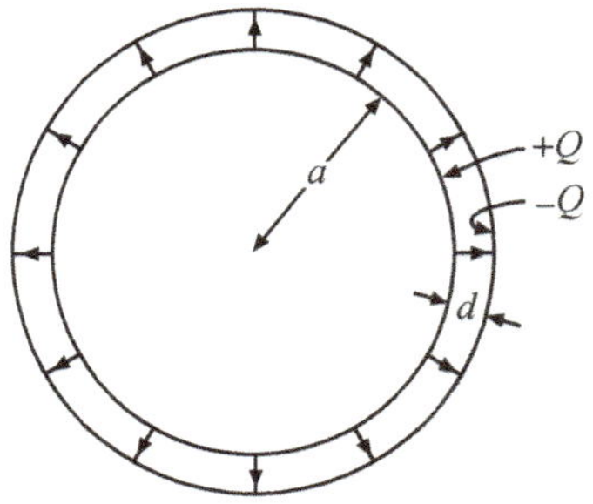

Fig. 4.2.14 Concentric sphere capacitor

An example - the concentric sphere capacitor

A battery connected briefly between two concentric spherical metal shells (hollow spheres) will transfer electrons from one shell to the other. Equal and opposite charges will therefore coat the facing surfaces of the shells, and there will be a field in the gap as shown in Fig. 4.2.14.

The electric field strength in the gap, at distance r from the centre of the spheres will simply be $E = \frac{Q}{4\pi\varepsilon_0 a^2}$ radially outwards. This is entirely due to the charged inner sphere; the charged outer sphere contributes nothing, as the gap is part of the space *inside* it (See Fig. 4.2.4).

Now suppose that the gap width d is much less than the radius of the inner sphere, that is $d << a$. We can then take the field as approximately constant across the gap and given by $E = \frac{Q}{4\pi\varepsilon_0 a^2}$. Using $\Delta V = -E\Delta x$, the pd between the shells is simply:

$$|\Delta V| = \frac{Q}{4\pi\varepsilon_0 a^2} d = \frac{Q}{\varepsilon_0 A} d$$

in which we have put $4\pi a^2 = A$, as it is the surface area of the inner shell, and almost that of the outer shell.

From the definition of *capacitance* (Section 4.1.1) $C = \frac{Q}{|\Delta V|} = \frac{\varepsilon_0 A}{d}$.

Study point

Our result for the capacitance of a concentric sphere capacitor is the same as that quoted for a parallel plate capacitor in 4.1.2. Perhaps this is not surprising, as our condition $d << a$ can be read as $a >> d$ (relatively large radius), but a proper derivation for a parallel plate capacitor is a little beyond A level.

4.2.6 Newton's law of gravitation

Isaac Newton gave convincing evidence for an inverse square law of forces between bodies with mass (so-called gravitational forces) roughly a century before Coulomb discovered the inverse square law of forces between charges. **Newton's law of gravitation** is given in modern language in Key terms.

- Newton showed that a consequence of the inverse square law would be that a spherical shell of evenly distributed mass would exert the same force on an external body as if all its mass were concentrated at its centre. No force would be exerted by the shell on any body placed *inside* it. We've already met similar results for charges in Section 4.2.4.

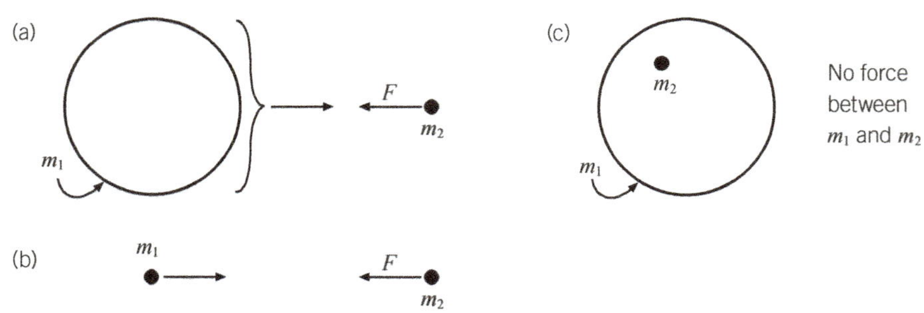

Fig. 4.2.15 Gravitational effect of a spherical shell – in (a) and (b) the forces between m_1 and m_2 are the same

Any spherical body with its mass distributed with spherical symmetry about its centre, can be thought of as an 'onion' of concentric shells, so we can deduce that *it*, too, attracts an external body as if all its mass were at its centre. That's a very useful result, because the bodies that exert appreciable gravitational forces are stars and planets, whose mass *is* distributed with roughly spherical symmetry.

- Newton's main evidence for the inverse square law of gravitation came from the heavens. He showed that an inverse square law of force from the Sun would account for the observed motion of the planets – more about this in Section 4.3.

Key terms

Newton's law of gravitation
Every body with mass attracts every other body with mass. For two 'point masses', m_1 and m_2, separated by distance r, the force is given by

$$F = \frac{Gm_1m_2}{r^2}$$

in which G is a constant called *Newton's gravitational constant* (or *big G*).

$G = 6.67 \times 10^{-11}$ N m^2 kg^{-2}

4.2.13 Knowledge check

For two protons at the same separation, calculate the ratio

$$\frac{\text{electrostatic repulsive force}}{\text{gravitational attractive force}}.$$

[If necessary look up the values for m_p and e – try and remember them!]

- Because Newton could only guess at the masses of astronomical bodies such as the Sun, he would not have been able to determine a reliable value for the proportionality constant in the law of gravitation, the constant we now call G. The problem was solved over a century later (in the 1790s) by Henry Cavendish (Fig. 4.2.16) who measured the force between lead spheres here on Earth. He used a torsion balance similar in principle to Coulomb's (Section 4.2.1), but much larger. The experiment was an astonishing feat of design and skill because of the low percentage uncertainty achieved in the result, even though the force to be measured was extremely weak.

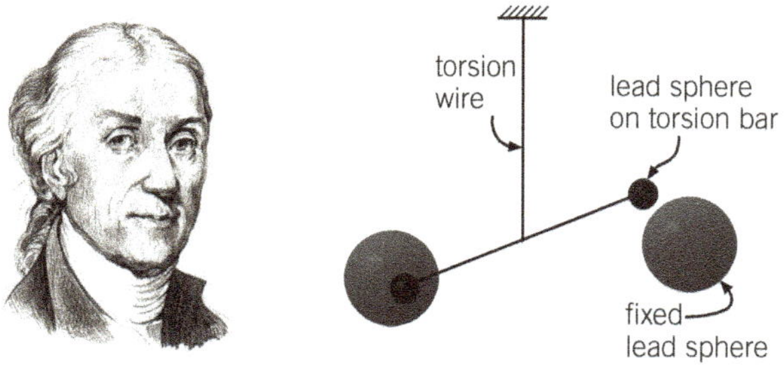

Fig. 4.2.16 Henry Cavendish and a simplified diagram of his apparatus

Knowledge check 4.2.14

Calculate the force between a movable sphere (mass 0.73 kg) and a fixed sphere (mass 158 kg) in Cavendish's experiment. The centres were 305 mm apart. [Assume a value for G!]

4.2.7 Gravitational field strength, g and potential, V

(a) Definition of g

Gravitational field strength is defined in a very similar way to electric field strength (see Section 4.1.6), but using a test *mass* rather than a test *charge*.

If the only force acting on a body is gravitational, then, using Newton's second law of motion, its acceleration is

$$a_{\text{grav}} = \frac{F}{m} = \frac{mg}{m} = g$$

This is a simple but non-trivial result (see Study point). It tells us that:

- Measuring a body's gravitational acceleration in a location amounts to measuring the gravitational field strength there. Indeed we shall now use g for both the field strength and the acceleration.
- A body's gravitational acceleration is independent of its mass; all bodies have the same gravitational acceleration in the same place. Galileo confirmed this by experiment, though almost certainly he never dropped cannon balls from the leaning tower of Pisa as the legend goes! The result has since been confirmed to very high accuracy.

Key terms

The **gravitational field strength**, g, at a point is defined as $g = \frac{F}{m}$, that is

$$g = \frac{\text{Force on test mass, } m}{m}$$

at that point. It is a vector.

UNITS: N kg^{-1}

Study point

The independence of a_{grav} from the *mass* of the accelerating body is due, in our derivation of $a_{\text{grav}} = g$, to the cancellation of m. On the top line m's role is gravitational: bigger m, bigger gravitational force; on the bottom line m's role is inertial: bigger m, smaller acceleration. In Newtonian physics this dual role of mass is a mystery.

(b) g outside a spherically symmetric mass distribution

If the mass of the sphere is M, and m is a test mass outside the sphere we can put $m_1 = M$ and $m_2 = m$ into Newton's law of gravitation, to give

$$F = \frac{GMm}{r^2} \quad \text{so} \quad \frac{F}{m} = \frac{GM}{r^2} \quad \text{that is} \quad g = (-)\frac{GM}{r^2}$$

in which r is the distance from the test mass to the centre of the sphere.

A minus sign is often included as shown; it signifies that g is directed *inwards* along a radius.

Stretch & challenge

Show that the gravitational field strength *inside* a sphere of uniform density at a distance r from its centre is given by

$$g = g_{\text{surf}}\frac{r}{r_{\text{surf}}}$$

4.2.15 Knowledge check

The mean radius of Mars is 3390 km, and its mean density is 3930 kg m^{-3}. Calculate a value for the gravitational field strength on its surface.

4.2.16 Knowledge check

By what factor would the length of the Earth's day have to decrease in order for the observed acceleration of free fall to be zero at the equator?

Example

Calculate a value for the mass of the Earth given that its mean radius is 6370 km and the mean field strength near its surface is 9.81 N kg^{-1}.

Answer

$$M = \frac{R_E^2 g}{G} = \frac{(6.37 \times 10^6 \text{ m})^2 \times 9.81 \text{ N kg}^{-1}}{6.67 \times 10^{-11} \text{ N kg}^{-2} \text{ m}^2} = 5.97 \times 10^{24} \text{ kg}$$

Dividing this mass by the volume $\frac{4}{3}\pi R_E^3$, gives a value for the mean density of the Earth as approximately 5.5×10^3 kg m^{-3}.

It's now time to confess that Cavendish never published a value for G as such. Instead, he used the results of his experiment to calculate a value for the mean density of the Earth, rather as we have just done. But a value for G can be derived from Cavendish's results.

Observed free fall acceleration on the Earth

On the Earth, at all latitudes except $\pm 90°$ (the poles), the *observed* acceleration of free fall, g_{obs}, is a little less than the gravitational acceleration, g. This is because g includes the centripetal acceleration, $r\omega^2$, needed to keep the body rotating about the Earth's axis (at a distance r from the axis), as well as g_{obs}. In our rotating frame of reference (the Earth's surface) we don't observe $r\omega^2$ directly.

Fig. 4.2.17 shows the easiest case: when a body is falling at a point P on the equator, so $r\omega^2$ like g, is directed towards the centre of the Earth. g_{obs} must therefore also be in this direction and

$$g = g_{obs} + r\omega^2 \qquad \text{that is} \qquad g_{obs} = g - r\omega^2$$

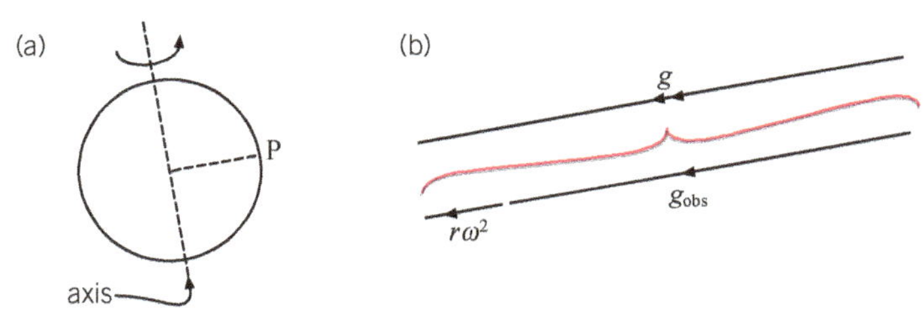

Fig. 4.2.17 g at a point P on the equator

At the poles $r\omega^2 = 0$ and $g_{obs} = 9.832$ m s^{-2}. At the equator, $r\omega^2 = 0.034$ m s^{-2} (as you should check!), but $g_{obs} = 9.780$ m s^{-2}. Clearly the different value of $r\omega^2$ doesn't account for all the 0.052 m s^{-2} discrepancy in g_{obs}. The rest is due to the Earth not being a perfect sphere. At the equator, g itself is actually very slightly less than at the poles – which are closer to the centre of the Earth.

Having just dragged out into the open the distinction between real and observed g, we must now admit that for many purposes (including exam questions unless specifically mentioned) we *don't* distinguish – for the simple reason that $r\omega^2 << g$.

Key terms

The **gravitational potential energy** of a test mass, m, at a point P in a gravitational field is the work done by the field on m if m goes from P to infinity.

(c) Gravitational potential, V

This concept is the gravitational twin of *electrical* potential, explained in Section 4.2.3(a), so we'll be briefer here.

We start from the definition of **gravitational potential energy** in Key terms. Here, the work done by the field on m is negative because the gravitational force on m will be back towards the body or bodies responsible for the field, that is in essentially the opposite direction to m's journey from P to infinity. This makes gravitational PE, E_P, a negative quantity, approaching zero as P approaches infinity. The zero of E_P at infinity – far enough for the gravitational force to be negligible – is contrived for mathematical convenience.

We now concentrate on the potential energy of m at a point P outside a spherically symmetric mass distribution, M. Fig. 4.2.18 shows how g varies with distance r from the centre of M.

Then, following our method in Section 4.2.3,

$$E_P \text{ of } m \text{ at P} = m \times \text{area 'under' } g\text{–}r \text{ graph from P to } \infty$$
$$= m \times \left(-\frac{GM}{r}\right) \quad \text{(see Study point)}$$

We can express this in terms of the **gravitational potential**, V, due to M at P (see Key terms):

$$E_P \text{ of } m \text{ at P} = mV \qquad \text{in which} \qquad V = -\frac{GM}{r}$$

Figs 4.2.19 (a) and (b) show how g and V vary with distance r from the centre of a star. See KC 4.2.17. Note the different effects of doubling r, in accordance with the inverse square law for g and the inverse law for V.

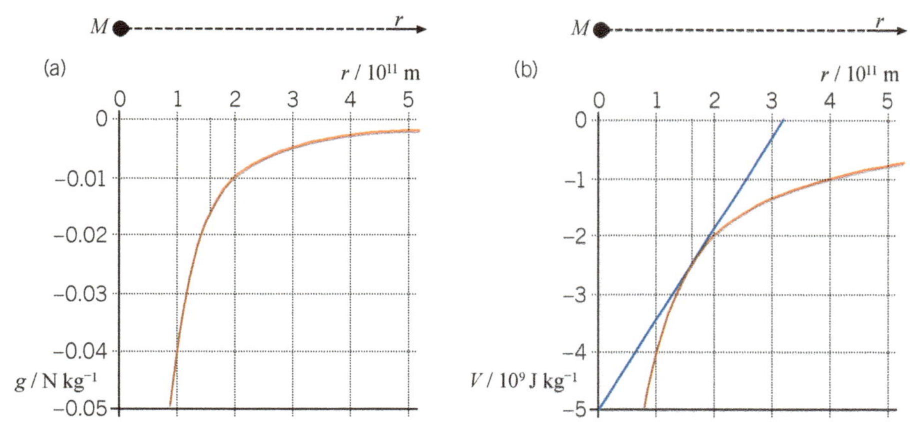

Fig. 4.2.19 (a) g against r (b) V against r, for the same star

(d) Gravitational potential gradient

A test mass, m, moving distance Δx in the direction of a gravitational field will have work $mg \times \Delta x$ done on it by the field,

so change in gravitational PE, $\Delta E_P = -mg\Delta x$ (see Study point)

and change in gravitational potential, $\Delta V = -\frac{mg\Delta x}{m}$

That is $\Delta V = -g\Delta x$ so $g = -\frac{\Delta V}{\Delta x} = -$ potential gradient.

For the field from a spherically symmetric mass distribution,

$$g = -\frac{\Delta V}{\Delta r} = -\text{ gradient of tangent to graph of } V \text{ against } r$$

See KC 4.2.18.

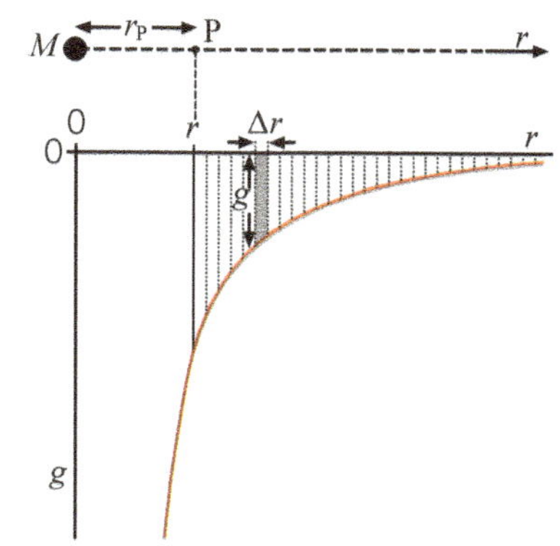

Fig. 4.2.18 Area under g vs. r

Study point

Similarly to the Study point in Section 4.2.3(a) …

Shaded area in Fig. 4.2.18 is

$$\int_{r_P}^{\infty} g\, dr = -GM\int_{r_P}^{\infty} \frac{1}{r^2}\, dr$$
$$= -GM\left[-\frac{1}{r}\right]_{r_P}^{\infty} = -GM\left[0 - \left(-\frac{1}{r_P}\right)\right]$$

which, having dropped the 'P' subscript, reduces to the result quoted in the main text.

Key terms

The **gravitational potential**, V, at a point P in a gravitational field is the work done by the field *per unit mass* on a test mass if it goes from P to infinity.

UNIT: J kg^{-1}

So, for test mass m, $E_P = mV$.

At distance r from centre of M:

$V = -\frac{GM}{r}$.

Knowledge check 4.2.17

Determine the mass of the star for which Fig. 4.2.19 (a) and (b) are drawn.

Study point

We can calculate ΔE_P from $\Delta E_P = -mg\Delta x$ only when Δx is small enough for g to be almost constant. We write it as $\Delta E_P = mg\Delta h$ for a body lifted through a height Δh near the Earth's surface ($\Delta h << R_E$). It won't usually be applicable to space-shots and the like.

4.2.18 Knowledge check

Determine the gradient of the tangent that has been drawn to the graph of V against r in Fig. 4.2.19 (b) and compare it with the value of g for the same value of r as read from Fig. 4.2.19 (a).

4.2.19 Knowledge check

Calculate the speed with which a rocket needs to be launched from the Moon's surface to reach a height of 8.70×10^5 m above the surface.

(Mean R_{Moon} = = 1.74×10^6 m,

g at surface = 1.62 m s^{-2}.)

Study point

The value of g at the surface, g_{surf} is given by

$$g_{surf} = \frac{GM}{r^2_{Mars}}$$

so we can write $g_{surf}r^2_{Mars}$ in place of GM.

4.2.20 Knowledge check

Locate the point P between the Sun and Jupiter at which their resultant g is zero. (See Fig 4.2.21.)

First determine the ratio $\frac{x}{y}$ and hence the values of x and y.

(Masses of Sun and Jupiter: 1.99×10^{30} kg, 1.90×10^{27} kg.)

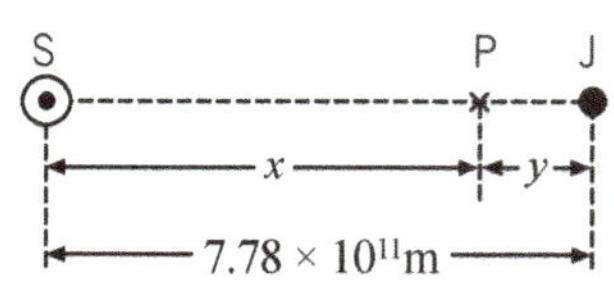

Fig. 4.2.21 Sun and Jupiter

Study point

A body placed at P in KC 4.2.20 wouldn't remain at rest (if P is a fixed point relative to the galaxy). For one thing, Jupiter (and the Sun) are in mutual orbit; they don't stay put.

Example: using gravitational potential

The mean radius of Mars is 3.39×10^6 m and the value of g at its surface is 3.71 N kg^{-1}. Neglecting any resistive effects, calculate:

(a) the distance that a rocket will rise if it is launched straight up from the surface of Mars with a speed, u, of 3200 m s^{-1}.

(b) the escape velocity of a rocket from the surface of Mars.

Answer

(a) Since energy is conserved,

initial E_K + initial E_P = E_K at highest point + E_P at highest point.

So $\frac{1}{2}mu^2 + \left(-\frac{GMm}{r_{Mars}}\right) = 0 + \left(-\frac{GMm}{r_{max}}\right)$

in which r_{max} is the distance from the centre of Mars at which the missile runs out of KE. Dividing through by GMm and rearranging:

$$\frac{1}{r_{max}} = \frac{1}{r_{Mars}} - \frac{u^2}{2GM} \quad \text{therefore} \quad \frac{1}{r_{max}} = \frac{1}{r_{Mars}} - \frac{u^2}{2g_{surf}r^2_{Mars}}$$

Note how we've just overcome a dirty trick: not being given the mass, M, of Mars! See Study point.

So $\frac{1}{r_{max}} = \frac{1}{3.39 \times 10^6 \text{ m}} - \frac{(3200 \text{ m s}^{-1})^2}{2 \times 3.71 \text{ m s}^{-1} \times (3.39 \times 10^6 \text{ m})^2} = 1.75 \times 10^{-7} \text{ m}^{-1}$

Thus $r_{max} = 5.72 \times 10^6$ m and $r_{max} - r_{Mars} = 2.33 \times 10^6$ m

(b) We use energy conservation as in (a), but with less effort, as this time not only the 'final' KE, but also the 'final' PE, is zero ($r = \infty$).

Initial KE + initial PE = KE at infinity + PE at infinity.

So $\frac{1}{2}mu^2 + \left(-\frac{GMm}{r_{Mars}}\right) = 0 + 0$

So $u_{esc} = \sqrt{\frac{2GM}{r_{Mars}}} = \sqrt{\frac{2g_{surf} \times r^2_{Mars}}{r_{Mars}}} = \sqrt{2g_{surf}\, r_{Mars}} = 5020$ m s^{-1}

(e) g and V due to more than one body

When the field at a point P arises from more than one body:

- The field strength, g, is the vector sum of the field strengths at P that would arise from each body by itself.
- The potential, V, is the scalar sum of the potentials that would arise from each body by itself.

Example

Fig. 4.2.20 shows the alignment of Sun, Moon and Earth during a total eclipse of the Sun. Calculate the resultant field strength and potential at the Moon due to the Earth and Sun.

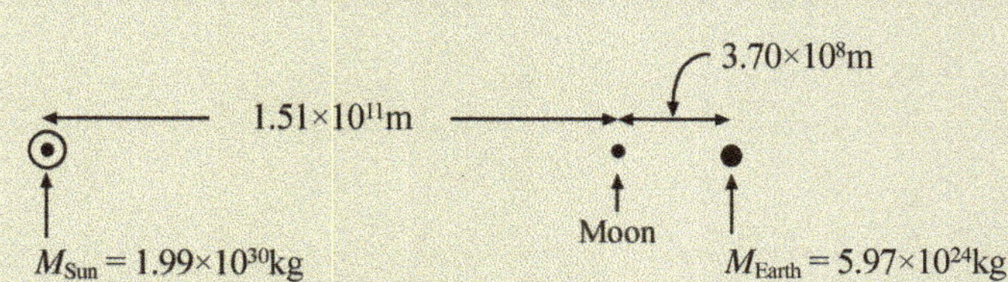

Fig. 4.2.20 Sun, Moon and Earth during a total eclipse

Answers

$$g = \frac{GM_E}{(3.7\times10^8\,\mathrm{m})^2} - \frac{GM_S}{(1.51\times10^{11}\,\mathrm{m})^2} \text{ towards the Earth}$$

$$= 6.67\times10^{-11}\,\mathrm{N\,kg^{-2}m^2}\left\{\frac{5.97\times10^{24}\,\mathrm{kg}}{(3.7\times10^8\,\mathrm{m})^2} - \frac{1.99\times10^{30}\,\mathrm{kg}}{(1.51\times10^{11}\,\mathrm{m})^2}\right\}$$

$$= -2.91\times10^{-3}\,\mathrm{N\,kg^{-1}} \text{ towards the Earth}$$

i.e. $2.91\times10^{-3}\,\mathrm{N\,kg^{-1}}$ towards the Sun.

and $V = -\dfrac{GM_E}{3.7\times10^8\,\mathrm{m}} + \left\{-\dfrac{GM_S}{1.51\times10^{11}\,\mathrm{m}}\right\}$

which similarly leads to $V = -8.80\times10^8\,\mathrm{J\,kg^{-1}}$.

4.2.8 Gravitational field lines and equipotentials

Allowing for the different type of field, **gravitational field lines** have the same properties as electric field lines (Section 4.2.5) except that they come from infinity and end on anything with mass. **Equipotential surfaces** are, again, pretty much self-explanatory.

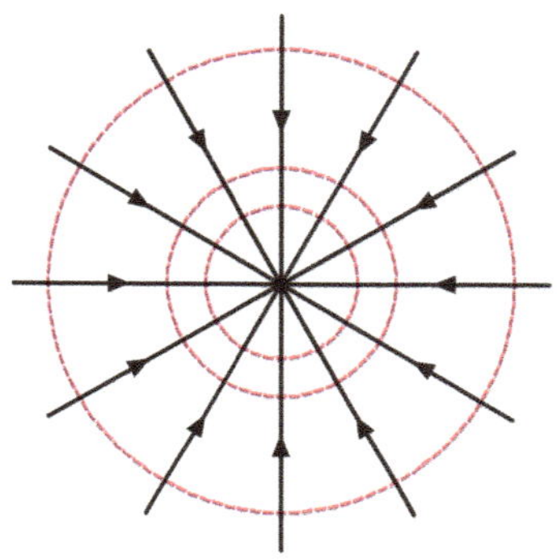

Fig. 4.2.22 Field lines and equipotentials for a small body with mass

Note that in Fig. 4.2.22 the equipotential surfaces are spheres of less and less negative potential as their radii increase.

Fig. 4.2.23 (a) shows the pattern of field lines and equipotentials for two spherically symmetric bodies of unequal mass. It is identical to the electric field pattern for two negative point charges such that $\dfrac{Q_1}{Q_2} = \dfrac{M_1}{M_2}$.

Figs 4.2.23 (b) and (c) are sketches of the field strength and potential along the line AB (the x direction, say). Note in particular the point between M_1 and M_2 at which the field strength is zero. It corresponds to a local maximum in the potential, in accordance with the relationship

$$g = -\frac{\Delta V}{\Delta x}$$

$$= -\text{ Gradient of tangent of } V\text{–}x \text{ graph}$$

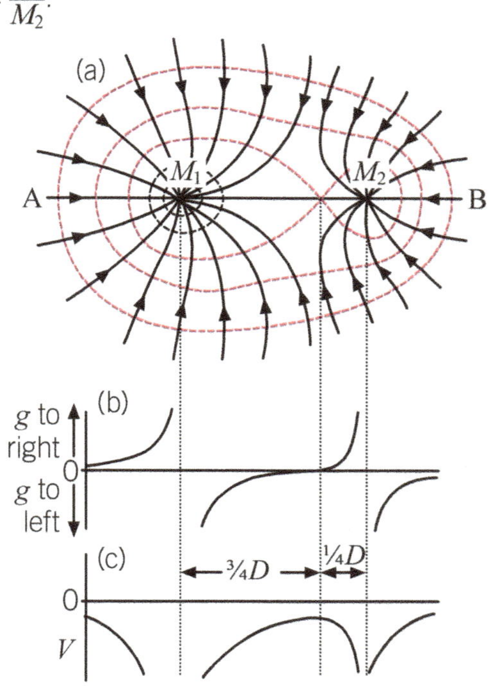

Fig. 4.2.23 Field and potential due to spheres of unequal mass

Key terms

A **gravitational field line** is a line whose direction at each point along it is the direction of the gravitational field at that point.

An **equipotential surface** in a gravitational field is an imaginary surface on which all points are at the same potential.

Knowledge check 4.2.21

In Fig. 4.2.23, the centres of M_1 and M_2 are distance D apart. Calculate the ratio $\dfrac{M_1}{M_2}$ from the additional data in the figure.

Summary of properties of electric and gravitational fields

Electric fields	Gravitational fields
The electric field strength, E, at a point is the force per unit charge on a small positive test charge placed at that point.	The gravitational field strength, g, at a point is the force per unit mass on a small test mass placed at that point
The inverse square law for the force between two point electric charges takes the form: $$F = \frac{1}{4\pi\varepsilon_0}\frac{Q_1Q_2}{r^2}$$ which is Coulomb's law	The inverse square law for the force between two point masses takes the form: $$F = G\frac{M_1M_2}{r^2}$$ which is Newton's law of gravitation
F can be attractive or repulsive	F is attractive only
$E = \frac{1}{4\pi\varepsilon_0}\frac{Q}{r^2}$ is the field strength due to a point charge in free space (a vacuum) or in air	$g = G\frac{M}{r^2}$ is the field strength due to a point mass
The potential, V, at a point is the work done per unit charge in bringing a small positive test charge from infinity to that point	The potential, V, at a point is the work done per unit mass in bringing a small test mass from infinity to that point
For a point charge: $V_E = \frac{1}{4\pi\varepsilon_0}\frac{Q}{r}$ and for two point charges: $PE = \frac{1}{4\pi\varepsilon_0}\frac{Q_1Q_2}{r}$	For a point mass: $V_g = -G\frac{M}{r}$ and for two point masses: $PE = -G\frac{M_1M_2}{r}$
Note that all the above equations also apply outside spherically symmetric charges	Note that all the above equations also apply outside spherically symmetric masses
The change of potential energy of a point charge displaced in an electric field is given by: $\Delta PE = q\Delta V_E$	The change of potential energy of a point mass displaced in an gravitational field is given by: $\Delta PE = m\Delta V_g$
Field strength at a point is given by: $E = -$ slope of the V_E–r graph at that point	Field strength at a point is given by: $g = -$ slope of the V_g–r graph at that point
$\varepsilon_0 = 8.854 \times 10^{-12}\ F\ m^{-1}$ But $\frac{1}{4\pi\varepsilon_0} \approx 9.0 \times 10^9\ F^{-1}\ m$ is an acceptable approximation	$G = 6.67 \times 10^{-11}\ N\ m^2\ kg^{-2}$

Test yourself 4.2

1. Diagrams (a) and (b) are of electrical field lines in different regions of space.

   In each case, draw an equipotential surface (line) through the point P and a second equipotential line which is at a higher potential.

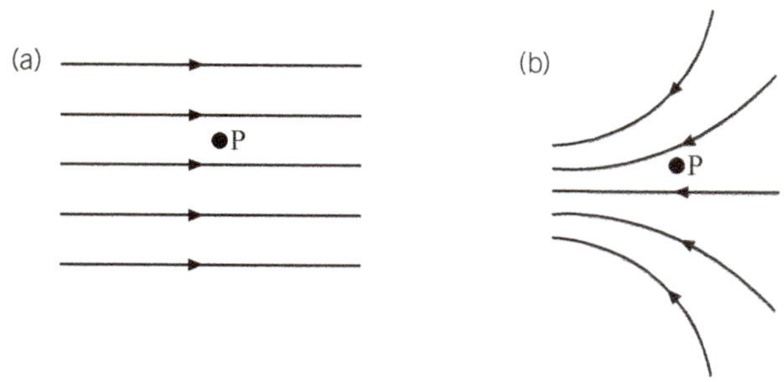

2. Two points, X and Y, are positioned as shown in a uniform electric field of intensity $150\ \mathrm{kV\ m^{-1}}$.

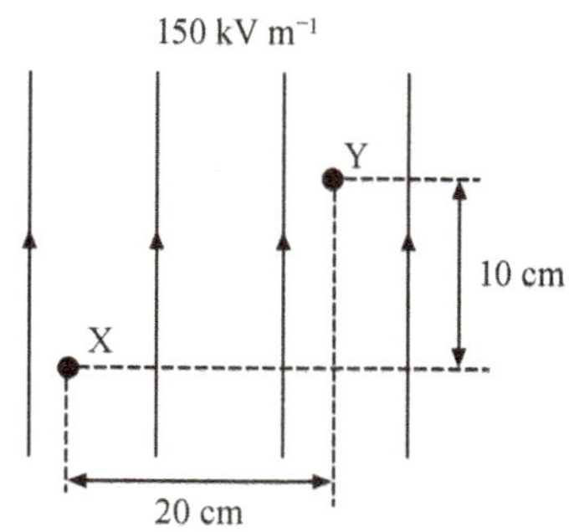

   (a) Explain which point is at the higher potential.
   (b) State the pd between X and Y.
   (c) A particle with a charge of $+10\ \mu\mathrm{C}$ is released at zero speed from X. Its path takes it subsequently through Y (presumably along a frictionless tube!). Assuming no friction, state its kinetic energy at Y.
   (d) Calculate the speed a proton would acquire in moving from X to Y.
   (e) A $^{7}_{3}\mathrm{Li}$ nucleus moves on a path that takes it through Y and X. Its kinetic energy at Y is $100\ \mathrm{keV}$. Calculate its speed at X. [You may need to use a data book for the mass and charge of a $^{7}_{3}\mathrm{Li}$ nucleus.]

3. A hollow conducting sphere of radius $10\ \mathrm{cm}$ carries a positive charge of $1.2\ \mu\mathrm{C}$. There are no other charges nearby.

   (a) Describe the distribution of charge.
   (b) Calculate the electric potential of the sphere and the electric field strength at its outer surface.
   (c) State the potential and field strength:
      (i) at a point $20\ \mathrm{cm}$ from the centre of the sphere,
      (ii) at the centre of the sphere.
      [One of these requires a trivial calculation.]

4. A small conducting sphere of mass $10\ \mathrm{mg}$ touches the outside of the charged sphere in Q3 and acquires a charge of $+10\ \mathrm{nC}$. It is observed to move rapidly away.

   (a) Explain this observation.
   (b) Calculate its initial acceleration.
   (c) Assuming no other significant forces act, calculate its final velocity.

5. Three charges are arranged in a right-angled isosceles triangle. The separation of the $-20\ \mu\mathrm{C}$ from each of the other charges is $10\ \mathrm{cm}$. The charges are not free to move.

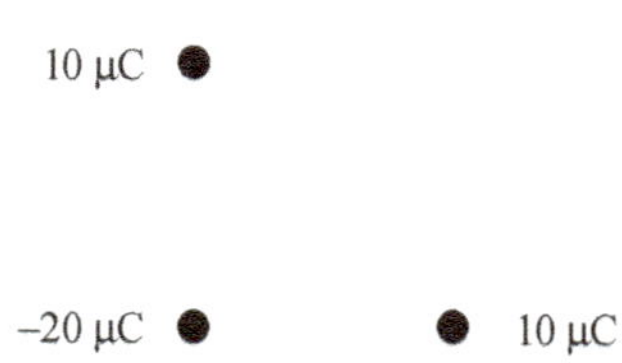

   (a) Calculate the electrical potential energy of the arrangement.
   (b) Calculate the resultant electrical force on each charge due to the other two charges.

6. The $-20\ \mu\mathrm{C}$ charge in Q5 is now free to move. Its mass is $1.00\ \mathrm{g}$.

   (a) Calculate its initial acceleration.
   (b) Describe its subsequent motion assuming no resistance to motion.
   (c) Calculate its greatest speed.

Questions 7–11 refer to a dwarf planet of radius 1000 km. The gravitational force on an object of mass 1.00 kg, due to the planet, is 1.00 N on the surface of the planet. Question 10 introduces a satellite (moon) of the dwarf planet with a radius of 100 km, of the same density as the planet and with an orbital radius of 20 000 km.

7. Calculate:

(a) the mass and mean density of the dwarf planet
(b) the acceleration due to gravity at the surface
(c) the gravitational potential at the surface
(d) the gravitational potential energy of a 100 kg object on its surface.

8. Calculate the gravitational potential and field strength:

(a) 5000 km from the centre and
(b) 1000 km above the surface of the dwarf planet.

9. An object of mass 10 kg is raised 100 km from the surface.

(a) Calculate the increase in gravitational potential energy using the uniform field approximation, $\Delta U = mg\Delta h$.
(b) Calculate the increase in gravitational potential energy using the equation derived from Newton's law of gravitation.
(c) Calculate the percentage error obtained by using the uniform field approximation.

10. For the satellite:

(a) Calculate the gravitational field strength at its surface.
(b) Calculate the gravitational potential at its surface due to:
   (i) the satellite alone, and
   (ii) the two bodies (i.e. dwarf planet and satellite).
(c) Discuss briefly to what extent the position on the satellite affects the answer to (b)(ii).

11. The satellite is in a circular orbit around the dwarf planet.

(a) Determine the gravitational acceleration due to the dwarf planet at the position of the satellite.
(b) By equating this to the centripetal acceleration of the satellite in its orbit, determine the period of the orbit. Give your answer in **days** (1 day = 86 400 s).

4.3 Orbits and the wider universe

Section 4.2 discussed Newton's law of gravitation. The investigation into the properties of gravity and gravitational mass and the motion of bodies under their influence has led to many important developments in our understanding of the universe. The story of Johannes Kepler's[1] development of these revolutionary laws, which not only debunked the philosophical notion that celestial (and therefore perfect) objects could only travel in perfect (i.e. circular) paths but also dethroned the Earth from its central position in the universe, is well worth looking into. The observational and experimental work of Galileo and mathematical work of Newton would eventually explain them and consolidate the idea that the whole of the universe was subject to the same physical laws.

Fig. 4.3.1 Kepler's portrait on a German postage stamp

4.3.1 Kepler's laws of planetary motion

Using the very accurate observations of the Danish astronomer, Tycho Brahe, Kepler was able to state three laws which sum up how planets move relative to the Sun. These were purely empirical laws, i.e. they did not explain the motion but they described it to a much greater precision than Copernicus. In fact, they also apply to comets, dwarf planets, objects further than Neptune from the Sun (that is trans-Neptunian objects (TNOs)) and asteroids, etc., i.e. all objects in orbit around the Sun; they also apply separately to objects in orbit around planets.

(a) Kepler's 1st and 2nd laws

As with Newton's laws we shall refer to Kepler's laws as K1, K2 and (later) K3. The first two laws can be written as:

K1. The orbit of a planet is an ellipse with the Sun at one focus.

K2. The line joining the planet to the Sun sweeps out equal areas in equal intervals of time.

Maths tip

See the Maths section for the terminology and mathematical properties of ellipses.

For most planets it is difficult to tell that the orbit is not circular. Mars has the most eccentric of any of the easily observed planets; its path drawn in Fig. 4.3.2 departs from circularity by about the thickness of the line!

In Fig. 4.3.2, the yellow and black discs represent the Sun and Mars (not to scale) and the red pecked line, the orbit of the Earth. The Sun is at F_1, one of the foci of the ellipse – the other focus is shown by the small grey circle.

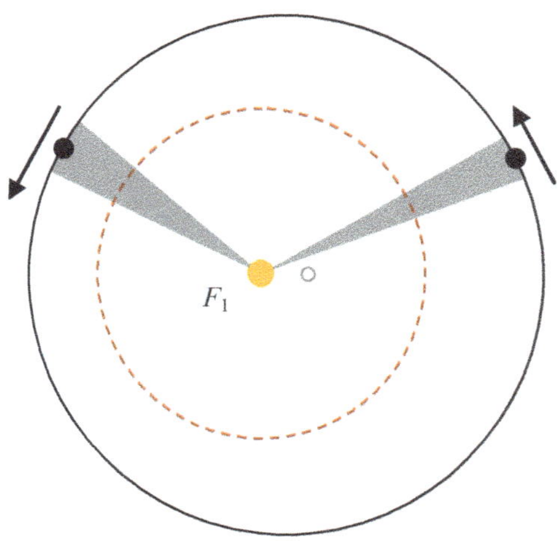

Fig. 4.3.2 Orbit of Mars

Study point

When applying Kepler's laws to moon systems, we replace *planet* with *moon* and *Sun* with *planet*. Thus:

The orbit of a moon is an ellipse with the planet at one focus.

Study point

Note that K2 refers to an individual planet (comet, TNO…) in its orbit. It doesn't compare the areas swept out by different bodies. The movement of different objects in orbit around the same central body is the subject of Kepler's third law (K3).

1 He was a mathematician, astronomer and astrologer who successfully defended his mother from a charge of witchcraft.

The elliptical nature of cometary orbits is much more obvious.

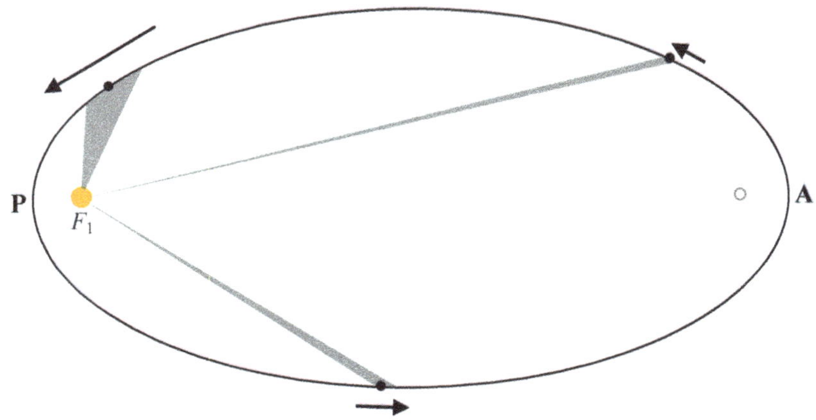

Fig. 4.3.3 Cometary orbit

The grey areas in the two figures illustrate K2. They represent equal time slices in the orbits. The comet in Fig. 4.3.3 travels much faster when it is close to the Sun, near its **perihelion**, P, than near its **aphelion**, A, but the areas swept out are the same.

Example

Comet 67P/Churyumov-Gerasimenko has a speed at perihelion of 34.23 km s^{-1} when its distance from the Sun is 1.2432 AU. The length of the semi-major axis of its orbit is 3.4630 AU. Using Kepler's 2nd law, calculate:

(a) the area swept out per second by the line joining it to the Sun

(b) its speed at aphelion.

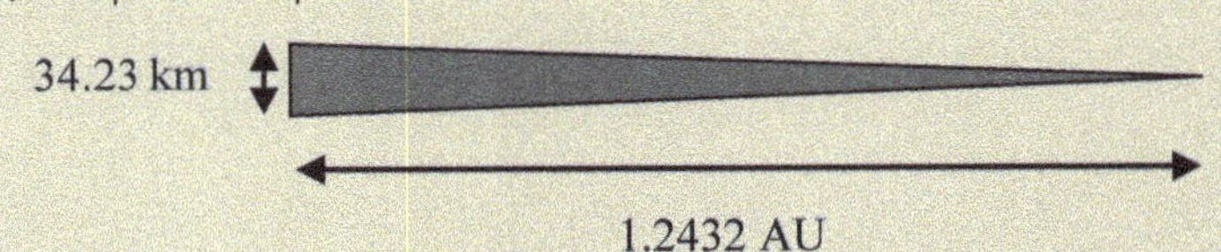

Answer

(a) The diagram shows the area swept out per second at perihelion.

$1.2432 \text{ AU} = 1.2432 \times 1.496 \times 10^8 \text{ km} = 1.86 \times 10^8 \text{ km}$

$\therefore$ Area swept out per second $= \frac{1}{2} \times 34.23 \times 1.86 \times 10^8 = 3.18 \times 10^9 \text{ km}^2 \text{ s}^{-1}$

(b) Distance $AF_1 = AP - PF_1$ (See Fig. 4.3.3)

$= 2 \times 3.4630 - 1.2432$

$= 5.6828 \text{ AU} = 8.50 \times 10^8 \text{ km}$

From K2: Area per second is constant, $\therefore 3.18 \times 10^9 = \frac{1}{2} \times 8.50 \times 10^8 \, v$

$\therefore$ Velocity at aphelion, $v = 8.3 \text{ km s}^{-1}$

Study point

The major axis of an ellipse is the longest diameter, i.e. the line joining P and A in Fig. 4.3.3. The length of the semi-major axis, $\frac{1}{2}$PA, plays a large role in orbit theory. It is equivalent to the radius of a circular orbit.

4.3.1 Knowledge check

Using your ruler to measure the maximum and minimum distances of the comet from the Sun, compare the gravitational potential energy of the comet at P and A.

Study point

Many calculations in astronomy are carried out in **astronomical units** (AU).

1 AU = semi-major axis of Earth's orbit

$= 1.496 \times 10^{11}$ m

The year, 3.156×10^7 s, is often used as a unit of time.

Another unit of distance is the light-year (ly) $= 9.46 \times 10^{15}$ m.

4.3.2 Knowledge check

The mass of 67P/Churyumov-Gerasimenko has been estimated as 1.0×10^{13} kg. Use this to calculate:

(a) the gravitational potential energy

(b) the kinetic energy

(c) the total mechanical energy of the comet at perihelion.

(Mass of Sun $= 1.99 \times 10^{30}$ kg)

Stretch & challenge

Assess whether the data in the example are consistent with the Principle of Conservation of Energy.

(b) Kepler's 3rd law

This law compares the motion of different bodies in orbit around the same central body.

K3. The period, T, of the orbit of a planet and the length, a, of its semi-major axis are related by $T^2 \propto a^3$.

This law can be used to compare the orbital motion of all the planets, dwarf planets, asteroids, comets and TNOs, which orbit the Sun. It can also be used for a family of extra-solar planets around an individual star or the satellites of a planet such as Jupiter. It *cannot* be used to compare objects which orbit different bodies, e.g. the moons of Saturn and Uranus.

Hence, for any two bodies, 1 and 2, orbiting the same central object, we can write:

$$\frac{T_1^2}{T_2^2} = \frac{a_1^3}{a_2^3} \quad \text{or} \quad \frac{T_1}{T_2} = \left(\frac{a_1}{a_2}\right)^{\frac{3}{2}}$$

This tells us that, as we might expect, the period increases with mean distance from the Sun (i.e. the semi-major axis) but more rapidly than a proportional relationship.

Example

The mean distance of the Earth from the Sun (the semi-major axis of its orbit) is 1.496×10^{11} m. Saturn's orbital distance is 1.427×10^{12} m. Show that the orbital period of Saturn is approximately 30 years.

Answer

From K3: $\frac{T^2}{a^3}$ = constant. The period of the Earth's orbit is 1 year.

$\therefore \frac{1^2}{(1.496 \times 10^{11})^3} = \frac{T_S^2}{(1.427 \times 10^{12})^3}$, where T_S = period of Saturn (in years).

$\therefore T_S = \left(\frac{14.27}{1.496}\right)^{\frac{3}{2}} = 29.46 \approx 30$ years

Study point

It follows from K3 that the period of a comet or planet doesn't depend on how long and thin its orbit is – just the length of the semi-major axis. Thus Comet 67P in the example opposite has almost the same period (6.5 years) as the asteroid Camilla which has an almost circular orbit of radius 3.49 AU.

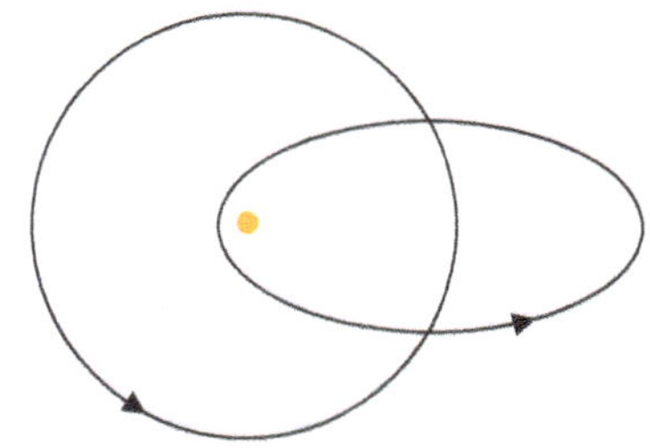

Fig. 4.3.4 Equal period orbits

Knowledge check 4.3.3

What does K3 imply about how the mean speed of an orbiting body depends upon the mean distance from the Sun?

Top tip

Any units of time and distance can be used when applying Kepler's 2nd law in the form $\frac{T_1^2}{a_1^3} = \frac{T_2^2}{a_2^3}$; e.g. distance in AU and time in years.

4.3.2 Newton's laws and orbital motion

To a very good approximation, bodies in orbit around the Sun move under the influence of just one force; that is the Sun's gravity. The gravitational force of other Solar System objects can usually be ignored because their masses are so much smaller than the Sun; the interplanetary medium exerts a negligible drag and the pressure of sunlight is miniscule.[2] In these circumstances, as Newton showed, objects move in paths which are elliptical, parabolic or hyperbolic. This section explores how Newton explained Kepler's laws in terms of his principles of force, mass and acceleration together with his universal law of gravitation, which we met in Section 4.2. We'll start with Kepler's 3rd law.

(a) Kepler's 3rd law – distance / period relationship

The proof of this law for an elliptical orbit is beyond the scope of A-level physics and maths. However, for a circular orbit, it builds on the work we have done on centripetal acceleration (Section 3.1.4) and Newton's law of gravitation (Section 4.2.6) and N2. **You need to learn this derivation.**

Knowledge check 4.3.4

The period of Mercury's orbit is 88 days. Calculate the semi-major axis of its orbit. Express your answer in AU.

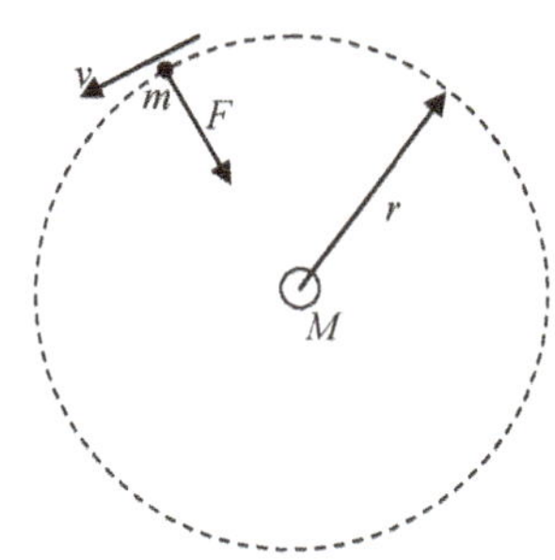

Fig. 4.3.5 Force on an orbiting body

2 In fact all these three forces can exert influences in some circumstances, e.g. when a comet approaches a planet, drag in the early stages of the formation of the Solar System and the effects of sunlight causing outgassing of comets.

Study point

An alternative derivation uses the angular velocity, $\omega = \frac{2\pi}{T}$.

From Section 3.1, the centripetal force is $mr\omega^2$. Equating this to the gravitational force gives

$mr\omega^2 = G\frac{Mm}{r^2}$, $\therefore \omega^2 = \frac{GM}{r^3}$.

Hence $T^2 = \frac{4\pi^2 r^3}{GM}$.

4.3.5 Knowledge check

How does the derivation of K2 depend upon Newton's laws of motion?

4.3.6 Knowledge check

The semi-major axis of the orbit of Miranda about Uranus is 129 400 km and the period is 1.413 days. Calculate the mass of Uranus.

$G = 6.67 \times 10^{-11}$ N m^2 kg^{-2}.

Fig. 4.3.5 shows an object in a circular orbit and introduces the symbols for the quantities involved.

The resultant force, F, on an object moving at constant velocity in a circle is directed towards the centre of the circle and is given by:

$$F = \frac{mv^2}{r} \qquad [1]$$

The only force on the object is the gravitational force $\frac{GMm}{r^2}$, so we can write

$$\frac{mv^2}{r} = \frac{GMm}{r^2}$$

$\therefore$ Simplifying: $v^2 = \frac{GM}{r}$. [2]

The velocity is related to the period, T, of the orbit by $v = \frac{2\pi r}{T}$.

Substituting for v in equation [2] and rearranging gives

$$T^2 = \frac{4\pi^2 r^3}{GM} \qquad [3]$$

In fact, as we mentioned in the Study point in Section 4.3.1(b), the orbital period turns out not to depend on the eccentricity of the elliptical orbit – just the length of the semi-major axis. Equations [2] and [3] give us very powerful tools for investigating the mass of distant objects – by measuring the orbital periods of bodies in orbit around them. We'll explore this further in Sections 4.3.3 and 4.3.4. We can also use it to work out the orbital energy in a system, as the example shows.

Example

A space probe of mass m orbits a planet of mass M. The orbital radius is r. Show that the energy needed for the space probe to escape the planet's gravitational field is $\frac{1}{2}\frac{GMm}{r}$

Answer

The gravitational potential energy, $E_P = -\frac{GMm}{r}$.

The kinetic energy, $E_k = \frac{1}{2}mv^2$; But $\frac{mv^2}{r} = \frac{GMm}{r^2}$, so $E_k = \frac{1}{2}\frac{GMm}{r}$

$\therefore$ Total energy $= E_k + E_P = \frac{1}{2}\frac{GMm}{r} - \frac{GMm}{r} = -\frac{1}{2}\frac{GMm}{r}$

$\therefore$ Energy required to escape $= \frac{1}{2}\frac{GMm}{r}$

The result of the example is quite interesting. It's also worth remarking that the energy of a body in an elliptical orbit of semi-major axis a is the same as if it were in a circular orbit of radius a. See Stretch & challenge.

(b) Kepler's 1st and 2nd laws

A derivation of these laws using Newton's law of motion is beyond the scope of this book but we'll point the way that Newton himself used to derive K2, leaving the details to a Stretch & challenge. The derivation only makes use of the fact that the force on the planet is always directly towards the star.

Fig. 4.3.6 shows a planet, **P**, moving near a star, **S**, with a velocity, u. Using the formula for the area of a triangle, in the short time Δt, as it approaches, the planet travels a distance $u\Delta t$, so the radius vector sweeps out an area $\frac{1}{2}uh\Delta t$. This means that the area swept out per second is $\frac{1}{2}uh$. Now we'll leave you to show in the Stretch & challenge that this area per second stays constant, just using one of Newton's laws of motion:

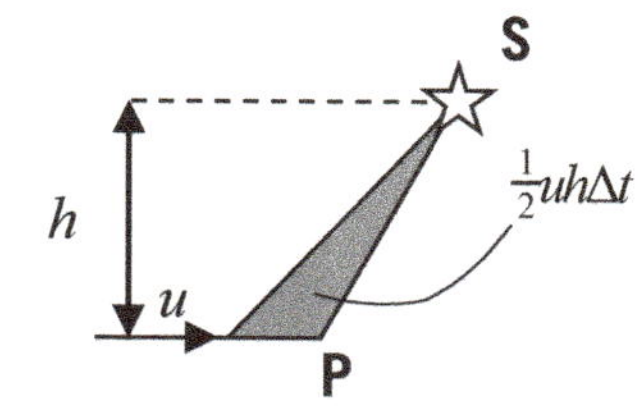

Fig. 4.3.6 Derivation of K2

Stretch & challenge

This is how Newton derived it: The star gives a tug to the planet towards it – along the dotted line **PS** – so that it changes direction and speed, as shown in Fig. 4.3.7. The tug changes the velocity by adding a velocity w to the planet towards S.

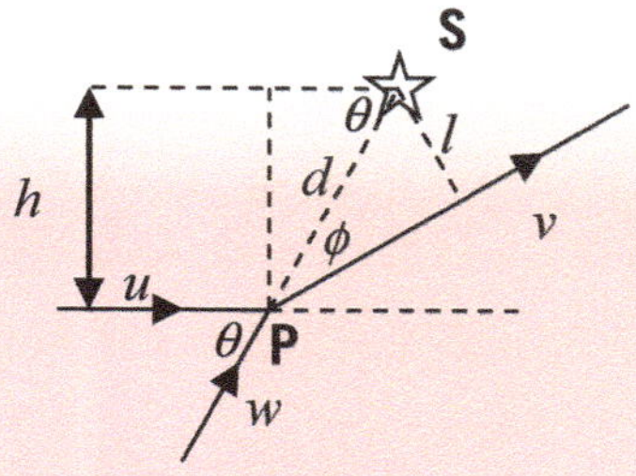

Fig. 4.3.7 Derivation of K2

1. Use the triangles including d to show that h and l are related by:

$$\frac{h}{\sin\theta} = \frac{l}{\sin\phi}$$

2. The component of velocity perpendicular to PS must be constant [N1]. Use this to derive a similar relationship between u and v.
3. Use the two equations to show that $\frac{1}{2}vl = \frac{1}{2}uh$, hence the area swept out per second is the same.

Newton then argued that a constant force towards the star could be regarded as lots of small impulses, after each of which the area per second was the same.

Kepler's 1st law is beyond the scope of A-level physics and maths to prove. It needs Newton's law of gravitation, plus a knowledge of differential equations. Suffice it to say that Newton showed that the planets move in ellipses – failing only to account for a slight departure in the case of Mercury, which was only solved by Einstein with his General Theory of Relativity. Objects with a high enough energy to escape from the solar system, have parabolic or hyperbolic orbits.

(c) The energy of orbiting bodies

The total mechanical energy of bodies which orbit in closed elliptical orbits is negative. The reason for this is that, by definition, the potential energy of an object an infinite distance from a gravitating body is zero. Thus bodies with a positive total mechanical energy would be able to escape to infinity.

First, we'll consider a body in a circular orbit, as in Fig. 4.3.5. The total mechanical energy, E, is given by:

$$E = \text{kinetic energy} + \text{gravitational potential energy}$$
$$= \frac{1}{2}mv^2 - \frac{GMm}{r}$$

From equation [2] in Section 4.3.2(a): $v^2 = \frac{GM}{r}$. Substituting for v^2 gives

$$E = \frac{1}{2}\frac{GMm}{r} - \frac{GMm}{r}$$

So the total mechanical energy of an object in a circular orbit is $-\frac{1}{2}\frac{GMm}{r}$.

Now we'll consider elliptical orbits. Remembering that the period of an elliptical orbit depends only on its semi-major axis (i.e. it is independent of the eccentricity – see Fig. 4.3.5 again) it should be no surprise that the total energy of an orbiting object also depends only on the semi-major axis, a. This means that we can write:

$$\text{Energy of an object in an elliptical orbit} = -\frac{1}{2}\frac{GMm}{a}.$$

The proof of this requires a bit of algebra and you do not need to know it for the examination. See Stretch & challenge.

Stretch & challenge

The diagram shows an object in an elliptical orbit. Show that its total energy is given by $E = -\frac{1}{2}\frac{GMm}{a}$.

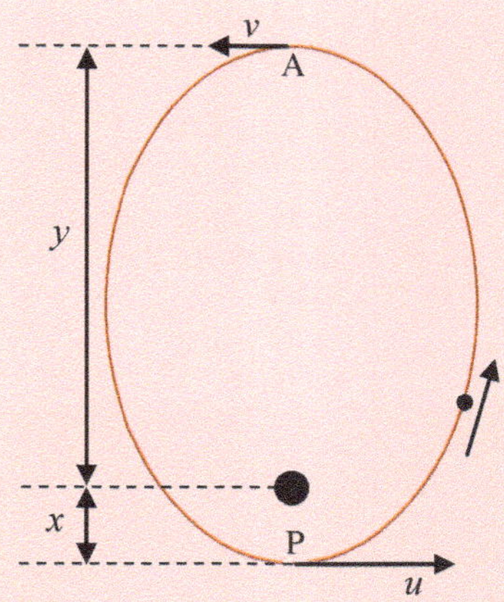

Hint: The energies at **A** and **P** are equal. Express the total energy in terms of u and x and separately in terms of v and y. Use K2 to eliminate one of the velocities and hence find the total energy in terms of $a = \frac{(x+y)}{2}$.

Knowledge check 4.3.7

Use the data in Knowledge check 4.3.2 and the value of the semi-major axis to calculate the orbital energy of the comet 67P/Churyumov-Gerasimenko.

Compare the two answers.

4.3.3 Using orbits to determine the mass of objects

In this section we'll look first at orbits around compact bodies, e.g. planets, stars and black holes. Then we'll examine the orbits of stars in galaxies.

Study point

Newton's law of gravitation is valid for point objects and also for spherically symmetric objects. Thus the orbit equations work well for stars and planets.

(a) Orbits around compact bodies

In Section 4.3.2(b) we obtained the following equations:

$$T^2 = \frac{4\pi^2 a^3}{GM} \quad [2] \qquad \text{and} \qquad v^2 = \frac{GM}{r} \quad [3]$$

We can apply equation [2] to any elliptical orbit. Equation [3] is only applicable to circular orbits but inserting the semi-major axis, a, instead of r will enable us to calculate an average figure of orbital speed [strictly the rms speed]. Hence, measuring the radius (or semi-major axis) of an orbit and either the orbital period or speed enables astronomers to determine the mass of the central object star or planet. If the diameter of the central body is known, its mean density can also be calculated, as in the following example. This enables planetary scientists to develop models of planetary composition.

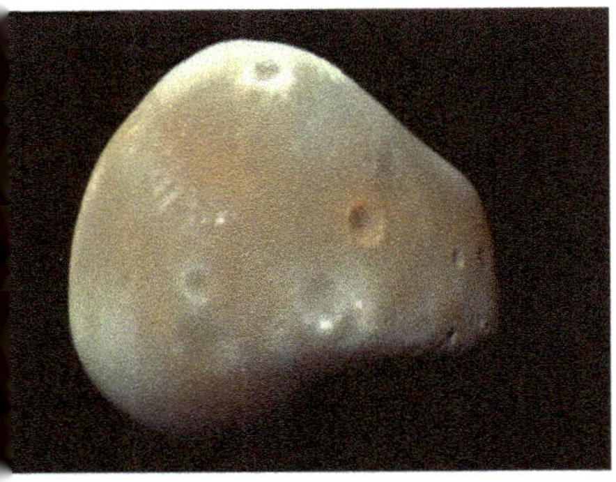

Fig. 4.3.8 Martian moon Deimos

Example

Deimos, the smaller of the two moons of Mars, has an orbit with a semi-major axis of 23 460 km and a period of 30.35 hours. The diameter of Mars is 6750 km. Calculate the mean density of Mars.

Answer

The mass, M, of Mars is given by $M = \frac{4\pi^2 a^3}{GT^2} = \frac{4\pi^2(23\,460 \times 10^3)^3}{6.67 \times 10^{-11} \times (30.35 \times 3600)^2}$

$= 6.40 \times 10^{23}$ kg

Density $\rho = \frac{M}{V} = \frac{M}{\frac{4}{3}\pi r^3} = \frac{3 \times 6.40 \times 10^{23}}{4\pi(3375 \times 10^3)^3} = 3970 \text{ kg m}^{-3}$

4.3.8 Knowledge check

The radius of the orbital of Phobos, the larger moon of Mars, is 9380 km.

Using the mass of Mars calculated in the example, determine the mean orbit speed of Phobos.

This figure for the density of Mars is much lower than the mean density of the Earth, which is 5520 kg m^{-3}. This suggests that Mars has proportionately much less iron than the Earth does. Also the absence of a significant magnetic field suggests that any iron core is completely solid. This fits in with the fact that small objects (Mars is about half the diameter of the Earth) cool more rapidly than large ones.

Fig. 4.3.9 Sagittarius A*

A more exciting example of the power of orbits to reveal mass is in the nature of the powerful radio source, known as Sagittarius A* (Sgr A*), at the centre of our galaxy – see Fig. 4.3.9. Various stars in the vicinity of Sgr A* have been tracked over the past 20 years. The positions of one of them, known as S2, are shown schematically in Fig. 4.3.10. The crosses are error bars; the circles have error bars which are too small to plot. Analysis of the data reveals an object of several million solar masses within a space of the order of the size of the solar system. The only possible object with this mass and size is a super-massive black hole. In the last few decades it has become clear, that virtually all spiral galaxies have super-massive black holes in their cores.

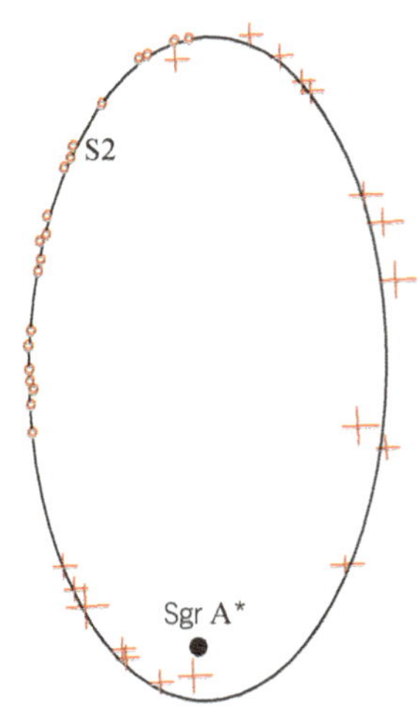

Fig. 4.3.10 The orbit of S2 around Sgr A*

Study point

The calculation of the mass of Sgr A* is in Test yourself 4.3.

(b) The rotation curves of the Solar System

A plot of orbital speed for objects circling the Sun is shown in Fig. 4.3.11. This sort of graph is called a **rotation curve**. The equation of the curve is

$$v = \sqrt{\frac{GM}{r}}$$

where M is the mass of the Sun, 1.99×10^{30} kg

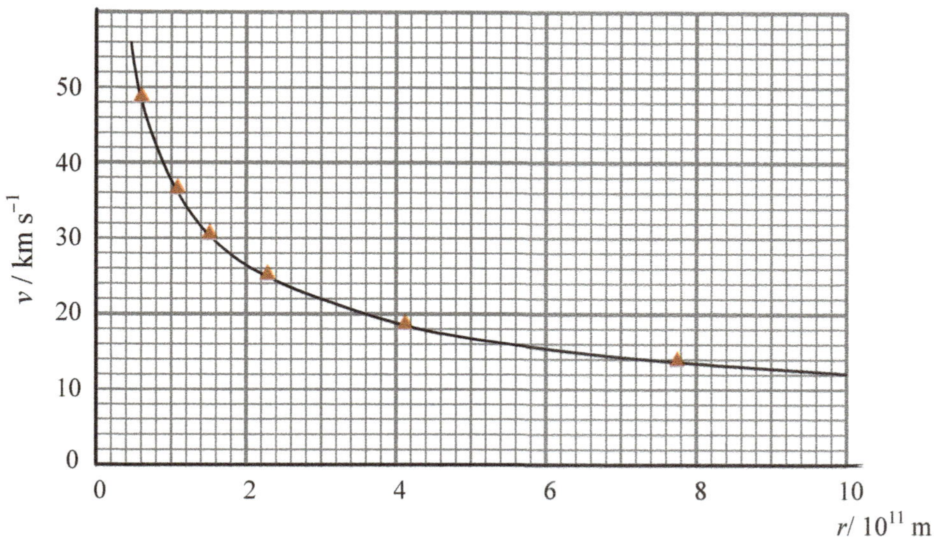

Fig. 4.3.11 Solar System rotation curve

The rotation curve has this simple form, in agreement with equation [3], because:

- the Sun contains the great bulk of the mass in the Solar System,
- the Solar System objects orbit outside the Sun and
- the Sun is almost spherically symmetric.

These features of the Solar System mean that we can treat the motion of any orbiting body as being subject only to the gravitational force due to a point mass in the centre of the Sun.

(c) The rotation curve of spiral galaxies

The stars in our galaxy orbit the centre of the galaxy. The image in Fig. 4.3.12 gives an idea of the appearance of our galaxy seen from 'above' the galactic plane and the position and the circular orbit of the Sun: radius ~2.7×10^{17} km, orbital speed ~220 km s^{-1}. Measuring the orbital speed of stars and gas clouds within our own galaxy is not easy because the Solar System itself is in orbit, but it has been done for a large number of other spiral galaxies using the Doppler effect (see Section 4.3.4). The result for a very similar galaxy called NGC 3198 is shown in Fig. 4.3.13.

Study point

The red triangles in Fig. 4.3.11 represent Mercury, Venus, Earth, Mars, Ceres and Jupiter.

Knowledge check 4.3.9

For Solar System data, what would be the gradients and y-intercepts of the following plots?

(a) v against $\frac{1}{\sqrt{r}}$

(b) v^2 against $\frac{1}{r}$

(c) $\ln v$ against $\ln r$.

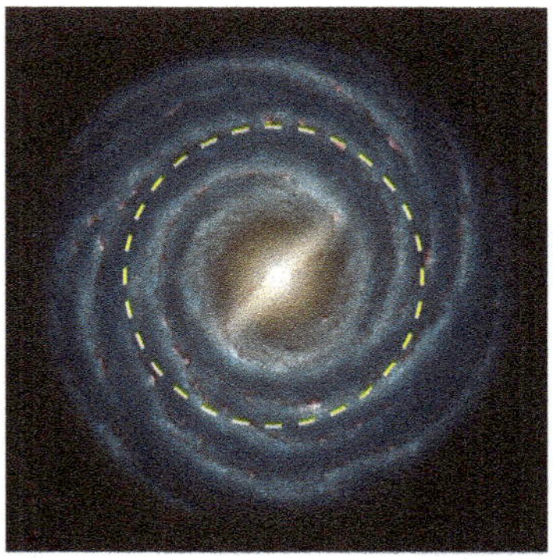

Fig. 4.3.12 The Sun's orbit about the centre of the galaxy

Study point

Astronomers usually express distance in *parsecs* (pc) rather than the more familiar light-years. A parsec is the distance at which the radius of the Earth's orbit subtends an angle of 1" $\left(\text{i.e. } \frac{1}{3600}^\circ\right)$. Distances within a galaxy are often expressed in kpc and intergalactic distances in Mpc or Gpc.

$1 \text{ pc} = 3.09 \times 10^{16} \text{ m}$

$1 \text{ kpc} = 3.09 \times 10^{19} \text{ m}$, etc

4.3.10 Knowledge check

(a) Express the distance of the Sun from the centre of the galaxy in light-years.

(b) Express the distance of the Sun from the centre of galaxy in parsecs.

Study point

Strictly speaking, the material outside a circular orbit can only be ignored for a spherically symmetric mass distribution. Clearly, the visible mass of a spiral galaxy is concentrated in a plane but, even so, ignoring extraneous mass should give quite a close result.

4.3.11 Knowledge check

Check the 160 km s^{-1} figure for the Sun's orbital speed, for a mass within the orbit of 1.0×10^{41} kg.

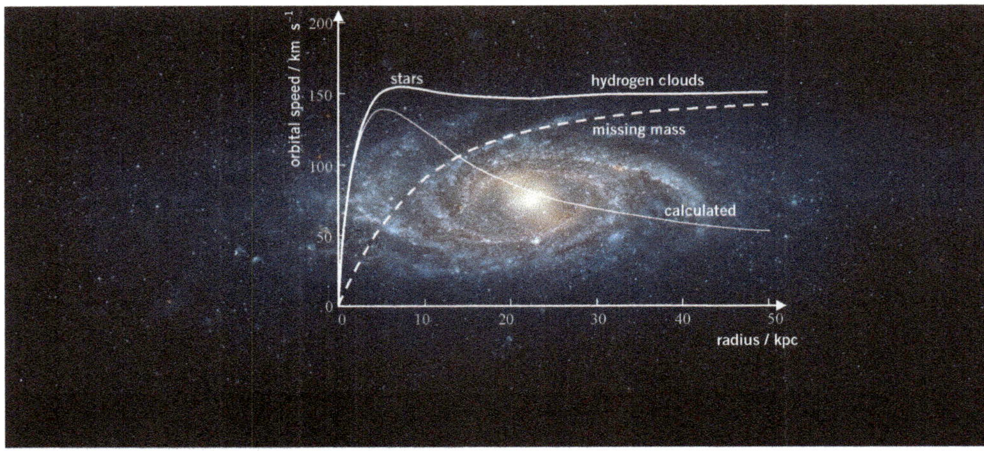

Fig. 4.3.13 Rotation curve of NGC 3198

The upper solid curve is the measured orbital speeds of objects. The inner part of the curve is obtained from the motion of stars within the visible disc of the galaxy, and the outer parts from the motion of hydrogen clouds. The galaxy extends a long way beyond its visible disc: there are large numbers of hydrogen clouds which are invisible but which emit characteristic 21 cm radiation (see AS/Year 12 book Section 1.6.5) the speed of which can be measured by radio astronomers, again using their Doppler shifts.

We saw in Section 4.2.6 that the gravitational effect of a spherically-symmetric mass is the same as that due to the same mass concentrated at its centre, as long as we are outside the mass (Fig. 4.2.15). So the Moon orbits the Earth just as it would an Earth-mass black hole. Also an object inside a uniform spherical shell of mass feels no gravitational effect due to the shell. We'll use features of the orbit of the Sun around the centre of the Galaxy (speed 220 km s^{-1}, radius $2.7 \times 10^{17} \text{ km}$) to determine the total mass within the orbit.

Rearranging $v^2 = \frac{GM}{r}$ gives $$M = \frac{rv^2}{G}$$

So, with the data converted to m and m s^{-1}:

$$\text{Mass within the Sun's orbit} = \frac{2.7 \times 10^{20} \text{ m} \times (220 \times 10^3 \text{ m s}^{-1})^2}{6.67 \times 10^{-11} \text{kg}^{-1} \text{ m}^3 \text{ s}^{-2}}$$

$$= 2.0 \times 10^{41} \text{ kg}$$

Missing mass

For decades, astronomers have made measurements of the mass of all the observable objects within the galaxy – stars, interstellar gas clouds, the central black hole. The maximum figure they can obtain for the mass of these objects within the orbit of the Sun is 1.0×10^{41} kg. They can only see half of the mass that's needed to hold the Sun in its orbit. And the further out we go from the centre, the bigger this discrepancy becomes. If this were the only matter towards the centre of the galaxy, the Sun would fly off into intergalactic space, along with all the other stars outside the central bulge of the galaxy. This appears to be true of all galaxies. Clusters of galaxies also apparently contain too little matter to hold themselves together.

Because galaxies are stable we know that the mass must be there. Astronomers often call this *missing mass*. Perhaps *missing matter* would be a better name – the mass is there but we cannot see the matter!

But just a minute: our calculation of mass is only valid if the galaxy is spherically symmetric. The visible disc certainly is not. However, in the next section we shall see that the missing matter, which accounts for most of the mass of the galaxy, is indeed spherically distributed.

Fig. 4.3.14 Coma galaxy cluster: gravitationally unstable without dark matter

(d) Dark matter holds galaxies together

The matter that the astronomers detect and 'weigh' is the normal everyday stuff we are familiar with. It is composed of quarks and leptons (with the addition of a tiny contribution from photons). By far the greatest contribution to this mass comes from protons and neutrons, which are baryons; so we call this matter **baryonic matter**. Apart from neutrinos, baryonic matter interacts with electromagnetic radiation – either emitting it or absorbing it – which we can detect with our multiwavelength astronomy.

The lowest curve in Fig. 4.3.13 represents the theoretical rotation curve calculated on the basis that the only matter in the galaxy is the observed baryonic matter. This curve drops off as roughly $r^{-0.5}$ beyond the visible galaxy. But the observed rotation curve shows an almost constant orbital speed in this region. Hence, assuming that Newton's law of gravitation is correct, there is a large amount of missing matter. How much? At 50 kpc the 'calculated' orbital speed is about one third of the observed value. But $M \propto v^2$, so the total mass within 50 kpc of the centre is about 9 times the observed mass, i.e. we are missing nearly 90% of the mass of the galaxy.

The entire galaxy appears to be embedded in a vast spherical cloud of matter, which is only detectable by its gravitational effect. It does not interact by the strong, weak or electromagnetic interactions. This matter is referred to as **dark matter**. Fig. 4.3.15 illustrates this. Notice that the density of the dark matter cloud decreases with distance.

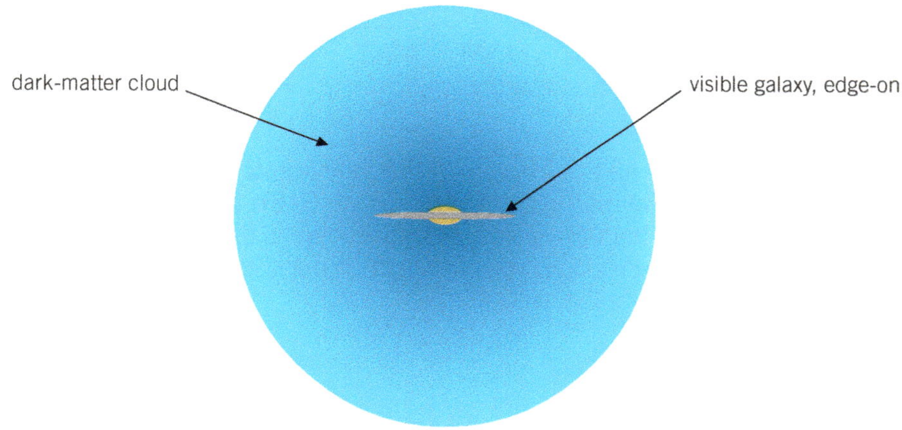

Fig. 4.3.15 Galaxy embedded in a dark matter cloud

Study point

At the time when the WJEC specification was being written, it was suggested that the Higgs boson could be the source of dark matter. The Higgs normally decays into bosons (such as the W or Z particles) or into fermions (such as $\tau^+ + \tau^-$ or $b + b^-$). Swedish theoretician Christopher Petersson has suggested that it could also decay into a photon plus a dark matter particle. This suggestion has now been largely discounted.

Stretch & challenge

1. If $\rho(r)$ is the variation of the density with radius, the mass of a spherical shell of radius r thickness δr is $4\pi r^2 \rho(r)\delta r$.
2. $v^2(r) = \dfrac{GM(r)}{r}$

Stretch & challenge

The 'missing mass' curve in Fig. 4.3.15 is the rotational speed due to the dark matter alone:

(a) Explain why the observed velocity is not equal to the sum of the calculated and dark matter velocities.

(b) What distribution of matter (dark + baryonic) could account for the constant velocity portion of the rotation curve?

4.3.4 Measuring radial velocity using the Doppler effect

The Austrian physicist Christian Doppler (1803–1853) predicted that the observed frequencies of sound waves depend on the motions of the source and observer[3]. He also predicted the same effect with light waves and this effect, together with the Newtonian analysis of Kepler's laws provides a very powerful tool for the investigation of the universe.

Consider the stationary source, **S**, of light waves of in Fig. 4.3.16. The light waves have a frequency f_0 and spread out with the velocity of light, c, producing a set of spherical wavefronts, centred on **S**. A stationary observer in any position will determine the frequency to be f_0 and wavelength to be λ_0, where $c = \lambda_0 f_0$.

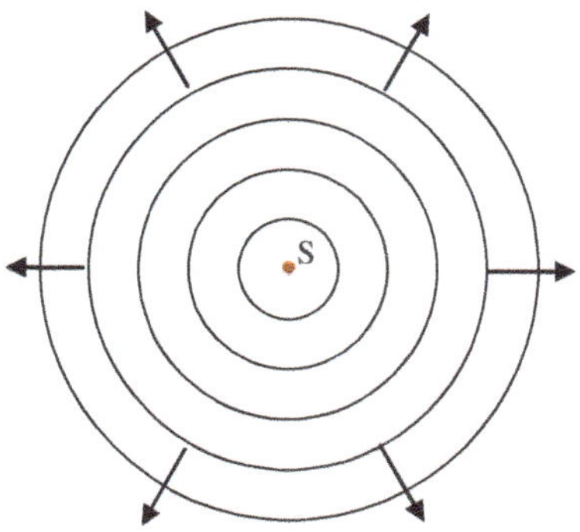

Fig. 4.3.16 Waves from a stationary source

Fig. 4.3.17 shows the same source moving away from the observer with a velocity, v. In the context of astronomy, this is called the **radial velocity**.

Key terms

The **radial velocity** of an object (e.g. a star) is the component of its velocity relative to an observer on the Earth, in the direction from the Earth to the object.

3 This was confirmed in 1845 by the Dutch meteorologist Buys-Ballot in an experiment involving a flat-bed train, a brass band and a group of listeners with perfect pitch! Like all scientists, Doppler didn't always get things right – he tried to explain the different colours of stars by assuming they were moving towards us (blue stars) or away from us (red stars).

Study point

The **radial velocity** of the source in Fig. 4.3.17 is positive. A source moving towards the observer has a negative radial velocity. Such an observer would observe a higher frequency and a shorter wavelength, e.g. if the observer were to the right of the source in Fig. 4.3.17.

Stretch & challenge

Derive the equation $\frac{\Delta f}{f_0} = -\frac{v}{c}$ from $\lambda = \lambda_0 + \frac{v}{c}\lambda_0$.

Hint: You will need to use the binomial approximation:

$(1 + x)^n \approx 1 + nx$ for $x \ll 1$

Study point

These equations are not valid for speeds approaching the speed of light. For more information, search for the 'relativistic Doppler effect'.

4.3.12 Knowledge check

The radial velocity of the Andromeda galaxy (M31) is given on a website as $-301 \pm 1\ \text{km s}^{-1}$.

The laboratory wavelength of Hα is 656.281 nm. Calculate the measured wavelength of Hα emissions from M31, together with its absolute uncertainty.

Study point

Note that the distance is given in km. Remember to convert to m for the calculation.

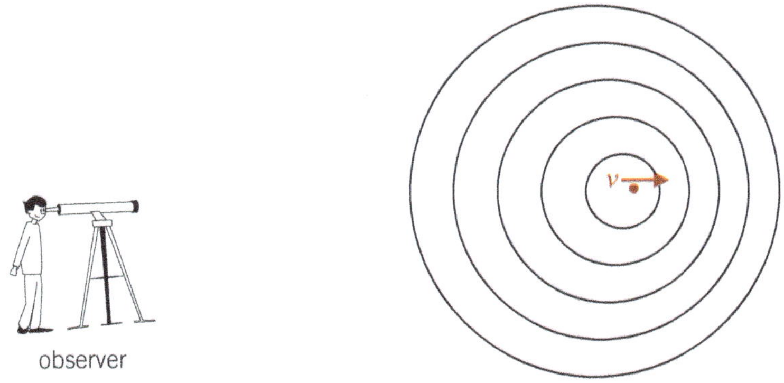

Fig. 4.3.17 Source moving away from observer

The wavefronts are still spherical but their centres move progressively to the right, so the waves which propagate towards the observer have a longer wavelength and therefore a lower frequency. We can derive the magnitude of the effect as follows:

Consider wavefronts which are one cycle apart. The time, T, between the emission of the wavefronts is given by $T = \frac{1}{f_0}$.

In this time the source has moved a distance to the right $= \frac{v}{f_0} = \frac{v}{c}\lambda_0$.

So the observed wavelength, $\lambda = \lambda_0 + \frac{v}{c}\lambda_0$

$\therefore$ The fractional change in wavelength, $\frac{\Delta\lambda}{\lambda_0} = \frac{\lambda - \lambda_0}{\lambda_0} = \frac{v}{c}$.

Similarly, the equation for the fractional frequency change, $\frac{\Delta f}{f_0} = -\frac{v}{c}$, which is valid as long as $v \ll c$.

Many astronomical objects have line spectra, either emission or absorption, and the wavelengths of the lines have been accurately measured in the laboratory (see AS book Section 1.6.4). Measuring the wavelengths (or frequencies) of these lines in the spectra of astronomical objects will tell us their radial velocities. In many cases this will enable us to determine a quantity of gravitating material.

Example

A hydrogen gas cloud is observed 8.5×10^{17} km from the centre of a spiral galaxy which is viewed edge on. The frequency of the 21.1 cm hydrogen line is observed to have a Doppler shift of -0.62 MHz. Estimate the mass of the galaxy within the radius of the gas cloud's orbit.

Answer

Laboratory frequency of 21.1 cm hydrogen line $= \frac{c}{\lambda} = \frac{3.00 \times 10^8}{0.211} = 1420$ MHz.

$\therefore$ Using $\frac{\Delta f}{f_0} = -\frac{v}{c}$, $v = \frac{-c\Delta f}{f_0} = \frac{3.00 \times 10^8\ \text{m s}^{-1} \times 0.62\ \text{MHz}}{1420\ \text{MHz}} = 131\,000\ \text{m s}^{-1}$

Using $v^2 = \frac{GM}{r}$, $M = \frac{rv^2}{G} = \frac{8.5 \times 10^{20}\ \text{m} \times (131\,000\ \text{m s}^{-1})^2}{6.67 \times 10^{-11}\ \text{kg}^{-1}\text{m}^3\text{s}^{-2}} = 2.2 \times 10^{41}$ kg

4.3.5 Bodies in mutual orbit

In many cases, one body in a system is overwhelmingly more massive than all the other bodies put together. In these cases we can do as we have done up to now and assume that the massive body is at rest and the light body (or bodies) orbits around it. For example, Mercury and Venus can be considered to orbit around a Sun which is metaphorically nailed to the spot; the stars orbiting Sgr A* – see Section 4.3.3(a) – don't tug the black hole noticeably out of position. However, there are many pairs of objects of similar mass, such the Pluto / Charon system and binary stars, which orbit each other. In order to analyse these we'll first need to develop the concept of the **centre of mass**.

(a) Centre of mass

We'll start with a thought experiment. The barbell in Fig. 4.3.18 is floating in space, away from gravitational influences, and is hit separately by various impulses, $I_1 - I_5$. Three of these, I_2, I_4 and I_5, cause the barbell to rotate as well as to move. Two of them, I_1 and I_3, only cause it to move in a straight line: I_1 makes it move to the right and I_3 upwards. This is because the lines of action of these two forces pass through the middle of the bar (the red dot). With an asymmetric barbell the 'no-rotation' point, called the **centre of mass (C)**, is shifted towards the more massive ball (Fig. 4.3.19). This time I_3 causes rotation and I_4 does not.

The position of the centre of mass for two small objects is calculated using the equation:

$$m_1 r_1 = m_2 r_2$$

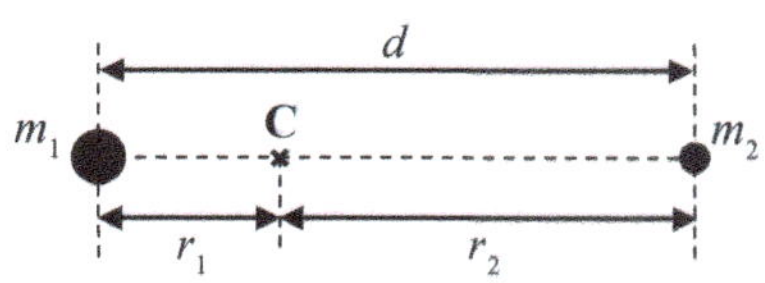

Fig. 4.3.20 Centre of mass

When we investigate binary stars or double planets we shall need to express the position of C in terms of the separation of the bodies. You should be able to show (see KC 4.3.13) that:

$$r_1 = \frac{m_2}{m_1 + m_2} d \quad \text{and} \quad r_2 = \frac{m_1}{m_1 + m_2} d$$

When we study the motion of binary systems, e.g. the orbit of a binary star system, we can consider the motion in two parts:

1 For the system as a whole, we consider the whole mass to be concentrated at the centre of mass.

2 For the motions within the system we consider the motions relative to the centre of mass.

So: the Earth and Moon orbit around their common centre of mass (see KC 4.3.13) and (strictly) it is this centre of mass that orbits in an elliptical orbit the centre of mass of the Earth-Moon-Sun system rather than the Earth orbiting the Sun. We'll now see how to apply this.

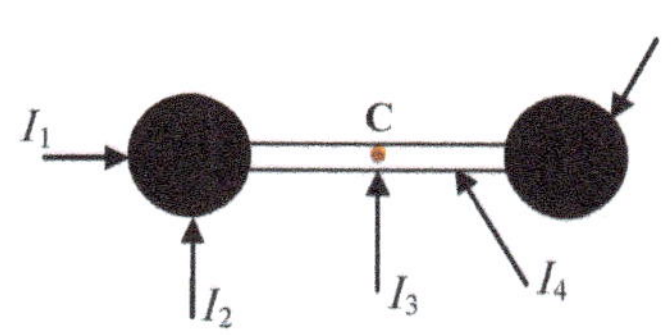

Fig. 4.3.18 Impulses on a barbell

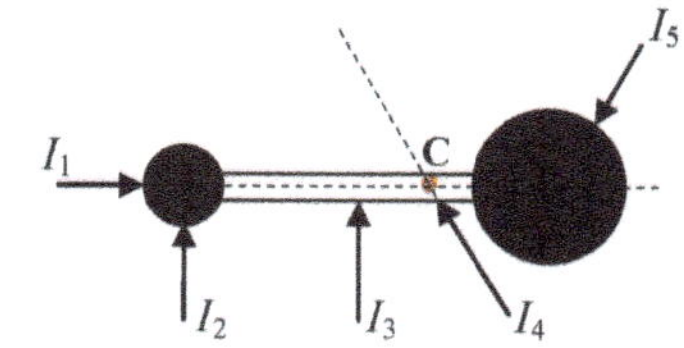

Fig. 4.3.19 Asymmetric barbell

Knowledge check 4.3.13

Use the equations

$m_1 r_1 = m_2 r_2$ and $r_1 + r_2 = d$

to derive the equation

$$r_1 = \frac{m_2}{m_1 + m_2} d.$$

Stretch & challenge

If we have N objects of mass m_i at positions (x_i, y_i), where $i = 1, 2, ...N$, the position of the centre of mass is given by

$$(x, y) = \left(\frac{\sum_{i=1}^{N} m_i x_i}{\sum_{i=1}^{N} m_i}, \frac{\sum_{i=1}^{N} m_i y_i}{\sum_{i=1}^{N} m_i} \right).$$

Show that the equations in the main text are consistent with this.

Knowledge check 4.3.14

The separation of the Earth and Moon is **400 000 km**. The mass of the Earth is **80 times** that of the moon. Find the position of the centre of mass of the Earth–Moon system.

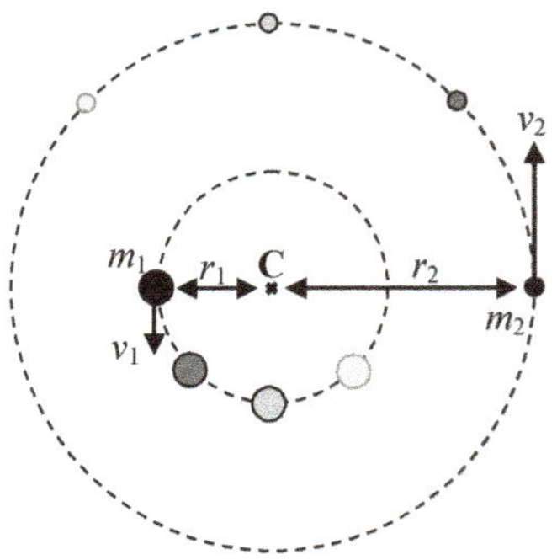

Fig. 4.3.21 Mutual orbit

(b) Orbits of binary objects

We'll consider circular orbits: as usual, the results also apply to elliptical orbits. The two objects from above are in mutual orbit – see Fig. 4.3.21. We are only considering motion within the system, so the centre of mass remains stationary. Starting off with the object in the black position, the velocities v_1 and v_2 must be such that $m_1v_1 = m_2v_2$, otherwise the centre of mass would be moving. Hence each orbits the centre of mass (fading into the future!) with the same period.

To calculate the period of the orbit, consider the motion of m_1:

The centripetal force on m_1 is provided by the gravitational force from the other mass, i.e.

$$m_1r_1\frac{4\pi^2}{T^2} = \frac{Gm_1m_2}{d^2} \quad \text{where we have used} \quad \omega = \frac{2\pi}{T}.$$

Substituting $r_1 = \frac{m_2}{m_1 + m_2}d, \rightarrow \frac{m_1m_2}{m_1 + m_2}d\frac{4\pi^2}{T^2} = \frac{Gm_1m_2}{d^2}$

Dividing by m_1m_2 and rearranging $\rightarrow T = 2\pi\sqrt{\frac{d^3}{G(m_1 + m_2)}} = 2\pi\sqrt{\frac{d^3}{GM_{tot}}}$

Top tip

You do not have to learn the derivation in (b). The equation for T is given in the Data booklet.

Example

A close pair of stars has a separation of 5.0 million km and orbit each other with a period of 2.5 days. Calculate the total mass of the system.

Answer

$T^2 = \frac{4\pi^2d^3}{GM_{tot}}$, so $\quad M_{tot} = \frac{4\pi^2d^3}{GT^2} = \frac{4\pi^2(5.0 \times 10^9\text{ m})^3}{6.67 \times 10^{-11}\text{ kg}^{-1}\text{m}^3\text{s}^{-2} \times (2.5 \times 86\,400\text{ s})^2}$

$= 1.6 \times 10^{30}$ kg

Study point

Notice the relationship between the formulae:

$T^2 = \frac{4\pi^2a^3}{GM}$ and $T = 2\pi\sqrt{\frac{d^3}{GM_{tot}}}$.

More information is needed to find the masses of the individual components of the binary system, e.g. the orbital speed of one of them (see KC 4.3.15).

If the separation of the bodies is not known, other information such as the orbital speeds of both objects also enables the system to be solved. This is explored below.

4.3.15 Knowledge check

The speed of one component of the binary pair in the example is 58 km s^{-1}. Calculate:

(a) the radius of its orbit, and

(b) the masses of the two stars.

4.3.6 Exoplanets and binary stars

An **exoplanet** is a planet which orbits a star other than the Sun. An observer on an exoplanet would not be able to image the Earth directly with current Earth technology because it would be lost in the glare of the Sun. However, there are two ways in which the existence of the planets or invisible low mass companions, such as red or brown dwarfs (stellar-type objects with too low a mass for hydrogen fusion), could be inferred and their properties investigated.

(a) The Doppler technique

The star and the unseen companion are in mutual orbit. The radial velocity of the star changes throughout its orbit; if our line of sight is in the plane of the orbit, the amplitude of the variation of radial velocity is the orbital speed of the star. This can be detected by Doppler shift in the wavelength of the star's Fraunhofer lines.

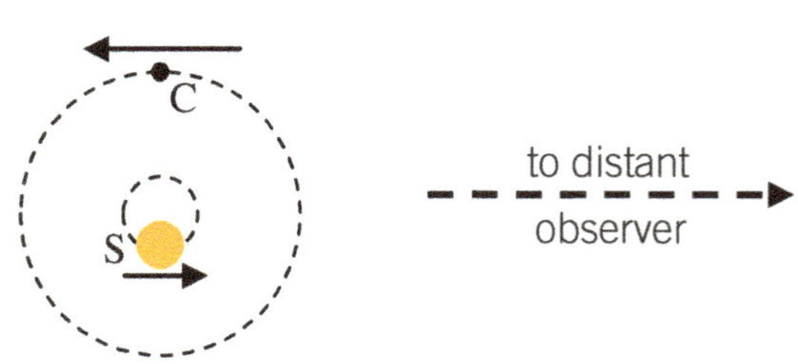

Fig. 4.3.22 Doppler method

4.3.16 Knowledge check

The plane of a star's orbit is at 15° to the line of sight to the Earth. The star's orbital speed is 150 ± 10 km s^{-1}. Compare any error caused by ignoring the angle of inclination with the uncertainty in the speed.

In Fig. 4.3.22 the radial velocity of the star, **S**, is negative, giving a negative $\Delta\lambda$. Half an orbit later $\Delta\lambda$ is positive; the companion, **C**, is detected only by this oscillating Doppler shift.

If the period and amplitude of the wavelength variation are determined, the mass of the planet can be estimated – even without using the *mutual orbit* formula in Section 4.3.5(b). This is illustrated by an example – the details of the calculations are left for KC questions.

Example

The wavelength of the sodium D-line (laboratory value 589 nm) in the spectrum of a star, **S**, of mass 1.0×10^{30} kg was found to vary by ± 0.20 pm with a period of 1.0×10^{6} s, due to the effect of an unseen orbiting companion, **C**. Estimate the orbital distance and mass of the companion.

Answer method

Start by assuming that the mass of the star, **S**, is much greater than that of **C**, so we can use the equations in Section 4.3.3, then:

1 Use $T^2 = \frac{4\pi^2 r^3}{GM}$ to find the orbital distance of **C**, then calculate its orbital speed using $v = \frac{2\pi r}{T}$.

2 Find the speed of **S** using the Doppler shift.

3 Finally use $M_s v_s = m_c v_c$ to estimate the mass of the unseen companion, **C**.

4 Check the initial assumption.

Now work through these steps – see Knowledge check 4.3.17.

Knowledge check 4.3.17

(a) Show that the orbital radius and orbital speed of **C** are 1.19×10^{10} m and 74.7 km s^{-1} respectively.

(b) Calculate the orbital speed of **S**.

(c) Calculate the mass of **C** and compare it to the mass of **S**.

Knowledge check 4.3.18

Alternative method (very similar):

(a) Calculate the orbital radius of **C**.

(b) Use the Doppler shift to calculate the speed, v_s, of **S** and therefore its orbital radius.

(c) Calculate m_c using $M_s r_s = m_c r_c$.

You should have found the mass of the companion to be about 1.4×10^{27} kg, making the star ~$700\times$ as massive. This makes the original assumption reasonable. It is worth checking that, putting in this value of m_C and using the given values of M_S and T give a value of $\Delta\lambda$ which is indistinguishable from the measured value.

If the two objects in a binary system, such as Pluto and Charon, have comparable masses, it is helpful to have extra information such as the orbital velocity of the two objects. The graphs in Fig. 4.3.23 are typical.

The whole system is moving away from us with a velocity v, derived from the Doppler shift. The two components, A and B, have orbital velocities given by $v_A = v_1 - v$ and $v_B = v_2 - v$ (assuming we are in the orbital plane). The masses of the two components can be determined from these and the orbital period, T – see the Study point and KC 4.3.19.

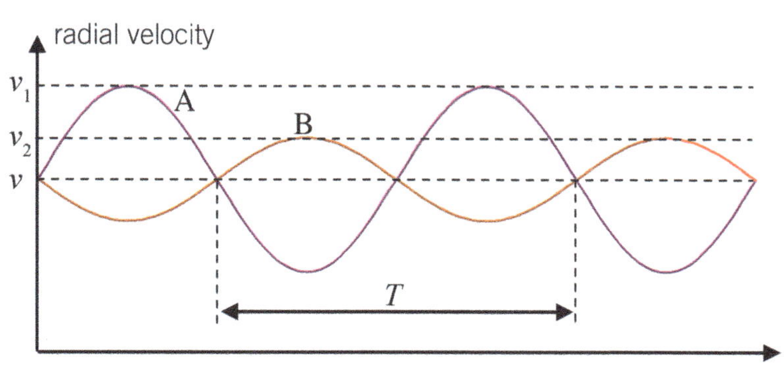

Fig. 4.3.23 Objects in mutual orbit

Study point

The radii of the orbits of components **A** and **B**, and hence their total separation, d, can be found using $v_A = \frac{2\pi r_A}{T}$ and $v_B = \frac{2\pi r_B}{T}$.

Then $T = 2\pi\sqrt{\frac{d^3}{GM_{tot}}}$ gives the total mass of the system; the individual masses can be found by comparing the orbital speeds – see Knowledge check 4.3.18

Knowledge check 4.3.19

The following data (see Fig. 4.3.23) are obtained for a binary star system.

$v = 110$ km s^{-1}; $v_1 = 150$ km s^{-1}
$v_2 = 135$ km s^{-1}; $T = 500$ days

Use the method given in the Study point to find the masses of the two stars in the system.

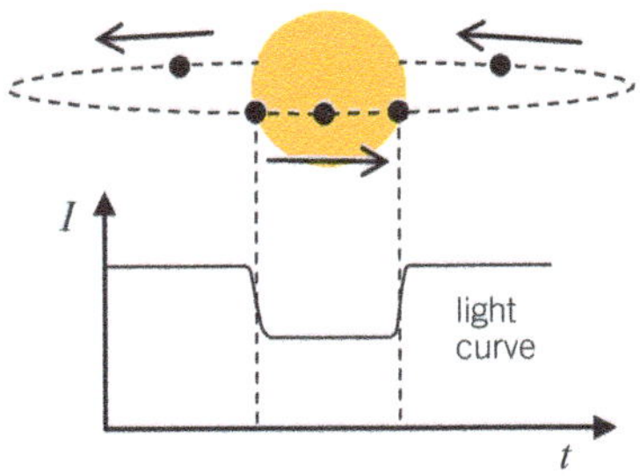

Fig. 4.3.24 Transit method

(b) The Transit method

If the observer is more or less in the plane of the orbit of the companion, the starlight will dim periodically as the planet passes in front of the stellar disk. This is shown schematically in Fig. 4.3.24. Search *Kepler mission* for more information. The duration of the eclipse and its depth provide information about the speed of the companion and its diameter. Both these properties depend upon knowledge of the diameter of the star itself. By the end of its mission on 30 October 2018 the Kepler Space Telescope had detected over 2600 planets using this technique. In conjunction with Doppler measurements, which deliver the mass of the planet, the mean density and therefore its nature can be investigated.

A close binary star system which we happen to view edge on, such as MY Camelopardalis [MY Cam] is called an *eclipsing binary*. Its components have 31 and 38 solar masses and have an orbital period of 1.17545 days. In fact, they are so close that their outer atmospheres overlap. Such pairs are also called *contact binaries*. The details of the light curve and Doppler measurements reveal a wealth of data. The variation in brightness shown in the light curve in Fig. 4.3.25 is due to the fact that at some periods in the orbit, one of the components is obscured by the other.

4.3.20 Knowledge check

A distant extra-terrestrial astronomer views a transit of the Sun by the Earth. Calculate the duration of the eclipse and the percentage dimming of the Sun's light. What assumption have you made?

Diameter of Earth = 12 700 km.

Diameter of Sun = 1 390 000 km.

Radius of Earth's orbit = 1.49×10^8 km.

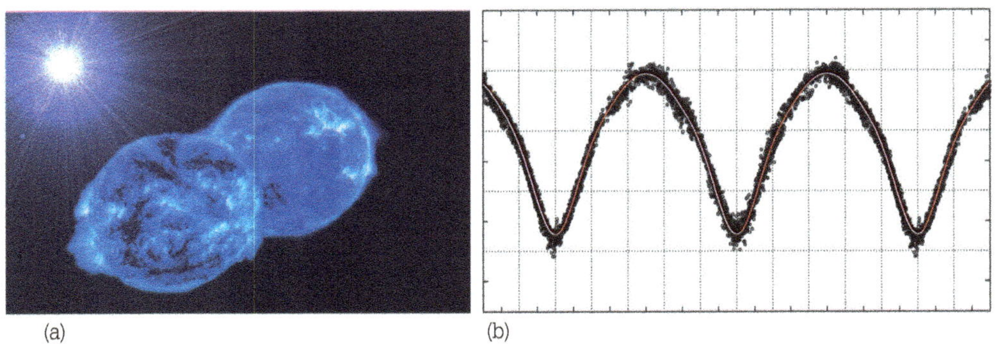

Fig. 4.3.25 MY Cam (a) artists impression and (b) light curve

Stretch & challenge

Sketch the brightness curve you would expect for two Solar-mass stars orbiting with a separation of 10 million km, when viewed edge on. How would the graph differ if:

(a) the orbit were tilted at 5° to the plane of observation, or

(b) one of the stars were twice as bright as the other but with the total mass remaining the same?

4.3.7 The expansion of the universe

(a) Hubble's law

Astronomers have discovered that all objects outside our local cluster of galaxies have a spectrum which is red-shifted, which can be interpreted as a Doppler shift showing a relative velocity away from our galaxy. This velocity, v, is found to be approximately proportional to their distance, up to a distance of several hundred megaparsecs. This relationship, known as **Hubble's law**[4], can be expressed as:

$$v = H_0 D,$$

where H_0 is the Hubble constant and D is the distance (strictly the **proper distance**) to the object.

Study point

Hubble's law states that the **recession velocity**, v, of deep-space objects is proportional to their proper distance.

Study point

A correct definition of the **proper distance,** D, of a deep-space object is out of the scope of this book but it is, in essence, the distance from us to the current position of the object.

4 In 2018 members of the International Astronomical Unit voted to call it the **Hubble-Lemaître law**. Georges Lemaître, a Belgian catholic priest and astronomer, was the first to attribute the observed recession of the galaxies to the expansion of the universe.

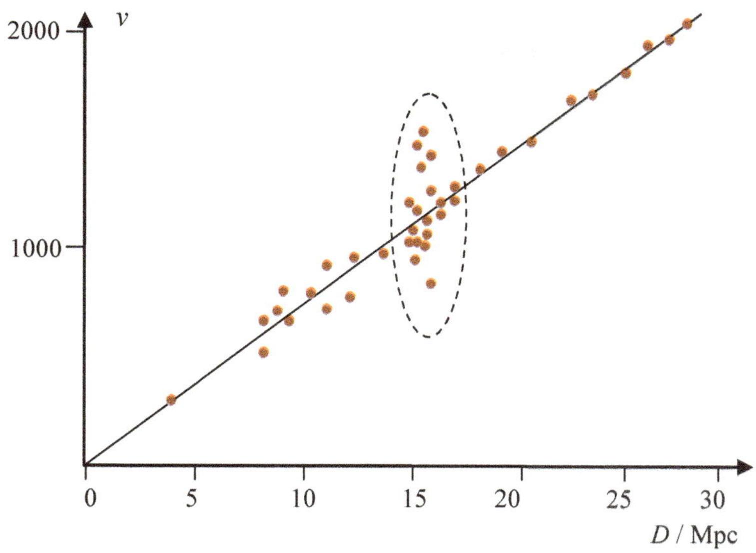

Fig. 4.3.26 Recession velocity–distance relationship

The value of the Hubble constant is not easy to determine. The recession velocity emerges from the red shift, but the distance to objects beyond about 100 kpc cannot be determined directly by parallax (see the AS book p73). A common method is to use the *apparent* brightness of objects of known intrinsic brightness such as Cepheid variables and type 1a supernovae. The details of this method are beyond the scope of this book but are well worth investigating.

The red dots in Fig. 4.3.26 represent nearby galaxies. The scatter about the line of best fit is due to the motion of the galaxies through space. The line of best fit itself represents the expansion of space. The galaxies within the pecked ellipse are members of the Virgo cluster of galaxies.

At the time of writing, the consensus value of H_0 is $67.9\ \text{km s}^{-1}\ \text{Mpc}^{-1}$, which means that a galaxy with a proper distance of 20 Mpc has a recession velocity of $20 \times 67.9 = 1360\ \text{km s}^{-1}$.

Why are all galaxies moving away from us? Are we the 'centre of the universe'? No. The whole of space is expanding and the galaxies are embedded in this expanding space. Fig. 4.3.27 is a schematic illustration of a region of space, the dots being clusters of galaxies. As space expands, the clusters recede from one another: an observer in *any* galaxy cluster observes all the other clusters receding, with a red-shift proportional to distance.

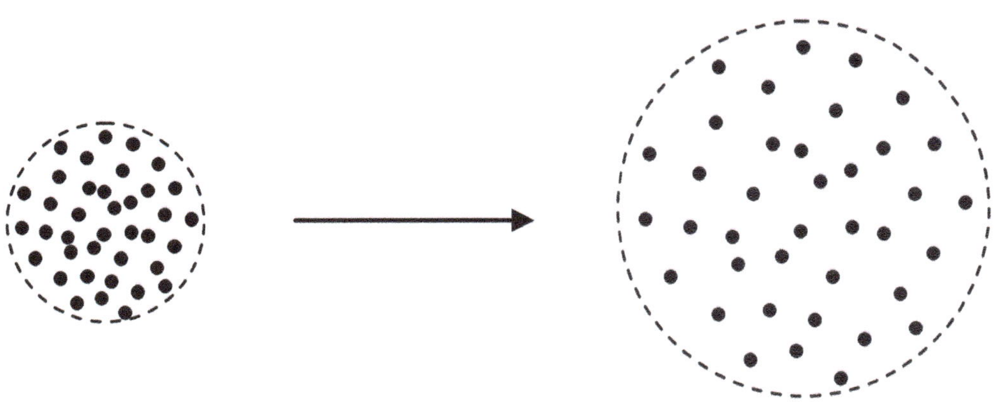
Fig. 4.3.27 The expansion of the universe

Study point

From its definition, the SI unit of the **Hubble constant** is s^{-1}. Cosmologists normally express it in $\text{km s}^{-1}\ \text{Mpc}^{-1}$.

Study point

A **standard candle** in astronomy is an object of known (or assumed!) luminosity, such as a type 1a supernova or a Cepheid variable star. Measuring the apparent brightness of such an object in a distant galaxy allows the distance to be calculated, using the inverse square law.

Knowledge check 4.3.21

Show that the value of the Hubble constant given in the main text ($67.9\ \text{km s}^{-1}\ \text{Mpc}^{-1}$) is equivalent to $2.20 \times 10^{-18}\ \text{s}^{-1}$.

[$1\ \text{pc} = 3.08568 \times 10^{16}\ \text{m}$]

Knowledge check 4.3.22

An Hγ emission line ($\lambda_{\text{lab}} = 434\ \text{nm}$) from a distant galaxy has a measured wavelength of 456 nm. Use the Hubble constant and the red shift to estimate the distance of the galaxy (a) in m and (b) in Mpc.

Stretch & challenge

A very neat way of considering the Hubble red-shift is to think of the light from a distance object as being 'embedded' in space: as space expands, so does the wavelength of the light. Calculate by what factor the universe has expanded since the light in Knowledge check 4.3.22 was emitted.

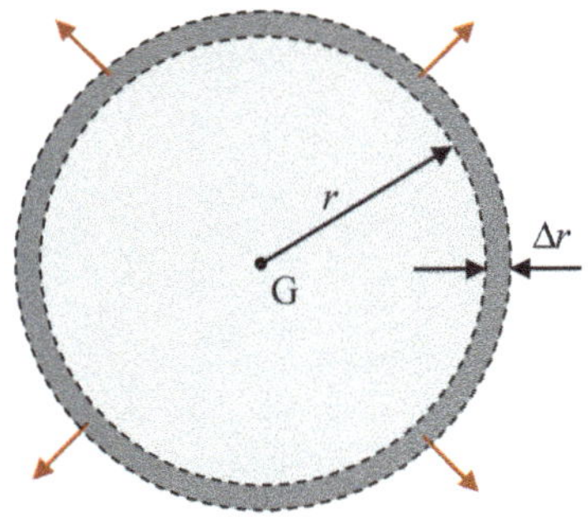

Fig. 4.3.28 Expanding shell

Study point

According to the General Theory of Relativity, the geometry of space itself relates to its energy density. If the energy density is negative, space curves in on itself; a positive density gives an outward curvature. If the energy density is zero then space is said to be *flat*, which means that Pythagoras' theorem is obeyed and angles in a triangle add up to 180°.

Key terms

The **Hubble time** is an estimate of the age of the universe and is defined as the reciprocal of the **Hubble constant**.

4.2.23 Knowledge check

Show that $t_H = 14.4 \times 10^9$ years.

(b) Will the universe expand for ever?

Treating the Doppler shift, for the moment, as representing the speed of recession of galaxies *through space* rather than with space, we can use Newton's law of gravitation to investigate this question. We'll consider the matter in a shell of space at a radius, r, from our galaxy, G. Assuming that the universe is homogenous:[5]

Let the mass of the shell between radii r and $(r + \Delta r)$ be m_S

The mass within the shell, $M_S = \frac{4}{3}\pi r^3 \rho$

where ρ = the mean density of the universe.

The recession speed of the shell, $v_S = H_0 r$.

The KE and PE_{grav} of the shell are: $E_k = \frac{1}{2}m_S v_S^2$ and $E_p = -\frac{GM_S m_S}{r}$.

So $E_{Tot} = E_k + E_p = \frac{1}{2}m_S v_S^2 - \frac{Gm_S M_S}{r}$

$= \frac{1}{2}m_S H_0^2 r^2 - Gm_S \frac{4}{3}\pi r^2 \rho$.

If $E_{Tot} > 0$, the shell will expand for ever; if $E_{Tot} < 0$, it will reach a maximum distance and then collapse; If $E_{Tot} = 0$ it will just come to a halt at an infinite distance and the universe has a *flat geometry* (see Study point). For a flat universe, then, setting the expression for E_{Tot} to zero and rearranging leads to the following expression for density:

$$\rho_c = \frac{3H_0^2}{8\pi G},$$ where ρ_c is called the *critical density*.

Current measurements of baryonic matter show that its density is much less than this. But the universe appears to have a flat geometry – more evidence for the existence of dark matter.

(c) How old is the universe

We have seen that the accepted value of the Hubble constant at the current time is $2.20 \times 10^{-18}\ s^{-1}$ (see Section 4.3.7(a) and KC 4.3.21). In other words, for every metre an object is away its recession speed is $2.20 \times 10^{-18}\ m\ s^{-1}$. *If this rate of expansion were constant* then the time, t_H, since all matter was gathered in the same infinitesimal spot is given by:

$$t_H = \frac{1}{2.20 \times 10^{-18}\ s^{-1}} = 4.55 \times 10^{17}\ s.$$

And more generally: $t_H = \frac{1}{H_0}$, where t_H is called the **Hubble time**.

5 .... meaning that matter is spread evenly through space. This is obviously not correct on the small scale [the Earth is a lot denser than interplanetary space, to say nothing of neutron stars] but it is a pretty good approximation on the scale of hundreds of Mpc

We might expect the expansion of the universe to be slowing down because of the effect of gravity. If this is true, then it was expanding more quickly in the past and therefore the time from the Big Bang is less than the Hubble time. This is illustrated in Fig. 4.3.29.

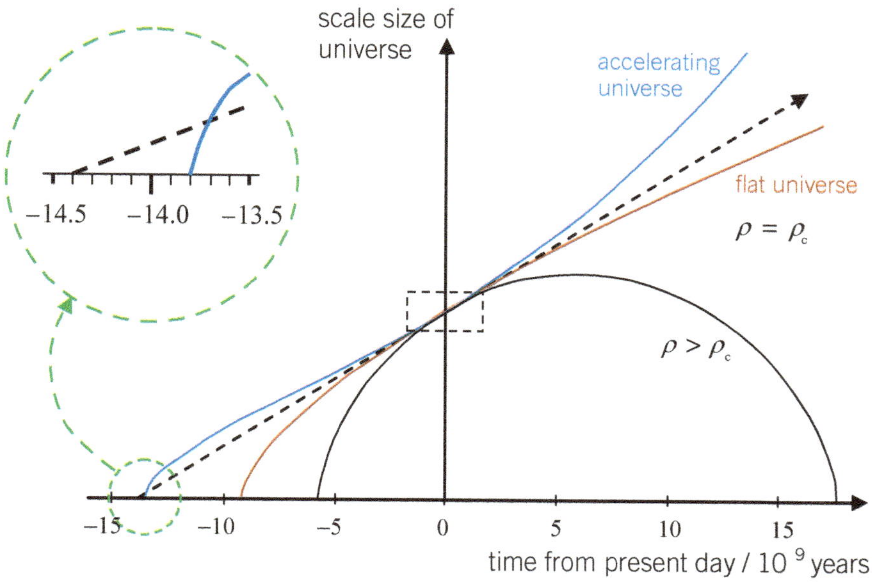

Fig. 4.3.29 Expansion models of the universe

The black dashed arrow represents a universe with a constant rate of expansion. This is a universe without any matter in it, so no gravitational deceleration. The red line shows a 'flat' universe, with the same observed value of the Hubble constant: the gradient within the pecked box is the same. Notice that this universe is 'only' 9 billion years old. A 'closed' universe, i.e. with enough gravity to stop and reverse the expansion would be even younger. However, our universe is none of these: recent measurements show that it has been *accelerating* for the last 5 billion years (blue line). Co-incidentally the resulting best age estimate for the universe (13.8 billion years) is very close to the **Hubble time** (14.4 billion years).

Study point

The 'scale size' plotted in Fig. 4.3.29 is any representative length which expands with the universe, e.g. the mean distance between clusters of galaxies.

Knowledge check 4.3.24

The wavelength of the H line measured in a laboratory is 410 nm. The same line in the spectrum of a distant galaxy is 820 nm. Estimate the current distance of this galaxy. Use Fig 4.3.29 and assume a constant expansion rate for the universe.

Stretch & challenge

The first indications that the expansion of the universe is accelerating came from measurements of the brightness and red-shift of distant type 1a supernovae; for a particular red-shift the supernovae are fainter than expected. Explain this.

[Hint: Consider your answer to KC 4.3.24]

Test yourself 4.3

1. A planet has two moons, **A** and **B**. Moon **A** has a period of 15 days; the period of moon **B** is 30 days. Find the ratio of their orbital radii.

2. The ratio of the orbital radii of two moons of a planet in circular orbits is 2.25. Calculate the ratio of the orbital periods.

3. The radius of the orbit of Io around Jupiter is 421 700 km. Its period is 1.769 days.

(a) Calculate the mass of Jupiter.
(b) The orbital distance of Callisto (another Jovian moon) is about 4.5 times that of Io. Estimate its orbital period.

4. The periods and orbital radii of the five major satellites of Uranus are given in the table to two significant figures. You will use these to investigate Kepler's 3rd law.

Satellite name	Miranda	Ariel	Umbriel	Titania	Oberon
Orbital radius / 10^3 km	130	190	270	440	580
Orbital period / day	1.4	2.5	4.1	8.7	13.5

(a) In view of the spread of data, explain why a plot of T^2 against r^3 is an inappropriate graph.
(b) What are the gradient and intercept of a graph of $\ln T$ against $\ln r$ expected to be?
(c) Steve claims that the following plots should give a straight line through the origin:
T against $r^{3/2}$; $T^{1/3}$ against $r^{1/2}$; $T^{2/3}$ against r.
(i) Explain why Steve is correct.
(ii) Of these, explain why $T^{2/3}$ against r is the most suitable plot with these data values [Hint: consider the spread of values of r in the data]

5. According to a website, the star S2 orbits the radio source Sgr A* with a period of 15.8 ± 0.11 years [see Section 4.3.3(a)]. The semi-major axis of the orbit has an apparent angular size of $(5.832 \pm 0.13) \times 10^{-7}$ rad and recent measurements give the distance of Sgr A*from the Earth as 7.94 ± 0.42 kpc. Newspaper reports identify the radio source as a black hole with a mass of 4 million Suns ($M_{Sun} = 1.99 \times 10^{30}$ kg).

(a) Ignoring the uncertainty values, calculate the semi-major axis of the orbit and hence the mass of the radio source. Does it agree with the newspaper reports?
(b) Use the uncertainty values to estimate the uncertainty in your value of the mass of Sgr A*.
(c)

Stretch & challenge

The perinigron [closest approach to the black hole] is estimated at 17 light-hours. Calculate the speed of S2 at this point. [Hint: consider the total orbital energy.] What fraction of the speed of light is this?

6. A neutron star of mass $1.4M_\odot$ and a white dwarf of mass $0.6M_\odot$ are in circular orbits about their centre of mass with a period of 15 hours $M_\odot$ = mass of Sun = 1.99×10^{30} kg. Calculate:

(a) the separation of the stars and hence the radii of their orbits
(b) the orbital speed of each of the two bodies
(c) the observed Doppler shift of the Hα spectral line (laboratory wavelength 656.281 nm) emitted by the white dwarf.

4.4 The magnetic field

We say that there is a **magnetic field**, or an electric field (Section 4.1.6), in any region where electric charges experience forces. The key difference between the fields is that only moving charges are affected by magnetic fields. With electric fields, both moving and stationary charges experience forces. See the Study point.

We'll be looking at the magnetic forces (forces due to applied magnetic fields) on free electrons moving in wires that are carrying electric currents, and on charged particles moving in empty space in particle accelerators. Later we'll study how currents in wires, straight or coiled, act as *sources* of magnetic fields.

Considering the name 'magnetic field', you might well ask: where do magnets fit into all this? The moving charges are orbiting and spinning electrons inside the magnets' atoms. The electron spins of neighbouring atoms in regions called 'domains' (Fig 4.4.1) align so that their effects re-inforce. In magnets the domains themselves are predominantly aligned roughly in one direction. All this is clearly quite complicated, or at any rate not exactly basic – which is why magnets aren't studied in depth at A-level. Nevertheless, by their very familiarity, magnets do give us an easy way into our work on magnetic fields.

Key terms

We say that a **magnetic field** exists in a region in which a moving electric charge experiences a force, but a stationary charge does not.

The **direction of a magnetic field** in a certain place is the direction in which the North pole of a small, freely pivoted magnet tends to point.

The **North pole (or North-seeking pole)** of a magnet is the end that tends to point roughly geographically North if only the Earth's magnetic field is present.

Study point

A charge that is stationary in our lab will be moving if viewed from a passing vehicle. Doesn't that make nonsense of the first paragraph in 4.4? No: it tells us that magnetic (and electric) fields depend on our frame of reference. At A-level we stick with one frame. Phew!

Fig. 4.4.1 False colour image of oppositely magnetised domains – image diameter ~ 0.2 mm

4.4.1 The direction of a magnetic field

A useful way to explore magnetic fields is with a small magnet, pivoted so it can point in any direction. A magnetic 'plotting compass' is usually good enough, though its magnet can point only in horizontal directions.

If we find that a freely-pivoted magnet points in a particular direction we know that it's in a magnetic field. We can use this to define the **direction of a magnetic field**.

(a) Magnetic field lines (lines of magnetic flux)

By moving a plotting compass so that its South pole is successively where its North pole was previously, we can trace or 'plot' a magnetic field line. If we start in different places we can plot (in theory) as many lines as we like in a region.

Fig. 4.4.2 shows what emerges when we investigate the region around a bar magnet in this way. The compass magnet's **North pole** points in the direction of the resultant magnetic field, in this case the resultant of the magnet's field and the Earth's field. As long as we stay near the magnet, its field is much stronger than the Earth's, so the pattern of field lines is almost entirely due to the magnet.

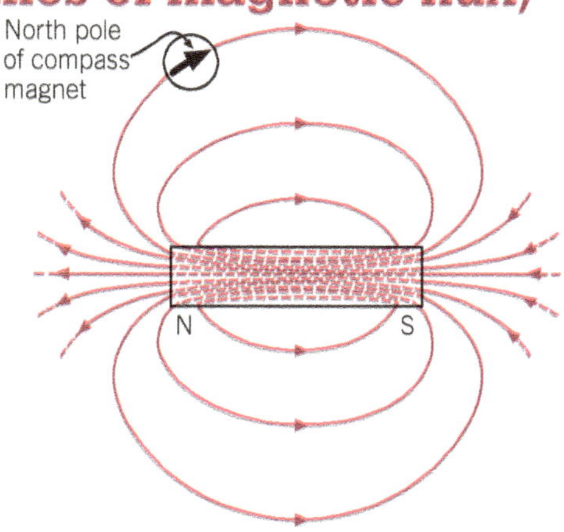

Fig. 4.4.2 Magnetic field lines of a bar magnet (horizontal section)

Knowledge check 4.4.1

Magnetic field lines *emerge* from the magnet's North pole. Deduce this from the 'like poles repel, unlike poles attract' rule for magnets. [Remember how the lines are plotted!]

Key terms

A **magnetic field line** (or **line of magnetic flux**) is a line whose direction at every point is the direction of the magnetic field at that point.

Study point

The properties of magnetic field lines are analogous to those of electric field lines, except that electric field lines due to stationary charges are not closed loops, but start on positive charges and end on negative. The concept of electric charge is indispensible, but arguably we could do without that of a magnetic pole.

(b) Properties of magnetic field lines

So what *is* a **magnetic field line**? The definition is very straightforward. Note the following properties of the lines, using Fig. 4.4.2 as an illustration:

1. Magnetic field lines never cross or meet. If they could, which way would a compass point at their intersection? In fact, it points in the direction of the *resultant* field, which is everywhere well defined.
2. Magnetic field lines are closed (end-less) loops. Although we can't show this with a plotting compass, the lines that arrive at the South pole of a magnet continue (from S to N) through the magnet, emerging at the North pole.
3. When field lines become closer together, the field is becoming stronger – as near the poles of a magnet.

Compare the lines with streamlines in a smoothly flowing liquid… The direction of a streamline shows the direction of liquid velocity at a point, and where streamlines become closer together, as when the pipe in Fig. 4.4.3 narrows, the magnitude of the velocity becomes greater. Indeed, magnetic field lines are sometimes called *lines of magnetic flux*, the primary meaning of the word 'flux' being 'flow'. Don't take the analogy too far: we don't believe anything is actually flowing along a magnetic field line.

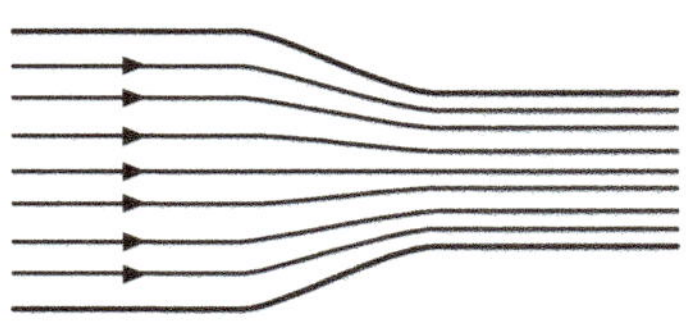

Fig. 4.4.3 Streamlines for liquid flowing in a narrowing pipe

4.4.2 The motor effect: force on a current-carrying conductor

We're now ready to move on to the central topic of this chapter: the forces experienced by charges moving in magnetic fields.

We find that a wire, carrying a current, placed in a magnetic field, at an angle to the field, experiences a force. This is 'the motor effect'. Fig. 4.4.4(a) shows one way of demonstrating it.

4.4.2 Knowledge check

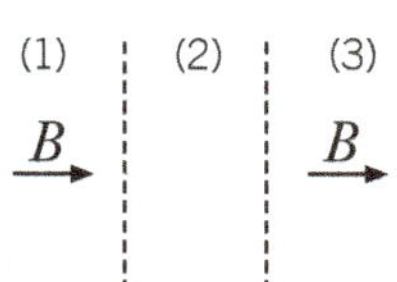

The arrows labelled 'B' denote magnetic field directions in two regions, (1) and (3). The field in (3) is stronger. Use the properties of magnetic field lines to show that the direction of the field can't be the same all over the transition region (2).

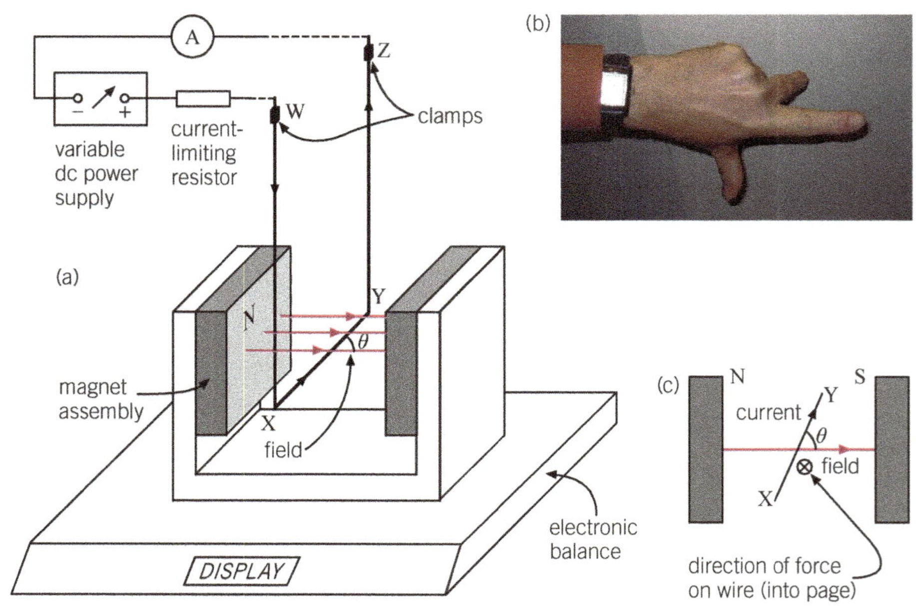

Fig. 4.4.4 The motor effect and Fleming's left hand motor rule

(a) The direction of the motor effect force

In the apparatus of Fig. 4.4.4 we are concerned with the force on the horizontal portion, XY, of the current-carrying wire, WXYZ. Note the position of XY, at an angle to the magnetic field direction.

With a current of a few ampère in the XY direction, XY experiences an appreciable downward force. The magnet assembly producing the field experiences an upward force of equal magnitude. This shows up on the electronic balance supporting the assembly.

The force on the wire isn't an attraction or repulsion, as it is at right angles to both the magnetic field and the current. **Fleming's left hand motor rule** (FLHMR) has been verified by countless observations and we will be using it to *predict* the direction of motor effect forces. Note the alliterative association of **F**irst finger with **F**ield, and so on. Fig. 4.4.4(b) shows how the rule works for XY in Fig. 4.4.4(a). Do try it with your own left hand; no injuries have been reported over more than a century of the rule's use …

The motor effect, then, has to be understood in three dimensions, but there's often a simple alternative to drawing a diagram that tries to project a third dimension on to a piece of paper. We draw a section, as in Fig. 4.4.4(c), adding dot or cross symbols:

⊙ based on an arrow coming towards us, means a current, field or force – we have to label it with which it is – coming out of the page;

⊗ based on the tail-feathers of an arrow, means something going into the page. Again it must be labelled.

The magnetic field direction is usually labelled 'B', the current, I, as in KC 4.4.3.

(b) The magnitude of the motor effect force

Again investigating this with the apparatus in Fig. 4.4.4, we find, when the other variables are held constant, the force is proportional to:

- the length, ℓ, of the wire (XY),
- the current, I,
- $\sin\theta$, in which θ is the angle between current and field.

The force must surely also increase with the strength of the field. But we can't even attempt to discover an exact relationship by experiment, as we haven't defined in a quantifiable way what we *mean* by the strength of a magnetic field! We turn the situation on its head and use the motor effect itself to provide us with a definition…

We define the magnitude, B, of the **magnetic flux density** (the official name for magnetic field strength) as

$$B = \frac{F}{I\ell\sin\theta}$$

in which F is the magnitude of the force on a short wire of length ℓ, carrying a current, I, at angle θ to the field direction.

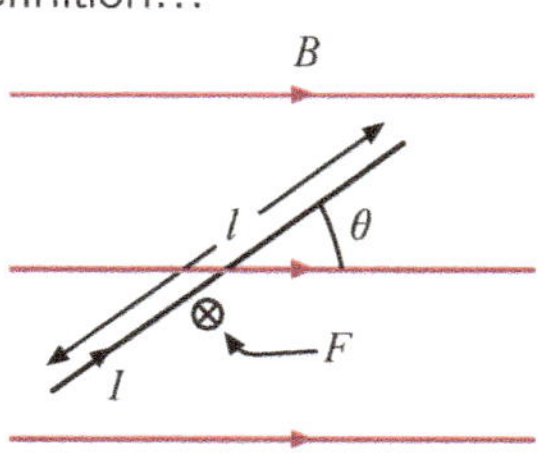

Fig. 4.4.5 Motor effect

[Note how, if we use twice the current, the force will be twice as much, so we'll get the same value for B, which is as it should be, because the field can't depend on what we're using to measure it. Similar points can be made about $\sin\theta$ and ℓ.]

Key terms

Fleming's left hand motor rule

Hold the left hand so that:

- the **F**irst finger points straight out from the palm and in the direction of the magnetic **F**ield,
- the se**C**ond finger points in the direction of the conventional **C**urrent
- the **Th**umb is at right angles to both these fingers and gives the direction of the force (or **Th**rust) on the wire.

Knowledge check 4.4.3

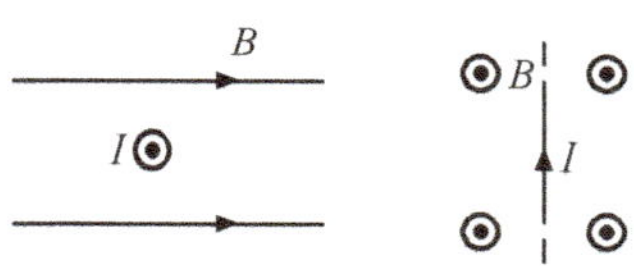

Determine the direction of the forces on the wires in the diagrams above.

Key terms

Magnetic flux density (**magnetic field strength**) is a vector of magnitude $B = \frac{F_{max}}{I\ell}$, where F_{max} is the force on a short wire of length ℓ, carrying current I, angled so as to maximise the force on it.

UNIT: N A^{-1} m^{-1} = tesla (T).

The direction of the vector is given by Fleming's left hand rule, knowing the force and current directions.

4.4.4 Knowledge check

If, in Fig. 4.4.5, $\ell = 2.5$ cm and $I = 3.5$ A and the wire experiences a maximum downward force of 0.0037 N, calculate the mean flux density in the region of the wire.

4.4.5 Knowledge check

A piece of wire 0.020 m long, carrying a current of 3.0 A, is placed horizontally in a horizontal field of 80 mT due North. Determine the force on the wire if it is angled so that the current is:

(a) due North
(b) at a bearing of 30° East of North
(c) due East
(d) South West.

Study point

When $\phi = 90°$, $\sin\phi = 1$, so the torque is a maximum, as should be clear from Fig. 4.4.8(a). When $\phi = 0$, $\sin\phi = 0$, and the torque is zero (Fig. 4.4.8(b)).

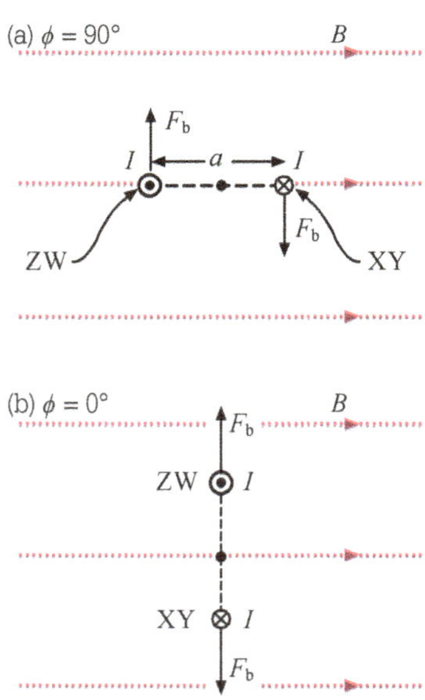

Fig. 4.4.8 (a) maximum torque and (b) zero torque

The equation tells us that the SI unit of the magnetic flux density, B, must be $\text{N A}^{-1}\text{ m}^{-1}$. We give this combination the name 'tesla', abbreviated to 'T', so $\text{N A}^{-1}\text{ m}^{-1} = \text{T}$.

Nikola Tesla (1856–1943) was a Serbian-American pioneer of alternating current technology, who revelled in high voltages and high frequencies.

Fig. 4.4.6 Nikola Tesla

Rearranging the defining equation for B,

$$F = BI\ell\sin\theta$$

(c) The magnetic flux density vector

Magnetic flux density is a vector. We can find its direction, as already explained, using a pivoted magnet. In the equation $F = BI\ell\sin\theta$, $B\sin\theta$ is the component of the vector perpendicular to the wire, sometimes written in one lump as $B_\perp$. See Fig. 4.4.5.

It may seem messy to define the direction of the vector in terms of a pivoted magnet, and its magnitude in terms of a current-carrying wire. It's fine to stick with these definitions, but see if you prefer the one-stop-shop approach given in Key terms on the previous page.

(d) Example of the motor effect: the current-carrying coil

We examine the forces on a rectangular current-carrying coil (of N turns) orientated with the normal to its plane at an angle ϕ to a uniform magnetic field.

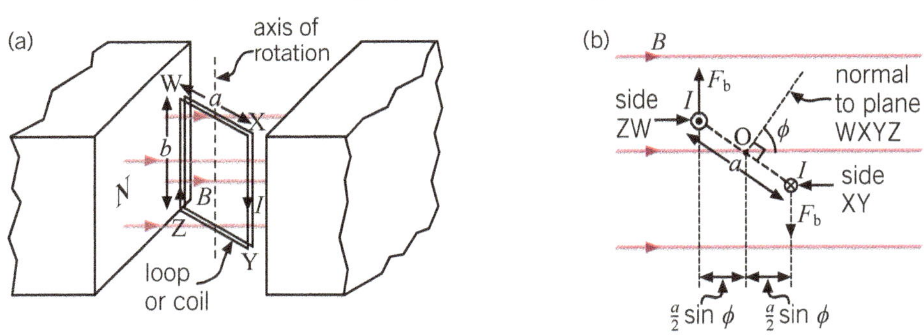

Fig. 4.4.7 Torque on a coil: (a) general view (b) horizontal section

First we note that, according to FLHMR, the force on side WX is upwards (towards the top of the page) and the force on YZ is downwards. These forces are not only equal and opposite, but in the same vertical plane – that of the coil. Their only effect is to stretch the coil – unnoticeably.

The forces (of magnitude F_b) on XY and ZW are also equal and opposite (see Fig. 4.4.7(b)), but this time they are 'staggered' apart by a distance of $a\sin\phi$, making up what is called a 'couple'. Adding the moments of these forces about the 'axis of rotation' (which goes through point O) …

Total torque, $\tau = 2F_b \times \frac{a}{2}\sin\phi = 2NBIb\sin 90° \times \frac{a}{2}\sin\phi$

Tidying up, $\tau = NBIA\sin\phi$

Here we have put $ab = A$, since ab is the area of the coil. There's more to this substitution than mathematical short-hand: it turns out that the last equation applies for a flat coil of *any* shape, A being the coil area. [This example has great practical importance, but the equations in it don't have to be learned for the exam!]

Note that the effect of having N turns on the coil is to make the effective current N times larger.

The torque on a coil is exploited in an electric motor. The current through the coil has to be reversed as the coil turns through the positions where the field through it changes direction, to prevent the torque direction reversing. Most motors have several coils at different angles so the total torque never even drops to zero. The torque is increased by winding the coils on iron 'cores'. See Fig. 4.4.9.

Fig. 4.4.9 Simple electric motor: coils, iron cores (green), magnets (blue and white)

Knowledge check 4.4.6

Insulated wire is wound around the edges of a door to make a flat, 250-turn coil, $0.75 \text{ m} \times 2.00 \text{ m}$. The open door is in the magnetic North–South plane. Taking the horizontal component of the Earth's field to be $20 \ \mu\text{T}$, calculate the current needed to start the door closing, against a frictional torque of 1.2 N m.

4.4.3 The effect of magnetic fields on moving charges

(a) The force on a moving charge

Since a current in a metal wire is made up of moving free electrons, it seems likely that the motor effect force on a current-carrying wire is simply equal to the sum of forces acting on the individual moving electrons. See second Study point. If so we can easily find the magnetic force on an electron with drift velocity v:

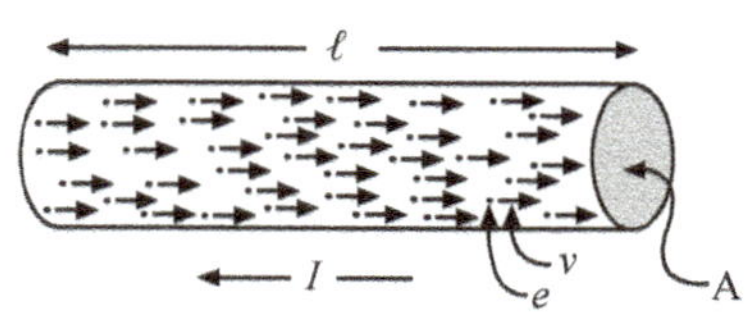

Fig. 4.4.10 Current in wire

First, recall that if the wire has cross-sectional area A and has n free electrons per unit volume, then the current will be $I = nAve$

We can therefore write the motor effect force on length ℓ of the wire as

$$F_{\text{wire}} = BI\ell \sin\theta = BnAve\ell \sin\theta$$

But the volume of the wire is $A\ell$, so there are $nA\ell$ free electrons in it.

Therefore force on each free electron $= \dfrac{BnAve\ell \sin\theta}{nA\ell} = Bev\sin\theta$

We now make the leap that this applies to any particle carrying charge q, provided we replace e by q. See Study points.

Study point

The magnetic force on a charge, q, moving in a magnetic field as shown has magnitude

$F = Bqv \sin\theta$

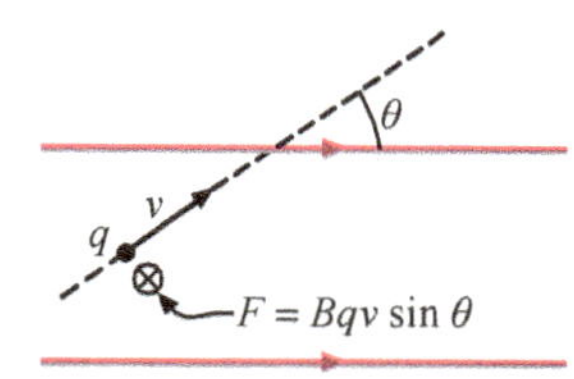

Fig. 4.4.11 Charge moving in magnetic field

The direction of the force is given by Fleming's left hand motor rule.

Study point

In $F = Bqv \sin\theta$, if $q < 0$, then $F < 0$, i.e. F acts in the opposite direction.

Study point

The magnetic forces don't push drifting electrons out of the sides of a wire, because opposing electric field forces come into play. (Strictly, the rest of the wire experiences the Newton 3 electric field 'partner' forces, rather than the magnetic forces themselves.)

(b) Path of a charged particle moving in a uniform magnetic field

Suppose that a charge, q, moving at speed v is 'injected' into a uniform magnetic field, B, with angle θ between the directions of the field and the particle velocity. If no forces other than the magnetic force act on the particle, how will it move?

First note that the magnetic ($F = Bqv \sin\theta$) force is always at right angles to the particle's direction of motion – see Fig. 4.4.11. Therefore the field can't do work on the particle. Its kinetic energy, and hence its speed, remain constant.

Instead, the force causes the particle's *direction* of motion to change at a constant rate. That's unless $\theta = 0°$, in which case there will be no force so the particle will carry on moving parallel to the field lines, at speed v.

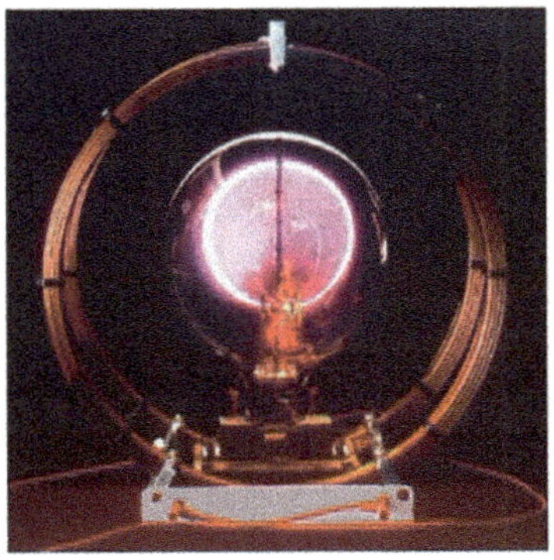

Fig. 4.4.13 Electrons' circular path in a uniform magnetic field

$\theta = 90°$

If, initially, the particle is moving at right angles to the field, it will continue to do so – in a circular path.

The magnetic force supplies the centripetal acceleration, so it's easy to find the radius of the circle, using $F = ma$:

Force on particle = mass × acceleration

So $Bqv = \frac{mv^2}{r}$. Hence $r = \frac{mv}{Bq}$

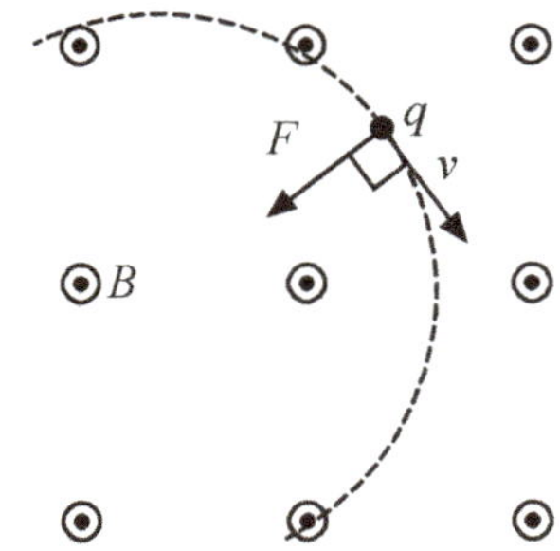

Fig. 4.4.12 Force on a + charge moving in a uniform magnetic field

Fig. 4.4.13 shows electrons moving in a circular path – the bright ring. The light comes from gas molecules excited by hits from electrons and de-excited by photon emission. [The gas pressure in the glass vessel is very low so that some electrons can complete a circle without hitting molecules!] The fast electrons are injected by an electron gun at the bottom of the ring. The magnetic field, almost uniform in the region of the ring, is made by passing current through the large (red) coils of wire.

Fig. 4.4.15 Particles trapped by Earth's magnetic field

θ neither 0° nor 90°

We can resolve the particle's velocity into a component $v_{\parallel} = v\cos\theta$ parallel to the field direction and a component $v_{\perp} = v\sin\theta$ at right angles to it. By themselves, $v_{\perp}$ would give circular motion, and $v_{\parallel}$, unaffected by the field, would give straight line motion parallel to the field lines.

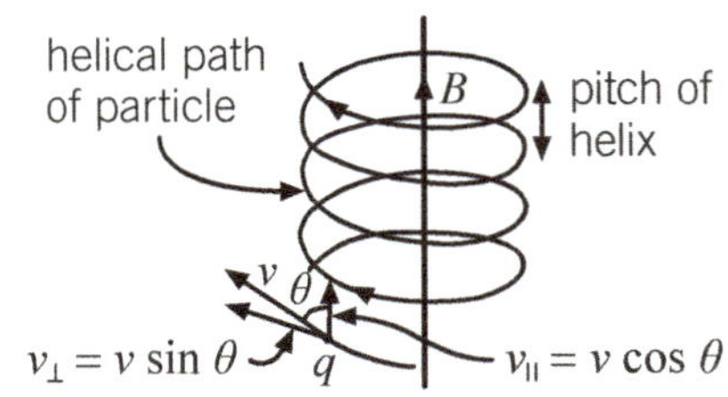

Fig. 4.4.14 Helical path of particle in uniform magnetic field

A combination of these two motions results: the particle traces out, at constant speed, a spiral staircase path called a helix (adjective: helical). It may help to imagine yourself viewing the particle from a platform moving with velocity v_2. From this viewpoint the particle would be moving in a circle with velocity, $v_{\perp}$.

Paths resembling helices are traced out by charged particles, mainly protons and electrons emitted by the Sun, that are trapped by the Earth's magnetic field and condemned to gyrate around field lines.

Fig. 4.4.16 Aurora borealis photographed in Manitoba

The Earth's magnetic field is believed to be generated by the motion of molten iron-rich material in the Earth's core. On a large scale the pattern of the Earth's field lines is quite similar to the pattern that would arise from a small but very strong magnet near the centre of the Earth and aligned at a few degrees to the Earth's axis of rotation. Clearly, then, the field is not uniform on a large scale, so the paths of the gyrating particles are not true helices.

Fig. 4.4.15 shows what happens in a stylised way. At near-polar latitudes the field lines are close together. They also pass through the upper atmosphere, where the fast-moving charged particles that they have captured may collide with molecules, exciting them. The light emitted during de-excitation is a major cause of the spectacular displays in the night sky that are the Northern and Southern 'lights' (aurora borealis and aurora australis).

4.4.7 Knowledge check

Electrons accelerated through a pd of 970 V are injected at right angles to the field direction into a magnetic field of strength 1.50 mT. The path radius is measured as 71 ±2 mm. Show whether or not this is consistent with the equations just developed.

[$e = 1.60 \times 10^{-19}$ C, $m_e = 9.11 \times 10^{-31}$ kg]

(c) Particle accelerators

From early in the 20th century, physicists have used high-speed particles as missiles to 'fire' at nuclei and, later, at smaller targets, such as nucleons. By noting how the missile particles are deflected, and what comes out of the target particles and in which direction(s), physicists have learned much about the structure of sub-atomic particles.

At first, α particles from radioactive sources were used as missile particles. This is how the very existence of the atomic nucleus was revealed. But for more detailed work, α particles (helium nuclei) have too much internal structure of their own, and not enough KE (usually a few MeV).

In the 1920s, physicists started to develop ways of accelerating particles such as electrons and protons. The electron gun (Section 4.1.7 (c)) is a simple form of accelerator – one in which the acceleration occurs in a single step. Unfeasibly enormous pds would be needed to give the particles enough KE for use as missiles in modern particle physics. The problem is overcome by accelerating the particle over and over again, through a large (but managable) pd.

The linear accelerator

Fig 4.4.17 shows a simplified linear accelerator (or linac). Suppose the particles to be accelerated are positive. They will be accelerated across the first gap into drift tube (1), when this is at a negative potential with respect to the particle production chamber (0). The particles will travel through the first drift tube at constant speed.

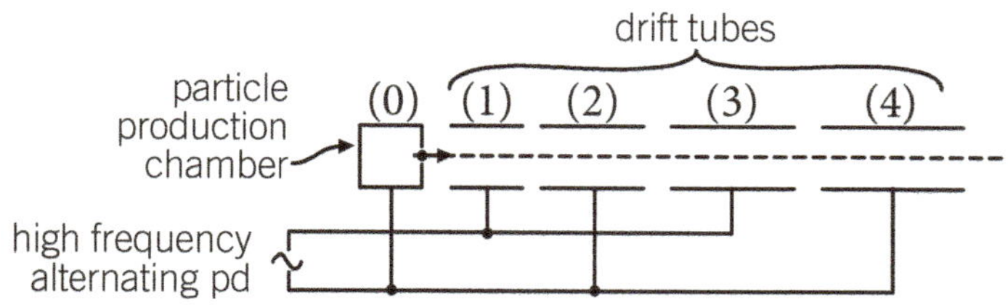

Fig. 4.4.17 Linear accelerator (simplified)

The pd needs to alternate at such a frequency that (2) is negative with respect to (1) when the particles reach the gap between (1) and (2), but the pd has reversed again so that (3) is negative with respect to (2) when the particles reach the gap between (2) and (3) – and so on. The increasing length of the drift tubes is designed to keep pace with the increasing speed of the particles, so that the frequency of the alternating pd can be roughly constant.

The longest linear accelerator in the world (3.2 km) forms a main part of the SLAC laboratory in California. It accelerates electrons or positrons up to kinetic energies of 50 GeV.

The cyclotron

The cyclotron is a particle accelerator that can take up much less space than the equivalent linear accelerator. This is because a cyclotron uses a magnetic field (at right angles to the particle velocity) to confine the charged particles to (semi)circular paths between the times when they acquire kinetic energy.

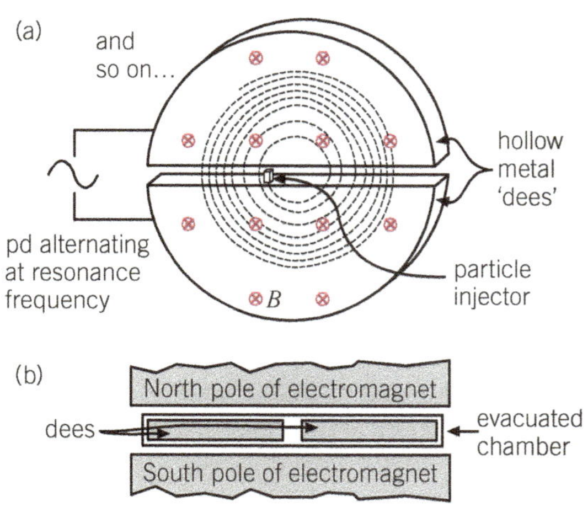

Fig. 4.4.18 A cyclotron: (a) path of particles (b) vertical section

Study point

When we speak of *acceleration* in the context of particle accelerators, we mean particles getting faster. This is a special type of acceleration. In a cyclotron or synchrotron there is also acceleration through rapid change of direction, though this doesn't add to the particles' kinetic energy.

Study point

Particles that have already been brought close to the speed of light can go on picking up kinetic energy in an accelerator, while only increasing their speeds slightly. That's why physicists usually refer to the energies of such particles, rather than their speeds.

Knowledge check 4.4.8

Assuming that, at each gap, the pd has a value V, show that the speed of particles (charge q, mass m) in successive drift tubes, (1), (2), (3) … is w, $w\sqrt{2}$, $w\sqrt{3}$, … in which

$$w = \sqrt{\frac{2qV}{m}}$$

Knowledge check 4.4.9

Deuterons (hydrogen nuclei of mass 3.343×10^{-27} kg) are brought to a high speed in a cyclotron that uses a magnetic flux density of 0.60 T. The pd applied between the dees is ±12 kV. Calculate:

(a) the frequency needed for the alternating pd,

(b) the path radius after the particle has performed 300 revolutions. [Neglect initial velocity.]

[$e = 1.60 \times 10^{-19}$ C.]

A particle sent out from the injector gets faster each time it crosses the gaps between the metal 'dees', because a high voltage is applied between them. In Fig. 4.4.18 a positive particle will travel anticlockwise – use that left hand rule! – so the top dee will need to be negative and the bottom positive when the particle crosses the gap on the right, but the other way round when the particle reaches the left-hand gap. Therefore the pd must alternate with a periodic time equal to the time, T_C, for the particle to make one revolution.

We now calculate T_C for when the particle has attained a speed v and a path radius r:

$$T_C = \frac{2\pi r}{v} = \frac{2\pi}{v} \times \frac{mv}{Bq} = \frac{2\pi m}{Bq}$$

The reciprocal of T_C gives us the frequency of revolution, the number of circles per second! It is sometimes called *the cyclotron resonance frequency*, f_C, but it applies for any charged particle moving at right angles to a uniform magnetic field, provided $v << c$ (the speed of light).

We have then: $f_C = \frac{Bq}{2\pi m}$

The really neat thing is that T_C and f_C are independent of v and r! The faster the particle the wider its circular path, and these factors cancel.

Therefore we should be able to stick with one frequency of alternating voltage for all particles of the same type anywhere in the cyclotron…

… If only life were that simple. The cyclotron was invented in 1932 and was brilliant for producing fast protons to use as missile particles in nuclear research. Cyclotrons are still used in hospitals to produce proton beams for cancer therapy and for bombarding nuclei to produce short-lived isotopes. Unfortunately, for cutting-edge experiments in Particle Physics the cyclotron simply can't give particles enough kinetic energy.

For one thing, as a particle's speed approaches the speed of light, its path radius increases faster than its speed, and the time per revolution increases. A cyclotron can be modified to deal with this, but a more serious issue is the very large path radii of high energy particles, which would require the magnetic field to be applied over a very wide disc-shaped region. The electromagnet would have to be vast.

The synchrotron

In a synchrotron the particles go round and round one circular path. Thus no magnetic field is needed at other radii. Even better, the field can be applied at regular stations along the path (Fig. 4.4.19) rather than all along it. All the same, the Super Proton Synchrotron (SPS) at CERN has 744 path-bending electromagnets along its 7 km circumference! In the SPS, the particles are given kinetic energy by electric fields in 'radio frequency cavities' at two places in the circular path, every time they go round it.

What is sacrificed in a synchrotron is the ability to produce high energy particles in a continuous stream. For the particles to remain in the same circular path as they gain KE, the magnetic field has to be increased continuously. The magnetic flux density that is needed for particles that have done many circuits of the synchrotron would be too large for particles at earlier stages in their ride. The only thing to do is to inject particles into the ring a batch at a time; in the SPS, typically around 10^{13} per batch, occupying less than the ring's circumference.

A by-product of a synchrotron's operation is 'synchrotron radiation': photons of up to X-ray frequencies emitted by the charged particles as they are deflected by the electromagnets and undergo radial acceleration. Although extra power has to be supplied to the beam by the electric fields to compensate for this energy loss, the synchrotron radiation may be used for other experiments.

4.4.10 Knowledge check

A mass spectrometer enables the masses of ions to be determined. Positive ions are accelerated by a known pd and a magnetic field is adjusted until the ions reach a detector by travelling round an arc of known radius.

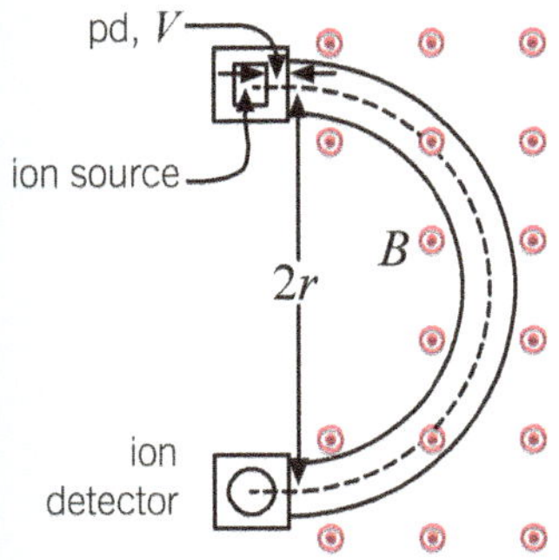

A field of 0.296 T enables Mg^+ ions, accelerated through 4500 V, to traverse a semicircle of diameter 0.320 m. Identify the nucleon number of the magnesium isotope.

[$e = 1.60 \times 10^{-19}$ C, 1 kg = 6.02×10^{26} u]

Study point

In a synchrotron we are usually dealing with particles near the speed of light, where Newtonian formulae for momentum and KE have to be modified. Particles can still go on acquiring KE from the electric fields, even though their speed increases very little!

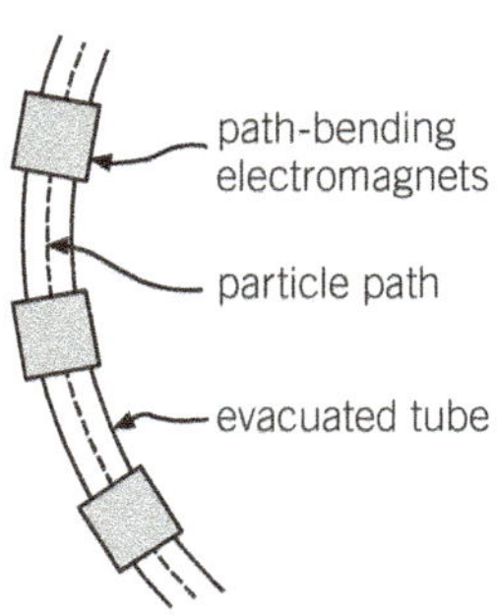

Fig. 4.4.19 Particle path in synchrotron

4.4.4 'Crossed' electric and magnetic fields

(a) In a vacuum

A charged particle in a uniform electric field moves in a straight or parabolic path with changing speed (See Section 4.1.7), but, a charged particle injected into a uniform magnetic field moves in a straight, helical or circular path at constant speed. If we apply *both* E and B fields over the same region, a particle's trajectory is likely to be complicated, but we can consider just one simple and useful case:

Suppose the E and B fields are at right angles to each other. A set-up for producing such fields is shown in Fig. 4.4.20. A charged particle is sent into the 'crossed fields' region travelling at speed v in a direction at right angles to both these fields (along the dotted line). If the fields are as shown (or if *both* fields are reversed!) the electric and magnetic forces on the particle will be in opposite directions. For zero resultant their magnitudes must be equal.

That is $\quad Bqv = Eq \quad$ so $\quad Bv = E$

This is therefore the – delightfully simple – condition for the particle to continue moving in a straight line at constant speed, even though there are fields present.

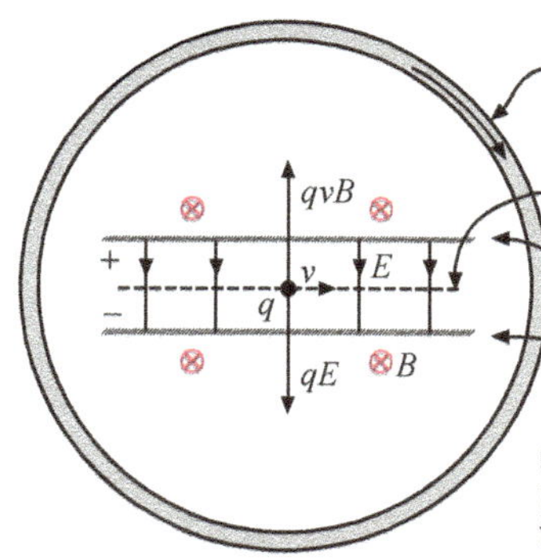

Fig. 4.4.20 Moving particle in crossed E and B fields

Knowledge check 4.4.11

What happens to a particle initially at rest in:

(a) a uniform E field,

(b) a uniform B field?

This gives us a neat way of measuring the speed of the particles in a beam of charged particles: send the beam normally through crossed fields and adjust the field strengths until the beam is undeflected. Then $v = E/B$. A very similar use is as a 'velocity filter': the ratio E/B is set at a desired value, so if particles of various speeds are sent normally through the crossed fields, only those with speed E/B will go straight through. [In one arrangement these can pass through a hole in a metal electrode while others, with the 'wrong' speed, are obstructed.]

In the next section we meet crossed fields inside a solid.

Knowledge check 4.4.12

Calculate the speed of a particle that passes undeflected through crossed E and B fields, if $B = 0.37$ T and E is provided by placing a pd of 3300 V between plates separated by 8.0 mm.

(b) In a conductor – the Hall effect

The Hall effect (discovered by Edwin Hall in 1879 – before the discovery of the electron) gave the first indication that the motor effect force on a current-carrying metal conductor arises from forces on negative charge-carriers moving inside it. The Hall effect results from the magnetic field exerting a force towards one side of the conductor on the carriers.

Although the effect occurs in metals, it is much greater in semiconductors, such as silicon. The thinner the specimen, the larger the effect. Thin pieces of semiconductor used to show the Hall effect are often called 'Hall wafers'. We shall first consider a wafer made of silicon doped (alloyed) with a small amount of phosphorus This material is called an 'n type' semiconductor for reasons we shall see.

Study point

Connections to the external current-supplying circuit are made right across the small end-faces of the wafer, rather than to single points.

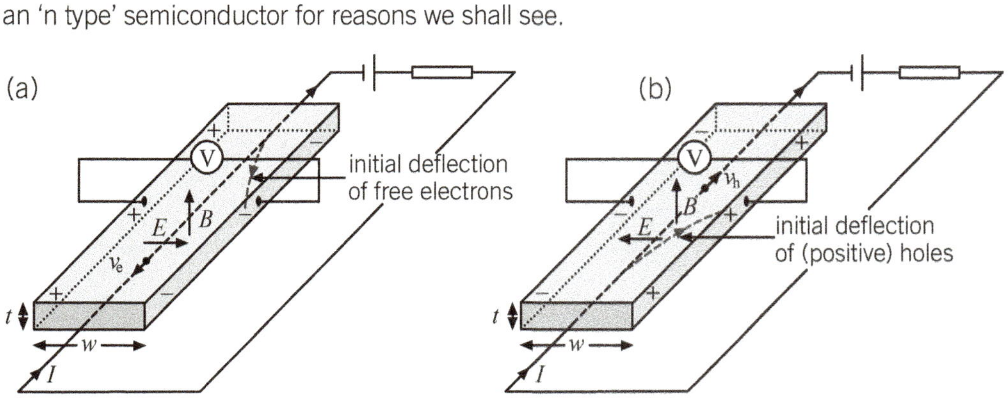

Fig. 4.4.21 Hall effect in: (a) n-type material (b) p-type material

Study point

There is a longitudinal electric field in the wafer, which causes the current. This means that the voltmeter will show a pd even with no magnetic field, if the pick-up points are not exactly opposite each other. This zero-error can be compensated for.

Study point

Pure silicon has equal numbers of electrons and holes, but the drift speed of the holes is lower, so there is still a resultant Hall voltage.

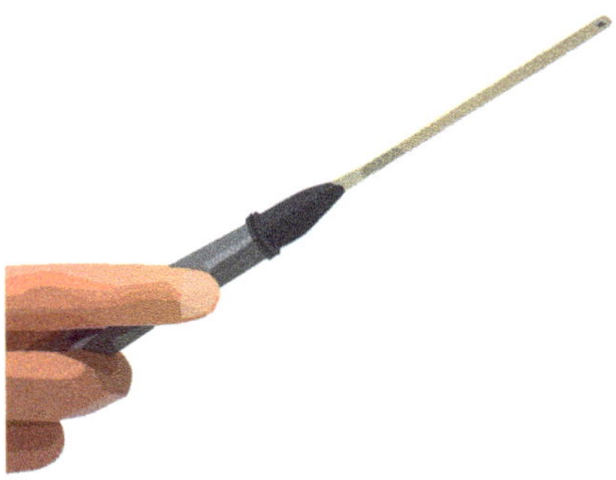

Fig. 4.4.22 Hand-held Hall probe (the wafer is near the end of the stick)

4.4.13 Knowledge check

In copper, $n = 8.5 \times 10^{28}\ \text{m}^{-3}$; in quite heavily doped n-type silicon n is typically ~ $1 \times 10^{22}\ \text{m}^{-3}$.

Explain in physical terms why the Hall effect is much greater in the silicon than the copper.

Stretch & challenge

Discuss how you would seek to maximise the Hall voltage across a copper specimen. Consider both the size and shape of specimen, and the circuit in which it is included.

A circuit is set up to send a current (usually a few mA) lengthways through the wafer (Fig. 4.4.21 (a)). See Study point. With no applied magnetic field there should be no pd between points exactly opposite each other on the side faces, so the voltmeter should read zero.

When a magnetic field is applied in the direction shown, across the whole of the wafer – not difficult, because the area of its largest face may well be less than that enclosed by this letter 'O' – the voltmeter indicates a small pd, V_H with the left-hand face at a higher potential.

This is just what we'd expect if free electrons in the wafer had been deflected, as shown in Fig. 4.4.21 (a), in accordance with Fleming's left hand motor rule (FLHMR), remembering their **n**egative charge! [Hence n-type]

Silicon doped with indium is an example of a 'p type' material. In such a material the main mobile charge carriers are positively charged; they are called **holes**. [They arise from covalent bonds that are missing an electron.] If we repeat the previous experiment with a wafer made of p-type material we find that the Hall voltage is the other way round; the right-hand face is at a more positive potential than the left-hand. We infer that the predominant charge carriers are indeed positive, deflected as shown in Fig 4.4.21 (b), again in accordance with FLHMR.

The magnitude of the Hall voltage

Because of the magnetic deflection of the charge carriers (electrons or holes) the side-faces of the wafer acquire charges as shown in Fig. 4.4.21 (a) or (b). These set up a transverse (width-wise) *electric* field. Within a fraction of a second of applying B, the force on the charge carriers due to the electric field is equal and opposite to the magnetic force, so there is no further build-up of charge.

Then $\quad (\pm e)E = (\pm e)vB \quad$ that is $\quad E = Bv$

This is the 'crossed fields' condition for the charge carriers, with drift speed v, to pass undeflected through the wafer. The magnitude of the Hall voltage, V_H, measured between points separated by the width, w, of the wafer with width-ways electric field strength, E, is (see Section 4.1.6)

$$V_H = Ew \quad \text{so} \quad V_H = Bvw$$

It helps to re-express v in terms of the current I through the wafer, using our old friend $I = nAve$. A is the cross-sectional area of the conductor normal to the current direction, that is the area, wt, of the small face.

So $\quad v = \dfrac{I}{nwte}$

Thus $\quad V_H = \dfrac{BIw}{nwte} \quad$ that is $\quad V_H = \dfrac{BI}{nte}$

in which n is the free electron or hole concentration (number per unit volume) and t is the thickness of the wafer (Fig. 4.4.21).

Note these features of the equation:

- V_H is proportional to B, the magnetic flux density. A Hall wafer, mounted in a 'probe' (Fig. 4.4.22), with a connecting cable to an external current supply and voltmeter (sometimes calibrated directly in mT), provides a practical way of measuring B, even when it varies over distances of a few mm. Measuring the force on a current-carrying wire isn't practicable in such cases (nor, indeed, we now confess, in most others!)
- The smaller the thickness, t, of the wafer and the smaller the free electron concentration, n, the larger V_H. This is because, for a given current, I, decreasing either of these factors increases the drift speed, v, and therefore the magnetic force on each charge carrier.
- The factor $\dfrac{1}{ne}$ in the equation for V_H is called the 'Hall coefficient'. It contains the quantities determined by the wafer *material*. A minus sign is sometimes put in front of it if the predominant charge carriers are negative.

4.4.5 The magnetic field due to a steady current

Almost all of this chapter so far has concerned the *effects* of magnetic fields. We've not discussed in any systematic way how the fields are related to their *sources*, though we've described the magnetic field around a bar magnet and mentioned the use of current-carrying coils to provide magnetic fields. It's time to put this right – by looking in more detail at the magnetic fields due to steady currents.

(a) The magnetic field due to a current in a circuit

A steady current requires a complete conducting circuit (including a cell or a power supply). We find that the circuit produces a magnetic field and that the lines of flux are linked with the circuit in the same topological way that handles of keys are linked with a key ring.

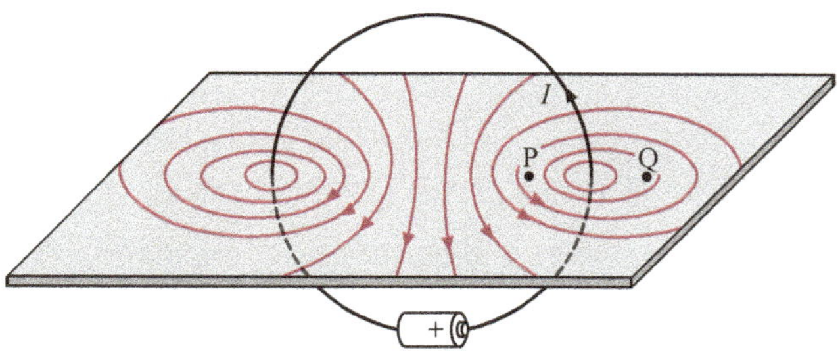

Fig. 4.4.23 Magnetic field due to current in a circuit

The direction of the field is given, as we can show with a compass, by the **right hand grip rule**. You need to be able to use it confidently.

The rule can be applied all round the circuit; it's natural to suppose that each part of the circuit is contributing to the magnetic field. The flux density at any point in the vicinity of the circuit is the vector sum of flux densities due to all the little bits of the circuit.

(b) The magnetic field due to a current in a long straight wire

The wire will have to be part of a circuit, of course, but we shall consider the field only in a region which is close to the wire and far from (i) the **ends** of the straight part, (ii) the rest of the circuit.

We can plot the field with a compass (if the wire is vertical). The current, I, needs to be large enough for the wire's field to be much greater than the Earth's. The field lines are circles around the wire (Fig. 4.4.25).

The magnitude of the field strength (flux density), B, can be investigated with a Hall probe, orientated at each point to give maximum Hall voltage (See Section 4.4.4 (b)). We find that for points at various distances, r, from the centre of the wire, $B \propto \frac{1}{r}$, so doubling r halves B, and so on: inverse proportionality. We also find that at any given point $B \propto I$.

Knowledge check 4.4.14

Give one example of a circuit which carries a current, but not a steady current.

Key terms

The **right hand grip rule**: Grasp the conductor in your right hand with your thumb sticking out in the direction of the current. The fingers will curl round in the direction of the field. See Fig. 4.4.24.

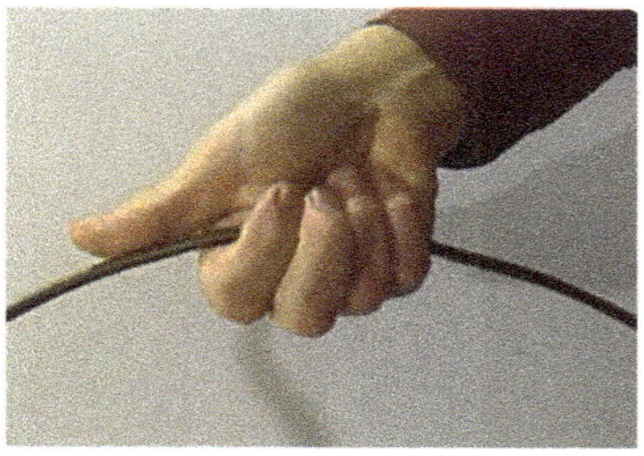

Fig. 4.4.24 Right hand grip rule

Knowledge check 4.4.15

Remembering that magnetic flux density is a vector, explain why we should expect the field lines to be closer at P than at Q in Fig. 4.4.23.

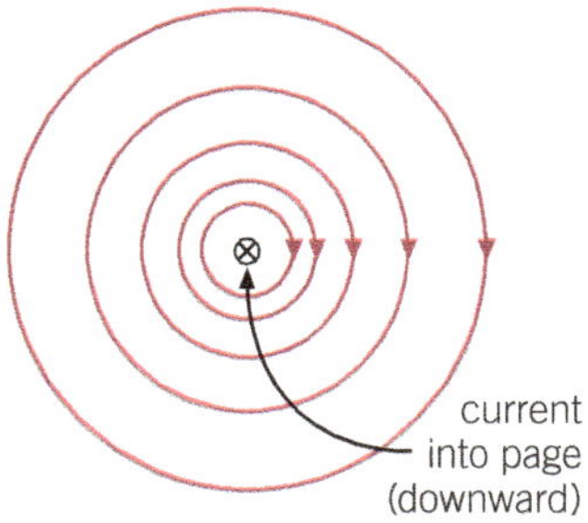

Fig. 4.4.25 Selected field lines due to current in long straight wire

>>> **Study point**

Keeping the 2π showing on the bottom line of the equation, and not dividing it into μ_0 until we put figures into the equation, isn't so silly: the $2\pi r$ is a reminder of the cylindrical symmetry of the field around the wire.

The dependences of B on both r and I are contained in the equation

$$B = \frac{\mu_0 I}{2\pi r} \text{ in which } \frac{\mu_0}{2\pi} \text{ is the constant of proportionality.}$$

μ_0 is an experimental constant officially called *the permeability of free space*. Blame the 19th-century pioneers! Most of us call it 'mu nought'. Its value is almost exactly

$$\mu_0 = 4\pi \times 10^{-7}\ \text{T m A}^{-1} = 4\pi \times 10^{-7}\ \text{H m}^{-1}\ (\text{henry metre}^{-1})$$

The henry (H), is the unit of self-inductance which is introduced in Section 4.5.6.

4.4.16 Knowledge check

Using the data in the example:

(a) determine the magnetic flux density at point Y, 35 mm due North of the wire in Fig. 4.4.26,

(b) determine the force on 0.20 m of the wire due to the Earth's horizontal field.

Example

If there is only a small current in a vertical wire, its field strength a few centimetres away will be comparable with the horizontal component, B_{EH}, of the Earth's magnetic field. B_{EH} varies from place to place and over time, but assume a value of 20 μT due North. The resultant magnetic field is shown in Fig. 4.4.26. Point X (due East of the wire) is a null point – where the (horizontal) field strength is zero.

How large a current must there be in the wire for X to be 35 mm from the wire?

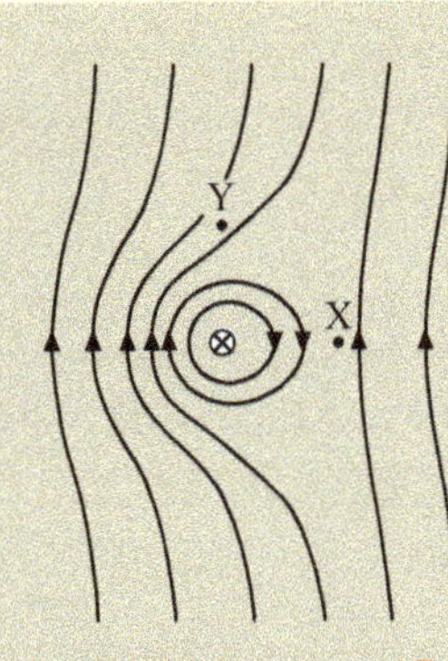

Fig. 4.4.26 Resultant of Earth's field and field due to wire

Answer

The magnitude of the wire's field at X is equal to that of the Earth's horizontal field.

$$\frac{\mu_0 I}{2\pi r} = B_{EH}, \text{ so } I = \frac{2\pi r\, B_{EH}}{\mu_0} = \frac{2\pi \times 0.035\ \text{m} \times 20 \times 10^{-6}\ \text{T}}{4\pi \times 10^{-7}\ \text{H m}^{-1}} = 3.5\ \text{A}.$$

>>> **Study point**

Some people like to imagine the field lines to the left of the wire in Fig. 4.4.26 acting like taut catapult strings pushing on the wire. You can use this analogy if it really appeals to you but Fleming's left hand motor rule and $F = BI\ell \sin\theta$ tell us all about the force on the wire.

(c) The magnetic field due to a current in a coil

Where the field lines pass through the plane of the loop in Fig. 4.4.23 the fields due to all parts of the loop are in the same direction, and vector addition leads to reinforcement. We can make the field here (and at every point) N times stronger still, by replacing the loop with a coil of N turns of insulated wire, giving, effectively, N times the loop current.

If we use a Hall probe to measure the field along the axis of a 'flat' circular current-carrying coil (Fig. 4.4.27(a)) we find that it varies with distance z from the centre of the coil as shown in Fig. 4.4.27(b).

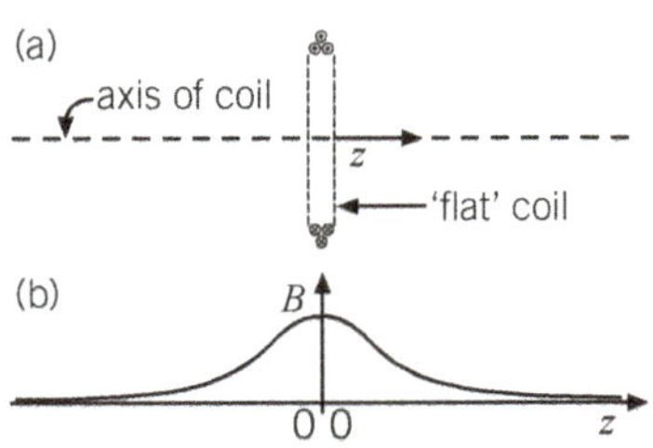

Fig. 4.4.27 (a) Section through a coil (b) Magnitude of B along coil axis

We now concentrate on the **solenoid**, a coil whose turns are wound in a close-packed helix, usually around a 'former' in the shape of a tube. [The word 'solenoid' means tube-like.] Field lines due to current in a **long solenoid** (length >> diameter) are shown in Fig. 4.4.28. Note these points:

>>> **Study point**

The field pattern of the solenoid makes sense if you imagine the solenoid sliced (like a baguette) into flat coils, each producing fields along the axis, as shown in Fig. 4.4.27(b). The solenoid's field is the sum of the 'staggered' fields due to the slices.

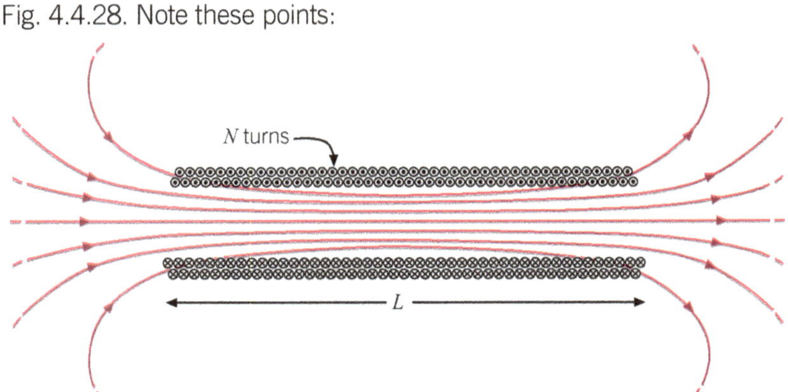

Fig. 4.4.28 Field due to current in long solenoid

- The field lines are closed loops, and the pattern resembles that of a bar magnet with its North pole near the solenoid's right-hand end. North because, with the current as shown, the field lines emerge from the right, according to the right hand grip rule.
- Inside the long solenoid, except near the ends, the field is almost uniform over the cross-section and along the solenoid. See first **Study point**. Its magnitude is given by

$$B = \mu_0 n I \quad \text{in which} \quad n = \frac{\text{number of turns}}{\text{axial length}} = \frac{N}{L}$$

- The same equation gives the flux density inside a toroidal coil (Fig. 4.4.29), provided that the 'tube' radius is much less than the overall radius of the torus. (A torus is an object shaped like an inflated inner tube.) A toroidal coil can be thought of as a solenoid with no ends – and no poles.

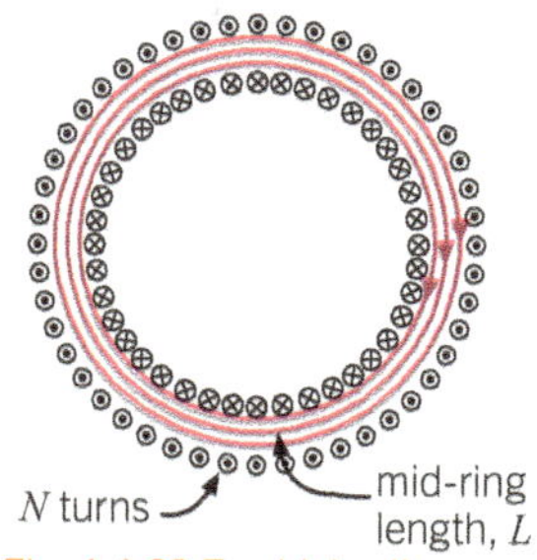

Fig. 4.4.29 Toroidal coil

4.4.6 Magnetic flux, Φ

A striking feature of the flux density inside a long solenoid is that, for given n and I, it doesn't depend on the cross-sectional area, A. But, in a way, with a larger A you do get more field for your money, even if the field is no stronger! We can make this idea precise by defining a 'new' quantity, the **magnetic flux**, Φ.

More flux is generated by the solenoid with larger A.

We can now see the sense in calling B the 'flux density', because, writing $B_\perp = B\cos\theta$ for the component of flux density normal to the area,

$$B_\perp = \frac{\Phi}{A} = \text{normal flux per unit area}$$

The properties of magnetic field lines (Section 4.4.1) are related to those of magnetic flux. For example, in Fig. 4.4.31 all the flux that passes to the right through the disc-shaped area A_{int} 'returns' leftward through area A_{ext}, an imaginary annulus (washer shape) surrounding the solenoid and extending indefinitely.

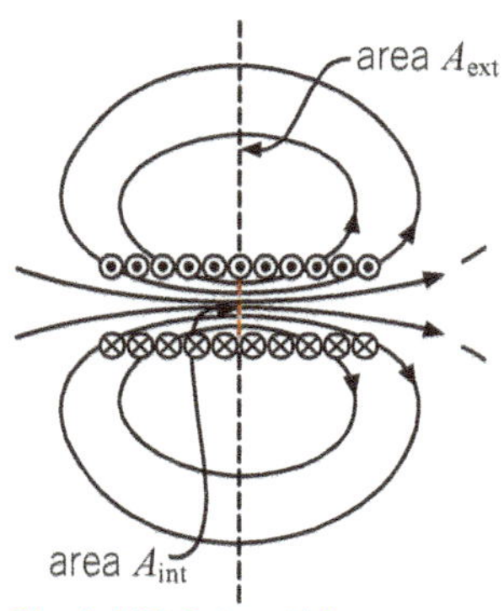

Fig. 4.4.31 Solenoid flux

Use of an iron core in a coil

We can greatly increase the flux density (and the flux), maybe by a factor of hundreds, in a solenoid or toroidal coil by including a soft iron core (Fig. 4.4.33(a)).

Soft iron is relatively pure iron. Its magnetic domains (see start of Section 4.4) are usually orientated so that their effects cancel, but the magnetic field produced by the current expands the domains that reinforce the field and shrinks those that oppose it. When the current is switched off the iron loses almost all its magnetism. (The domains are no longer aligned.)

An **electromagnet** is an iron-cored coil designed to produce a strong field beyond the ends of the core (the poles). In the electromagnet of Fig. 4.4.33(b) the field will be roughly uniform in the gap between the poles, though there is some bulging out of field lines near the edges. The flux crossing the gap will be the same as that passing through a cross-section of the core.

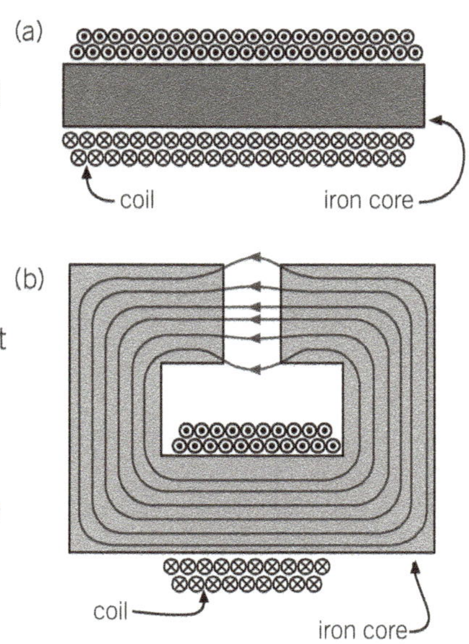

Fig. 4.4.33(a) Sections through (a) iron-cored coil and (b) electromagnet (one type)

Study point

Do note carefully the meaning of n in the long solenoid equation $B = \mu_0 n I$.

Knowledge check 4.4.17

Calculate the current needed to produce a flux density of 0.0050 T well inside a 450-turn solenoid of length 0.60 m.

Key terms

The **magnetic flux**, Φ, through an area, A is defined by $\Phi = BA\cos\theta$, in which θ is the angle between the normal to the area and the flux density (Fig. 4.4.30). It is a scalar.

UNIT T m^2 = Wb (weber)

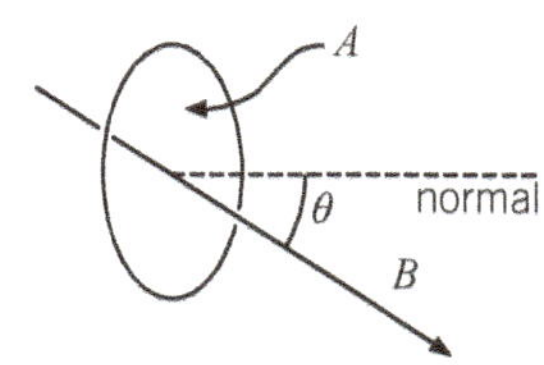

Fig. 4.4.30 Magnetic flux

Knowledge check 4.4.18

Calculate the flux passing through the central cross-section of the long solenoid in Fig. 4.4.32 when there is a current of 3.5 A.

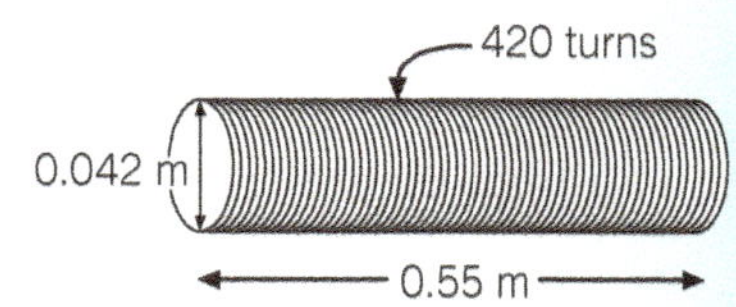

Fig. 4.4.32 Solenoid

Knowledge check 4.4.19

Explain in terms of flux and areas how you would expect the magnitude of the flux density in the air gap in the electromagnet of Fig. 4.4.33 to compare with that in its core.

4.4.7 Forces between current-carrying wires

Knowing the magnetic flux density around a long straight wire carrying a current I_1, we can calculate the motor effect force on a nearby parallel wire of length ℓ carrying current I_2.

The force on ℓ will be the same whether the set-up is that shown in Fig. 4.4.34 (a) or (b); we're not concerned with the forces on the rest of the circuit that carries I_2.

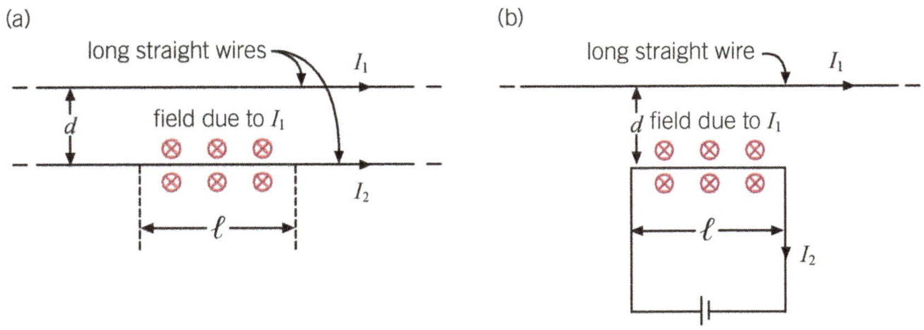

Fig. 4.4.34 Current-carrying wires that exert forces on each other

In the vicinity of ℓ, the field due to I_1 is into the page, as shown (using the right hand grip rule). So ℓ experiences a force towards I_1 (using Fleming's left hand motor rule). The magnitude of the force on ℓ is

$$F = BI_2\ell \quad \text{so} \quad F = \frac{\mu_0 I_1}{2\pi d} \times I_2\ell \quad \text{that is} \quad F = \frac{\mu_0 I_1 I_2 \ell}{2\pi d}$$

Newton's third law tells us that ℓ exerts an equal and opposite force on the I_1 wire *as a whole* (because the field that ℓ sits in is due to the current in this wire as a whole).

We see that 'like currents attract'. Check that unlike currents repel.

The ampère and μ_0

Until May 2019 the ampère was defined as follows:

'The ampère is that current which, in two long parallel wires exactly 1 metre apart in a vacuum, would produce a force between the wires of exactly 2×10^{-7} newton per metre of their length.'

Putting the values from this definition into the equation just derived gives

$$2 \times 10^{-7}\ \text{N} = \frac{\mu_0 \times 1\ \text{A} \times 1\ \text{A} \times 1\ \text{m}}{2\pi \times 1\ \text{m}} \quad \text{so} \quad \mu_0 = 4\pi \times 10^{-7}\ \text{N A}^{-2}\ (\equiv \text{H m}^{-1})$$

The ampère is now (since May 2019) defined as a current of exactly 1 coulomb per second, and the coulomb as the total charge of $6.241\,509\,074 \times 10^{18}$ protons. This number was chosen so that the new ampère is as close as possible to the same current as the old. So μ_0 is still almost exactly $4\pi \times 10^{-7}\ \text{N A}^{-2}$, but the value can be checked and, if necessary, corrected by experiment. What set-up might be used, and what would be measured?

4.4.20 Knowledge check

Calculate the resultant force on the square wire loop WXYZ in Fig. 4.4.36 due to the current in the long straight wire, and state the direction of the force. **Note**: there is no need to attempt to calculate the forces on XY and ZW, as they cancel.

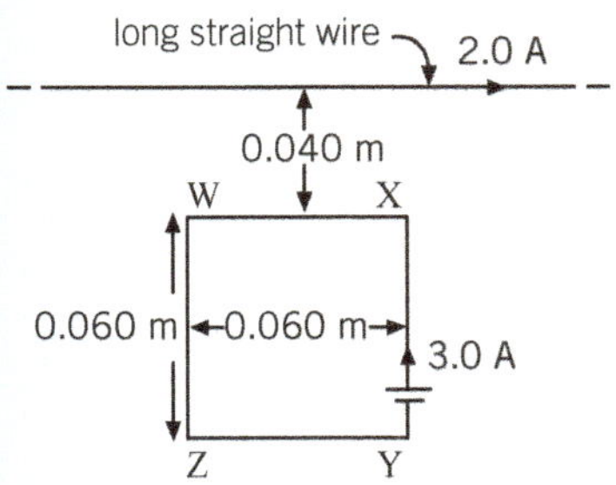

Fig. 4.4.35 Straight wire and square loop

Study point

The equation $F = \frac{\mu_0 I_1 I_2 \ell}{2\pi d}$ is not in the Data booklet, but there is no requirement to learn the equation itself.

Stretch & challenge

For the two long parallel wires of Fig. 4.4.34(a), the force on a length ℓ of the I_1 wire is equal and opposite to that on length ℓ of the I_2 wire.

(a) Why is this not a case of Newton's third law?

(b) Give a *valid* argument for these forces being equal and opposite.

4.4.21 Knowledge check

A hollow conductor in the form of a flimsy tube made of aluminium foil will implode when a very large current is sent through it from end to end. Explain why this happens.

4.4.8 Specified practical work

(a) Investigation of magnetic flux density using a Hall probe

A Hall probe contains a wafer connected as in Fig. 4.4.21 and often mounted in an enclosure at the end of a stick with a handle. The Hall pd (perhaps with a zero-adjusting facility) is fed to a meter that may read mV or μV or may have been calibrated directly in mT or μT.

Here are two possible investigations...

The field inside a long solenoid

The solenoid is included in a circuit to provide a known current. The probe needs to be of a type with wafer orientated to measure fields that are *parallel* to its stick. The reason should be clear from Figs 4.4.28 and 4.4.36.

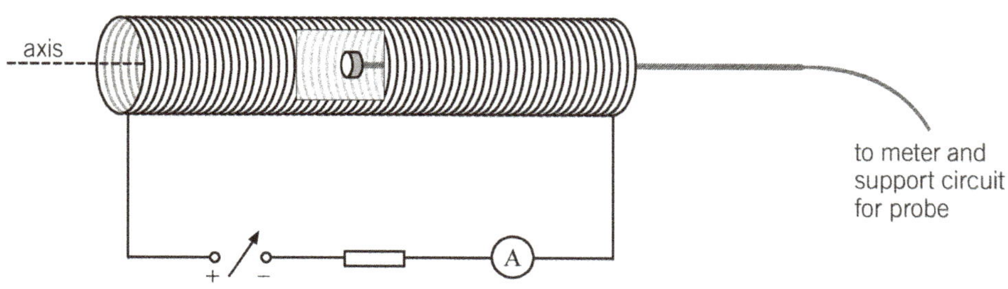

Fig. 4.4.36 Investigating B in a long solenoid with a Hall probe

(a) The flux density at the centre of a solenoid whose length is many times its diameter is given – see Section 4.4.5(c) – by

$$B = \mu_0 n I$$

Knowing n and having read the current I, we can calculate B. This enables us to check the calibration of a probe that gives a direct reading of B, or to calibrate a voltmeter wired to a special 'Hall chip'. We need to check for zero error, and having eliminated or allowed for this, to test for linearity: doubling the solenoid current should give twice the reading.

(b) With the probe near the centre, lengthways, of the solenoid we measure the field at three or four places across a diameter, to see if the field is uniform over a cross-section of the interior.

(c) The main investigation would be of the flux density, B, at different distances, z, along the solenoid axis from the centre ($z = 0$) to a little way beyond one end of the solenoid, with the goal of plotting a graph of B against z. Near the end, B decreases rapidly with z, so readings need to be at closer z-intervals than are needed at smaller z. Readings can also, of course, be taken for negative z values.

Study point

A cheap self-assembled Hall probe (Fig. 4.4.36) uses a special integrated circuit (SS495A1 chip) with built-in Hall wafer. When wired to a battery the chip produces a pd linearly related to the flux density. The pd may be fed to a digital multimeter on a *volts* range; a resolution of 0.1 mV is preferred.

Knowledge check 4.4.22

The following readings are taken from a voltmeter connected to a Hall probe correctly orientated at the centre of a long solenoid with 2200 turns per metre of its length.

I_{sol} / A	0	0.50	1.00
V_{Hall} / mV	8.2	49.8	91.0

(a) Determine the zero error, which has not been corrected.

(b) Show that the probe's response seems to be linear or close to linear, and calculate a value for the probe's sensitivity in V T^{-1}.

Stretch & challenge

For a truly long solenoid (length >> diameter), the field at the point on the axis at the very end of the solenoid is expected to be

$$B_{end} = \frac{1}{2} B_{middle}.$$

Prove this without intricate mathematics by using symmetry. Hint: half a long solenoid is still a long solenoid!

4.4.23 Knowledge check

What is the theoretical gradient of a graph of B against $\frac{1}{r}$ for a long straight wire?

Study point

'In the region of the probe, the other three sides of the coil produce almost no field *in this direction*.' (fourth bullet in main text). Check this claim using the *right hand grip rule*.

Stretch & challenge

Because of the thickness of the bundle of wires forming the coil, it may be difficult to decide where to position the zero of the scale used to measure r. For a zero error in r simply to produce a non-zero intercept we need to plot a different straight line graph from the one recommended in the main text. What should be plotted against what?

4.4.24 Knowledge check

The lower parts of the vertical portions, WX and YZ, of the wire shown in Fig. 4.4.4 experience magnetic forces. Why don't these forces affect the balance reading?

Study point

To increase the forces we are trying to measure, we could take the wire several times round the path WXYZW. This will be equivalent to increasing the current in a single wire this number of times.

4.4.25 Knowledge check

If the gradient of a graph of F against I for $\theta = 90°$ is $(4.2 \pm 0.1) \times 10^{-4}$ N A^{-1}, and ℓ is measured to be 20 ± 1 mm, calculate a value for B in the region of the wire, together with its absolute uncertainty.

The field due to a long straight wire

Fig. 4.4.37 shows a practical set-up (not the only one possible) for doing the investigation. The main problem it addresses is that a long, straight, current-carrying wire can't exist in isolation!

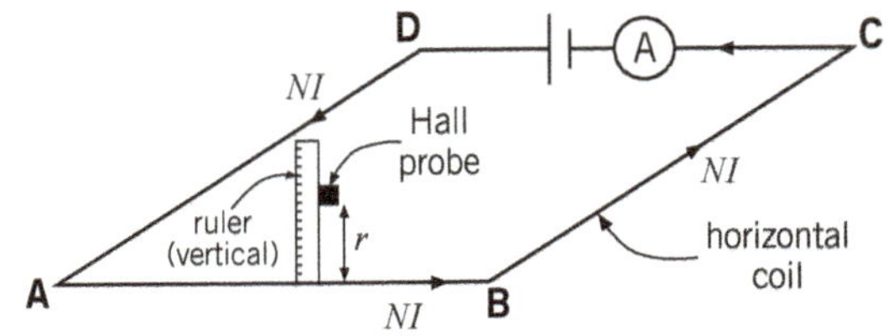

Fig. 4.4.37 Set-up for investigating B due to a long straight wire

Fig. 4.4.37 shows a large plane square coil, **ABCD**, consisting of N turns of wire, which carries a current, I. [80 cm × 80 cm fits nicely onto a laboratory bench.]

- We investigate the field due to one side, **AB**, of the coil, along a vertical line from the middle of **AB**.
- We use the ruler to measure the distance, r, of our Hall probe from the wire.
- The probe is orientated to measure magnetic fields in the direction parallel to **AD** or **BC**. To see why, use the right-hand grip rule on **AB**.
- Close to **AB**, the fields in this direction due to the other three sides of the coil are negligible.
- r shouldn't exceed about 50 mm. If r is too large the field becomes too weak to measure accurately, and **AB** won't be 'very long' compared with r.

Readings of B are taken, using the Hall probe for values of r up to about 50 mm. A graph of B against $\frac{1}{r}$ is plotted (see Stretch & challenge). It should be a straight line and the gradient could be compared with its theoretical value (see Knowledge check 4.4.23).

(b) Investigating the force on a current-carrying conductor

The apparatus is shown in Fig. 4.4.3, and indeed the investigation itself is outlined in Section 4.4.2. Here are a few practical points:

- The digital balance will no doubt be calibrated in grams, but it actually responds to force. A balance made for use in the UK registers 1.000 gram for a force of 9.81×10^{-3} N on its top-pan.
- The main investigation is of how force, F, varies with current, I, for a fixed length, ℓ, of wire (the length of the horizontal portion, XY) at a fixed angle θ (the obvious choice is 90°). Currents of up to a few ampère will probably be needed to produce forces large enough to measure reliably (see Study point).
  We plot F against I. A value for B can be found from the gradient, if ℓ has been measured with a ruler. See KC 4.4.24. Note that the value of B that emerges will be an average over XY; the field in the gap is unlikely to be very uniform.
- Another investigation could be of the dependency of F on θ. Some ingenuity is needed to measure θ reliably with a protractor. F should be plotted against $\sin\theta$, hoping for a straight line through the origin.
- It should be possible to check the dependency of F on ℓ for constant I and θ, but this may not be easy.

Test yourself 4.4

1. State the direction of the force on each red wire for the diagrams given in (a)-(e). The dot and cross symbols relate to the direction of the current or magnetic field as indicated. The circle in part (e) is a plane current-carrying coil.

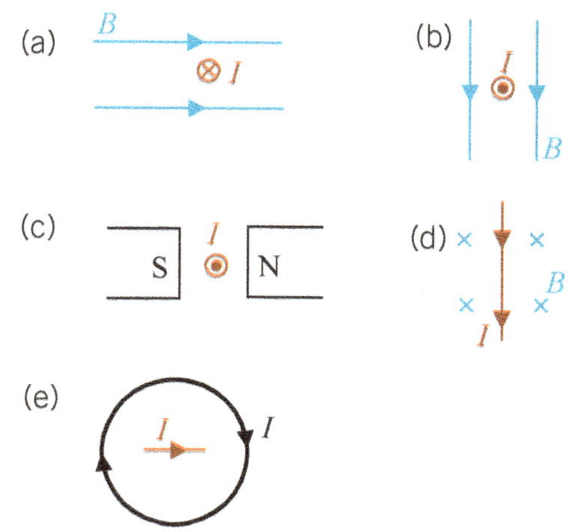

2. A horizontal wire of length $10\ \mathrm{m}$ is oriented W–E and carries a steady current of $25\ \mathrm{A}$ in the eastward direction. The flux density of the Earth's magnetic field at the position of the wire is $50\ \mu\mathrm{T}$ and its direction is downwards at an angle of $70°$ to the horizontal in a NS plane. Calculate:

   (a) the magnitude and direction of the force on the wire due to the Earth's magnetic field;
   (b) the horizontal and vertical components of the force on the wire.

3. The wire in question 2 now carries an alternating current given by $I = 50\cos 100\pi t$, where I is in A and t in s. Sketch a graph to show the variation in the horizontal component of the force on the wire over a period of $40\ \mathrm{ms}$. [Hint: first find the period of the AC.]

4. A square coil of 500 turns measures $7.07\,\mathrm{cm} \times 70.7\,\mathrm{cm}$ and carries a current of $3.00\ \mathrm{A}$. It is placed in a magnetic field of $0.200\ \mathrm{T}$ so that the sides XY and ZW are always at right angles to the field (Fig. 4.4.6). By first calculating the forces on XY and ZW, determine the torque on the coil when its plane is (a) parallel to the magnetic field (b) at $60°$ to the magnetic field.

5. Electrons of kinetic energy $1.0\ \mathrm{keV}$ are travelling at right angles to a uniform magnetic field. They are observed to travel in circular paths of radius $5.0\ \mathrm{cm}$. Calculate:

   (a) the momentum of the electrons,
   (b) the flux density of the magnetic field and
   (c) the gyrofrequency of the electrons (i.e. the number of orbits per unit time).
   [Required data: electron charge and mass]

6. Electrons in the upper atmosphere have a gyrofrequency of $1.7\ \mathrm{MHz}$. This results in radio wave absorption due to resonance. Calculate:

   (a) the flux density of the Earth's magnetic field and
   (b) the gyrofrequency of protons.
   [Required data: electron and proton charge and mass]

7. A cyclotron operates with a magnetic flux density of $0.50\ \mathrm{T}$. It is used to accelerate protons. When the protons cross the space between the dees, the pd is $1000\ \mathrm{V}$.

   (a) Calculate the frequency of the pd required.
   (b) State the energy of the protons after 2000 complete orbits.
   (c) Calculate the radius of protons' path after 2000 complete orbits.
   (d) State why a cyclotron could not be used in this way to accelerate electrons to this energy.

8. Monoenergetic charged metal ions are passed through a velocity selector. The magnetic field has a flux density of $0.10\ \mathrm{T}$. The electric field is provided by a pair of parallel plates $4.0\ \mathrm{cm}$ apart. The ions are observed to pass undeviated through the selector when the pd between them is $116\ \mathrm{V}$.

   (a) Calculate the speed of the ions.
   (b) The ions (assumed to be singly ionised) are produced by acceleration through a pd of $100\ \mathrm{V}$. Identify them by calculating their mass.
   [Data required: atomic masses]

9. The diagram shows the principle of a simple mass spectrometer, a device for separating atoms of different masses. Singly ionised atoms, which are initially of negligible energy, are accelerated through a pd, V, pass through a slit, S, and enter a uniform magnetic field.

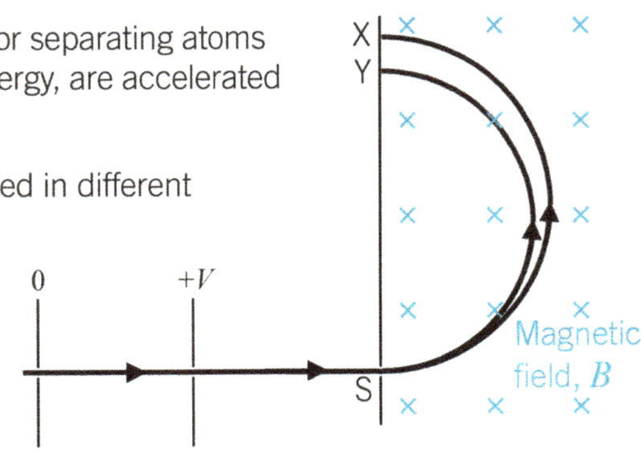

The ions travel in circular paths, and those with different masses are detected in different places, **X** and **Y**.

(a) State the direction of the magnetic field.
(b) There are two isotopes of masses, m_X and m_Y.
Give an expressions for the velocities of the two isotopes as they enter the magnetic field.
(c) Use your answer to part (b) to derive expressions for the radii of the circular paths, r_X and r_Y.
(d) A mass spectrometer uses a magnetic field of flux density 0.050 T. It is used to separate isotopes of copper. The $^{63}Cu^+$ ions were detected at a distance of 25.0 cm from S.
(i) Calculate the potential, V.
(ii) Calculate the separation (**YX** on the diagram) of $^{63}Cu^+$ and $^{65}Cu^+$ ions.

10. A Hall probe uses a p-type Silicon chip with a charge carrier concentration of 1.0×10^{15} cm^{-3}. It has a resistivity of $13.5\ \Omega$ cm. The dimensions, $w \times t \times l$ of the chip are 2 mm $\times$ 0.1 mm $\times$ 5 mm. It is supplied with a current of 5.0 mA.

(a) Calculate the resistance of the chip.
(b) Calculate the longitudinal pd required.
(c) Calculate the ratio of the Hall voltage, V_H, to the magnetic flux density, B, when the chip is placed at right angles to a magnetic field. [Care units]
(d) The contact points for measuring the Hall voltage are not exactly opposite each other on the chip. The difference is 0.01 cm. Estimate the pd between these points with zero magnetic field.

Stretch & challenge

In order to achieve the (almost) uniform magnetic field in the deflection tube shown in Fig. 4.4.12 a Helmholtz coil arrangement is used. This is two plane coils of the same radius, a, and number of turns, N, arranged on the same axis and separated by a. The coils are connected in series with the currents in the same direction. For a single plane coil, the variation of axial flux density with position, shown in Fig. 4.4.26 has the formula:

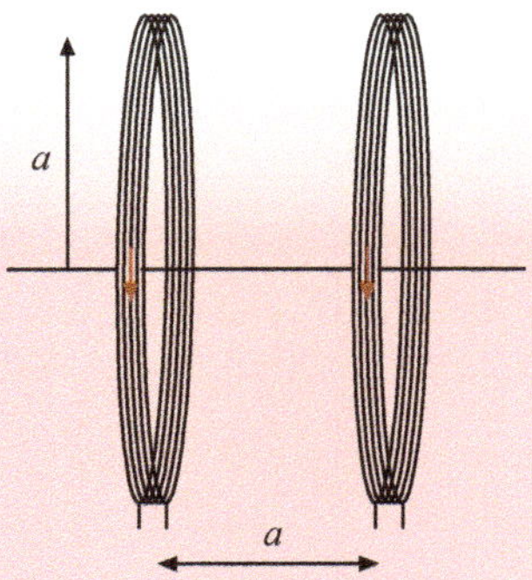

$$B = \frac{\mu_0}{2}\frac{NIa^2}{(a^2 + x^2)^{3/2}}$$

1. Plot a graph of this function from $x = -2a$ to $x = 2a$. We are only interested in the shape of the graph so you can set μ_0 to 2 and N, I and a to 1. So you should plot the function $\frac{1}{(1 + x^2)^{3/2}}$ between $x = -1$ and $x = 2$. [Hint: Use a spreadsheet]
2. Superimpose a second graph shifted to the right by a [i.e. 1], to represent the field of the second coil.
3. Add the two fields of the two coils together and comment on the value of the field between the coils.
4. The field is roughly constant because, in the middle region, the field from one coil decreases at the same rate as the other increases. This is achieved when the points of inflection of the two graphs coincide. By differentiating the flux density formula twice, show that the points of inflection are at $x = \pm\frac{1}{2}a$.

4.5 Electromagnetic induction

Electromagnetic induction is the physical effect made use of in electrical generators (for example, in power stations and in cars – to charge their batteries). Transformers (as used in electrical energy distribution and to produce low voltages from 'mains' supplies) also depend for their operation on electromagnetic induction, though the effect involved is really rather different. We explain how one equation covers both cases.

4.5.1 The emf induced in a moving conductor

An emf is 'induced' in a conductor when it moves in such a way that it cuts (moves across) lines of magnetic flux. Fig. 4.5.1 shows one way of demonstrating this: we pull the horizontal wire XY sharply upwards and observe that the voltmeter gives a reading (usually a few millivolts) while the wire is moving.

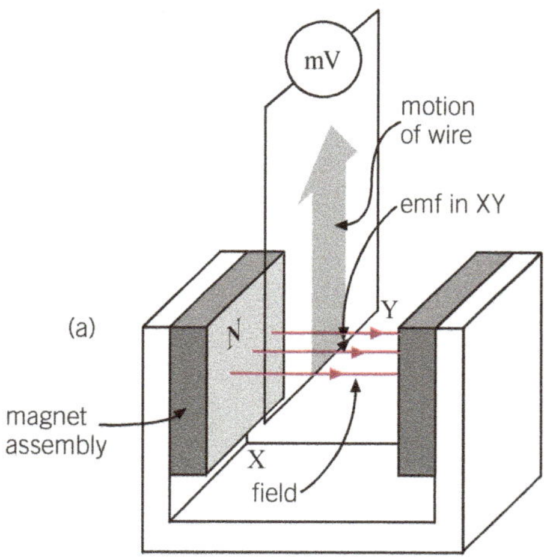

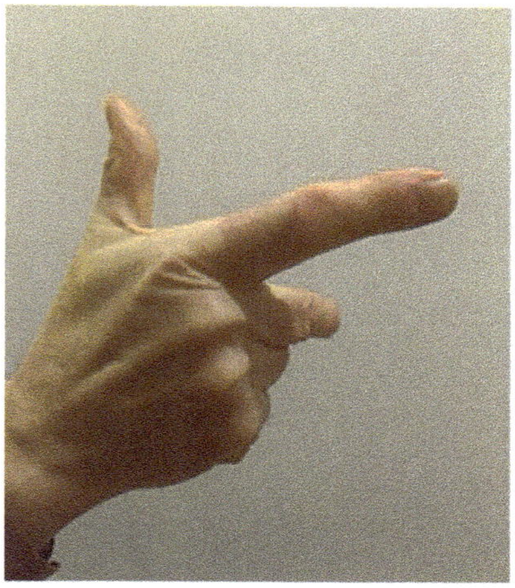

Fig. 4.5.1 (a) Wire cutting flux (b) Fleming's right hand generator rule

Note these points:

- There is no reading if we move the wire XY in Fig. 4.5.1(a) from left to right; it is not *cutting* lines of flux, but slipping between them.
- If the moving conductor is part of a complete circuit, the emf will give rise to a current. An emf in the X to Y direction in the wire XY urges a current from X to Y in the wire and therefore from Y to X in the rest of the circuit! From the voltmeter's point of view, Y acts like the positive terminal of a battery. If the circuit is *not* complete there will still be a pd between X and Y equal in magnitude to the emf. See Study point.
- The direction of the emf is given by **Fleming's right hand generator rule** (FRHGR). It's the third and last 'hand rule'!
- Strengthening the magnetic field, lengthening the wire within the field, and moving the wire faster all increase the emf. This much can be discovered using apparatus like that in Fig. 4.5.1 (a). We can derive a simple equation for the emf using a thought-experiment.

Key terms

Fleming's right hand generator rule

Hold the right hand so that:

- the **F**irst finger points straight out from the palm and in the direction of the magnetic **F**ield,
- the Thu**M**b points in the direction of **M**otion of the wire,
- the se**C**ond finger is held in towards the palm…

The se**C**ond finger points in the direction of the conventional **C**urrent in the conductor (and so in the emf direction).

Knowledge check 4.5.1

The rectangular loop, PQRS in Fig. 4.5.2 is moved steadily to the right. The loop is partly in a magnetic field that is negligible outside the dotted boundary.

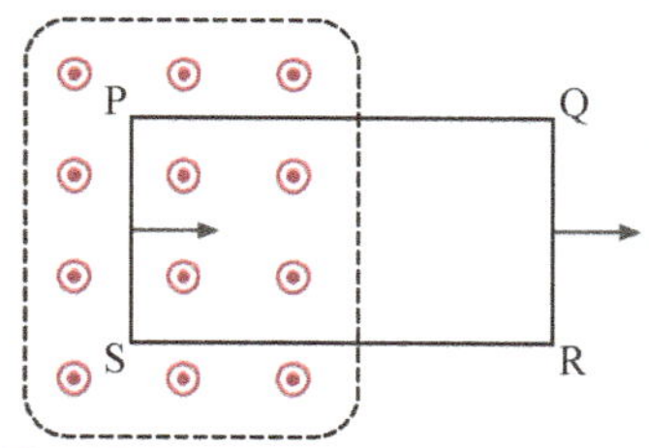

Fig. 4.5.2 Cutting of flux

(a) Explain in terms of cutting of flux why no emf is induced in sides PQ, QR or RS.

(b) State the direction of the emf induced in SP.

(c) State the direction of current in QR.

Study point

If the circuit is not complete, electrons urged towards X will pile up there, and there'll be an electron deficit at Y. The process is quickly self-limiting, stopping when the pd due to the charge separation is equal to the emf.

Key terms

Faraday's law

When the magnetic flux, Φ, through a circuit changes, an emf is induced in the circuit, proportional to the rate of change of flux linkage. For a circuit with N identical turns

$$\mathcal{E} = -\frac{\Delta(N\Phi)}{\Delta t}.$$

$N\Phi$ is the **flux linkage**. If the circuit is a coil of N turns, with the same flux, Φ, through each, the emf is the same as if there were N times the flux through a single turn!

4.5.2 Knowledge check

Show that the equation $\mathcal{E} = B\ell v$ is homogeneous in terms of units.

Study point

The minus sign in the Faraday's law equation relates to the direction of the emf: see last Study point of 4.5.3. It's no great sin to leave out the minus sign, as we shall be using the equation to find only the magnitude of the emf.

4.5.3 Knowledge check

The wire XY in Fig. 4.5.1(a) is 0.020 m long and is at right angles to the magnetic field. When it is moved upwards at 3.0 m s^{-1} the meter reads 1.3 mV. Calculate B.

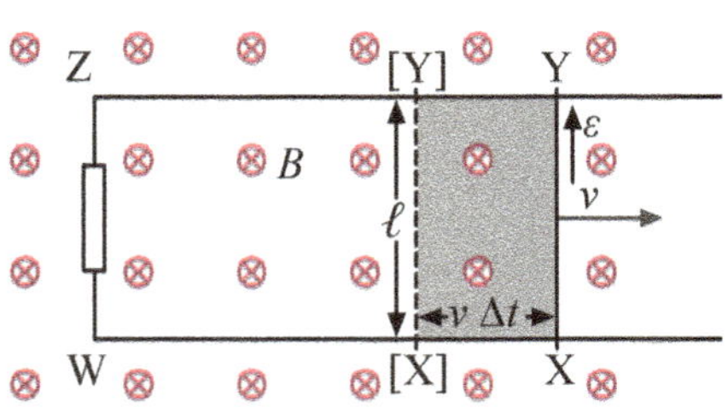

Fig. 4.5.4 Rod on rails again

How the emf arises when a conductor cuts magnetic flux

We consider a metal rod of length ℓ moving at constant velocity, v, cutting a uniform magnetic field, B. The set-up is shown in Fig. 4.5.3.

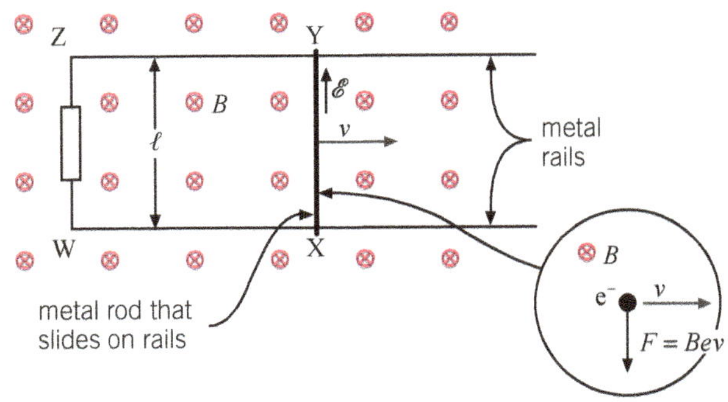

Fig. 4.5.3 Rod on rails: thought-experiment

Each free electron in the rod is carried along with the rod and so has the rod's velocity, v. It therefore experiences a magnetic force Bev as shown in the bubble. See Section 4.4.3. Free electrons in the rod are, then, urged to drift in the direction YX, making a conventional current in direction XY, (so XYZWX around the circuit). This is just as predicted by FRHGR.

We can go further... The emf, $\mathcal{E}$, is the energy per unit charge given to electrons passing through the rod, that is from Y to X. Therefore

$$\mathcal{E} = \frac{\text{work done on electron}}{e} = \frac{\text{force on electron} \times \text{distance moved in force direction}}{e}$$

So $\mathcal{E} = \dfrac{Bev \times \ell}{e}$ that is $\mathcal{E} = B\ell v$

This equation applies to the rather limited case of a straight wire, when wire, field and wire's velocity are all at right angles to each other.

4.5.2 Faraday's law

Faraday's law gives the emf induced in a circuit in the *general* case. The concept of magnetic flux, needed to make sense of the law, was introduced in Section 4.4.6, where the defining equation, $\Phi = BA\cos\theta$, is explained.

How does the law work for the rod on rails? In time Δt the rod sweeps across an area $\ell \times v\Delta t$ (shaded in Fig. 4.5.4), so, because the flux density is normal to the area ($\theta = 0$) the flux linked with the circuit has increased by an amount

$$\Delta\Phi = B \times \ell \times v\Delta t.$$

This much more flux goes *through* WXYZ than when it was W[X][Y]Z. Since $N = 1$, Faraday's law gives the induced emf magnitude as

$$\mathcal{E} = \frac{\Delta\Phi}{\Delta t} = \frac{B\ell v\Delta t}{\Delta t} = B\ell v$$

We now apply Faraday's law to some cases where using $\mathcal{E} = B\ell v$ is either not wholly straightforward – or quite impossible.

Case 1: Emf in a loop moving in a uniform magnetic field

In Fig. 4.5.5 side ZW of the wire loop WXYZ has an induced emf, Bbv, in the direction WZ and XY has an emf, Bbv, in the direction XY. These emfs are in opposite 'senses', meaning that, as we go (either way) round the circuit the emfs encountered are equal and opposite. The resultant emf is zero.

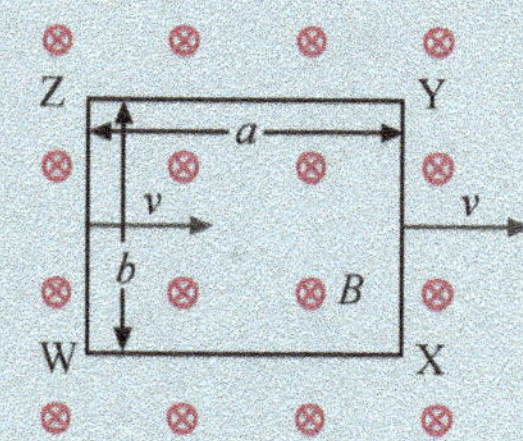

Fig. 4.5.5 Translating loop in uniform B

Faraday's law makes even lighter work of things:

As the whole loop is moving, and the field is uniform, the same amount of flux is always passing through it, so $\mathscr{E} = \dfrac{\Delta\Phi}{\Delta t} = 0$

Case 2: Shrinking band cutting flux

Suppose a round balloon with a band of metallic paint around its equator deflates from a radius of 0.16 m to a radius of 0.11 m in a time of 0.18 s. What would be the mean emf induced in the band if the Earth's field component field at right angles to the balloon's equatorial plane is $30\ \mu$T?

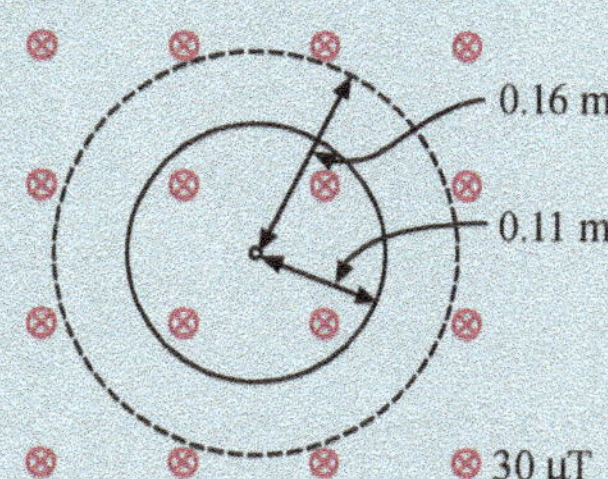

Fig. 4.5.6 Balloon deflating in uniform B

The area enclosed by the band decreases, so the flux linked with it decreases. $N = 1$ and the flux density is always normal to the area, so $\theta = 0$. We therefore have

$$\mathscr{E} = \frac{\Delta\Phi}{\Delta t} = \frac{\Delta(BA)}{\Delta t} = \frac{B\Delta A}{\Delta t} = \frac{30 \times 10^{-6}\ \text{T} \times \pi((0.16\ \text{m})^2 - (0.11\ \text{m})^2)}{0.18\ \text{s}} = 7.1\ \mu\text{V}$$

Note that this is the mean emf. See Study point.

Case 3: Mean emf in a loop turning in a uniform field

Suppose that, in a time Δt, we turn a metal loop through $180°$ in a magnetic field so that the normal to its plane shown in Fig. 4.5.7(b) changes from being parallel to a uniform magnetic field as in Fig. 4.5.7(a)) to being 'antiparallel' as in Fig. 4.5.7(c).

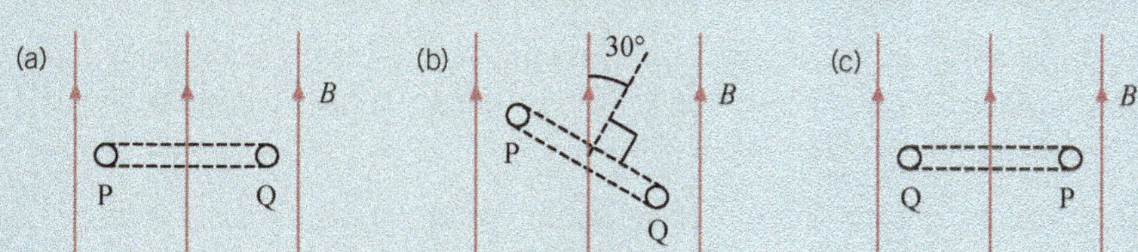

Fig. 4.5.7 Loop flipped through 180° in uniform B field (section)

Initially the flux through the loop is BA, in which A is the area enclosed by the ring. Finally, flux of the same magnitude, BA, threads through the ring in the opposite direction. This counts as a flux change of $2BA$.

So the mean emf is ..... $\mathscr{E} = \dfrac{\Delta\Phi}{\Delta t} = \dfrac{2BA}{\Delta t}$

Study point

$\dfrac{\Delta(N\Phi)}{\Delta t}$ really gives us the mean emf induced over the time interval Δt. If the flux linkage is changing at a constant rate (as in the rod on rails), the emf is constant and we can drop the word 'mean'. It's not seriously misleading to do so even when the rate of change isn't constant, provided that we understand $\dfrac{\Delta(N\Phi)}{\Delta t}$ to be applied to a small enough time interval Δt.

Knowledge check 4.5.4

The wingspan of the Airbus A380 is 80 m.

(a) Calculate the emf generated between the wingtips when the plane is flying horizontally at $260\ \text{m s}^{-1}$ in a vertical magnetic field of $45\ \mu$T.

(b) Explain why you would not expect to read this voltage on a voltmeter inside the cabin, connected to the wingtips by wires.

Knowledge check 4.5.5

Take the 'loop' in Fig. 4.5.7 to be a ring of *diameter* 30.0 cm, initially with the normal to its plane parallel to a B field of $25\ \mu$T (Fig. 4.5.7(a)). Calculate the mean emf induced in the ring when it is turned through $30°$ (see Fig. 4.5.7(b)) in a time of 0.24 s.

4.5.6 Knowledge check

Calculate the emf induced in a 75-turn coil wound tightly around the outside of the middle of the solenoid in Case 4, for the same variation of solenoid current.

Case 4: Mutual induction

This occurs when a changing current in one circuit causes an emf to be induced in another circuit, because some or all of the (changing) magnetic flux produced by the first is linked with the second.

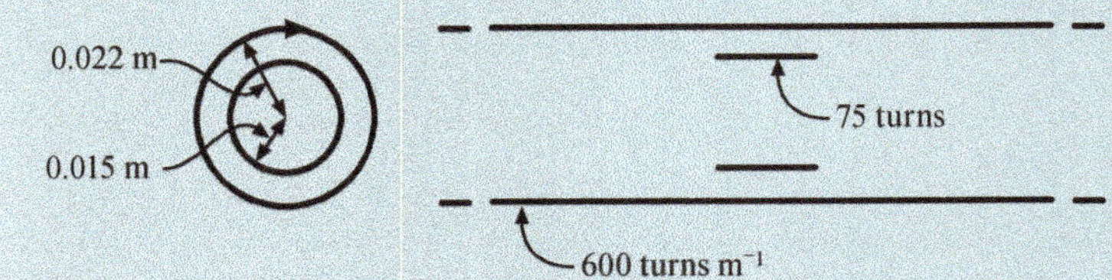

Fig. 4.5.8 'Small' coil inside long solenoid. Sectional diagrams.

In Fig. 4.5.8, suppose that (by some deft action with a variable power supply or using a sawtooth wave from a signal generator?) the current in the long solenoid is reduced at a constant rate from 3.0 A to zero in 0.25 s. What emf would be induced as a result in the 75-turn coil inside the solenoid?

In the central region of the solenoid the flux density is uniform and parallel to the axis of the solenoid. Its magnitude is $B = \mu_0 n I_{sol}$.

So the initial flux linkage for the 75 turn coil is $75\Phi = 75\ BA = 75\ (\mu_0 n I_{sol})A$ in which A = area of small coil = $\pi(0.015\text{ m})^2$

So the emf in the 75 turn coil is

$$\mathcal{E} = \frac{\Delta(N\Phi)}{\Delta t} = \frac{75 \times (4\pi \times 10^{-7} \times 600 \times 3.0\text{ T}) \times \pi(0.015\text{m})^2 - 0}{0.25\text{ s}} = 0.48\text{ mV}$$

Study point

Mutual induction is an example of when the induced emf is *not* due to magnetic forces on free electrons in a moving conductor. There *is* no moving conductor! In this case the emf arises from forces on the electrons due to an *electric* field around a closed path through which there is changing magnetic flux. Faraday's law covers both kinds of e-m induction!

4.5.3 Lenz's law

Lenz's law gives the direction of the induced emf in the *general* case. The wording of the law is deliberately rather open, as it can often be applied in more than one way. Examples will show this. We first re-visit an old friend (Sections 4.5.1, 4.5.2)… pushing it!

Key terms

Lenz's law = The direction of an induced emf is such that its effects oppose the change producing it.

(a) Lenz's law applied to the rod on rails

What counts, in this case, as the *change* producing the emf? Is it the rod moving to the right (due to some external agency pushing it)? Or is it the consequent increase in flux linked with the circuit? Relax: either will do.

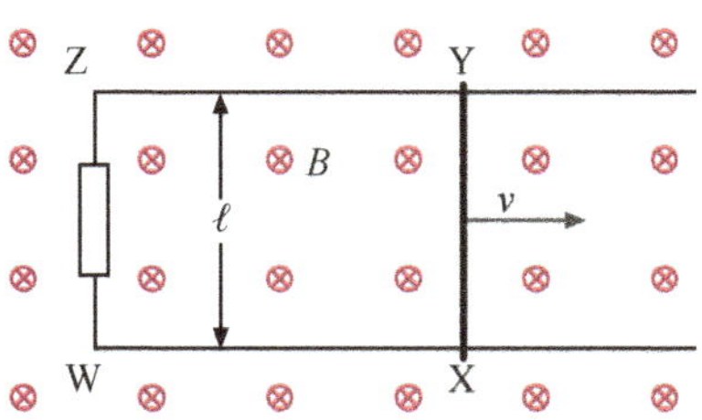

Fig. 4.5.9 Rod on rails (for the last time!)

What are the *effects* of the emf? The emf drives a current round the circuit and this in turn has two effects of interest:

1. There will be a motor effect force on the rod XY. Lenz's law demands that this must be to the left to oppose the rod's motion. Using Fleming's left hand motor rule we see that the current in the rod is in the direction XY, so the emf must be in the direction XY (see Fig. 4.5.9).
2. The current in the loop will set up its own magnetic field. Across the area WXYZ this must be out of the page, opposing the increase of 'original' flux into the page due to the increasing area. According to the right hand grip rule, to produce flux out of the page the loop current has to be in the sense WXYZW, so the emf in the rod is in the direction XY. The same conclusion as before! Choose whichever method you prefer.

Study point

If there's a break in the circuit WXYZ, there is still an induced emf in XY even though it can't drive a current round. We use Lenz's law to find the direction of the emf by *imagining* the circuit to be closed!

(b) Lenz's law and energy conservation

Why *should* the effects of the emf oppose the change producing it? Suppose the emf in the rod XY were in the other direction: then, once the rod were given an initial push to the right, the motor effect force, now to the right, would take over the rod's propulsion, while at the same time we'd be getting free heat from the resistor! Marvellous, but a violation of the Principle of Conservation of Energy.

In the real world, with the direction of emf required by Lenz's law, the external agency has to do work against the opposing motor effect force, and it is this work input which supplies the resistor's thermal energy (via the action of an electric current). In other words, Lenz's law is required by the Principle of Conservation of Energy.

Knowledge check 4.5.7

In the thought experiment of the rod on rails, let the induced current be I.

(a) Express the electrical power dissipation in the resistor in terms of B, ℓ, v, I.

(b) Show that the power input to the system is the same, by calculating the work done per second pushing the rod against the motor effect force on it.

(c) More applications of Lenz's law

We go back to Case 2 in Section 4.5.2. The emf in the band must be in the clockwise sense in Fig. 4.5.6 so that the current is in this sense and – according to the right hand grip rule – produces flux directed into the page through the area bounded by the band. This opposes the decrease, due to the shrinking of the bounded area, in the original flux.

In Case 4 in Section 4.5.2 (Fig. 4.5.8) we can similarly deduce that a decrease in the 'clockwise' current in the solenoid will induce an emf in the clockwise sense in the 75-turn coil. Check this!

Next we look at a favourite way of demonstrating electromagnetic induction: plunging a bar magnet into a coil connected to a voltmeter. Fig. 4.5.10 (a) – (c) are 'snapshots' of a magnet moving at constant speed through a coil.

In (a) the sense of the emf and current in the coil is such that flux produced by the coil goes from right to left through the coil, opposing the increase in left to right flux from the magnet. And the coil's left-hand end acts as a North pole repelling the advancing magnet!

Check that you understand the emf in (c) and the lack of one in (b).

By studying the magnet's field lines in relation to the coil in (a) – (c), observe how the sketch-graph in (d) shows the time variation of the flux linkage ($N\Phi$). The gradient of this graph gives us (minus) the induced emf, in accordance with Faraday's law. See graph (e) and Study point.

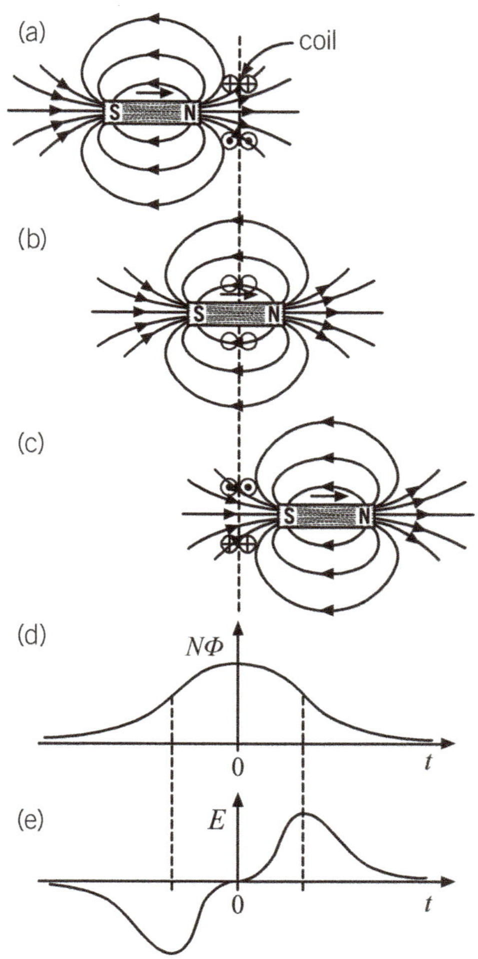

Fig. 4.5.10 Magnet passing through coil

Stretch & challenge

Explain in clear steps how the sense (clockwise) of the emf in the band in Case 2 in Section 4.5.2 can be deduced by applying Lenz's law to the motion of any small part of the band.

Knowledge check 4.5.8

Work out the direction of current in the loop in Fig. 4.5.7 as it turns through the position in (b). Should P show a current into or out of the page?

Study point

We can now justify the minus sign in $\mathcal{E} = -\frac{\Delta(N\Phi)}{\Delta t}$. By convention $\mathcal{E}$ and $\Delta\Phi$ would have the same sign if the current urged by $\mathcal{E}$ produced flux in the direction of $\Delta\Phi$. But, according to Lenz's law, the flux due to the current urged by $\mathcal{E}$ is in the opposite direction to $\Delta\Phi$!

If you find this confusing just stick with the methods of applying Lenz's law explained in the main text. They're fine for A-level.

4.5.4 The simple alternating current generator

This is a 'flat' coil that is turned by some external means in a uniform magnetic field. As it rotates, the flux linkage changes, so an emf is induced. For simplicity, we consider a rectangular coil, WXYZ, represented as a single turn in Fig. 4.5.11. The first Study point deals briefly with a practical matter.

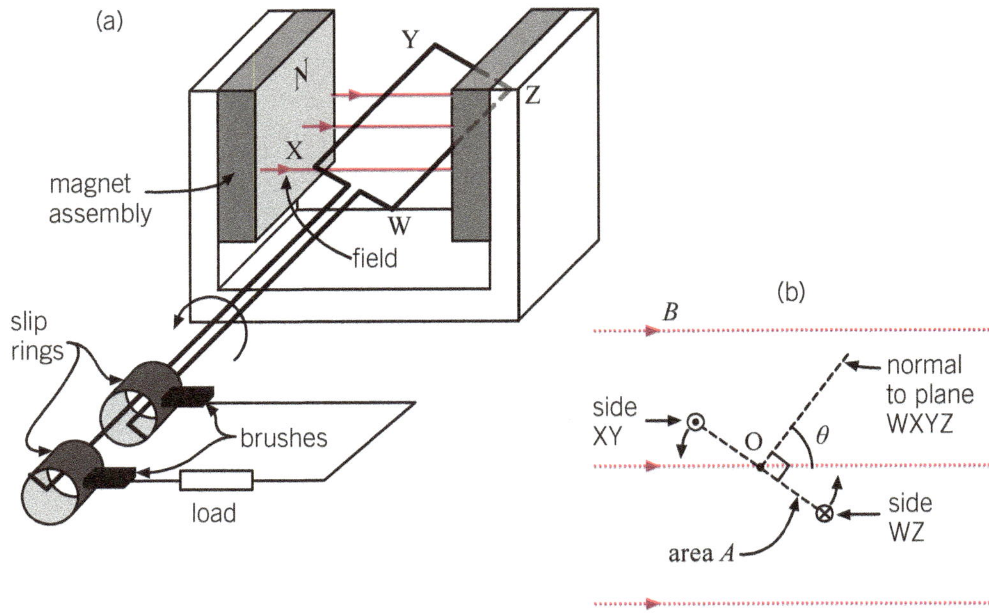

Fig. 4.5.11 (a) Simple ac generator (b) Section through its coil

When the normal to the coil (see second Study point) is at angle θ to the field, the magnetic flux, Φ, (Section 4.4.5) through the coil is

$\Phi = BA\cos\theta$ [A = side length XY × side length YZ]

For a coil of N turns, the flux *linkage* is

$$N\Phi = NBA\cos\theta$$

$A\cos\theta$ is the area that the coil presents normally to the flux. (For example, when $\theta = 0$, the coil presents 'full frontally' and $A\cos\theta = A$, but when $\theta = 90°$, the coil presents 'edge-on' and $A\cos\theta = 0$.)

Suppose that we rotate the coil at a constant angular velocity, ω. In Fig. 4.5.12 the top graph shows how $N\Phi$ varies over one revolution. The corresponding coil orientations are shown every quarter of a revolution to help you make sense of the graph. We have chosen $t = 0$ to be when the normal to the coil's plane is parallel to the field. The time scale on the graphs is given in terms of T, the time for one revolution.

The emf at any instant is, from Faraday's law, minus the gradient of the graph of $N\Phi$ against time at that instant. Hence the bottom graph. Check its shape!

You'll see that the maximum emf, $\mathscr{E}_{max}$, during the cycle has been labelled on the $\mathscr{E}$ axis as $BAN\omega$. This follows from a mathematical treatment (given in Section A.1 of the Alternating current option). For the examination you don't need to know that treatment – unless you're doing that option! Nor, strictly, do you need to know that $\mathscr{E}_{max} = BAN\omega$. However, you should appreciate that the rate of change of flux linkage is proportional to each of B, A, N and ω. Hence, by Faraday's law, $\mathscr{E}_{max}$ is proportional to each of these – which you should know.

Study point

How do we connect a rotating coil to a stationary 'load' without twisting the connecting wires to destruction? One solution, shown in Fig. 4.5.11 (a), is to connect the coil to metal 'slip rings' which rotate with the coil, and the load to carbon brushes pressing against the slip rings, but allowing slipping.

In a serious AC generator (or 'alternator'), brushes and slip rings are used to supply a rotating electromagnet. The coil in which the emf is generated (the *stator*) is stationary and wound on a soft iron core.

Study point

It was an arbitrary choice for our normal to stick out as shown, rather than in the opposite direction. But we stay with this choice. As the coil rotates, the normal rotates with it.

Study point

With the coil as shown in Fig. 4.5.11 and turning so that θ is increasing, $N\Phi$ in the direction of the chosen normal is decreasing. According to Lenz's law, the sense of the emf is as shown by ⊙ and ⊗, so any current will produce flux through the coil in the direction of this normal, opposing change in flux.

Study point

Recall from Section 3.2.3 that periodic time, T, and frequency, f, are related to ω by:

$T = \frac{2\pi}{\omega}$ and $f = \frac{1}{T} = \frac{\omega}{2\pi}$

Remember that ω must be in radians per second.

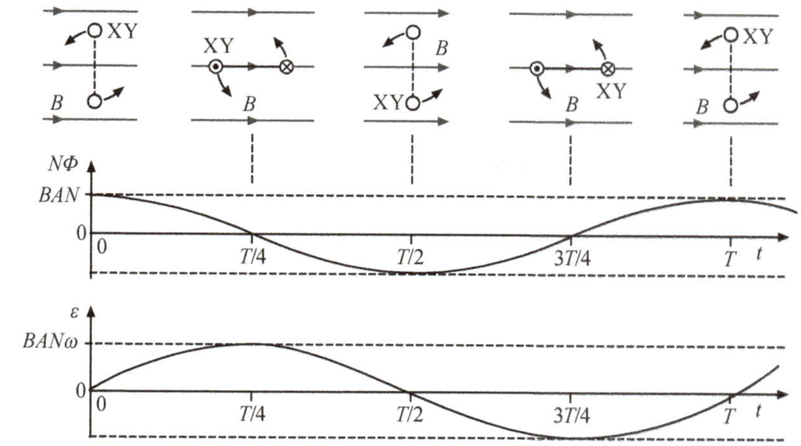

Fig. 4.5.12 Flux linkage and emf for coil turning in uniform field

Emf given by a simple AC generator: summary

The emf varies sinusoidally with time, being zero when the normal to the coil is parallel to the magnetic field (maximum flux linkage). The emf has its peak value (positive or negative when the coil normal is at right angles to the field (zero flux linkage).

As well as its dependence on angle, at any instant the emf is proportional to each of these factors:

- the number of turns, N, on the coil
- the area, A, of the coil
- the magnetic flux density, B, in which the coil rotates
- the angular velocity, ω, of the coil.

4.5.5 Eddy currents

Fig. 4.5.13 (a) shows a magnet moving towards a suspended aluminium ring. An emf is induced in the ring, as in the coil in Fig. 4.5.10 (a). The ring is effectively a single turn coil with its ends connected together ('short-circuited'), so the emf drives a current round it. The ring is repelled by the magnet in accordance with Lenz's law.

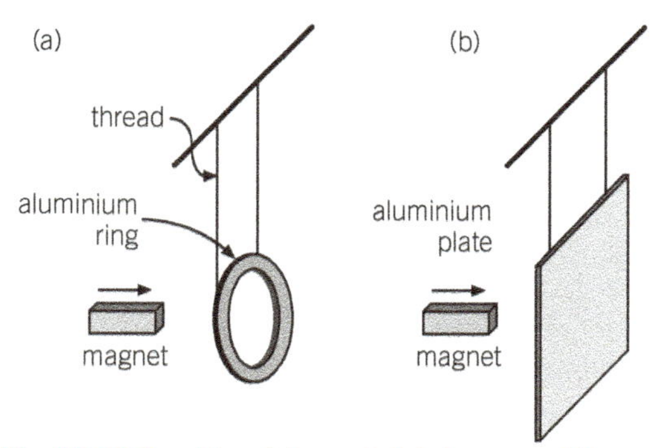

Fig. 4.5.13 Repulsion of ring and plate by approaching magnet

The plate in Fig. 4.5.13 (b) is also repelled. We can think of the plate as *containing* a ring like that in (a) – and other rings besides. The constraint of an *actual* ring or loop is not needed for the current to take such a path. Currents that follow closed loops in conducting material in bulk are called **eddy currents** (after the eddies that can occur in fluids).

A striking demonstration of the effect of eddy currents uses the apparatus in Fig. 4.5.14 (a).

Knowledge check 4.5.9

Refer to the instant shown in Fig 4.5.11.

(a) By considering the rate of cutting of flux by XY and WZ explain whether $|E|$ is increasing or decreasing.

(b) By considering the directions of forces on XY and WZ dictated by Lenz's law, explain the sense of the emf in the coil.

Knowledge check 4.5.10

A coil rotated in a uniform magnetic field of 0.060 T has a peak emf of 9.0 V induced in it when it is rotated with a time per revolution of 0.020 s.

(a) Determine the emf induced in the same coil rotated in a field of 0.040 T, with a time per revolution of 0.015 s.

(b) State the law of Physics on which your answer rests.

Top tip

The terms 'Eddy current and Electromagnetic damping' are not in the WJEC specification. However, the concepts in this section are and can form part of examination questions.

Knowledge check 4.5.11

When a bar magnet is dropped down a copper pipe, held vertically, it soon reaches a terminal velocity. The magnet takes much longer to reach the bottom than if dropped through a plastic pipe of the same length and diameter.

Explain how an electromagnetic force arises opposing the magnet's motion.

Key terms

Eddy currents are currents that take closed loops in conducting material in bulk. They arise from electromagnetic induction.

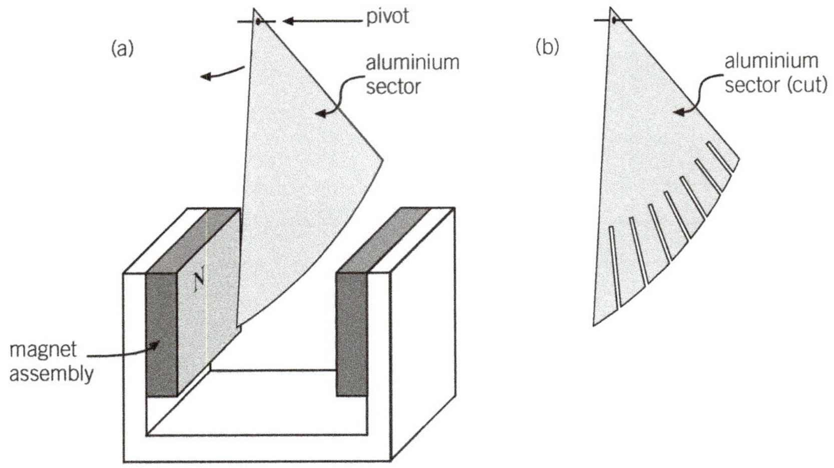

Fig. 4.5.14 Aluminium sector swinging in a magnetic field

We displace the sector through an angle from its lowest position, and release it. As it swings through the magnetic field its motion is very noticeably damped. Eddy currents in the metal experience motor effect forces that oppose the motion: **electromagnetic damping**. A convincing control experiment is to substitute a sector with radial slots (Fig. 4.5.14 (b)). The electromagnetic damping is greatly reduced, as there can no longer be large eddy current loops.

Key terms

Electromagnetic damping is the slowing of motion due to forces experienced by induced currents, in accordance with Lenz's law.

Energy dissipation through eddy currents

Eddy currents heat conductive material, just as a current heats a wire. The atoms of the material gain extra random energy of vibration – and the temperature rises. We say that energy is *dissipated*.

'Induction hobs' use eddy currents in the bottoms of suitable saucepans to heat them and hence, by thermal conduction, and (usually) convection, the contents of the saucepan. The eddy currents arise from emfs generated by rapidly alternating (and therefore changing) currents in coils underneath the pans. See Study point.

Study point

Energy is also dissipated (by domain realignment) in iron if its magnetisation is constantly changing. This is called 'hysteresis loss'. It contributes to the heating of suitable saucepans on an induction hob. It would be a nuisance in transformer cores, which are therefore made of 'soft' (low hysteresis) iron.

Eddy currents are often unwanted. For example, if the cores of coils (Section 4.4.5) were lumps of solid iron, they would heat up if the coils carried alternating currents. Changing magnetic flux in the core would be linked with closed paths in the core (as in Fig. 4.5.15).

Energy dissipation in the iron cores of transformers and of AC motors is greatly reduced by *laminating* the cores, that is making them out of thin sheets of iron, insulated from each other, so preventing large eddy current loops.

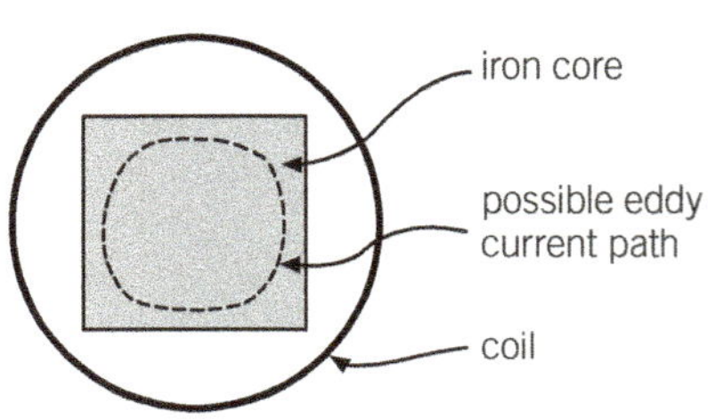

Fig. 4.5.15 Eddy currents in iron core

4.5.12 Knowledge check

The power dissipation *per unit volume*, Π, near a point P in a conductive solid is given by $\Pi = J^2\rho$ in which ρ is the resistivity and J is the *current density* near P, that is the current per unit area. Show that the equation is homogeneous in terms of units.

4.5.6 Self-induction

A changing current in one coil induces an emf in another coil, if flux from the first is linked with the second. See **Case 4**, Section 4.5.2. But flux produced by any coil is linked with the very coil producing it (Fig. 4.4.27). So we must have **self-induction**.

A striking demonstration uses an *inductor* made of many turns of wire wound around an iron core. (In Fig. 4.5.16 (a) the two coils can be connected in series so that their fields are in the same sense round the core.) The inductor is included in the circuit shown in Fig. 4.5.16 (b), which features the circuit symbol for an inductor. (The straight line next to the coil symbol represents an iron core.)

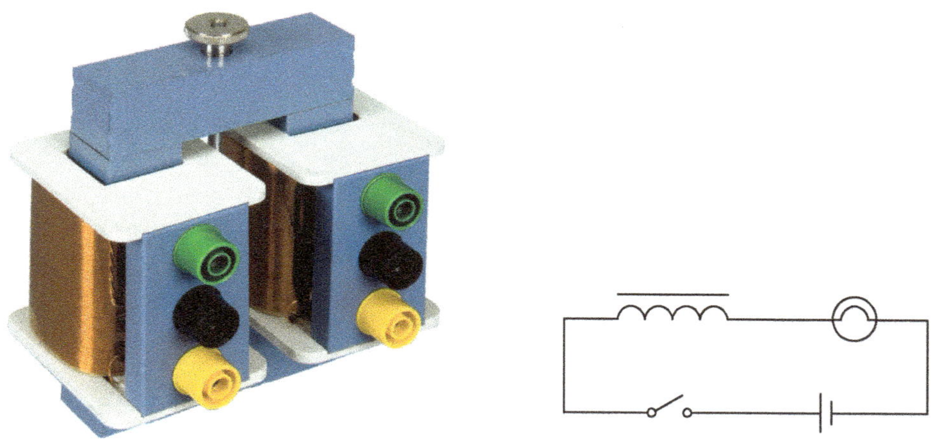

Fig. 4.5.16 (a) An inductor (b) Circuit to demonstrate effects of self-induction

When the switch is closed the lamp is slow to light, because of the emf in the inductor (often called a *back-emf*) opposing the battery emf, in accordance with Lenz's law. **When the switch is opened it is important for no-one to be touching any part of the circuit**. This is because the current, and therefore the magnetic flux, decreases to zero in a very short time, and a large emf (maybe several hundred volts) is induced in the coil. This produces a fat spark or 'arc' at the switch contacts (prolonging the current through motion of ions in the arc, thereby temporarily opposing the decrease of current!).

A very nice control experiment is possible with the two-coil inductor of Fig. 4.5.16 (a). We simply reverse the connections to one of the coils so that the fields of the coils oppose, but the resistance stays the same. The lamp now comes on with little delay when the switch is closed, and the spark is much less noticeable when the switch is opened.

Top tip

The terms Self-induction and Self-inductance don't appear in the WJEC specification, but you could meet the concepts in parts of examination questions. If you have chosen the AC option, this section contains useful introductory material.

Study point

Self-induction is the induction of an emf in a coil (or circuit) because of a changing current in that coil (or circuit).

Self-inductance

We now look quantitatively at the emf induced in a coil of N turns by self-induction, that is by a changing current, I, in the coil.

From Faraday's law, $\mathcal{E} = -N\frac{\Delta\Phi}{\Delta t}$

But $N\Phi \propto I$

So $\mathcal{E} \propto -\frac{\Delta I}{\Delta t}$ which we can write as $\mathcal{E} = -L\frac{\Delta I}{\Delta t}$

in which L is a constant for the coil, called its **self-inductance**. Indeed, in the context of electric circuits, a coil is often called an *inductor*.

The last equation may also be written in the form $V = L\frac{\Delta I}{\Delta t}$

One way to see the equivalence of the last two equations is to compare an inductor with a cell or battery (Fig. 4.5.17).

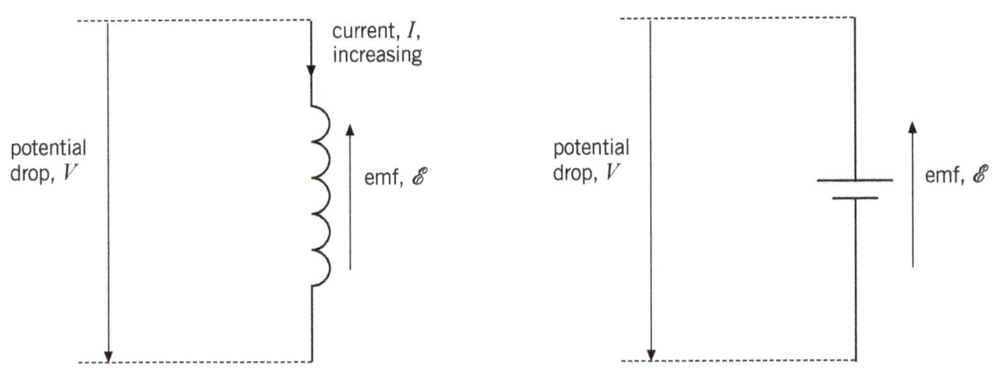

Fig. 4.5.17 V and $\mathcal{E}$ for inductor and battery

The Stretch & challenge invites you to derive a formula for the inductance of a long solenoid in terms of its dimensions and number of turns. An iron core increases the inductance of a coil (maybe by a factor of hundreds if the core is a closed loop) because of the extra flux due to the alignment of magnetic domains – see Section 4.4.6. This extra flux is not strictly proportional to the current, so the inductance of an iron-cored coil is not really a constant. We can usually make do with some sort of mean value.

Study point

When the current through a coil changes, the potential difference across it is given by

$$V = L\frac{\Delta I}{\Delta t}$$

in which L is a constant for the coil called its **inductance** (or **self-inductance**).

UNIT: Henry (H) ≡ V A^{-1} s ≡ Ω s

4.5.13 Knowledge check

Calculate the time in which a current of 2.0 A through an inductor of 4.2 H must fall to zero in order to induce an emf of 280 V.

Stretch & challenge

A long air-cored solenoid has N turns, length, l and internal cross-sectional area A.

(a) Show that the flux linkage for a current, I, is approximately

$$N\Phi = \frac{\mu_0 N^2 A I}{l}.$$

(b) Hence write an expression for the inductance, L of the solenoid.

4.5.14 Knowledge check

A pd of 3.0 V is applied to a 2.0 H inductor (assumed to have zero resistance). What happens to the current?

Test yourself 4.5

Questions 1–4 relate to the diagram of the magnet and copper loop. The magnet is moving to the **left**. We'll apply Lenz's law in two ways to predict the direction of the induced current in the loop.

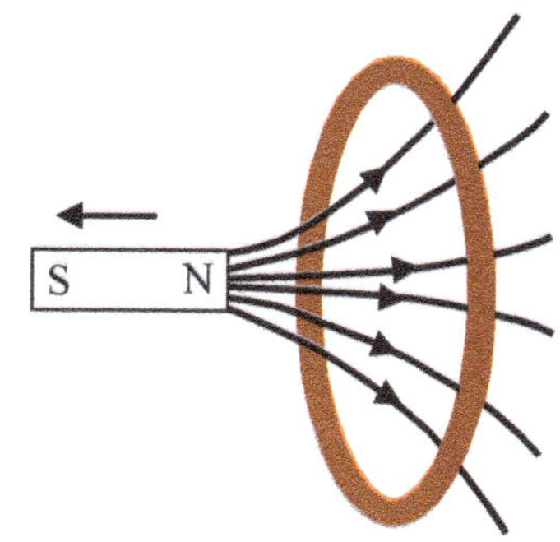

1. In terms of Faraday's law, why is there an induced current in the loop?

2. (a) What is happening to the flux density of the magnetic field in the loop?
(b) What is happening to the flux linkage in the loop?
(c) What direction does Lenz's law predict for the direction of the magnetic flux produced by the induced current? Explain your answer.
(d) Apply the right hand grip rule to predict the direction of the induced current in the loop.

3. Given that there is an induced current in the loop:

(a) What does Lenz's law predict about the direction of the motor effect force on the coil due to the interaction between the magnet's field and the induced current? Explain your answer.
(b) Use Fleming's left hand motor rule to predict the direction of the current in the top of the loop so that there is a component of the force in the direction you predicted in part (a).
(c) How does this apply to the whole loop?

4. Explain why the direction of the induced current would be the same as you have predicted in questions 2 and 3 if the magnet were the other way around **and** moving towards the loop.

5. A small upright magnet is dropped through a horizontal coil of wire, which is connected to a data logger. A graph of the induced emf against with time is shown.

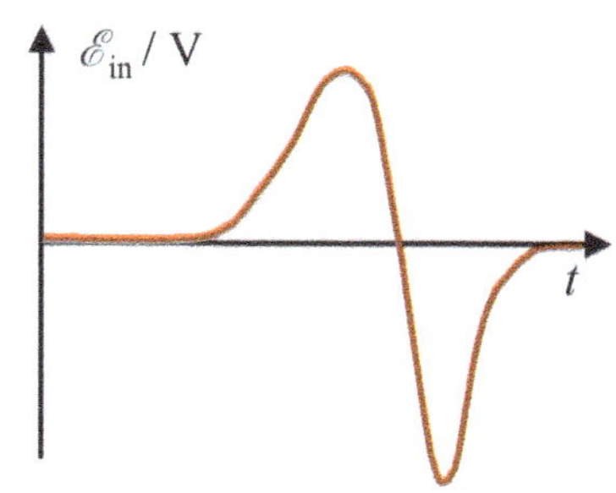

(a) Briefly explain the overall shape of the graph.
(b) Give a reason why the negative part of the graph is shorter and taller.
(c)

Stretch & challenge

Compare the areas between the positive and negative parts of the graph and explain your answer

6. A Boeing 787 Dreamliner, with a wingspan of $60\ \text{m}$ is cruising at $250\ \text{m s}^{-1}$ through the Earth's magnetic field which is $60\ \mu\text{T}$ at an angle of $70°$ to the horizontal.

(a) Calculate the pd between the wingtips because of the induced emf.
(b) Give a reason why this pd does not give rise to a current.

Questions 7–10 relate to a rectangular coil of 100 turns of copper wire, measuring $50\ \text{cm} \times 40\ \text{cm}$ that is rotating at a steady rate of 2.0 revolutions per second about an axis in the plane of the coil, parallel to a long side. At time $t = 0$ the coil's normal is parallel to a uniform field of flux density $0.050\ \text{T}$.

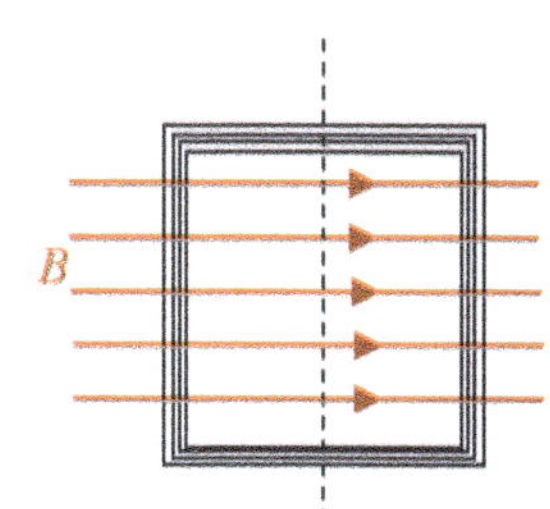

7. Calculate the flux linkage when the angle, θ, between the normal to the coil and the magnetic field is:

(a) $0°$ (b) $60°$ (c) $90°$

8. Calculate the mean emf induced in the coil as it turns from $\theta = 0°$ to $\theta = 90°$.

9. (a) Show that the speed of motion of the long sides of the coil is $2.5\ \text{m s}^{-1}$ (2 sf).
(b) State the value of θ when the long sides of the coil are cutting flux at the maximum rate.
(c) Making use of the equation $E = Blv$, calculate the maximum emf induced in the coil.
(d) Explain why we don't have to include emfs induced in the short sides of the coil.

10. The graph is of the output pd of a generator, consisting of a rotating plane coil in a uniform magnetic field.

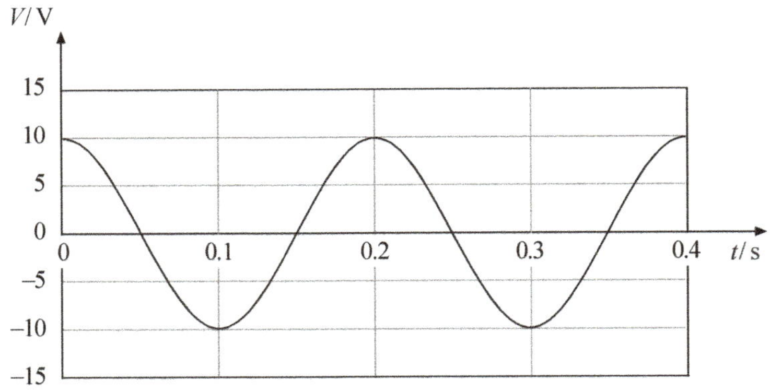

(a) State, giving reasons, the times at which the generator coil is (i) at right angles to and (ii) parallel to the magnetic field.
(b) With only trivial calculations sketch a graph to show the variation of output of the same generator with a rotational frequency 1.5× as high. Explain your answer without using algebra.

Stretch & challenge

Questions involving inductance, L, won't appear in the examination, except in the AC option.

An inductor has both inductance and resistance. It can be considered to be a series combination of a pure inductor (i.e. one with zero resistance) and a resistor. This is rather like the emf and internal resistance in a power supply.

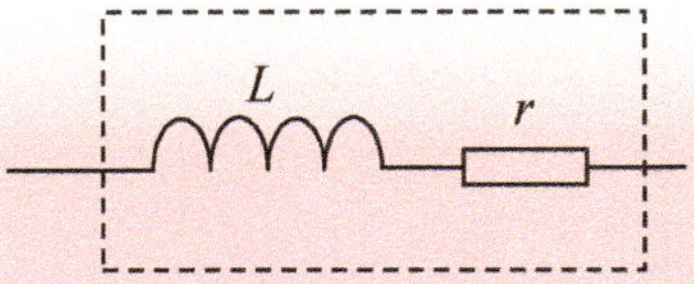

We are going to use this idea to derive the I-t relationship when an inductor is switched into a DC circuit. We'll start by considering a 10 mH inductor which has a resistance of 2Ω. The cell has a negligible internal resistance.

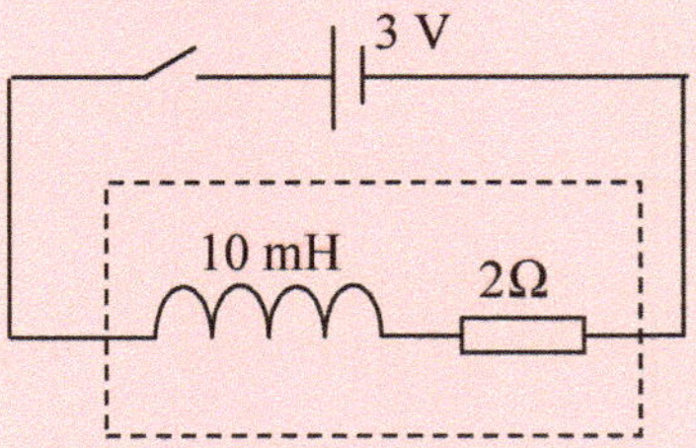

1 At the instant the switch is closed, the current is zero, so the pd across the 2Ω 'resistor' is zero and the pd across the 10 mH 'pure' inductance is 3 V.
   Use $V = L\frac{dI}{dt}$ to calculate the rate of change of current.
2 A (very) short time later, the current is 0.1 A. Calculate the new rate of change of current.
3 Repeat part 2 for currents of 0.2 A, 0.3 A, 0.4 and 0.5 A
4 What can you say about time intervals between the 0, 0.1, 0.2, 0.3, 0.4 and 0.5 A currents?
5 When the current is 1.5 A, state the pd across the inductor. What is the rate of change of current?
6 Estimate the time taken for the current to increase from 0 to 0.5 A.
7 Sketch a graph I against t.
   Now we'll move on to the general case, with a power supply V_0, an inductance L and a resistance R.
8 When the current is I, show that $\frac{dI}{dt} = \frac{1}{L}(V_0 - IR)$
9 By rearranging to give $\int \frac{dI}{V_0 - IR} = \int \frac{dt}{L}$, integrate and apply the initial condition $I(0) = 0$ to show that the relationship between I and t is $I = \frac{V_0}{R}\left(1 - e^{-(R/L)t}\right)$.
10 Show that the ratio $\frac{L}{R}$ has units of time and work out the value of this 'time constant' for the 10 mH, 2Ω inductor. Discuss how this relates to your answers to 6 and 7.

Option A: Alternating currents

Electrical currents and potential differences in circuits are very often alternating. They are generally referred to as AC [or ac] which stands for *alternating current*, even when referring to voltages! Energy is transmitted from power stations to homes by alternating currents because the voltages can easily and efficiently be stepped up and down by transformers to save energy and for safety reasons. Radio aerials convert em radiation into very high frequency electrical signals, which are controlled using resistors, capacitors and inductors. This optional unit investigates the generation, properties and control of sinusoidal electrical oscillations.

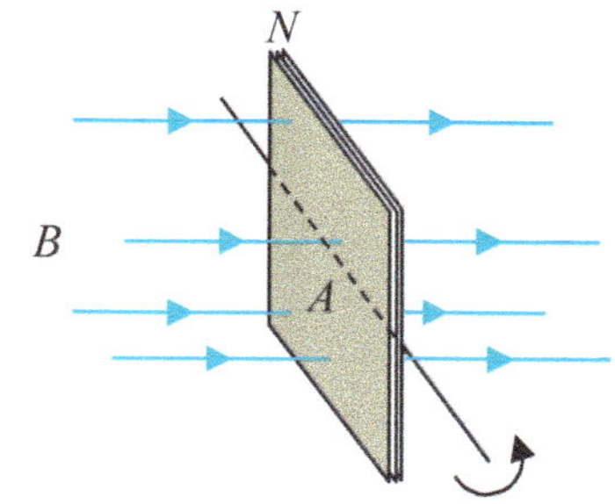

Fig. A.1 Plane coil in a magnetic field

A.1 The generation of sinusoidal voltages and currents

Fig. A.1 shows a rectangular coil with N turns and area A, placed at right angles to a magnetic field of flux density B. The flux linkage, $N\Phi$, is defined by:

$$N\Phi = BAN$$

If the coil is made to rotate about the axis shown, then the area the coil presents to the field changes and so the flux linkage changes, which induces an emf in the coil.

The coil in Fig. A.2 is shown edge on. It is rotating with angular velocity ω.

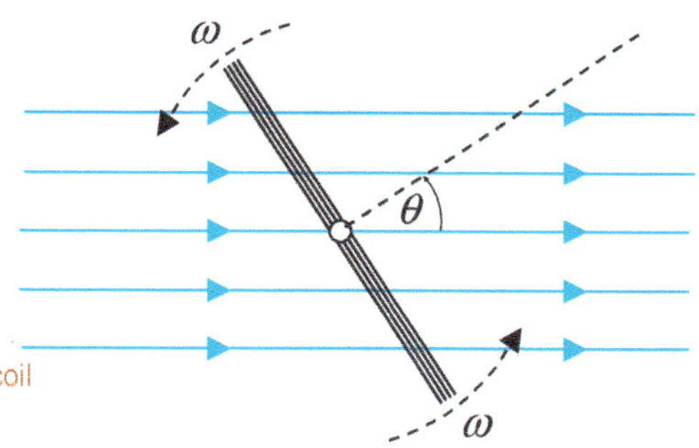

Fig. A.2 Rotating coil

The area the coil presents to the magnetic field is reduced by a factor of $\cos\theta$.

$$\therefore N\Phi = BAN\cos\theta$$

If $\theta = 0$ when $t = 0$, then $\theta = \omega t$

$$\therefore N\Phi = BAN\cos\omega t$$

Faraday's and Lenz's laws state that the induced emf, $\mathscr{E}_{\text{IN}}$, is equal to and opposed to, the rate of change of the flux linkage.

$$\mathscr{E}_{\text{IN}} = -\frac{\Delta(N\Phi)}{t} = -\frac{\text{d}}{\text{d}t}(N\Phi)$$

We saw in Section 3.2, Vibrations, that

if $x = A\cos\omega t$ then $v = -A\omega\sin\omega t$.

Remembering that v is the rate of change of x we deduce that the induced emf is given by

$$\mathscr{E}_{\text{IN}} = \omega BAN\sin\omega t.$$

If an opening is made in the coil, this induced emf results in an output pd across this opening. If no current is taken, or the resistance of the coil is negligible, the output pd, $V_{\text{OUT}} = \mathscr{E}_{\text{IN}}$, so we can write:

$$V_{\text{OUT}} = \omega BAN\sin\omega t.$$

Knowledge check A1

For the coil in Fig. A.1, state the flux linkage if:

$B = 2.0$ T; $A = 150$ cm^2;

$N = 1000$ turns.

Study point

Lenz's law: the direction of the induced emf is such as to tend to oppose the change that produces it.

Hence the '–' sign in $\mathscr{E}_{\text{IN}} = -\frac{\Delta(N\Phi)}{t}$.

Study point

Note that $\mathscr{E}_{\text{IN}}$, V_{OUT} and I are all proportional to all the significant quantities: B, A, N and ω (and remember that $\omega = 2\pi f$).

A2 Knowledge check

What emf is produced by a 1000-turn coil of area $10\ \text{cm}^2$ rotating at 3000 rpm in a magnetic field of flux density 1.5 T?

A3 Knowledge check

Sketch a labelled graph of $\mathscr{E}_{\text{IN}}(t)$ from Knowledge check A2. Add a second curve for $\mathscr{E}_{\text{IN}}(t)$ if the rotational frequency is halved.

Study point

Many commercial generators have a stationary coil and rotating magnets. This allows fixed connections to the coil (rather than slip rings).

A4 Knowledge check

From Fig. A.4, find the first time for which the pd is 0 and increasing.

A5 Knowledge check

State 3 values of θ for which $\sin\theta = 0$ [and increasing as θ gets larger].

Study point

Most of the time the phase angle θ can be taken as 0 or $\pm\frac{\pi}{2}$, so the sinusoids will be of the form

$V = V_0 \cos \omega t$,

$V = V_0 \cos(\omega t \pm \frac{\pi}{2})$,

$V = V_0 \sin \omega t$.

Study point

When taking oscilloscope measurements it is often useful to measure the *peak-to-peak* voltage, which is from top to bottom of the graph, i.e. $2V_0$.

Recapping from Section 4.5, the relationship between the coil orientation, $N\Phi$ and $\mathscr{E}_{\text{IN}}$ is shown in Fig. A.3 in which the magnetic field is horizontal and the thick black bars represent the coil, shown edge on.

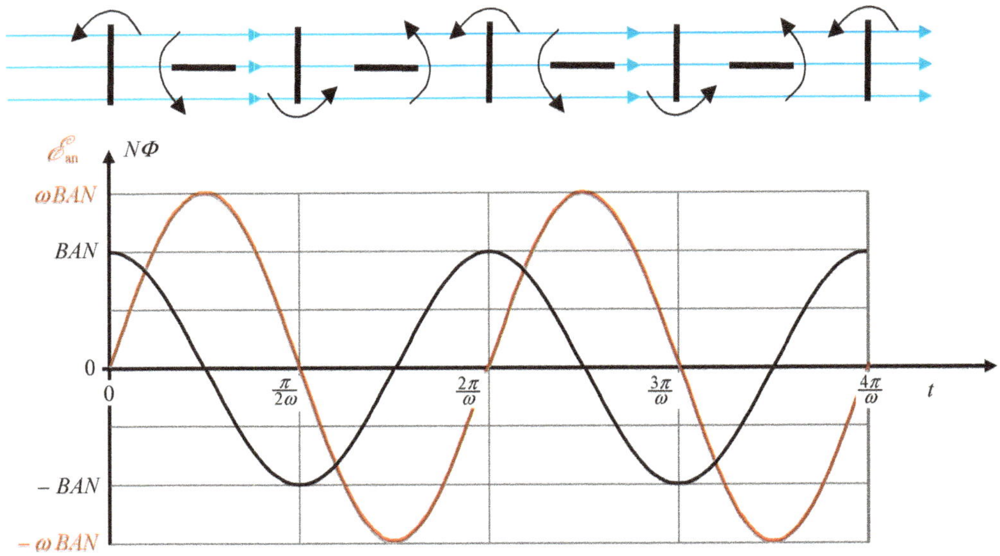

Fig. A.3 Coil position and emf

A.2 Basic AC terms

If you connect a 12 V AC supply to an oscilloscope[1] you'll see a trace which appears something like the one in Fig. A.4 [but without the axes and scales!]. It is a typical voltage–time graph for AC.

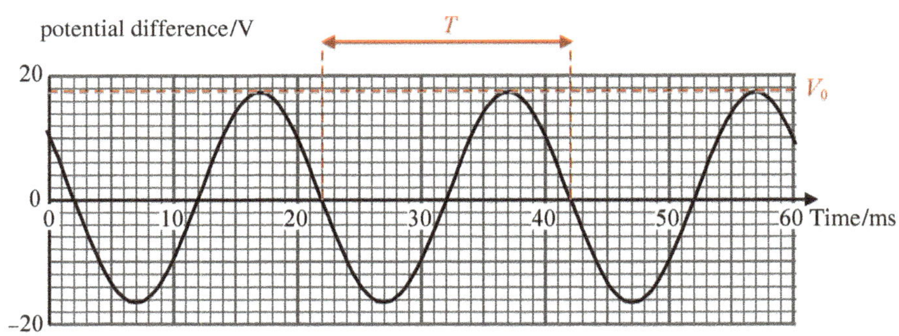

Fig. A.4 AC V–t graph

This graph has just the same form as those in simple harmonic motion. It is a sinusoid. In other words we can represent it by an equation of the form:

$$V = V_0 \cos(\omega t + \phi)$$

where V_0 is the maximum voltage, which we call the *peak voltage*.

Similarly the relationship between current, I, against time is expressed as:

$$I = I_0 \cos(\omega t + \phi)$$

and I_0 is called the *peak current*.

1 See Section A9 for oscilloscopes.

The quantity ω is called the *angular frequency* (or *pulsatance*) and is related to the *frequency*, f, and the period, T, by the equations:

$$\omega = 2\pi f = \frac{2\pi}{T}.$$

The angle ϕ is a *phase angle*. The graph in Fig. A.4 corresponds to the equation:

$$V = 17.0 \cos (100\pi t + 0.3\pi).$$

Example

Find the equation for the graph in Fig. A.4 in the form

$$V = V_0 \cos (\omega t + \phi).$$

Answer

1 The peak voltage = 17.0 V, $\therefore V_0 = 17.0\ V$

2 **2 cycles** of the graph take 42.0 – 2.0 ms = 40.0 ms. $\therefore T = 20$ ms

$$\therefore \omega = \frac{2\pi}{20 \times 10^{-3}} = 100\pi \text{ s}^{-1} [= 314 \text{ s}^{-1}]$$

3 V_{max} occurs when $t = -3$ ms. This is $\frac{3}{20}$ of a cycle

$$\therefore \phi = \frac{3}{20} \times 2\pi = 0.3\pi$$

$$\therefore V = 17.0 \cos (100\pi t + 0.3\pi)$$

A.3 rms values

At the beginning of Section A.2, we said that the graph was for a 12 V supply. But the voltage has a peak value of 17 V. What's going on?

The answer is to do with power dissipation. The 12 V figure is the voltage of the constant voltage DC power supply that would dissipate the same power through a resistor. Let's do the maths. Consider the circuit in Fig. A.5:

The power dissipated in the resistor at any instant is given by:

$$P = \frac{V^2}{R}$$

If V varies with time the mean value of the power, $\langle P \rangle$ is given by

$$\langle P \rangle = \frac{\langle V^2 \rangle}{R}.$$

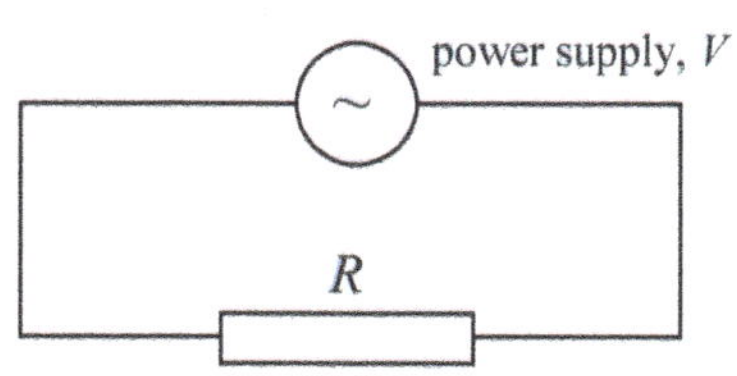

Fig. A.5 AC power supply

We **define** the **root mean square voltage V_{rms}** as $\sqrt{\langle V^2 \rangle}$.

Then the mean power dissipated is given by: $\langle P \rangle = \frac{V_{rms}^2}{R}$

So the rms voltage V_{rms} is the voltage of a smooth DC supply that would produce the same power dissipation. In the case of a sinusoidal variation, $V = V_0 \cos \omega t$, so $V_{rms} = V_0 \sqrt{\langle \cos^2 \omega t \rangle}$

Knowledge check A6

Use the answers to Knowledge checks A4 and A5 to write the equation of the graph in Fig. A.4 in the form: $V = V_0 \sin(\omega t + \varepsilon)$.

Maths tip

$\langle \cos^2 \omega t \rangle = \frac{1}{2}$. Why?

Graphs of $\cos \omega t$ and $\cos^2 \omega t$:

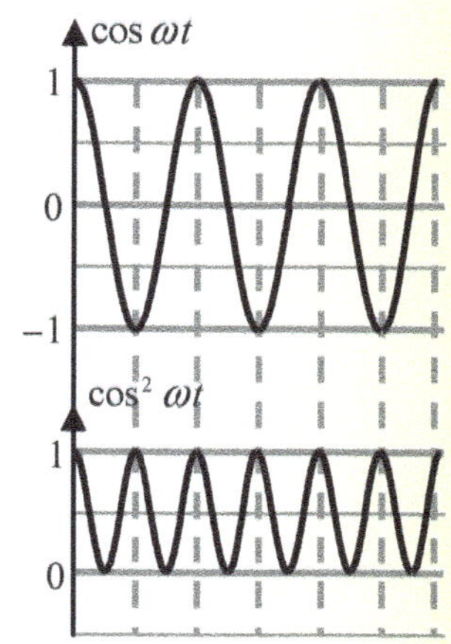

Using the trig relationship

$$\cos 2\theta = 2 \cos^2 \theta - 1$$

you should be able to show that the $\cos^2 \omega t$ graph is as shown. You need to be able to sketch this graph.

But $\langle \cos 2\theta \rangle = 0$

$\therefore\ 2\langle \cos^2 \theta \rangle - 1 = 0$

$\therefore\ \langle \cos^2 \theta \rangle = \frac{1}{2}$.

Key terms

RMS pd, V_{rms}: the square root of the mean value of the pd (taken over a whole number of cycles), i.e. $V = \sqrt{\langle V^2 \rangle}$

Similarly: $I_{rms} = \sqrt{\langle I^2 \rangle}$

Knowledge check A7

Show that the rms voltage of the supply graph shown in Fig. A.5 is 12.0 V.

A8 Knowledge check

(a) Sketch graphs of $\sin \omega t$ and $\sin^2 \omega t$.

(b) Explain why $\langle \sin^2 \omega t \rangle = \frac{1}{2}$.

Hint: See the Maths tip on the previous page.

A9 Knowledge check

An AC supply delivers an rms current of 0.2 A through a $47\ \Omega$ resistor. Calculate:

(a) the rms pd of the supply.

(b) the mean power dissipated.

A10 Knowledge check

For the graphs in Fig. A.6 calculate:

(a) the value of the resistance

(b) the frequency and angular frequency

(c) the rms voltage and current

(d) the mean power dissipated.

Study point

The power dissipation at any instant is the product IV at that moment.

The value of $\cos \omega t$ oscillates between -1 and $+1$ so its mean value is 0, but the value of $\cos^2\omega t$ is always positive and it oscillates between 0 and 1. The symmetry of the graph shows that its mean value is $\frac{1}{2}$.

$$\therefore V_{\text{rms}} = \frac{V_0}{\sqrt{2}}$$

Example

A student in Britain uses a CRO and measures the peak mains voltage, V_0 as 340 V. Calculate the value of V_{rms}.

Answer

$V_{\text{rms}} = \frac{V_0}{\sqrt{2}} = \frac{340}{\sqrt{2}} = 240$ V.

Note: Historically the nominal domestic mains (rms) voltage was 240 V in Britain and 220 V in mainland Europe. In 1995, they were 'harmonised' to 230 V with a tolerance of $+10\% / -6\%$. Both 220 V and 240 V are thus within the allowed range, and following the harmonisation, the actual domestic voltages remain as they were!

Similarly, the rms current, I_{rms}, defined as $\sqrt{\langle I^2 \rangle}$ can be used to calculate the mean power dissipated using:

$$\langle P \rangle = I^2_{\text{rms}}R = I_{\text{rms}}V_{\text{rms}} = \frac{V^2_{\text{rms}}}{R}$$

Unless otherwise specified, a quoted value of voltage or current is almost always the rms value. For power and current calculations, in circuits containing only resistors (i.e. no capacitors or inductors), the rms values can be used in just the same way as DC voltage and current values.

Example

A $15\ \Omega$ resistor is connected across a 6.0 V (rms) power supply. Calculate: (a) the rms current, (b) the peak current and (c) the mean power dissipated.

Answer

(a) $I_{\text{rms}} = \frac{V_{\text{rms}}}{R} = \frac{6.0}{15} = 0.40$ A

(b) $I_0 = \sqrt{2}I_{\text{rms}} = \sqrt{2} \times 0.40 = 0.57$ A

(c) $\langle P \rangle = I^2_{\text{rms}}R = 0.40^2 \times 15 = 2.4$ W

Note: for part (c) we could also use $\langle P \rangle = I_{\text{rms}}V_{\text{rms}}, \frac{V^2_{\text{rms}}}{R}$

or even $\langle P \rangle = \frac{I_0^2 R}{2}, \frac{I_0 V_0}{2}$, or $\frac{V_0^2}{2R}$.

A.4 Alternating currents through resistors, inductors and capacitors[2]

(a) Resistors

The current through a **resistor** at any instant is directly proportional to the potential difference at that instant. Because of this, the current and voltage oscillate **in phase** with each other. Fig. A.6 gives an example of this.

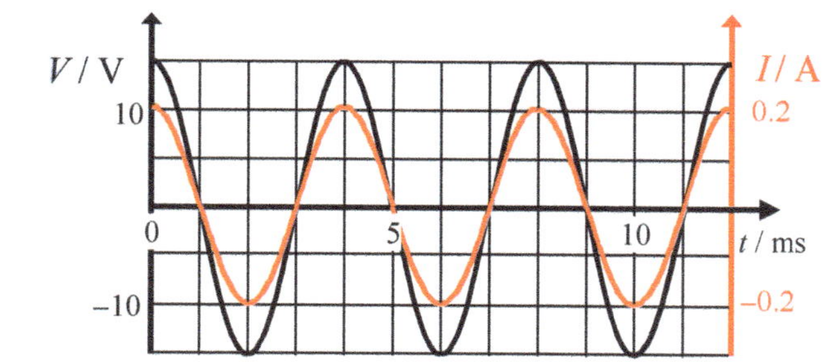

Fig. A.6 Current and pd variation for a resistor

For inductors and capacitors, there is a phase difference between V and I, so the situation is more complicated. We'll examine this phase difference qualitatively before looking at it mathematically.

(b) Capacitors

Consider a **capacitor**, C, which is initially uncharged. Suppose it is being charged up by a current, as shown in Fig. A.7.

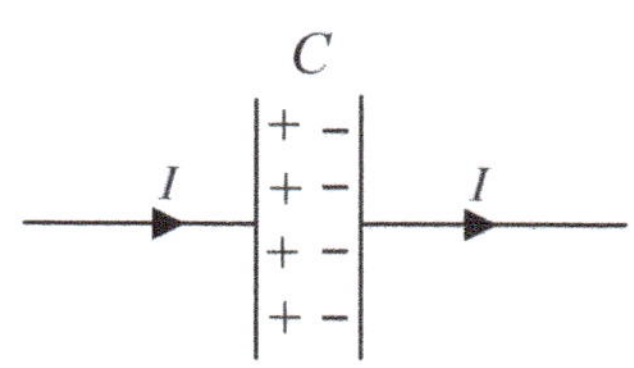

Fig. A.7 Charging capacitor

Positive charge builds up on the left-hand plate and negative charge on the right plate, so a pd becomes established.

We see from this description that, for a capacitor, current causes a change in voltage – the current leads the voltage.

Taking the conclusions from the Stretch & challenge box, the current and voltage graphs for a capacitor in an AC circuit are as in the example in Fig. A.8. The current graph is $\frac{1}{4}$ of a cycle ahead of the voltage graph, i.e. it is ahead in phase by $\frac{\pi}{2}$.

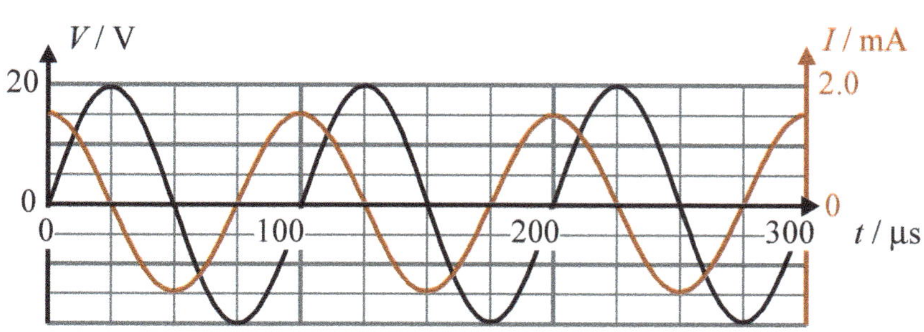

Fig. A.8 Current and pd variation for a capacitor

2 It is slightly strange to talk about the current **through** a capacitor – charge cannot flow through the dielectric, which is an insulator. The current is in the wires on either side of the capacitor!

Knowledge check A11

For the graphs in Fig. A.6 calculate the power at:

(a) $t = 0$

(b) $t = 2$ ms

(c) $t = 5$ ms.

Knowledge check A12

Sketch a graph of the variation of power with time between 0 and 6 ms for the graphs in Fig. A.6 and state the mean power.

Stretch & challenge

For a capacitor $Q = CV$.

The current, I, is the rate of change of charge, $I = \frac{dQ}{dt}$.

(a) By differentiating the first equation write I in terms of V.

(b) Let $V = V_0 \sin \omega t$. Use your answer to (a) to show that:

$I = \omega C V_0 \cos \omega t$

$I = \omega C V_0 \sin (\omega t + \frac{\pi}{2})$.

Conclusions

(1) The current is $\frac{\pi}{2}$ **ahead** of the voltage and

(2) the peak current and voltage are related by $I_0 = \omega C V_0$.

Knowledge check A13

For the graphs in Fig. A.8:

(a) Determine the values of V_0, I_0, V_{rms}, I_{rms}, T, f and ω.

(b) Use the equation $I_0 = \omega C V_0$ to calculate the capacitance, C.

(c) Confirm that the rms pd and current are related by $I = \omega C V$.

Study point

The **inductance** of an inductor is defined by the equation:

$V = L\frac{\Delta I}{\Delta t}$ or, in calculus, $V = L\frac{dI}{dt}$.

The unit is the henry (H).

Stretch & challenge

For an inductor $V = L\frac{dI}{dt}$.

Let $I = I_0 \cos \omega t$.

By differentiating this equation, show that:

$V = \omega L I_0 \cos (\omega t + \frac{\pi}{2})$

and hence that:

(1) the voltage is $\frac{\pi}{2}$ **ahead** of the current and

(2) the peak current and voltage are related by

$V_0 = I_0 \omega L$.

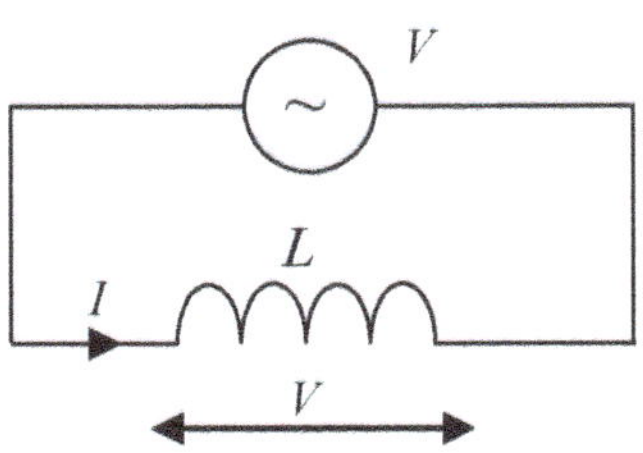

Fig. A.10 Inductor in a circuit

A14 Knowledge check

For the graphs in Fig. A.11:

(a) Determine the values of V_0, I_0, V_{rms}, I_{rms}, T, f and ω.

(b) Use the equation $V_0 = I_0 \omega L$ to calculate the inductance L.

(c) Inductors

Another component that is often found in an AC circuit is an **inductor**. It consists of a coil of wire often wound on a soft iron core. Fig. A.9 shows some images of inductors and the circuit symbol. L is the **inductance** of the inductor, defined in the box.

Fig. A.9 Inductors

The coil of wire usually has a very low resistance and its effect on the circuit is due to its magnetic properties. A current in an inductor produces a magnetic field and so there is a magnetic flux linking the circuit. If a pd is applied across the inductor so the current starts to grow, this produces a change in the flux linkage which produces an emf which, by Lenz's law, opposes the growth in current. The inductor therefore acts as a brake on a change of current: a reduction in current induces a supporting emf, which slows down the drop in current.

So, for a pure inductor (i.e. one with zero resistance) the **rate of change of current** is proportional to the voltage applied across it. For example, if a voltage of 10 V is applied across a 2 H inductor, the current grows at a rate of 5 A s^{-1}.

From the above discussion, it should be clear that an inductor behaves in just the opposite manner to a capacitor. A pd across an inductor produces a changing current, so in an AC circuit the voltage leads the current for an inductor – by $\frac{\pi}{2}$ for a pure inductor (see Stretch & challenge).

Consider an inductor connected across an AC power supply as shown in Fig. A.10. If the current varies as $I = I_0 \cos \omega t$, i.e. is at a maximum when $t = 0$, the voltage maximum is ¼ of a cycle before this, i.e. at $-\frac{T}{4}$. An example of this is shown in Fig. A.11.

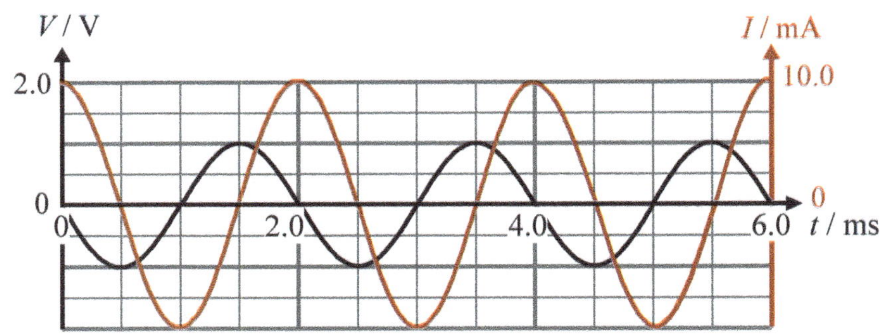

Fig. A.11 Current and pd variation for an inductor

A useful mnemonic

CIVIL: In a **C**apacitor, the current (**I**) leads the **V**oltage; the **V**oltage leads the current (**I**) in an inductor (**L**).

A.5 Capacitors and inductors dissipate no power

We have seen that **resistors** in AC circuits dissipate a mean power, $\langle P \rangle$, given by

$$\langle P \rangle = IV = I^2R = \frac{V^2}{R}$$

where V and I are the rms values. Let us examine the $V(t)$ and $I(t)$ graphs over one complete cycle to explain why this is not the case for capacitors and inductors. The graphs in Fig. A.12 are the current (red) and voltage graphs for a capacitor.

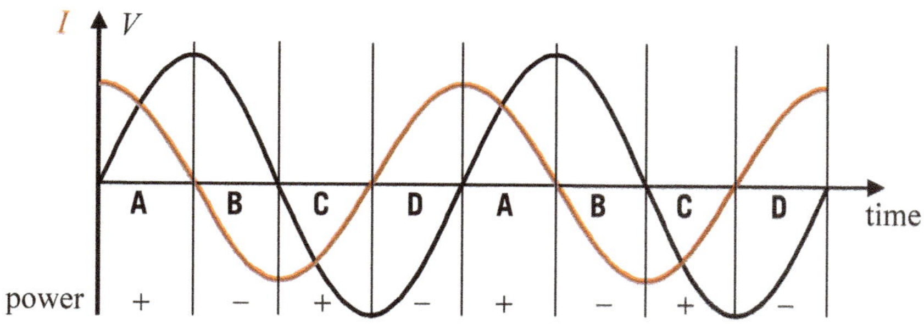

Fig. A.12 $I(t)$ and $V(t)$ graphs for a capacitor

Consider the energy exchange between the power supply and the capacitor. The instantaneous energy transfer, P, **from** the power supply is given by $P = IV$.

In the quarter-cycle intervals labelled **A**, the current and pd are both positive, so P is positive and the capacitor takes energy from the supply. In the intervals labelled **C**, both I and V are negative, so again P is positive. In **B** and **D**, one of I and V is positive and the other negative so, in these two intervals, P is negative, i.e. the capacitor returns energy to the supply. From the symmetry of the curves energy given by the power supply in **A** and **C** is the same as that returned in **B** and **D**. Thus, over a complete cycle, the mean energy taken is zero.

Let us now consider the energy balance over 1 cycle for an **inductor**. From the previous discussion, it should be clear that, because of the $\frac{\pi}{2}$ phase difference between I and V, the mean power taken from the power supply over a complete cycle is zero. It doesn't matter whether the voltage or current leads, the instantaneous power will be sometimes positive and sometimes negative.

In order to behave like this, the inductor must store energy in its magnetic field. It can be shown that the energy density is $\frac{B^2}{2\mu_0}$ in a vacuum but you do not need to know or recall this.

Stretch & challenge

Use the trigonometrical identity

$2\sin\theta\cos\theta = \sin 2\theta$

to write an expression for the instantaneous power taken from the supply by a capacitor or inductor, in terms of $\sin 2\omega t$. Deduce that the mean power over a cycle is zero.

Study point

How does the capacitor return energy to the power supply?

Unlike a resistor, a charged capacitor possesses energy in the form of electrical potential energy. This is because it has separated + and – charges. In parts B and D of the cycle, these charges are re-combining – hence releasing their potential energy.

A different, but equivalent, way of looking at it is that there is energy stored in the electric field between the plates of a charged capacitor.

Stretch & challenge

The energy density of an electric field, i.e. the energy per unit volume is $\frac{1}{2}\varepsilon_0 E^2$. Use this to derive the formula for the energy stored by a capacitor, $W = \frac{1}{2}CV^2$.

What assumption must you make?

Knowledge check A15

Show that the expressions for the energy density in electric and magnetic fields are dimensionally correct.

A.6 Resistance and reactance

We have seen that, for resistors, capacitors and inductors, the rms (and peak) voltages and currents are proportional. We have also seen that, although the current is in phase with the voltage for a resistor, this is not the case for capacitors and inductors.

For these components, or their combination, the ratio $\frac{V_{rms}}{I_{rms}}\left(=\frac{V_0}{I_0}\right)$ is referred to as the **impedance**, with the symbol Z. If the current and voltage are in phase, we call this ratio, the **resistance**, R. If they are $\frac{\pi}{2}$ out of phase, we call it the **reactance**, X. Like resistance, both impedance and reactance are expressed in ohms (Ω). Table A1 summarises the values of Z, R and X.

Component	R	X	Z	Notes
Resistor, R	R	–	R	I and V in phase
Capacitor, C	–	$X_C = \frac{1}{\omega C}$	$\frac{1}{\omega C}$	The current leads by $\frac{\pi}{2}$
Inductor, L	–	$X_L = \omega L$	ωL	The voltage leads by $\frac{\pi}{2}$

Table A1 Resistance, reactance and impedance

For individual components, we can use the values of reactance to relate I and V (rms or peak) just like resistances in a DC circuit.

Example

A capacitor, connected to a 6 V, 50 Hz supply has a reactance of 80 Ω.

(a) Calculate the current.

(b) If the frequency of the supply were changed to 100 Hz what would the current be?

Answer

(a) $I = \frac{V}{X_C} = \frac{6}{80} = 0.075$ A

(b) The reactance of a capacitor is inversely proportional to the frequency. The reactance at 100 Hz is thus half that at 50 Hz [40 Ω] so the current is doubled to 0.15 A.

We can now handle *individual* components in a circuit. The next problem is how to relate the current and voltage in a circuit with both resistive and reactive components. For example, consider the arrangement in Fig. A.13.

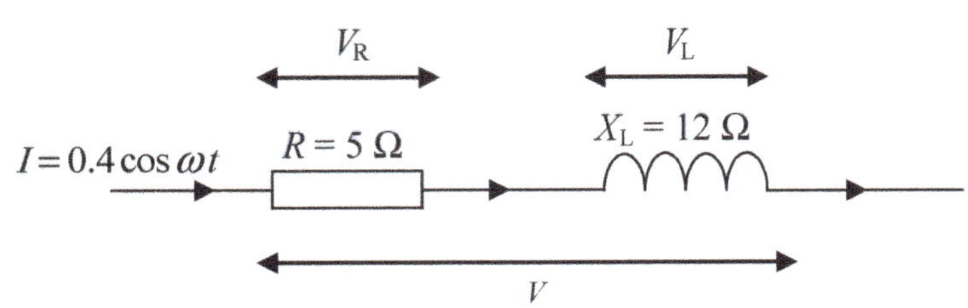

Fig. A.13 Series RL combination

From the above information, you should be able to show that V_R and V_L are given by:

$$V_R = 2.0\cos\omega t \text{ and } \quad V_L = -4.8\sin\omega t \text{ [or } 4.8\cos(\omega t + \tfrac{\pi}{2})\text{]}.$$

We'll now see how to calculate V, the pd across the combination.

Study point

The reactance, X_L, of an inductor is **directly** proportional to the frequency; X_C is **inversely** proportional to the frequency: ω (or f).

Key terms

Impedance, Z: The ratio $\frac{V}{I}$ for a component (or circuit), for any phase relationship between current and pd, where V and I are the rms pd and current [or peak pd and current] respectively of sinusoidal AC.

Resistance, R: The ratio $\frac{V}{I}$ for a component (or circuit) for which the current and pd are in phase, where V and I are the rms pd and current [or peak pd and current] of sinusoidal AC.

Reactance, X: The ratio $\frac{V}{I}$ for a component (or circuit) for which the current and pd are $\frac{\pi}{2}$ (90°) out of phase, where V and I are the rms pd and current [or peak pd and current] of sinusoidal AC.

Study point

In terms of the frequency, f:

$X_C = \frac{1}{2\pi fC}$ and $X_L = 2\pi fL$.

Study point

Note that the question in the **Example** does not specify rms or peak values. It doesn't matter: 6 V rms will produce a current of 0.075 A rms; if $V_0 = 6$ V, $I_0 = 0.075$ A.

A16 Knowledge check

Calculate the capacitance of the capacitor in the **Example**.

A.7 Solving RCL circuits

(a) Introduction to phasors

This is the technique most commonly used to solve AC circuits in A-level Physics. Consider a vector of amplitude I_0 rotating **anticlockwise** about the origin, **O** as in Fig. A.14. The x-projection of the vector is given by:

$$x = I_0 \cos \theta$$

If the vector is rotating with angular velocity ω, then

$$\theta = \omega t + \varepsilon,$$

where ε is the value of θ when $t = 0$. So $x = I_0 \cos(\omega t + \varepsilon)$.

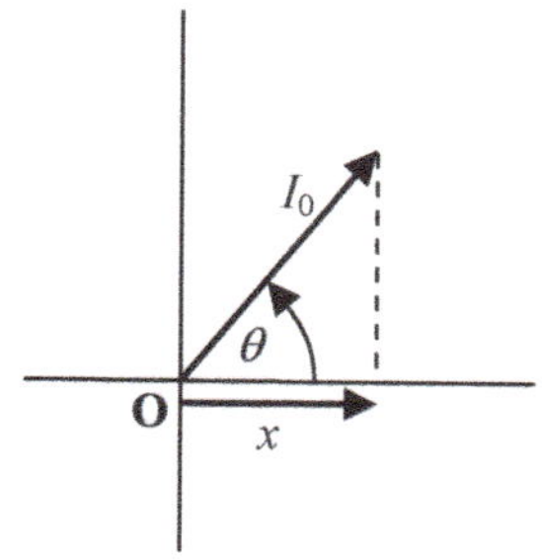

Fig. A.14 Rotating vector

This means that x is varying just like the current in a circuit: one which has a peak value of I_0 and angular frequency of ω. This suggests that we treat the vector I_0 as *representing* the current in magnitude and phase – and that the instantaneous value of I that we measure **is** the projection of this vector – a.k.a. **phasor**.

The top diagram in Fig. A.15 shows the voltage and current phasors for a resistor at a particular instant (the current phasor is in red). The directions of the two phasors are the same because the voltage and current are in phase for a resistor. The values of V and I at the instant of the diagram are the projections on the x-axis.[3] In the first diagram the values of V and I are positive and decreasing. One quarter of a cycle later the phasors have moved round $\frac{\pi}{2}$ anticlockwise. The values of V and I are now both negative and decreasing. Remember that a negative quantity that is become more negative is decreasing even though its magnitude is increasing.

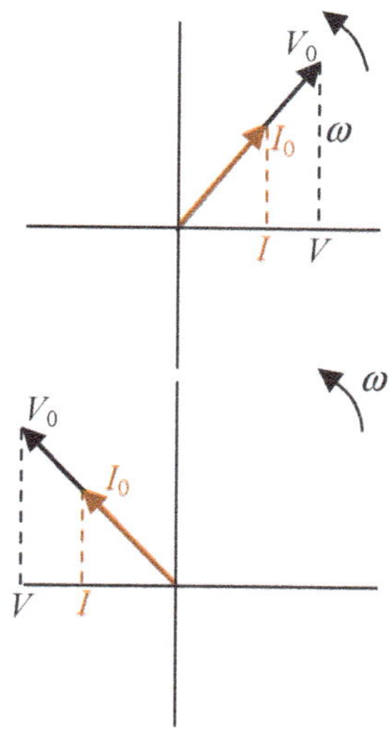

Fig. A.15 I and V phasors for a resistor

In I-V phasor diagrams for capacitors and inductors, the current and voltage are $\frac{\pi}{2}$ (90°) apart. We can draw the diagrams, with the correct phase difference, by using CIVIL. Remember that the phasors rotate anti-clockwise. Fig. A.16 shows a sequence of phasor diagrams for a capacitor. Notice how the diagrams show that the **current leads the voltage**.

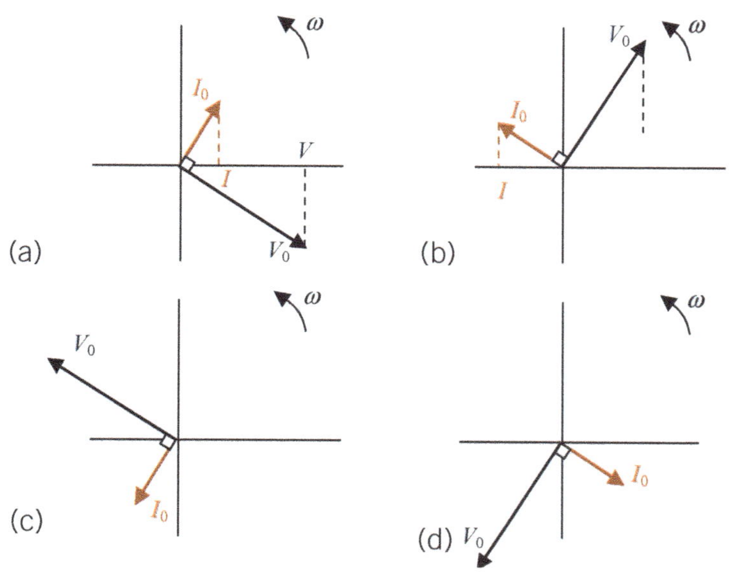

Fig. A.16 Capacitor phasor diagrams

Study point

In Fig. A.16(a), as the voltage phasor rotates anti-clockwise, its projection onto the x-axis, i.e. the measured voltage, V, is positive and increasing towards its maximum value; at the same time I is positive and decreasing towards 0.

Key terms

Phasor: A vector rotating at the frequency of the alternating current. It represents a voltage or a current (rms or peak) in magnitude and phase.

Knowledge check A17

In the same way as the Study point above, identify whether the values of I and V in Fig. A.16(b) – (d) are positive or negative and increasing or decreasing.

3 Note that the apparent lengths of the current and voltage phasors are not significant – the lengths depend upon different scales.

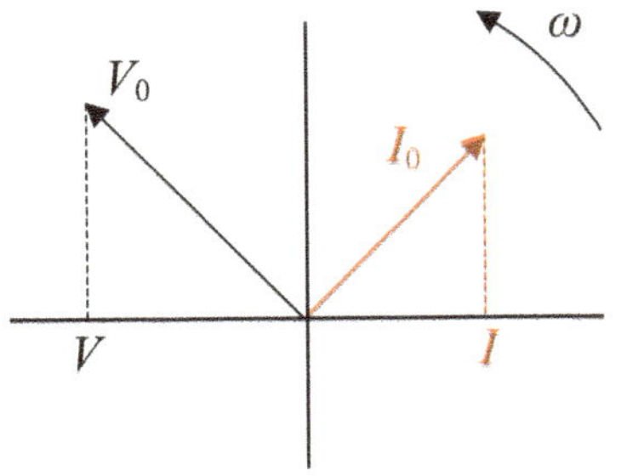

Fig. A.17 I and V phasors for an inductor

Fig. A.17 is a phasor diagram for an inductor, at a particular point in the cycle. The voltage is ahead of the current.

(b) Using phasors in AC circuits

RC and *RL* circuits

Phasors are useful in solving AC circuits involving combinations of resistors, capacitors and inductors. We shall restrict ourselves to series circuits, such as the RC combination in Fig. A.18. The current and voltages shown are **not** the instantaneous values. For the moment we'll take them to be the peak values.

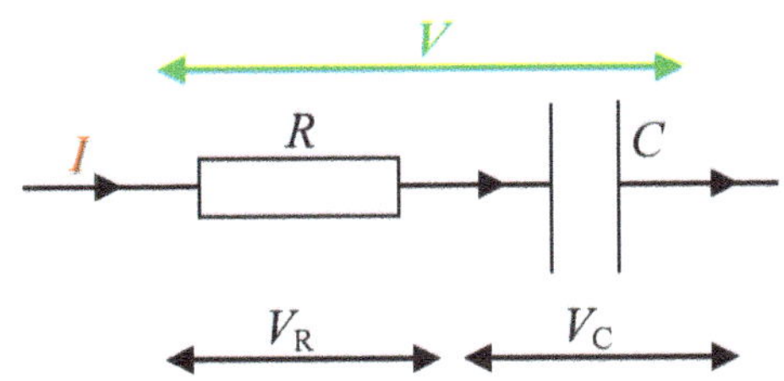

Fig. A.18 RC combination

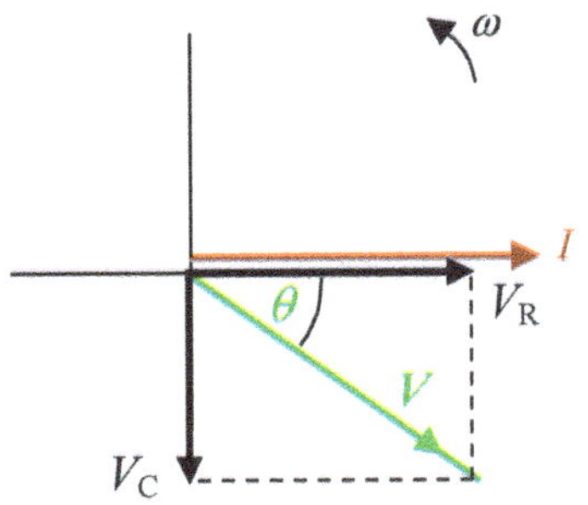

Fig. A.19 V_R and V_C phasors

The two components are in series, so they have the same current in them. When we're analysing these series circuits, we'll always draw the phasor diagram with the current phasor, I, along the x-axis. We are going to use the phasor diagram in Fig. A.19 to determine the total voltage, V, in terms of V_R and V_C. The V_R phasor is in phase with I; the V_C phasor is $90°$ behind.

The phasors are vectors, so we can add the two voltage phasors, V_R and V_C, according to the parallelogram law to give the total voltage phasor, V, as shown.

Hence: $V = \sqrt{V_R^2 + V_C^2}$ and the phase angle, $\theta = \tan^{-1}\left(\frac{V_C}{V_R}\right)$.

We can use the same technique with RL circuits:

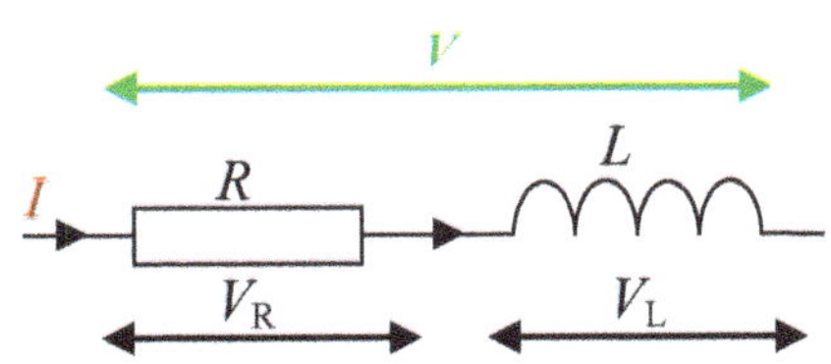

Fig. A.20 RL combination

A18 Knowledge check

State the voltage across the capacitor at the instant of Fig. A.19.

A19 Knowledge check

In terms of V_R and V_C, state the value of V:

(a) at the instant of Fig. A.19;

(b) $\frac{1}{4}$ of a cycle after Fig. A.19.

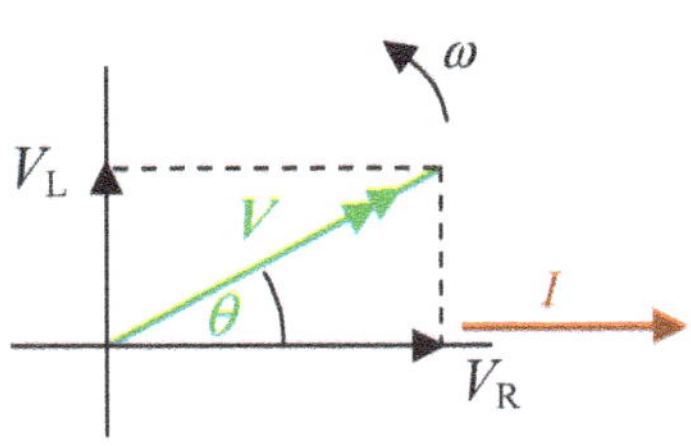

Fig. A.21 V_R and V_L phasors

Note that, again the diagram (Fig. A.21) is drawn for the instant when I is along the x-axis. Using the parallelogram law again:

The total voltage, $V = \sqrt{V_R^2 + V_L^2}$ and the phase angle, $\theta = \tan^{-1}\left(\frac{V_L}{V_R}\right)$.

The impedance of *RC* and *RL* circuits

In the above equations, $V_R = IR$, $V_C = IX_C$ and $V_L = IX_L$. So we can write the above relationships as:

$$V = I\sqrt{R^2 + X^2} \text{ and } \theta = \tan^{-1}\left(\frac{X}{R}\right).$$

where X is the reactance, either $X_C = \frac{1}{\omega C}$ or $X_L = \omega L$.

Impedance, Z, is defined by $V = IZ$, so $Z = \sqrt{R^2 + X^2}$

Hence and $Z_{RC} = \sqrt{R^2 + \frac{1}{\omega^2 C^2}}$ and $Z_{RL} = \sqrt{R^2 + \omega^2 L^2}$

A20 Knowledge check

Calculate the impedance of an RC combination with

$R = 100\ \Omega$ and $X_C = 100\ \Omega$.

Example

A variable frequency AC power supply with negligible internal impedance is connected across: (a) a series RL combination and (b) a series RC combination. Sketch graphs of I against ω.

Answer

(a) RL combination: $I = \dfrac{V}{\sqrt{R^2 + \omega^2 L^2}}$

When $\omega = 0$, $I = \frac{V}{R}$. As ω increases I decreases and tends to 0 as $\omega \to \infty$.

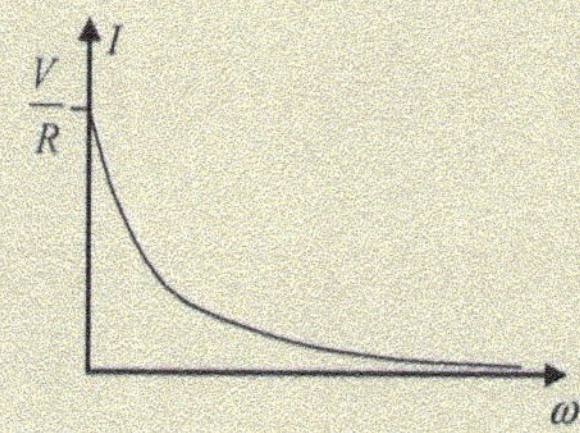

(b) RC combination: $I = \dfrac{V}{\sqrt{R^2 + \frac{1}{\omega^2 C^2}}}$

As $\omega \to 0$, $\frac{1}{\omega} \to \infty$, $\therefore I \to 0$.

As $\omega \to \infty$, $\frac{1}{\omega} \to 0$, $\therefore I \to \frac{V}{R}$

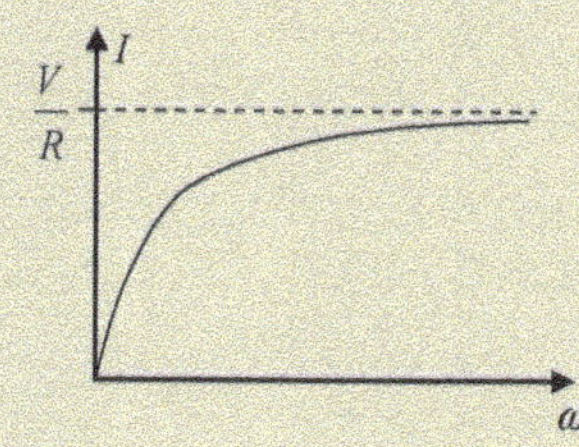

RCL circuits

We are now in a position to use phasors to solve the problem of series circuits containing all three components – resistors, capacitors and inductors.

Consider Fig. A.22

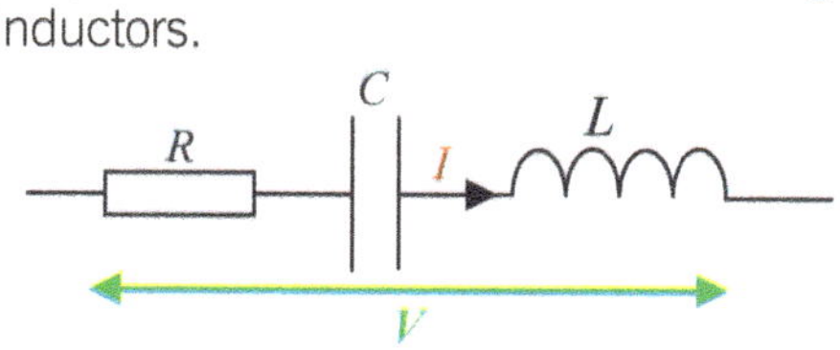

Fig. A.22 Series RCL combination

As usual, we'll draw the phasor diagram for the moment when the current phasor is along the x-axis. There are now three voltage phasors:

V_R in phase with I $\therefore$ along the x-axis

V_C $\frac{\pi}{2}$ behind I, $\therefore$ along the negative y-axis

V_L $\frac{\pi}{2}$ before I, $\therefore$ along the positive y-axis.

These phasors are shown in Fig. A.23. For the purposes of the diagram, we have assumed, for the moment, that $V_L > V_C$. Actually, the relative length of the phasors depends upon the frequency.

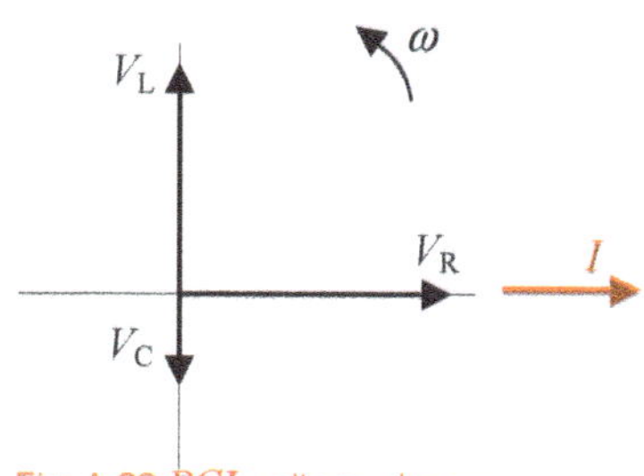

Fig. A.23 RCL voltage phasors

To add these phasors we first combine V_L and V_C, i.e. we find $V_L - V_C$ and then use Pythagoras' theorem to add this phasor to V_R. This is shown in Fig. A.24.

From the diagram:

$V = \sqrt{V_R^2 + (V_L - V_C)^2}$

From which, as before, $V = I\sqrt{R^2 + \left(\omega L - \frac{1}{\omega C}\right)^2}$.

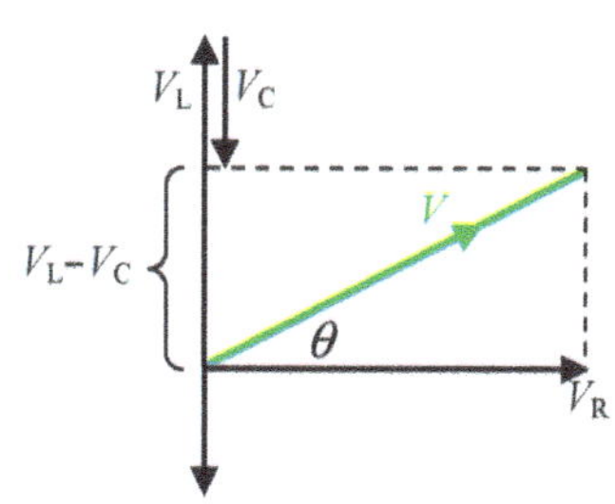

Fig. A.24 RCL phasor analysis

So the impedance, Z, of this circuit is given by:

$Z = \sqrt{R^2 + \left(\omega L - \frac{1}{\omega C}\right)^2}$ which is again of the form $Z = \sqrt{R^2 + X^2}$ but $X = X_L - X_C$.

The reason for the '–' sign should be obvious. The voltage across the inductor and capacitor are π out of phase, i.e. in *antiphase*.

Another useful form of this equation for Z is

$Z = \sqrt{R^2 + (X_L - X_C)^2}$ which may also be written as $Z = \sqrt{R^2 + (X_C - X_L)^2}$

Knowledge check A21

Calculate the phase angle between the current and voltage for the combination in Knowledge check A20.

Knowledge check A22

What value capacitor would have $X_C = 100\ \Omega$ at a frequency (f) of 1 kHz?

Knowledge check A23

Confirm that, $V_0 = 5.2$ V and $\theta = 1.18$ rad for the RL combination in Fig. A.13. Hence, write V as a function of time.

Knowledge check A24

A 0.127 H inductor and a 30 Ω resistor are connected in series across a 12 V, 50 Hz supply. Calculate:

(a) the reactance of the inductor

(b) the impedance of the combination

(c) the current

(d) the phase angle between the current and voltage.

Knowledge check A25

Sketch a phasor diagram for the circuit in Knowledge check A24 with values of V, V_L and V_R shown.

Study point

The wire of a real inductor must possess resistance (unless it is a superconductor). We can consider a real inductor to be composed of a resistor and an inductor in series. The I-ω graph for a real inductor is therefore as shown in the **Example** with R being the resistance of the inductor wire.

We can also calculate the phase angle, θ, as follows:

$\theta = \tan^{-1}\dfrac{V_L - V_C}{V_R}$, which leads to $\theta = \tan^{-1}\dfrac{X_L - X_C}{R}$ and so to $\theta = \tan^{-1}\dfrac{\omega L - \frac{1}{\omega C}}{R}$.

A26 Knowledge check

A resistor of resistance $10\ \Omega$ is connected across a power supply in series with an inductor and a capacitor of reactances $50\ \Omega$ and $30\ \Omega$ respectively.

Calculate:

(a) the total impedance

(b) the phase angle between the voltage of the supply and the current.

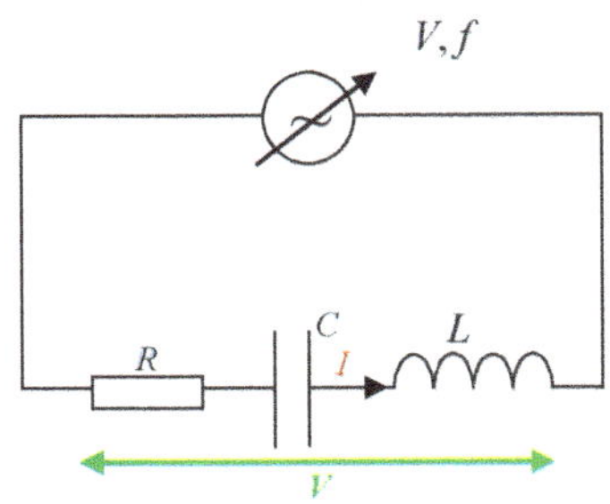

Fig. A.25 Series resonance circuit

Stretch & challenge

Reproduce Fig. A.26 using a spreadsheet program and then investigate the effects of varying R, C and L.

Study point

The lower the value of R, the sharper the resonance curve. Unlike for mechanical resonance, the frequency, f_0, for the peak current doesn't change with resistance.

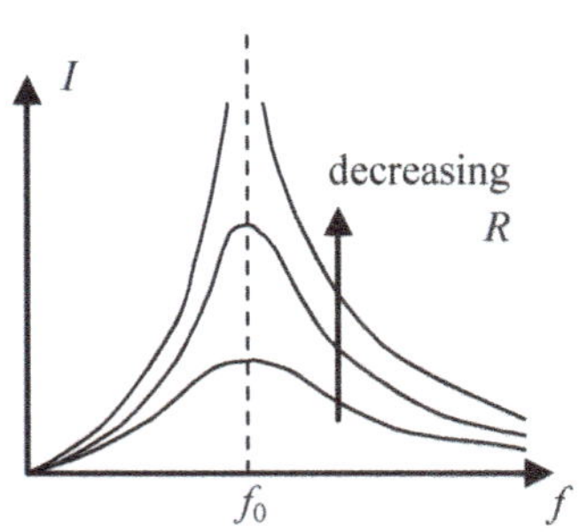

Fig. A.27 Effect of R on the resonance curve

A.8 Resonance in *RCL* circuits

(a) Variation of current with frequency

As we have seen, the relationship between V and I for the circuit in Fig. A.25 is:

$$I = \frac{V}{\sqrt{R^2 + \left(2\pi fL - \frac{1}{2\pi fC}\right)^2}}$$

The graph of I against f is plotted in Fig. A.26 for the component values $12\ \mathrm{V}$, $100\ \Omega$, $0.10\ \mathrm{H}$ and $1.0\ \mu\mathrm{C}$.

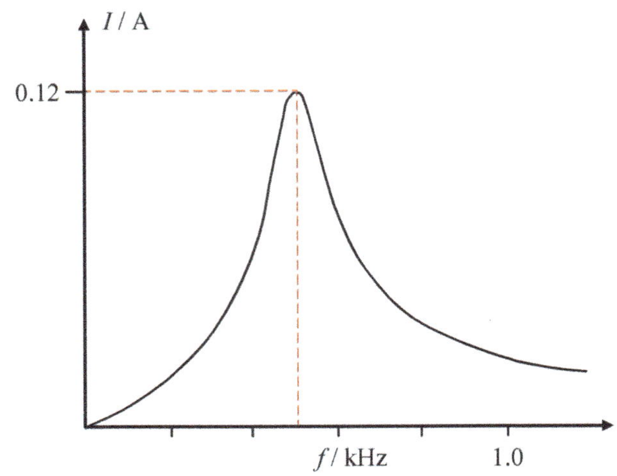

Fig. A.26 Resonance curve for a series *RCL* circuit

The curve – called a **resonance curve** – has the following properties:

1 The current $I \rightarrow 0$ as $f \rightarrow 0$ or $f \rightarrow \infty$.

2 The current rises to a peak value, for $f \sim 500\ \mathrm{Hz}$ in this case. This value of f is called the resonance frequency and is given the symbol f_0.

3 The current has a peak value of V/R which, in this case, is $0.12\ \mathrm{A}$.

Why does I vary in this way? For the following reasons:

1 If we try and insert $f = 0$ or let $f \rightarrow \infty$ in the equation for I, the denominator becomes infinite and hence I is zero.

2 The expression $\left(2\pi fL - \frac{1}{2\pi fC}\right)^2$ on the bottom line has a minimum value of zero. If $\left(2\pi fL - \frac{1}{2\pi fC}\right)^2 = 0$, then $f_0 = \frac{1}{2\pi\sqrt{LC}}$, which in this case is $503\ \mathrm{Hz}$.

3 When $\left(2\pi fL - \frac{1}{2\pi fC}\right)^2 = 0$ the equation simplifies to $I = \frac{V}{R}$, which in this case is $I = \frac{12\ \mathrm{V}}{100\ \Omega}$, i.e. $0.12\ \mathrm{A}$.

(b) The Q factor

The resonance curve in Fig. A.26 is quite sharp. If you've done the Stretch & challenge task above, you'll have found that the sharpness depends on the value of the resistance in the resonance circuit, as shown in Fig. A.27. Lower values of R give sharper resonance curves.

Electrical and electronic engineers need to consider the sharpness of resonance curves when designing systems. In radio receivers, for example, the peak needs to be wide enough to capture all the important frequencies in a station's signal but not so wide that it picks up neighbouring stations. The sharpness of the curve is expressed by the Q factor.

The **Q factor** is a measure of the sharpness of the resonance curve – the larger the Q factor the sharper the curve.

The Q factor is defined by $Q = \frac{\omega_0 L}{R}$, where ω_0 is the angular frequency at resonance.

In fact there are several other formulae for calculating Q, all giving the same answer, but this is the one in the WJEC Data Booklet. Other formulae are:

$$Q = \frac{1}{R}\sqrt{\frac{L}{C}} \qquad Q = \frac{V_C}{V_R} \quad \text{and} \quad Q = \frac{V_L}{V_R},$$

where V_C, V_L and V_R are the rms pds across the components at resonance.

(c) Understanding the resonance circuit

The phase angle, θ, between the voltage and current also varies with f, as shown in Fig. A.28. Why is this? To understand the circuit more completely it may help to stand back from the details of the maths and appreciate its physics qualitatively. As the frequency varies, ask what the controlling influence on the circuit is in various frequency ranges.

1. For **very low frequencies** we can forget about the inductor – its reactance approaches zero as we lower f to zero. The capacitor's reactance becomes larger, the smaller the frequency so at very low frequencies it dominates the circuit. For a capacitor the voltage is $\frac{\pi}{2}$ behind the current and at low frequencies the current is very small.
2. For **very high frequencies** the opposite argument applies: the inductor dominates, so V is $\frac{\pi}{2}$ ahead of the current and again I is very small.
3. When $f \sim f_0$, the effects of the capacitor and inductor cancel out because the voltages across these components are equal in magnitude but opposite in phase. This leaves a circuit which is dominated by the resistor. Hence $I \sim V/R$ and $\theta \sim 0$.

A.9 Using oscilloscopes

(a) Introducing oscilloscope controls and traces

An oscilloscope is a device for drawing a voltage–time graph of an electrical signal. Traditionally it is based upon a cathode-ray tube, like old-fashioned televisions, but computers can also detect and display electrical signals via a signal acquisition board which feeds in to the computer sound card via a USB port.

The pictures in Fig. A.29 show (a) a simple CRO and (b) a typical computer-based oscilloscope screenshot.

Top tip

When sketching resonance curves for RCL circuits with the same C and L remember that, for different values of R:

1. The resonance frequencies are all the same.
2. The graphs never cross [they meet at (0,0)].
3. The graphs all tend to the same gradient as $f \to 0$.

Knowledge check A27

Show that the formulae:

$Q = \frac{1}{R}\sqrt{\frac{L}{C}}$, $Q = \frac{V_C}{V_R}$ and $Q = \frac{V_L}{V_R}$,

are all equivalent to $Q = \frac{\omega_0 L}{R}$

Knowledge check A28

State why V_C and V_R in the equation

$$Q = \frac{V_C}{V_R}$$

could also be the peak but not the instantaneous voltages.

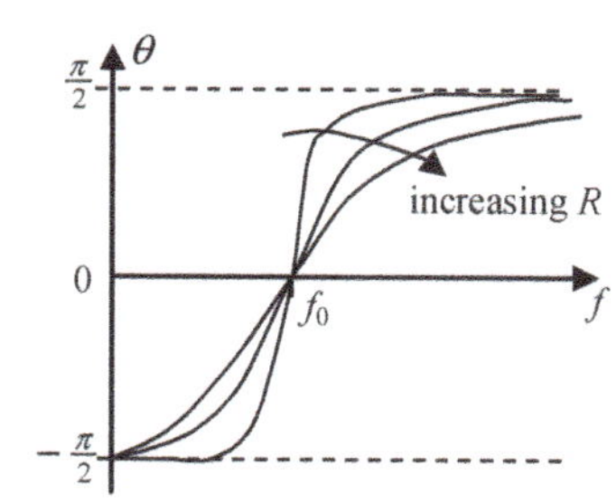

Fig. A.28 Variation of phase angle with frequency

Knowledge check A29

Calculate the Q factor for the circuit corresponding to Fig. A.26.

Knowledge check A30

How would the graph in Fig. A.26 be affected if the **DC resistance** of the inductor were 20 Ω?

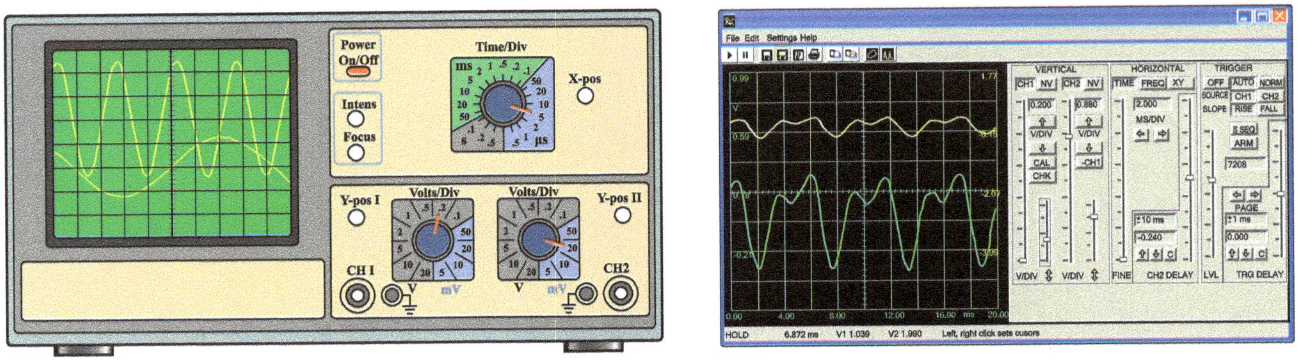

Fig. A.29(a) A CRO and (b) A computer CRO program in action

An oscilloscope, whether hardware or software based, has one or more input channels, which act in the same way as the terminals of a voltmeter. Both the devices in Fig. A.29 have two input channels labelled CH1 and CH2; each input channel produces one of the wave traces. The graph-grid on the screen has a vertical scale and a horizontal scale.

- The vertical scale is called Y-gain: for CH1, this is set at $0.2\ \text{V}$ per division on both machines. The two input channels' Y-gains can be separately adjusted.
- The horizontal scale is controlled by the **time base**. The setting for this on the CRO is marked Time/Div and can be adjusted between $0.5\ \text{s/div}$ to $0.5\ \mu\text{s} / \text{div}$. The time base setting for the two traces on an oscilloscope must be the same – this allows comparisons of the two signals to be made.

The vertical and horizontal positions of each trace can be adjusted [Y-Pos and X-Pos]. This helps to measure the amplitude and period against the fine scale divisions. The oscilloscope also has a **trigger** which acts to begin each sweep of the trace at the same point in the cycle of one of the inputs [usually CH1] to provide a steady display.

A31 Knowledge check

State the vertical scale of CH2 on the CRO in Fig. A.29.

A32 Knowledge check

State the horizontal scale on the CRO.

Example

Assuming that the lower-frequency signal on the CRO in Fig A.29 is on CH1, give the frequency and peak voltage of this signal.

Answer

Period ~ 8.4 divisions (by eye); time base = $5\ \mu\text{s} / \text{div}$, from Fig. A.29 (a)

$\therefore$ Period = $8.4 \times 5 = 42\ \mu\text{s}$

$\therefore$ frequency = $\dfrac{1}{42\mu\text{s}} = 0.024\ \text{MHz} = 24\ \text{kHz}$

Vertical height of wave [peak to peak] = 2.4 divisions (by eye). Y-gain = $0.2\ \text{V/div}$

$\therefore$ Peak voltage = $\frac{1}{2} \times 2.4 \times 0.2\ \text{V} = 0.24\ \text{V}\ [V_{\text{rms}} = \dfrac{0.24}{\sqrt{2}} = 0.17\ \text{V}]$

A33 Knowledge check

Find the frequency and peak voltage of the high frequency (CH2) signal on the CRO in Fig A.29.

Study point

X-shift and Y-shift are alternative names for X-Pos and Y-Pos.

Fig. A.30 shows how the Y-gain, X-Pos and Y-Pos controls can be used to facilitate measuring the voltage and period of an oscillation [see Study point].

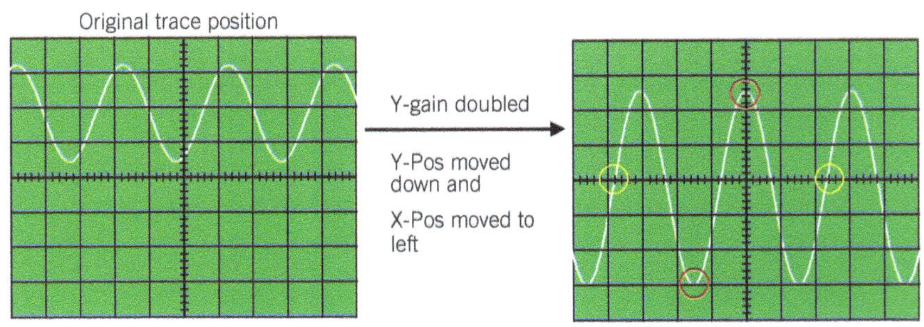

Fig. A.30 Adjusting the trace to help taking readings

A34 Knowledge check

Use the positions marked in Fig. A.30 with the red circles to measure the peak voltage.

Use the positions marked with yellow circles to measure the period of the oscillations.

Y-gain = $0.5\ \text{mV} / \text{div}$

Time base = $1\ \text{ms} / \text{div}$

- Changing the Y-gain gives a greater number of divisions so the fractional uncertainty in the vertical distance is reduced.
- Moving the Y-Pos down positions the bottom of the traces on a grid line.
- Moving the X-Pos to the left puts the top of the trace on the vertical scale. The period can be measured from the points where the trace crosses the central scale (again the distance should be as large as possible – so measure multiple cycles or adjust the time base).

Knowledge check A35

Use the answers to Knowledge checks A31 and A32 to find the rms voltage and frequency of the oscillations.

(b) Using oscilloscopes to investigate AC circuits

A dual-beam oscilloscope is very useful for investigating AC circuits but it does have two restrictions:

1. It cannot measure current directly. However, it can measure the pd across a known resistor, which enables the current to be calculated. Remember that I and V for an ohmic resistor are in phase and related by $V = IR$.
2. Both input channels are earthed. This means that the earthed sides must be connected to the same point in the circuit. If the circuit in Fig. A.31 were set up, the resistor R would be short-circuited by the two earths.

There is an additional problem that the output of signal generators is also often earthed on one side – all three earths must be connected to the same point in the circuit. Fig. A.32 shows how a circuit should be set up to investigate a series RC combination.

The use of the oscilloscope in such a circuit is most easily illustrated by working through an example.

Study point

Circuit symbol for an oscilloscope.

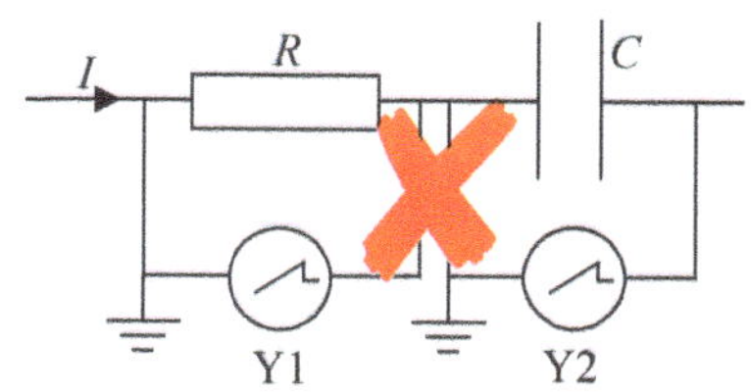

Fig. A.31 Incorrect connections

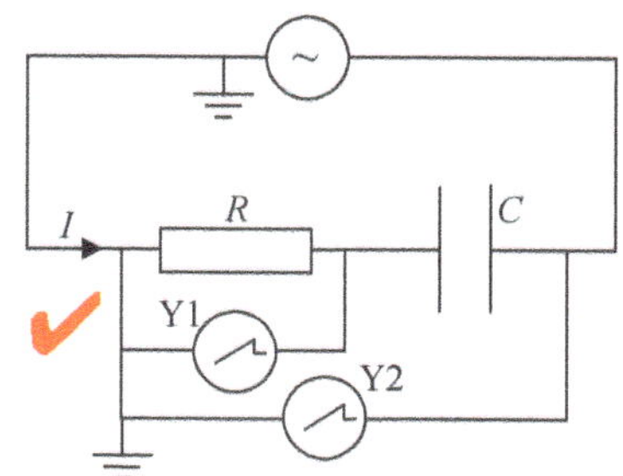

Fig. A.32 Correct earth connections

Example

A 10 Ω resistor is placed in series with a capacitor of nominal value, $C = 33$ nF, in an AC circuit. The oscilloscope is connected as in Fig. A.32 and the result displayed in Fig. A.33. Determine an accurate value for C.

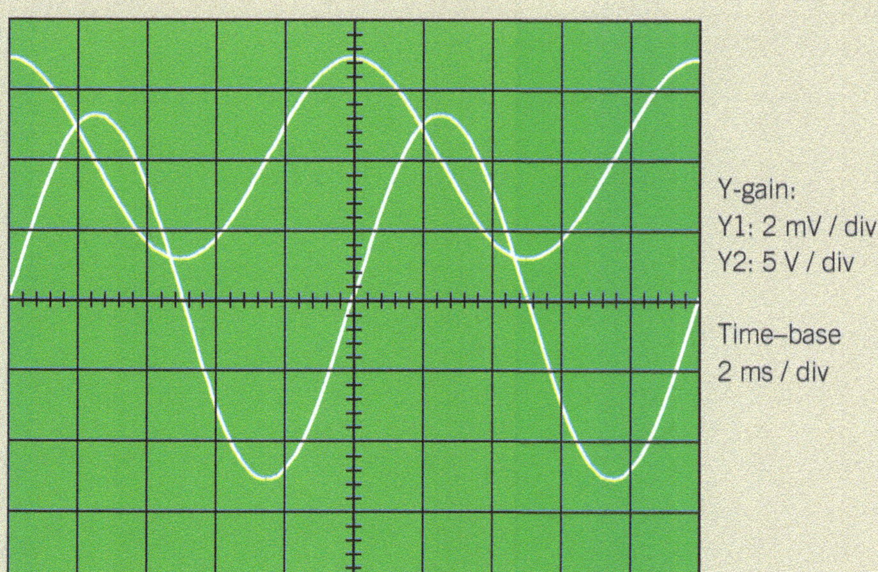

Fig. A.33 RC circuit traces

Knowledge check A36

For Fig. A.33, calculate the rms values of the voltages and currents. Calculate the capacitance using rms values in $V_C = \frac{I}{2\pi fC}$. The answer should be the same.

Study point

The pd across the resistor leads the pd across the capacitor, so the top trace is the pd across the resistor (Y1).

Study point

Notice that the resistor value in the example was chosen so that $R << X_L$. This enabled us to write $V_C = V_{RC}$ to a very good approximation.

1 Frequency: 2 cycles = 10 divisions = 20 ms. ∴ Period = 10 ms
∴ Frequency = 100 Hz.

2 V_R = peak resistor voltage $= \frac{1}{2} \times 2.9 \text{ div} \times 2 \text{ mV/ div} = 2.9 \text{ mV}$

3 I = peak current $= \frac{V_R}{R} = \frac{2.9 \text{ mV}}{10 \ \Omega} = 0.29 \text{ mA}$

4 V_{RC} = peak voltage across RC combination
$= \frac{1}{2} \times 5.2 \text{ div} \times 5\text{V/ div} = 13.0 \text{ V}$

5 Because $V_R << V_{RC}$ [3mV against 13 V]. We can write $V_C = 13.0$ V.

6 X_C = capacitor reactance $= \frac{V_C}{I} = \frac{13.0 \text{ V}}{0.29 \text{ mA}} = 44.8 \text{ k}\Omega$

7 So the capacitance, $C = \frac{1}{2\pi f X_C} = \frac{1}{2\pi \times 100 \times 44.8 \times 10^3} = 36 \text{ nF}$

Study point

In an RC circuit if V_{RC} and V_R are comparable in size, then we can use $V_{RC}^2 = V_R^2 + V_C^2$ to calculate V_C.

The final example gives a method of calculating the reactance (and hence the capacitance or inductance) using the phase difference between the V_R and the V_{total} signals. It is useful when V_X and V_R are comparable in size. Alternatively, see the Study point opposite.

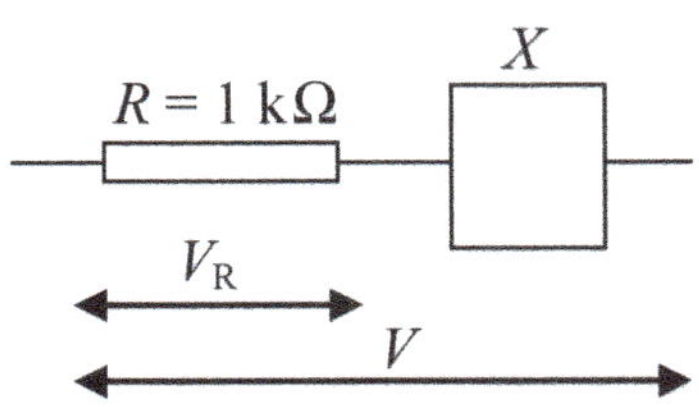

Fig. A.34 RX combination

Example

CRO traces for V_R and V were examined for the combination shown in Fig. A.34. X was known to be a capacitor, a resistor or an inductor.

The period of the AC voltage was found to be 10 ms and the V trace was 1 ms ahead of the V_R trace. Determine the nature and value of component X.

Answer

$1 \text{ ms} = \frac{1}{10} = 0.1$ cycle. ∴ Phase difference, $\theta = 0.1 \times 2\pi = 0.628$ rad.

$\tan\theta = \frac{X}{R}$. ∴ $X = R\tan\theta = 1 \times 10^3 \tan 0.628 = 726 \ \Omega$.

Because V is ahead of I [which is in phase with V_R] then X must be inductive.

∴ $2\pi fL = 726$. But $f = \frac{1}{T} = \frac{1}{0.01 \text{ s}} = 100 \text{ Hz}$ ∴ $L = \frac{726}{2\pi \times 100} = 1.2 \text{ H}$

A37 Knowledge check

Let the value of V_R in the example be 1.5 V.

Calculate V_L and V.

Test yourself Option A

1. The coil in a generator has N turns and area A. It is rotated in a magnetic field, B, with an angular velocity ω. Explain, in terms of Faraday's law, why the output voltage is proportional to all the values, N, A, B and ω.

2. The graph shows the variation of flux linkage with time for a generator coil rotating in a magnetic field.

   (a) Determine the angular frequency of the coil and the peak output voltage.

   (b) Sketch a graph of the output voltage with time, over the same time period.

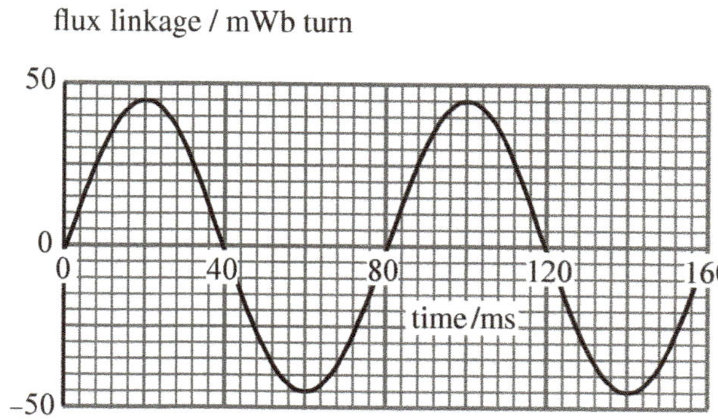

3. An alternating pd with $V_{rms} = 9.0$ V and $f = 50$ Hz is applied across a $10\ \Omega$ resistor.

   (a) Calculate (i) the rms current and (ii) the mean power dissipation.

   (b) If the pd is at a maximum at $t = 0$, write the variation of pd with time in the form $V = V_0\cos(\omega t + \varepsilon)$.

   (c) Write the variation of current with time in the form $I = I_0\cos(\omega t + \phi)$.

   (d) Sketch graphs of the variation of V, I and the power P with time between $t = 0$ and $t = 50$ ms.

4. A 12 V (rms), 1 kHz power supply is connected separately across (a) a 1 μF capacitor and (b) a 100 mH inductor. Calculate the rms current in each case. [Hint: see KCs A8(c) and A9(c).]

5. Three components, **A**, **B** and **C**, are connected in series. The graphs show the current (red) through the components and the pd across each. Identify the components in as much detail as possible.

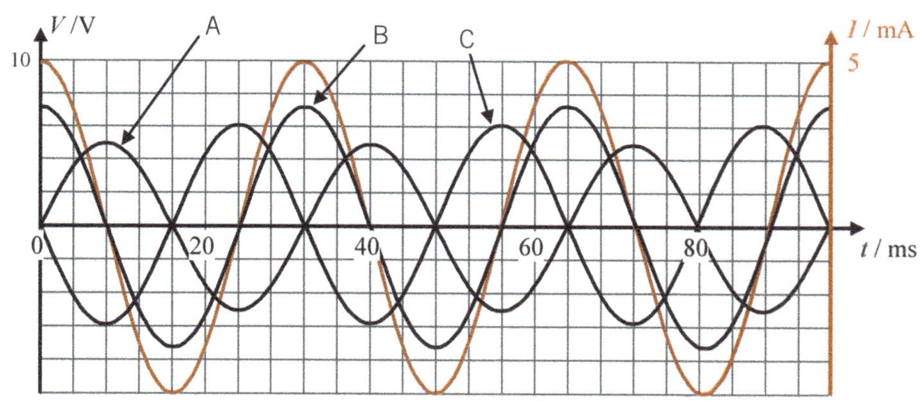

6. A resistor, capacitor and inductor are connected in series across a power supply with a frequency of 50 Hz. The rms pds across the components are 5.0 V, 20.0 V and 10.0 V respectively and current through the components is 2.0 A.

   (a) With the help of a phasor diagram, determine the rms pd of the supply and hence the impedance of the combination.

   (b) Determine the resistance and reactance of each component and hence the value of each component.

   (c) Determine (i) the phase difference between the current and voltage and (ii) the mean power dissipated.

7. The frequency of the power supply in question 6 is doubled but the rms voltage is unchanged. Explain, using a phasor diagram, why the rms current is unchanged and (without further calculation) state the phase difference between the current and voltage.

8. The pd, in V, of a power supply is given by $V = 24\cos 400\pi t$, in which t is the time in s. The supply is connected across a 47 Ω resistor and a 12 μF capacitor connected in series. Show that the rms current is approximately 0.2 A and determine the phase difference between the pd and current.

9. The oscilloscope screen shows two traces, A and B. The Y-gain of A is 2.0 V/div and that of B is 1.0 mV/div. The time base is 20 μs /div.

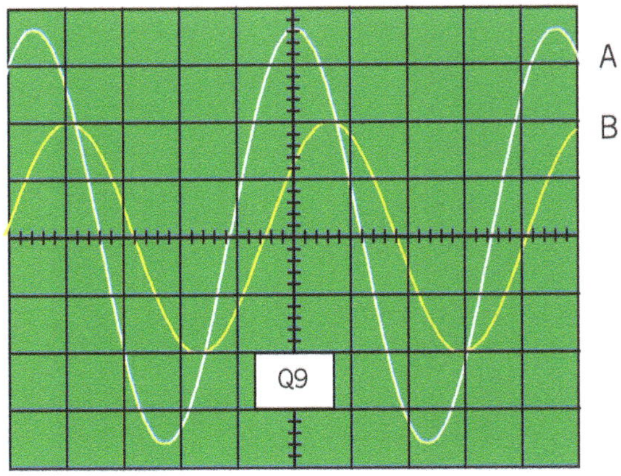

(a) Determine:
   (i) the frequency of the voltages
   (ii) the rms values of V_A and V_B
   (iii) the phase difference between V_A and V_B.

(b) Damian suggests that V_B could be the voltage trace across the resistor if V_A were applied across a series combination of a resistor and an inductor. Evaluate this claim.

10. An electronics student is designing a series *RCL* circuit to have a resonance frequency of 100 kHz and a Q factor of 10. A 1.0 mH inductor is available.

(a) Calculate the values of the capacitor and resistor the student should use.
(b) The circuit is connected across a 10 V rms supply. Calculate:
   (i) the power dissipated in the circuit at resonance
   (ii) the power dissipated in the circuit at 95 kHz
(c) The student reads that an alternative relationship for Q is $Q = \frac{f_0}{\Delta f}$, where Δf is the half-power bandwidth, i.e the range of frequencies where the power is at least equal to the power at resonance. Comment on whether your answers to part (b) agree with this.

Option B: Medical physics

Medical diagnosis and treatment increasingly rely on physics and technology enabled by physics knowledge. Even though the technology may be operated by technicians, it needs to be set up and calibrated by medical physicists, e.g. the dose delivered by radioactive sources needs careful checking.

This topic covers four distinct areas of medical physics:

- X-rays
- Ultrasound
- Magnetic resonance imaging (MRI)
- Nuclear medicine.

Each of these is covered in its own section.

B.1 X-ray medicine

This is the largest of the topics and covers the production of X-rays as well as their use in diagnosis and treatment.

Study point

X-rays are high-energy, ionising electromagnetic radiation produced by accelerating electrons or by high-energy atomic transitions.

Soft X-rays have photon energies from ~100 eV to ~10 keV.

Hard X-rays have photon energies between ~10 keV and a few hundred keV.

(a) X-ray nature and properties

With the possible exception of visible light, the division of the electromagnetic spectrum into different regions (infra-red, visible, X-ray, etc.) is not, for the most part, an objective one. The black body emission spectrum (see AS textbook Section 1.6.1) of a 2000 K object is mainly in the IR with a small amount of visible. As its temperature is raised to 6000 K (roughly the temperature of the Sun's photosphere) the proportion of IR decreases and there is more visible and significant quantities of UV. There is also no scientific consensus on how to distinguish between X-rays and γ-rays. For the purposes of this book we shall adopt the definition that X-rays are produced by electrons and γ-rays by nuclear transitions, and that their ranges overlap.

The wavelength of the hard X-ray photons used in medicine is so small (see KC B1) that their particle nature is much more significant than their wave properties. The ionisation energy of atoms and molecules is in the eV range so medical X-rays can easily ionise biological molecules, with potentially serious consequences, e.g. causing cell damage through DNA mutations.

Knowledge check B1

(a) Show that the wavelength of a 100 keV photon is approximately 10 pm.

(b) Give the approximate wavelengths of a 10 keV and a 1 keV photon.

(b) X-ray production

X-rays in hospitals are produced by firing high energy (up to ~ 100 keV) electron beams at heavy metal targets, usually tungsten or a tungsten-molybdenum alloy.

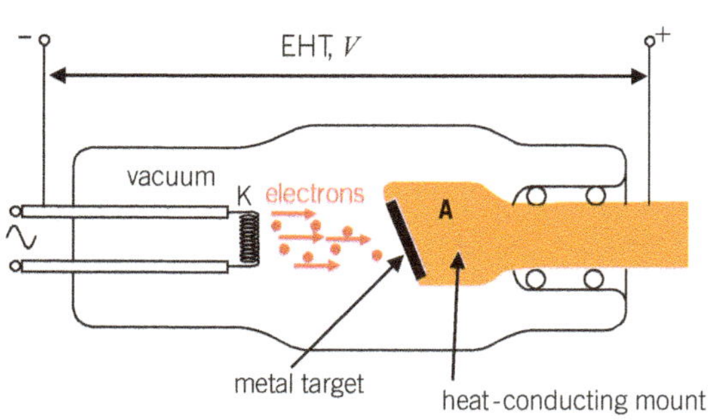

Fig. B.1 Schematic diagram of an X-ray tube vacuum

Study point

Fig. B.1 is only a schematic diagram. Modern hospital tubes have rotating anodes and focus the electron beam onto a small area of the target.

These are the principles involved:

1. The hot cathode (K) emits electrons. This process, called thermal emission, is analogous to photoemission: the electrons are able to overcome the work function because of thermal motion.
2. The thermal electrons are accelerated by the pd between the anode (A) and cathode (K).
3. The electrons which hit the target are rapidly decelerated and emit radiation doing so. This process is called Bremmstrahlung [German: braking radiation] and occurs whenever charged particles undergo acceleration (or deceleration). See Study point.

Study point

Synchrotron radiation is a form of Bremmstrahlung: it is emitted when electrons follow curved paths in magnetic fields. The Diamond Light Source in Oxfordshire uses this to produce very bright X-ray beams for research. See Section 4.4.3(c).

For some of the electrons, the loss of kinetic energy occurs over many collisions and for others in only a few. Because of this variation, a continuous spectrum is formed.

A second process occurs when an energetic electron collides with an inner electron in the metal target atoms, knocking it out of the atom and leaving a vacant inner energy level. An electron from a higher energy level falls down into the vacant one, releasing the energy difference as a photon. This gives rise to a line spectrum that is characteristic of the target element(s).

Study point

The X-ray intensity is sometimes plotted against photon energy rather than wavelength. In this case the graph is as in Fig. B.3. Note the difference in the appearance of the continuum spectrum.

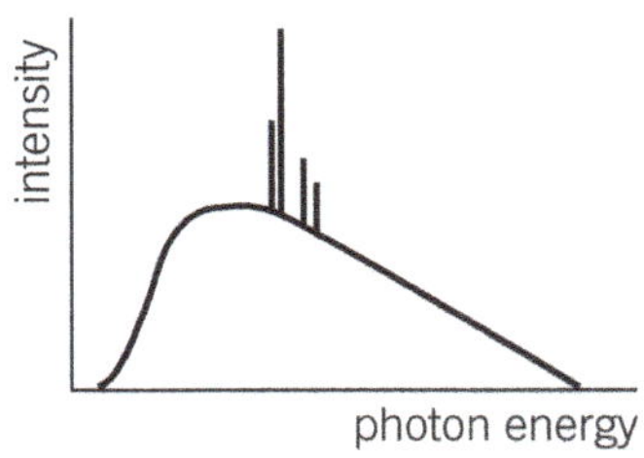

Fig. B.3 Intensity v photon energy

Fig. B.2 gives the spectra for two different accelerating voltages, $V_1 > V_2$, with the same target element.

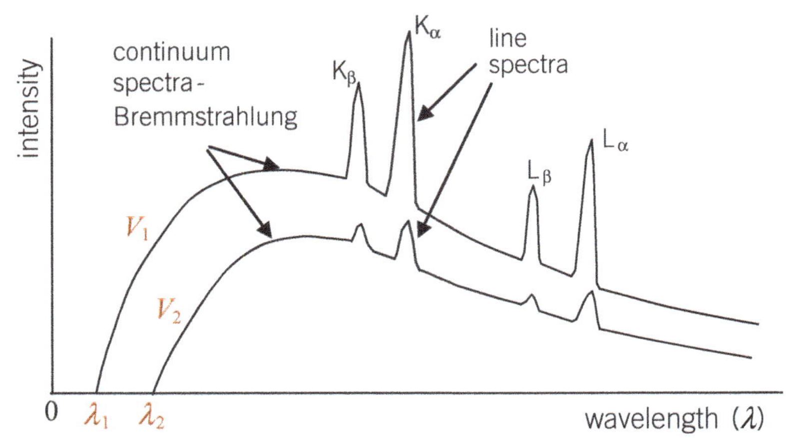

Fig. B.2 X-ray spectra for two different voltages

The simplified atomic energy level diagram of tungsten in Fig. B.4 can be used to illustrate the process of the production of the characteristic spectrum. If an incident electron ionises a tungsten atom by ejecting one of its K shell electrons, one of the L electrons can replace it by emitting a 59.3 eV photon (the difference in the levels).

Study point

The lines in the X-ray line spectrum are labelled K, L, ... with subscripts α and β. The capital letters refer to the electron shell which the electrons drop down into and the α and β to energy levels within the shell.

Example

What other photons could result from the ejection of a K electron from a tungsten atom?

Answer

After the L electron has replaced the K electron, its place could be taken by an M electron with the emission of a $10.2 - 2.5 = 7.7$ keV photon.

Also an M electron could directly replace the K electron with the emission of a $69.5 - 2.5 = 67.0$ keV photon.

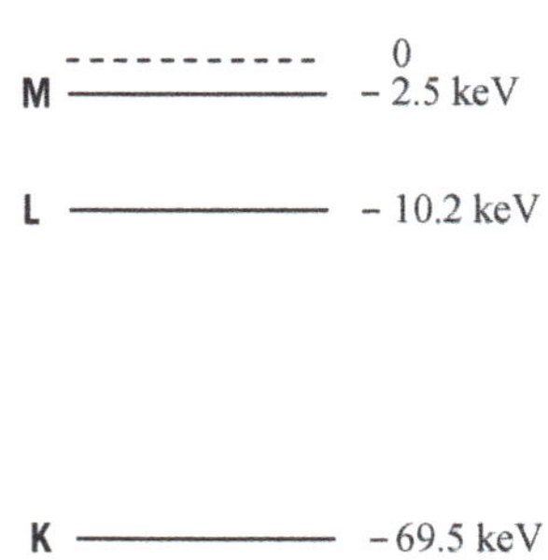

Fig. B.4 Tungsten atomic energy levels

(c) Efficiency of X-ray production

Considered in purely energy terms, the X-ray tube is very inefficient. Typically only a very small fraction of the electron beam energy is converted to X-ray energy. This is because most of the electrons lose energy gradually, producing large numbers of IR photons. The efficiency increases with accelerating voltage: at $50\ \text{keV}$ the efficiency is typically $\sim 0.5\%$; at $100\ \text{keV}$ it is $\sim 1\%$.

Example

Estimate the number of X-ray photons produced per second by a $100\,\text{kV}$ tube operating with an anode current of $10\,\text{mA}$. State your assumptions.

Answer

Assuming an efficiency of 1% and a mean photon energy of $50\,\text{keV}$:

$$P_{\text{OUT}} = 0.01 \times P_{\text{IN}} = 0.01\ VI = 0.01 \times 100\,\text{kV} \times 0.01\ \text{A} = 10\,\text{W}$$

$$\therefore \text{No. of photons per second} = \frac{P_{\text{OUT}}}{E_{\text{ph}}}$$

$$= \frac{10\,\text{W}}{50\ \text{keV} \times 1.60 \times 10^{-16}\,\text{J keV}^{-1}} \sim 1 \times 10^{15}\,\text{s}^{-1}$$

This low efficiency means that a lot of heat needs to be conducted away from the anode – hence the conducting mount in Fig. B.1.

(d) Photon energy and beam intensity

Notice that the continuum spectrum is cut off at a minimum wavelength. This is because no radiation can be emitted with a photon energy greater than that of the incident electrons. For an EHT voltage, V, the kinetic energy of the electrons is eV, so the minimum wavelength, λ_{min}, is given by

$$eV = \frac{hc}{\lambda_{\text{min}}},$$

from which we can see that λ_{min} is inversely proportional to the accelerating voltage.

For a particular accelerating pd, the beam intensity is proportional to the number of electrons per second which hit the metal target. This is a function of the heater temperature: the greater the temperature, the more electrons are emitted per second and hence the greater the beam intensity. To obtain high definition X-ray images a combination of a high intensity beam and a small source spot on the anode is needed. The rotating anode tube has been developed to allow this (Fig. B.5). The X-rays are produced from a small spot on the anode. The temperature of a stationary anode would be too great, so the anode is made to rotate, allowing time for the heat to be conducted away.

(e) X-ray attenuation

Section 3.5.2 dealt with the absorption of nuclear radiation. X-rays are absorbed in a similar way to γ-rays. If a mono-energetic beam of X-rays passes through a material, the probability of a photon being absorbed in any distance is constant, so the intensity, I, of the X-rays decreases according to an exponential decay law: $I = I_0 e^{-\mu x}$, where μ is called the **attenuation** (or absorption) **coefficient** or **attenuation** (or absorption) **constant**. Different materials have different attenuation coefficients and these also depend upon the photon energy.

Knowledge check B2

A 100 kV X-ray tube is operating with an anode current of 10 mA. Estimate the rate at which heat needs to be conducted away from the anode.

Knowledge check B3

(a) Calculate λ_{min} for an X-ray tube operating at 40 keV.

(b) What can you say about the pd needed to produce L photons from a tube with a tungsten target? See Fig. B.4.

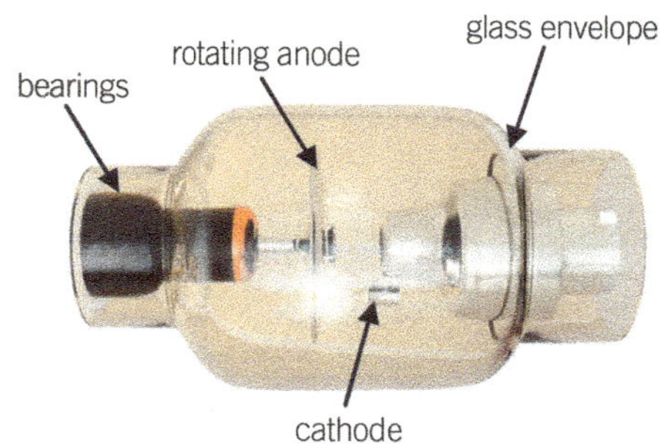

Fig. B.5 Rotating anode X-ray tube

Study point

If an X-ray fraction f_1 gets through object 1 and a fraction f_2 gets through object 2, the fraction which penetrates the two is $f_1 f_2$.

B4 Knowledge check

In an investigation into X-ray attenuation, 60% of a beam of monoenergetic photons penetrates a 2.0 cm thickness of a material. Calculate:

(a) the percentage getting through 4.0 cm of the material

(b) the attenuation coefficient of the material.

Study point

Notice that in the example, the brain absorbs most of the X-rays, even though μ is lower for the brain. This is because the brain tissue is much thicker than the two layers of bone.

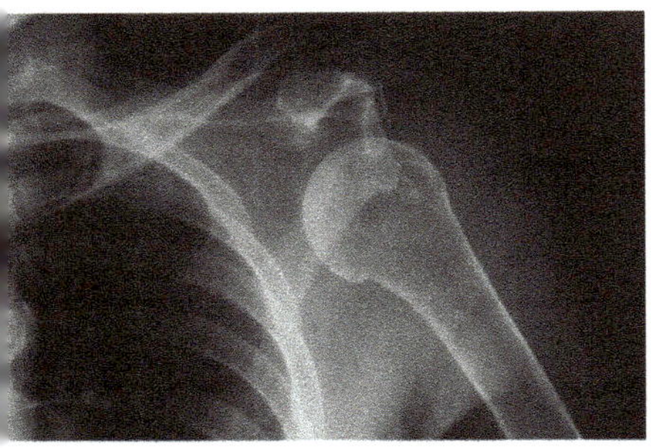

Fig. B.6 Broken left clavicle (collar bone)

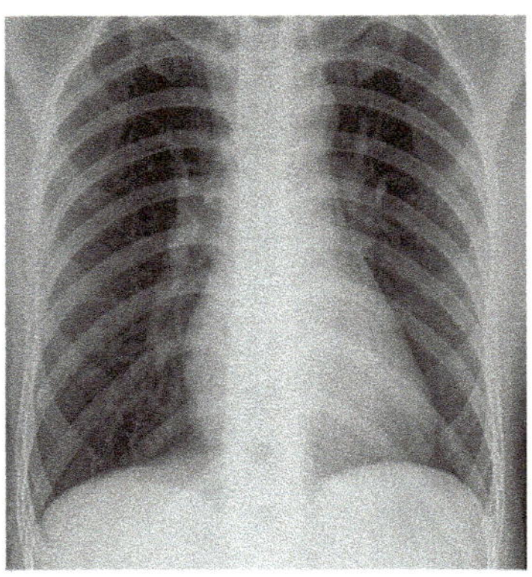

Fig. B.7 Thoracic cavity X ray

Key terms

X-ray fluoroscopy: real-time digital images produced of soft tissues using image intensifiers.

Contrast medium: a substance introduced into a part of the body (e.g. gut, blood vessels) to give a greater contrast with surrounding tissues.

E_{phot} / keV	Attenuation coefficient / cm^{-1}					
	bone	muscle	fat	blood	soft tissue	brain
10	18	5	3	6	7	5
30	1.6	0.40	0.38	0.50	0.30	0.40
60	0.45	0.20	0.19	0.21	0.21	0.18

Table B1 Attenuation coefficients of different materials

Example

The thickness of the human skull averages 6.3 mm. The width of the brain is 140 mm. Use the data in Table B1 to calculate the fraction of a beam of 60 keV X-ray photons which passes through the human head.

Answer

Total bone thickness = 12.6 mm = 1.26 cm

$\therefore$ Fraction penetrating the bone $= \frac{I}{I_0} = e^{-\mu x} = e^{-0.45 \times 1.26} = 0.57$

And: fraction penetrating brain $= e^{-0.18 \times 14} = 0.080$

$\therefore$ Total penetration fraction $= 0.57 \times 0.080 = 0.046$.

The differences in attenuation coefficient are important in medical X-ray examinations.

(f) Low-energy X-rays in radiography

Bones attenuate low-energy X-rays much more than soft tissue because of their greater density; in particular their concentrations of mineralised calcium compounds. Fig. B.6 shows a typical X-ray image of a broken collar bone, the sort of image which allows medical practitioners to decide on treatment – in this case, perhaps an intramedullary pin, to hold the two halves of the bone together to allow them to knit.

Soft tissues can also be revealed by X-rays as the chest X-ray (Fig. B.7) shows. Although the muscles of the heart have a lower attenuation coefficient than the ribs, the heart has absorbed more X-rays because it is thicker. Even blood vessels are visible on the image.

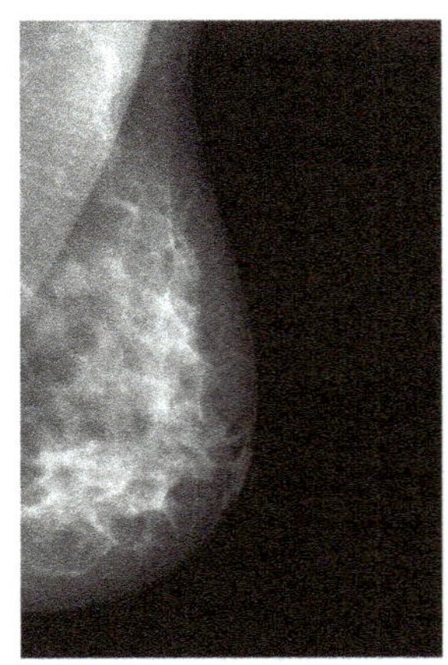

Fig. B.8 Normal breast tissue

In mammography, low energy X-rays are used to exploit the differential absorption of various tissues. The mammogram is examined for abnormal areas of density, which can indicate the presence of a cancer.

If real-time images are needed, e.g. of the blood vessels of a beating heart (cardiogram), **fluoroscopy** is used with a **contrast medium** and an *image intensifier*. Barium sulfate is commonly used in alimentary canal imaging; iodine compounds in angiograms. There are many images available on the web showing fluoroscopic images of the alimentary canal and heart.

Images are acquired using digital image receptors: tiny sensors, integrated in an array, each of which responds to X-rays by giving a digital output representing the X-ray intensity for that pixel. The advantages include rapid image acquisition, great sensitivity, image enhancement possibilities and the ability to transmit and store the image.

(g) X-ray computed tomography (CT scan)

In this imaging technique, the patient has many X-ray images taken from different angles when passing through the circular machine, as shown in Fig. B.9. The X-ray tube and camera rotate around the patient to build up the images.

The cross-sectional images obtained are virtual slices of the body and the use of digital techniques allows structures with very little contrast to be distinguished. In the upper abdomen section in Fig. B.10 you can probably recognise some ribs, a vertebra, the liver and the two kidneys – other structures in the image are the stomach and duodenum.

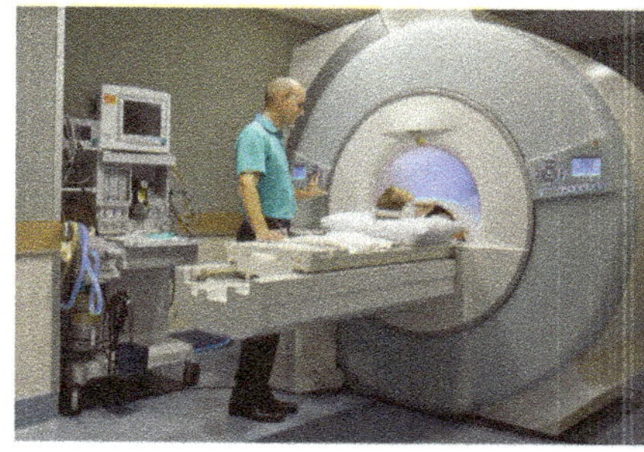

Fig. B.9 CT scanner in action

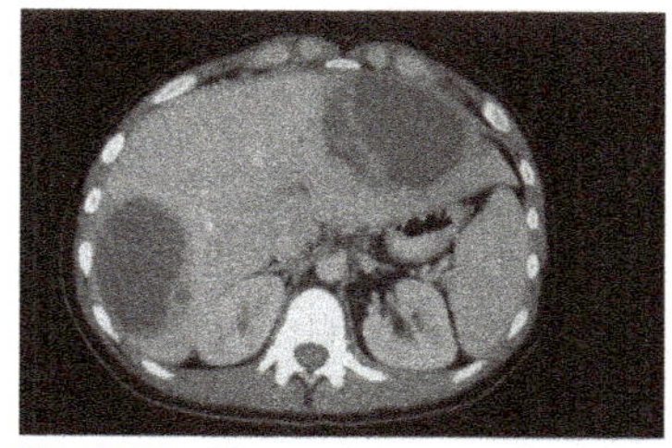

Fig. B.10 – Upper abdomen CT section

(h) X-ray radiotherapy

X-rays are also used to treat cancers. The DNA of dividing cells is more sensitive to ionising radiation than that of quiescent cells. Hence, even though X-rays are themselves carcinogenic, they can be used to kill cancer cells. In order to concentrate the dose on the tumour and to mitigate the damage to healthy tissue, several beams of X-rays are used. These are aimed at the tumour from several directions and the X-ray apparatus rotated around the tumour (or the patient rotated).

High energy X-rays are used for this treatment: for shallow cancers (e.g. in the skin), so-called *superficial* X-rays ($50–200$ keV) are used, with *megavoltage* X-rays ($1–20$ MeV) being deployed for deep cancers. These last are usually produced using particle accelerators rather than standard X-ray tubes. Radiation from cobalt-60 is often used (in which case it should be called γ-ray radiotherapy).

Knowledge check B5

Suggest why radiotherapy is not used for dispersed cancers such as leukaemia and for cancers which have metastasised, that is, secondary cancers have formed in other parts of the body.

B.2 Ultrasound medicine

Sound with frequencies above 20 kHz, the upper limit for human hearing, is called ultrasound. Ultrasonic diagnostic and treatment procedures generally use ultrasound in the range 500 kHz – 5 MHz. As with X-rays, we'll start by considering how ultrasound is generated and detected.

(a) Generating and detecting ultrasound

Some crystals, e.g. lead zirconate titanate, generate electrical charge when stressed. This phenomenon is called *piezoelectricity*. Conversely, they deform when an electric field is applied. These piezoelectric crystals are used to generate and detect ultrasound. The principle of an ultrasound generator/detector is quite simple. Fig. B.11 gives the structure of an ultrasound transducer.

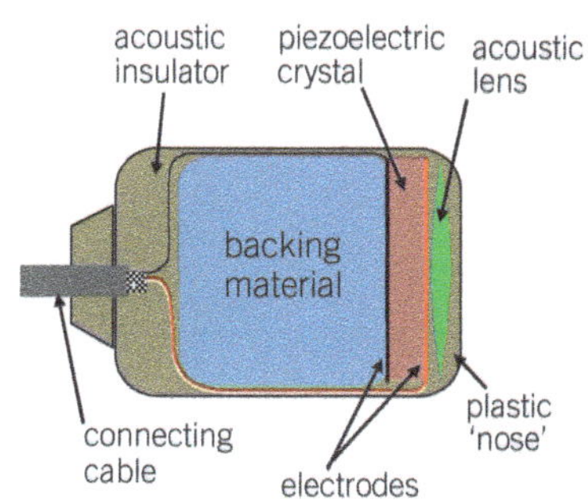

Fig. B.11 Ultrasound transducer

A high-frequency oscillating pd from from an external source, is fed to the crystal through the connecting cable, which generates the ultrasound sound waves (of the same frequency). These are emitted to the right: the backing material absorbs sound waves which propagate to the left and thus avoids reflected waves which could introduce spurious signals. Incoming ultrasound waves from the right are absorbed by the crystal, distorting it and causing it to generate an oscillating pd which is fed out through the same cable, to be analysed.

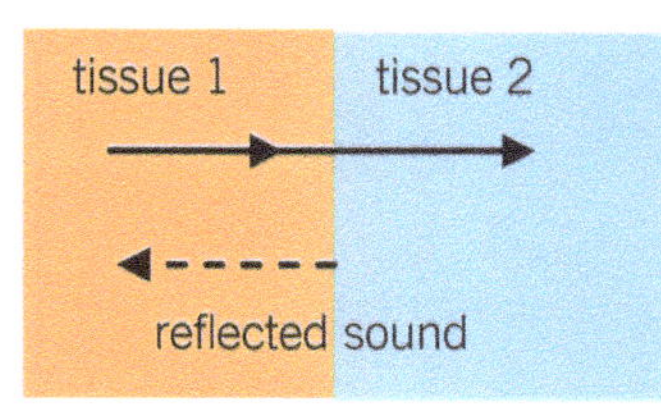

Fig. B.12 Ultrasound reflection

Key terms

The **acoustic impedance**, Z, of a material is defined by

$$Z = c\rho$$

where ρ is its density and c the speed of sound in the material.

Tissue	c	ρ	Z
Air	340	1.3	0.44
Lung	650	400	260
Fat	1470	920	1350
Water	1520	1000	1520
Brain	1560*	1030	1600*
Kidney	1558	1040	1620
Liver	1570	1060	1660
Muscle	1600*	1070	1700*
Bone	3400*	1600*	5400*

Table B2 Acoustic impedances

Units in the table:
c m s^{-1}; ρ kg m^{-3}; Z 10^3 kg m^{-2} s^{-1}

* These values vary a lot within and between tissues

B6 Knowledge check

A medical dictionary gives the acoustic impedance of human skin as 1.6×10^6 kg m^{-2} s^{-1}. Calculate the fraction of an ultrasound wave which would be reflected between skin and air.

B7 Knowledge check

Calculate the frequency of a sound wave with a wavelength of 1.0 mm in water. For data refer to Table B2.

B8 Knowledge check

Calculate the time taken for a sound wave to travel 15 cm in water.

(b) The reflection of ultrasound

Ultrasound scans work in a similar way to echo location. The generator sends out a pulse of waves and receives the echo from an object – in this case the boundary between two different tissues. The time delay between transmission and echo reception gives us twice the distance to the boundary.

The important property of the tissues, which determines the fraction of the incident sound that is reflected, is the **acoustic impedance**, Z, of the two tissues. The bigger the difference in Z for the two materials, the greater the fraction reflected. The equation is:

$$\frac{I_r}{I_i} = \frac{(Z_2 - Z_1)^2}{(Z_2 + Z_1)^2}$$

where I_i and I_r are the incident and reflected intensities respectively.

Example

Use the values of Z in Table B2 to calculate the fraction of ultrasound reflected at the boundary between back muscle and the kidney.

Answer

$$\text{Fraction reflected} = \frac{(1700 - 1620)^2}{(1700 + 1620)^2} = 0.00058\ (0.058\%)$$

Notice that it doesn't matter which way the sound wave is travelling because $(1700 - 1620)^2$ is the same as $(1620 - 1700)^2$.

Notice the low acoustic impedance of air. You should be able to show (Knowledge check B6) that over 99% of the sound will be reflected at an air/tissue boundary. If an ultrasound transducer is placed in contact with dry skin, there will always be a (very thin) layer of air between the two, i.e. there will be **two** air boundaries to cross and virtually no signal will be passed into the body. Hence examiners use a **coupling agent** – a jelly with a similar acoustic impedance to that of the body to achieve maximum signal transmission.

(c) Frequency, pulses and repetition frequency

The reflections of continuous sound waves cannot be analysed, so the wave is broken up into a series of pulses.

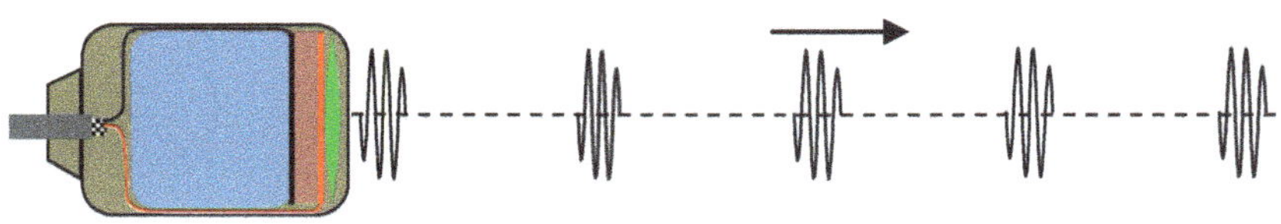

Fig. B.13 Pulsed ultrasound wave train

The resolution of the system is limited by the wavelength of the sound wave. For a resolution of 1 mm, the wavelength cannot be greater than this, so frequencies in the MHz range are employed (Knowledge check B7). Each pulse needs to travel through the body and back to the detector before the next is sent out. This takes approximately 0.1 ms (Knowledge check B8) so the gap between pulses must be at least this – so the repetition frequency (or pulse frequency) must be lower than 10 kHz.

(d) Ultrasound A scans (Amplitude scans)

These are the simplest kind of scans. They are commonly used by optometrists to make measurements on the eye, especially the length of the eyeball. They can also be used to measure the position of a known object inside the body, e.g. a cancer, but they do not produce an image.

The output of the scan is connected to a CRO (see Option A) which displays the time of arrival of the reflected pulses. In Fig. B.14 we see return pulses from the cornea, the front and back surfaces of the crystalline lens and from the retina at the back of the eye. The distances to these structures can be calculated from the time delays of the pulses.

Study point

Ultrasound waves also suffer attenuation. This limits the depth of penetration especially at high frequencies. Penetration of 10–15 MHz waves is limited to 2–3 cm.

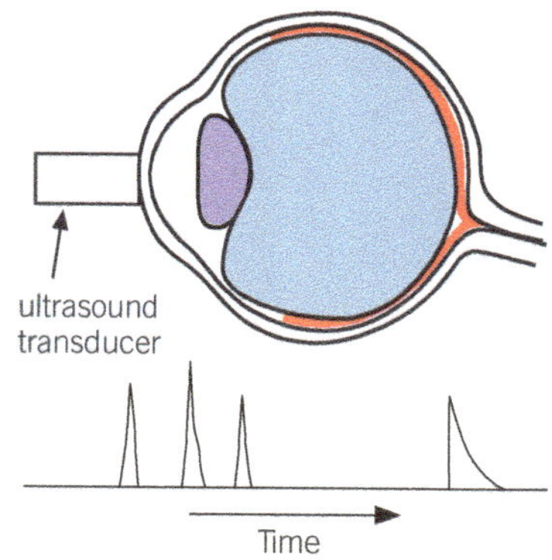

Fig. B.14 A-scan of the eye

(e) Ultrasound B scans (Brightness scans)

These scans give us the familiar images of foetuses in the uterus. They are produced by moving the scanner transducer over the area. The return signals from the different directions are combined electronically to produce the image on the display (Fig. B.15).

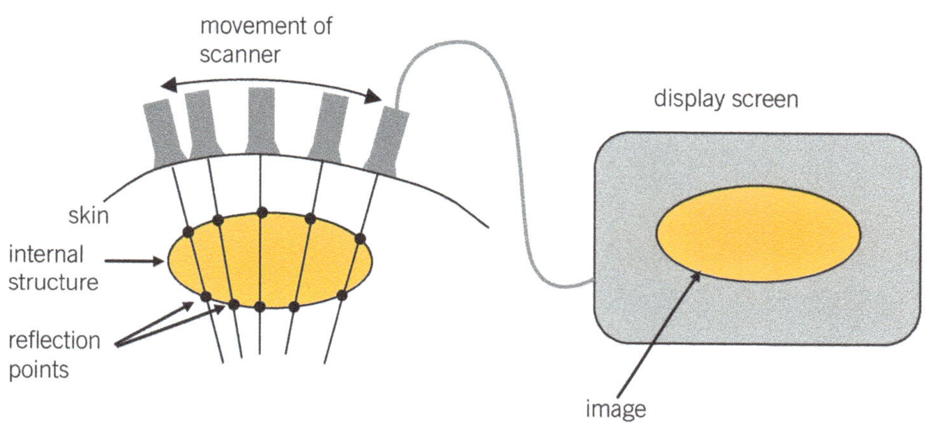

Fig. B.15 Ultrasound B scan of an internal structure

Heart scans, such as in Fig. B.16, need contrast agents because the blood/tissue boundary doesn't reflect well. A common agent is a liquid containing air microbubbles, which is injected into a vein.

Knowledge check B9

In the scan of an eyeball, the reflection from the back of the eye occurred after 32 μs. Assuming a speed of sound in the eye of 1550 m s^{-1}, what is the diameter of the eyeball?

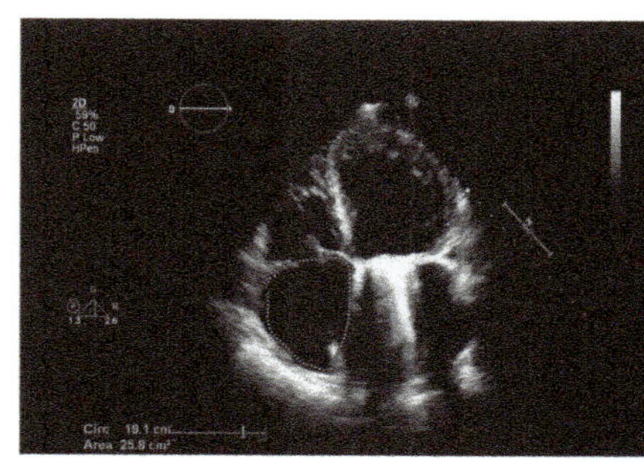
Fig. B.16 Heart scan

(f) Doppler scans

Ultrasound is used in conjunction with the Doppler effect, which we met in Section 4.3.4, to study the speed of flow of blood cells. The basic principle is illustrated in Fig. B.17.

For this application, the basic Doppler equation $\frac{\Delta f}{f} = \frac{v}{c}$ needs to be amended in two ways.

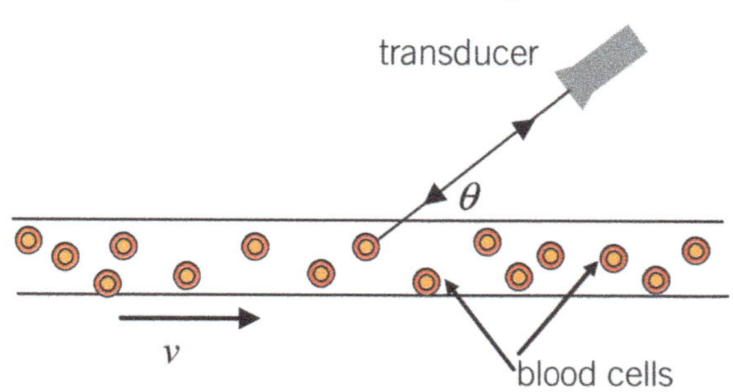

Fig. B.17 Doppler scan

1 There are **two** Doppler shifts involved. The red blood cells 'see' a Doppler-shifted sound wave from the transducer, which they see as coming towards them. The reflected wave is further Doppler shifted because the source of the waves (the cells) is moving towards the transducer (in receive mode).

Knowledge check B10

Explain why bubbles of air in the blood are good reflectors of ultrasound.

Study point

Police radar speed traps work on the same principle as Doppler scans, only using electromagnetic waves rather than ultrasound.

B11 Knowledge check

Estimate the Doppler shift of a 1 MHz ultrasound wave of blood in a vein. Assume that the angle θ is small and use the following data:

Speed of flow $10\ \text{cm s}^{-1}$

Speed of sound in blood $1550\ \text{m s}^{-1}$

2 Only the radial component of the velocity of the cells is significant.

So the relevant equation is: $\dfrac{\Delta f}{f} = \dfrac{2v\cos\theta}{c}$,

where θ is shown in Fig. B.17 and c is the speed of sound in the tissue.

For small θ, the equation becomes $\dfrac{\Delta f}{f} = \dfrac{2v}{c}$

Stretch & challenge

Therapeutic uses of ultrasound

It has been mentioned that ultrasound is attenuated in tissues. This means that there is some energy transfer and that there are potential biological and therefore medical effects. These are insignificant at the intensities used for imaging but high power ultrasonic beams can be used in medical treatment. Here is a selection of medical uses:

- Warming muscles and joints prior to manipulation by physiotherapists, or to promote healing by the warming effect.
- Descaling teeth by dental hygienists.
- Breaking up kidney stones and gall stones ('lithotripsy').
- Tumour ablation using high intensity focused ultrasound (HIFU).
- Assisting liposuction.
- Tooth and bone regeneration using low intensity pulsed ultrasound.

B.3 Magnetic resonance imaging (MRI)[1]

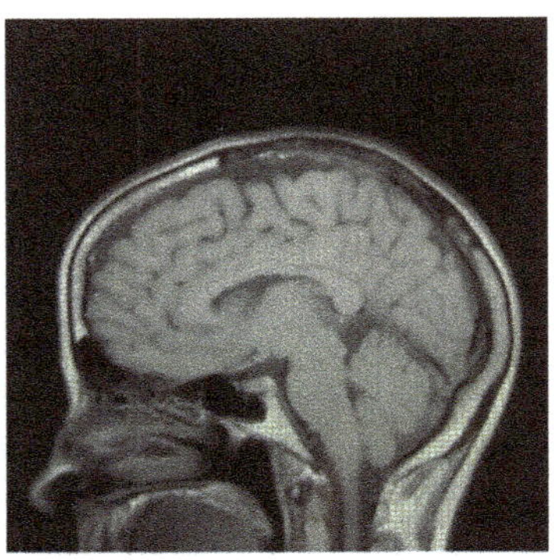

Fig. B.18 MRI head scan

This technique, which produces the most exquisitely detailed images of the internal tissues of the body, employs advanced physics and some very expensive kit.

It involves lying still and being passed through an intense non-uniform magnetic field (Fig. B.19) and having radio waves fired at you. However, as far as is known, the procedure carries very low risk (unless you have a heart pacemaker or forget to take your jewellery off!).

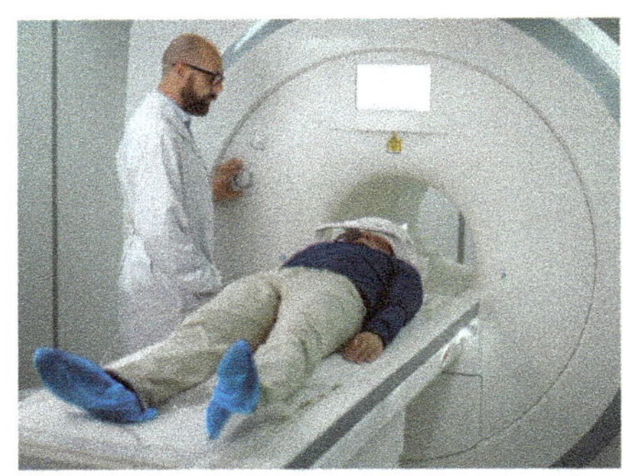

Fig. B.19 MRI scan preparation

So what's going on?

Study point

Remember that protons and neutrons are composite particles, consisting of quarks that are charged. So a spinning nucleon consists of circulating charges and so it behaves like a small magnet.

(a) Magnetic nuclei

All nuclei spin. For nuclei such as ${}^{1}_{1}\text{H}$, with an odd number of protons or neutrons (see Study point), this results in the nucleus behaving like a small magnet. In the absence of a magnetic field, the direction of the spin of hydrogen nuclei in a material (such as water) is random. In a magnetic field, the nuclei line up approximately with the magnetic field lines. Actually, they can be pictured like gyroscopes, with the direction of the rotation precessing around the field lines (Fig. B.21). There are two energy states available to the nuclei: parallel (low energy) and anti-parallel (high energy) to the magnetic field.

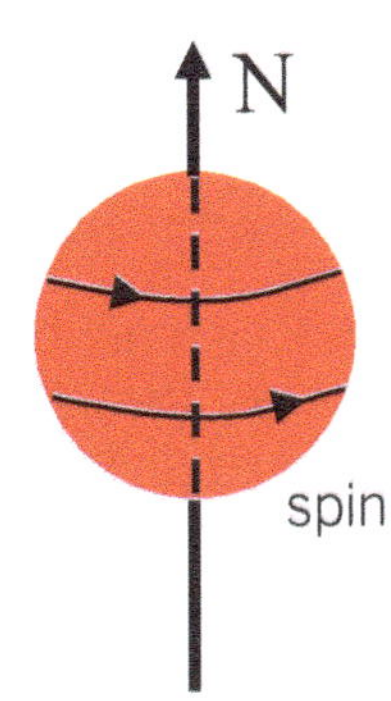

Fig. B.20 Spinning nucleus

B12 Knowledge check

Calculate ΔE for a 2.5 T magnetic field (typical MRI flux density). Express your answer in (a) J and (b) eV.

1 MRI scans used to be called NMR (nuclear magnetic resonance) scans. Rumour has it that the name was changed because 'an NMR' sounds like 'enema,' which is quite a different medical procedure!

The energy difference, ΔE, between the two states is very small even by atomic standards: it is proportional to the magnetic field:

$$\frac{\Delta E}{B} = 2.821 \times 10^{-26}\ \text{J T}^{-1}$$

For a 1 T field, ΔE is orders of magnitude less than the Boltzmann energy, kT, at room temperature (see Knowledge check B13), so the numbers of nuclei in the two states are very nearly equal.

Very nearly – but not exactly. For a 1 T field there is a roughly 4×10^{-6} excess of the nuclei in the lower energy state, i.e. for every million nuclei, about 500 002 will be in the lower state and 499 998 in the upper state! So, in a magnetic field, water (because it contains hydrogen) will behave like a very weak magnet. This is true for lots of materials.[2]

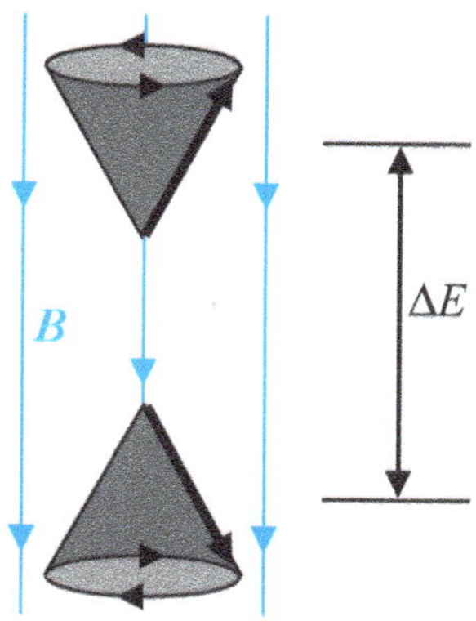

Fig. B.21 Precessing nuclei

Knowledge check B13

Calculate the mean translational kinetic energy of a gas molecule at room temperature (300 K) and compare this with your answer to Knowledge check B12.

Knowledge check B14

Use $E_{phot} = hf$ find the frequency of photons with an energy of $1.010\ 168 \times 10^{-26}$ J.

(b) Nuclear magnetic resonance – Larmor frequency

Suppose we put a hydrogen-rich piece of material (e.g. living cells) into a magnetic field of flux density 1 T. Consider the effect of flooding this material with a beam of photons of energy 2.821×10^{-26} J. From our study of lasers (see AS book, Section 2.8.1) we expect to see absorption, spontaneous and stimulated emission of photons of this energy. As the lower energy state is more populated, absorption will predominate, increasing the population of the higher (up) state. When we switch off the photon beam the hydrogen atoms will 'relax' into their previous state, with slightly more nuclei in the 'down' state than the 'up' state, i.e. by giving out a small burst of radiation. This emitted radiation shows the presence of hydrogen.

If you've done Knowledge check B14, you know that the frequency of photons needed to provide just the right energy to promote electrons from the lower to the higher state is 42.6 MHz. The exactness of the frequency needed makes the absorption akin to a very sharp resonance, and the effect is called nuclear magnetic resonance. The frequency needed is called the **Larmor frequency** or sometimes the resonance frequency. The Larmor frequency is also the frequency of precession of the nuclei about the magnetic field direction. Generally for hydrogen in a magnetic field with flux density B (in tesla), the Larmor frequency is given by:

$$f_{\text{Larmor}} = 42.6 \times 10^{6}\ \text{B Hz}$$

Hence the Larmor frequency for hydrogen in a magnetic field with $B = 1.5$ T is $1.5\ \text{T} \times 42.6\ \text{MHz T}^{-1} = 63.9$ MHz.

Key terms

The **Larmor frequency** is the photon frequency which resonates with the energy level difference between the magnetic moment states of a nucleus in a magnetic field.

The **relaxation time**, T, for the relaxation emission is the time for the intensity of the radiation to drop to $1/e$ of its original intensity.

(c) Relaxation

The return of the nuclei to their pre-excited state is called *relaxation*. This occurs as a burst of radiation, which decays exponentially with a characteristic **relaxation time**, T, of a few hundred milliseconds. The value of T depends upon the concentration of hydrogen nuclei in the material. This is because the magnetic fields of the hydrogen nuclei are weakly linked – the greater the concentration of hydrogen nuclei the longer the relaxation time. This is shown in Table B3: the greater the water concentration, the longer the relaxation time.

Tissue	T / ms
Fat	180
Liver	270
White matter	390
Spleen	480
Grey matter	520
Muscle	600
Blood	800
Cerebrospinal fluid	2000
Water	2500

Table B3 Relaxation time for body tissues in a 1 T field

2 See the Institute of Physics 2011 schools lecture 6 for this demonstrated for a grape and a piece of Blu-tak.

Study point

Although the detail is not great, the granulation of the liver showing cirrhosis is clearly visible in Fig. B.22. The white mass to the lower right is an enlarged spleen due to portal hypertension.

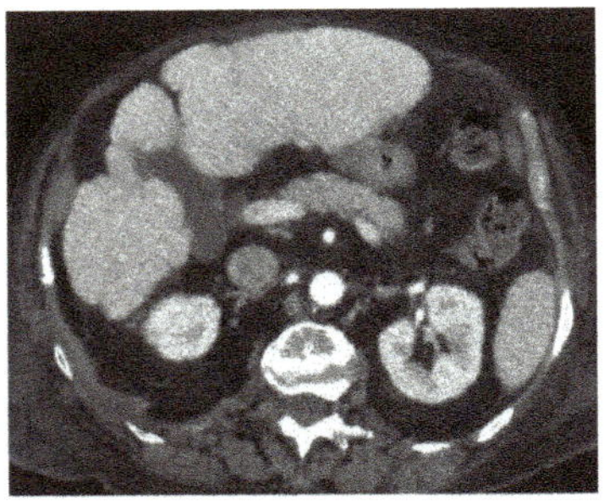

Fig. B.22 Liver with cirrhosis (top), and enlarged spleen (bottom right)

(d) Application to imaging

There are two requirements for forming an image of the internal organs of the body:

1. The MRI scanner must 'know' where the signal is coming from. If all hydrogen nuclei radiate with the same frequency, we cannot tell where a particular photon has come from.
2. The MRI scanner must be able to distinguish different kinds of tissue.

The 'where' problem is solved by arranging for the magnetic field to be non-uniform. If the field increases in strength along the axis, the Larmor frequency will also increase. The subject is then flooded with a band of radio frequencies including all Larmor frequencies. Received photons must therefore come from the slice of the body with the appropriate field strength. In addition a small, non-uniform rotating field is applied across the subject. This enables the point of origin of the photons to be known across the slice.

The 'tissue' problem is solved by exploiting the different relaxation times of the tissues. If the signal is read and integrated over (say) the first 300 ms, the blood will not have given out much of its energy but the fat and liver will have given out most of theirs. Thus different tissues will have different image densities. In Fig. B.22, the enlarged spleen is paler than the liver because it has a longer relaxation time.

Stretch & challenge

Functional MRI (fMRI)

This is one of the most exciting recent developments in nuclear magnetic imaging. Using it, neurologists can identify the activity of different brain areas.

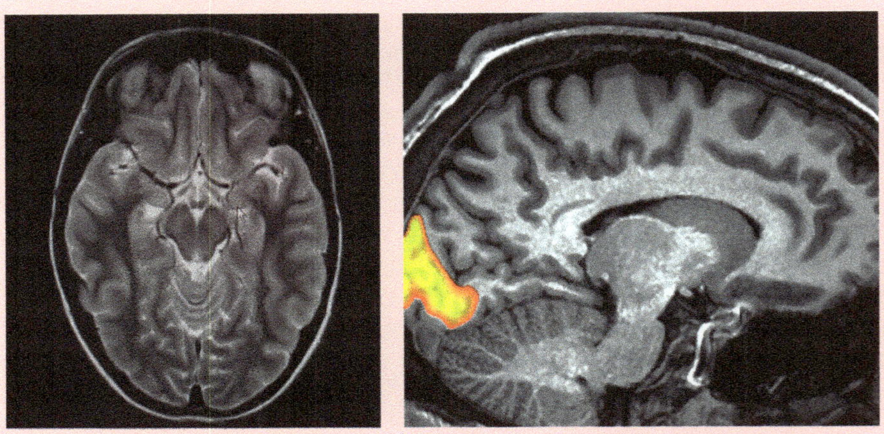

Fig. B.23 FMRI showing the response to visual stimulation

The basis of the technique is that the active areas of the brain have an increased supply of oxygenated blood that has different MRI properties from de-oxygenated blood. Fig. B.23 picks out the areas of the visual cortex that are active.

B.4 Comparison of Ultrasound, X-ray, CT scan and MRI

All four of these imaging techniques rely on sending waves (sound or electromagnetic) into a person and examining the reflected, transmitted or emitted radiation. Having looked at all four techniques, a brief comparison is called for.

Property	Ultrasound	Standard X-ray	CT scan	MRI
Radiation exposure	No ionising radiation	Exposure to ionising radiation	Significant exposure to ionising radiation (2–10 mSv) equivalent to up to 5 years' background radiation	No ionising radiation
Application	Generally soft tissue including foetal; skeletal joints	Mainly bone breakages; with contrast agent can also be used for soft tissue	Bone injuries, lung and chest imaging, cancer detection, A&E investigations	Different kinds of soft tissue imaging, e.g. injuries, tumours
Biological effects	No known hazards in imaging	Carcinogenic effects and developmental defects in embryos	As X-ray	No known radiation hazards; allergic reaction to contrast agents
Cost	Low cost	Low cost	About half the cost of MRI	High cost
Conditions	Short time; relatively painless (probe, e.g. rectal, may be required)	Very short time	Quite short time (5 min), ideally no movement but less of a problem than MRI	Long time; uncomfortable (no movement allowed); noisy; claustrophobia
3D imaging	Not possible	Not without moving the patient	Possible using a helical scan	Yes
Definition	Not high – depends on the skill of practitioner	High definition of bony structures	High definition of bony structures; moderate definition of soft structures (specially with contrast agent)	High definition (but requires stationary subject)
Contra-indications	None in imaging	Pregnancy	Weight limit of ~200 kg because of space and strength of moving table	Some metal implants; heart pacemaker; weight limit ~ 150 kg (space / strength of table)

B.5 Nuclear medicine

The topic of nuclear medicine includes the use of radioactive tracers for the investigation of organ function and body scanning as well as the treatment of medical conditions, mainly cancer. We shall start by discussing how nuclear radiation affects the human body.

(a) The effect of nuclear radiation on living material

As we discussed in Section 3.5, nuclear radiation (α, β and γ) is ionising. As it passes through living material, it ejects electrons from molecules, which can include structural and control molecules in cells as well as genetic material. This can damage or (in extreme cases) kill the cells. Cells that are undergoing mitosis (cell division) are particularly susceptible to damage. A large dose of radiation will therefore have consequences for the immune system, the reproductive apparatus and the epithelia (skin, lining of lungs, alimentary canal, vagina …).

The **acute effects** are largely due to failure of the immune system to deal with the consequences of cell death due to the heating effect of a large radiation dose. The **short-term effects** are observed during high dose radiotherapy and include the frequently observed hair loss. The lung lining is very susceptible to radiation damage, which can result in inflammation and difficulty in gaseous exchange.

These effects are very strongly dependent on radiation exposure.

Key terms

The effects of nuclear radiation can be divided into:

Acute effects (radiation poisoning) – illness and death due to a sudden large dose of radiation.

Short-term effects – damage apparent in the few weeks following exposure, mainly to the skin, lungs, ovaries and testes.

Long-term effects (or stochastic effects) – cancers, cataracts, development problems (e.g. foetal).

Study point

It is important to distinguish between the activity of a radioactive source, which is the number of emissions per second and dose of radiation which is a measure of the radiation received by a body.

(b) Radiation exposure

There are several units of radiation exposure, each expressing a different aspect. Unfortunately there are also obsolete units which are also still in use. The fundamental quantity is the **absorbed dose**, D, which is the energy deposited per kilogram of tissue. The unit is the gray (Gy) which is equivalent to $\mathrm{J\,kg^{-1}}$. (See Study point for the rad.)

We saw in Section 3.5 that α radiation is much more strongly ionising than β and γ, so its range is shorter. This results in greater damage to cells, i.e the damage in 1 kg of material due to the absorption of 1 mJ of α particles is much greater than that due to the same energy of β and γ. This is expressed in the *radiation weighting factor*, W_R – see Table B4 – and the **equivalent dose**, H.

Radiation	Weighting factor, W_R
X, γ	1
β	1
n	5–20
p	5
α	20
heavy nuclei	20

Table B4 Radiation weighting factors

$$\text{Equivalent dose, } H = DW_R,$$

The unit of equivalent dose is the sievert, Sv.

Study point

You may come across the rad as a unit of absorbed dose.

1 rad = 0.01 Gy

This is not quite the end of the story because different tissues (e.g. the lung lining, bone marrow) are more sensitive to radiation damage than others. In order to estimate the total **stochastic effect** on the body, the contribution from the radiation absorbed by the different tissues is expressed by the *tissue weighting factor*, W_T – see Table B5.

Tissue/organ	Tissue weighting factor, W_T
stomach, colon, lung, red bone marrow, breast, remaining tissues*	0.12
gonads	0.08
bladder, oesophagus, liver, thyroid	0.04
bone surface, brain, skin, salivary glands	0.01

* The 'remaining tissues' means all the tissues of the body not specified in the table, e.g. skeletal muscles, heart muscles.

Table B5 – Tissue weighting factors

B15 Knowledge check

For each line of Table B5, calculate W_T × the number of tissues listed. Hence show that the sum of the products is 1.0.

The procedure for calculating the **effective dose**, E, is to multiply the equivalent dose, H, received by each tissue by the relevant tissue weighting factor, W_T, and then add these together. The unit of effective dose is also the sievert, Sv. The result gives a way of comparing the likelihood of a cancer (or other stochastic condition) arising from the dose of radiation received.

Example

A patient is irradiated in the course of treatment. His lungs receive an equivalent dose of 4 mSv; the thyroid 3 mSv; the remaining tissues 2 mSv. Calculate the total effective dose and explain what it means.

Answer

$$E = 0.12 \times 4 \text{ mSv} + 0.04 \times 3 \text{ mSv} + 0.12 \times 2 \text{ mSv}$$
$$= 0.84 \text{ mSv}$$

This means that the probability of the patient getting cancer from the treatment is the same as if his whole body was uniformly irradiated with an equivalent dose of 0.84 mSv, i.e. an absorbed dose of 0.84 Gy (J kg^{-1}) assuming the radiation was β or γ.

Study point

The rem is an obsolete unit of equivalent and effective dose. It is still in use in the US.

1 rem = 0.01 Sv.

Knowledge check B16

In a treatment for a stomach cancer with γ rays, a patient's organs receive the following doses:

Stomach	3 mGy
Liver	2 mGy
Lung	2 mGy
Colon	0.5 mGy
Oesophagus	0.5 mGy

Calculate the total effective dose.

(c) Gamma-emitting radio tracers

Radio tracers are chemical compounds in which one of the atoms has been replaced by a radioactive isotope. Such tracers have a wide industrial application, e.g. finding the location of cracks in buried gas pipelines. They are also used in medicine for locating medical problems and for imaging. One of the most common radiotracers used is the gamma emitting metastable isotope of technetium, ^{99m}Tc.

Technetium is not found in nature as it has no stable isotopes. ^{99m}Tc has a half-life of only 30 minutes, so it is not possible for hospitals to obtain it directly. Instead it is extracted in hospitals from a molybdenum generator (known colloquially as a moly cow) which contains the radioactive molybdenum isotope, ^{99}Mo, which has a half-life of 67 hours and decays by β^- emission.

$$^{99}_{42}\text{Mo} \rightarrow {}^{99m}_{43}\text{Tc} + {}^{0}_{-1}\beta$$

The m in nuclide ^{99m}Tc indicates that this nuclide is in an excited state. It decays to ^{99}Tc with the emission of a 0.14 MeV photon (γ), which can be used for imaging.

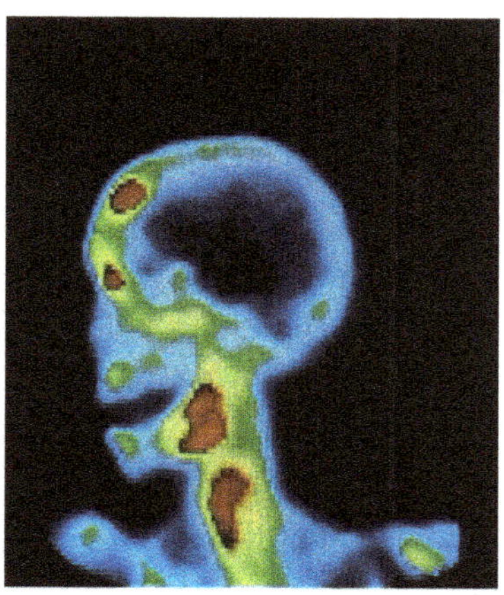

Fig. B.24 Gamma camera image of a patient showing secondary bone cancers (in red).

The image in Fig. B.24 is taken by a gamma camera (see below). The subject has been injected with a solution of sodium pertechnate (Na^+ TcO_4^-) which is absorbed more strongly by tumours because of their higher metabolic rates than normal tissue. ^{99m}Tc is also frequently used in myocardial perfusion scans which investigate heart function and blood flow.

Iodine-123 ($^{123}_{53}$I – half-life 13 hours) is also used as a medical tracer to investigate the functioning of the thyroid gland. Again it results in γ-emission (see Knowledge check B17) and, in addition to the short half-life, has the advantage over the previously used iodine-131 that there is no beta emission.

Study point

^{99}Mo is prepared either by neutron bombardment of ^{98}Mo or extracted from the fission products of a ^{235}U nuclear reactor.

Knowledge check B17

(a) Identify the missing particle in the equation for the decay of ^{99}Mo.

(b) Write the equation for the decay of ^{99m}Tc.

(c) Calculate the difference in mass of ^{99m}Tc and ^{99}Tc. Express your answer in u.

Knowledge check B18

^{123}I is prepared by bombarding ^{124}Te with protons. The resulting excited nucleus immediately emits two neutrons. ^{123}I decays by electron capture to ^{123m}Te, which decays almost immediately to ^{123}Te with the emission of a 157 keV photon, which can be detected by a gamma camera. Write equations for these reactions.

B19 Knowledge check

Explain, using a relevant conservation law, why two photons are produced from the annihilation of an electron and a positron.

(d) Positron emission tomography (PET scan)

In PET scans, a biologically active molecule is tagged with a positron-emitting atom, such as fluorine-18. An example of such a molecule is fluorodeoxyglucose, which is a glucose molecule with a hydroxyl group being replaced by an ^{18}F atom. The two gamma photons produced when the positron and an electron mutually annihilate can be detected outside the body using a gamma camera.

Cancers contain actively dividing cells and so have a high requirement for energy and tend to concentrate glucose. Hence PET scans are useful for cancer imaging.

The patient in Fig. B.25 has testicular cancer which has metastasised (spread) to the lungs. The bright areas in the left image indicate concentrations of glucose which show the presence of the secondary cancers. The image on the right shows the same patient after a 6 month course of chemotherapy, which resulted in complete regression.

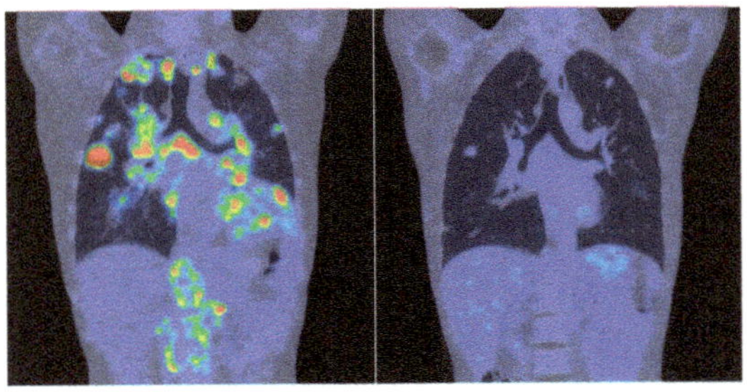

Fig. B.25. PET scans during cancer treatment

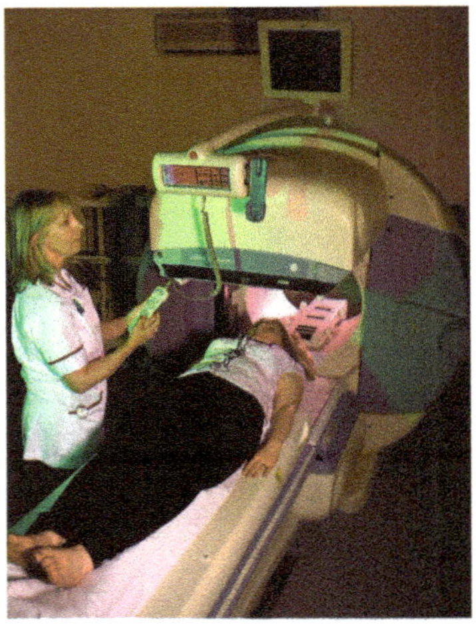

Fig. B.26 Gamma camera in use

(e) Gamma camera

Both the above two tracer procedures involve the detection of gamma rays and the production of images using a gamma camera. Fig. B.26 shows a gamma camera in use and its principles are illustrated in Fig. B.27.

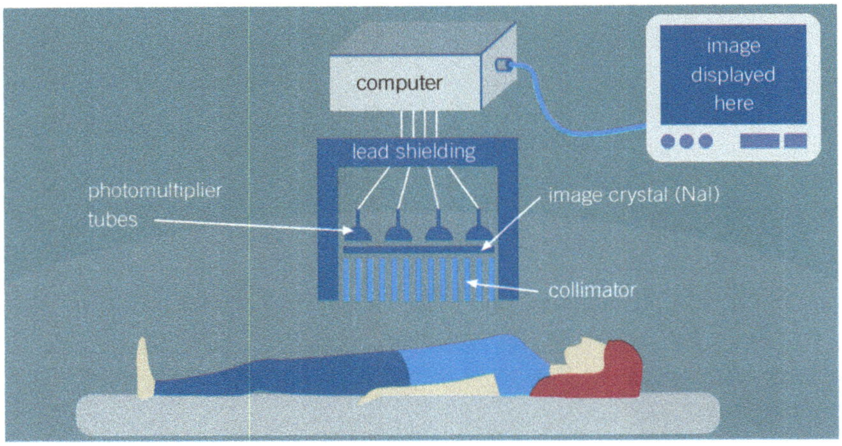

Fig. B.27 Principle of operation of a gamma camera

Study point

The compound eyes of insects work on the same principle as the collimator in a gamma camera. They have no lenses but only allow light rays to hit the retina which are travelling parallel to the axis of each tiny eye. The brain then constructs the image. The praying mantis even manages 3D images using these eyes.

There is no way of focusing γ-rays, so a lead collimator is used to ensure that only the γ rays travelling vertically upwards (in Fig. B.27) hit the crystal.

The sodium iodide crystal flashes each time a photon hits it, the faint light is magnified by the photomultiplier tubes and the number of flashes in each pixel registered on a scintillation counter (not shown). The signal is fed to the computer which generates the image. In PET scans, the patient is surrounded by the detector (as in Fig. B.26) and the computer only registers when it detects two photons on opposite sides at almost the same time. This allows it to reject stray background radiation photons. The very small time difference in the arrival of the two photons is used to determine the position of their source.

Test yourself Option B

1. The wavelengths of the K_α and K_β lines in the X-ray spectrum of copper are 154 pm and 139 pm respectively.

   (a) Calculate the differences in the atomic energy levels which give rise to photons with these wavelengths. Express your answers in eV.

   (b) Comment on the minimum tube pd which would be required to produce both K_α and K_β photons.

2. The spectrum is of a silver-target commercial X-ray tube without a filter (black curve) and with a thin aluminium filter (red curve). Note that the horizontal axis is of photon energy, not wavelength.

Mini-X-Ag Output Spectrum with and without 80 mil (2mm) Aluminum (Al) Filter

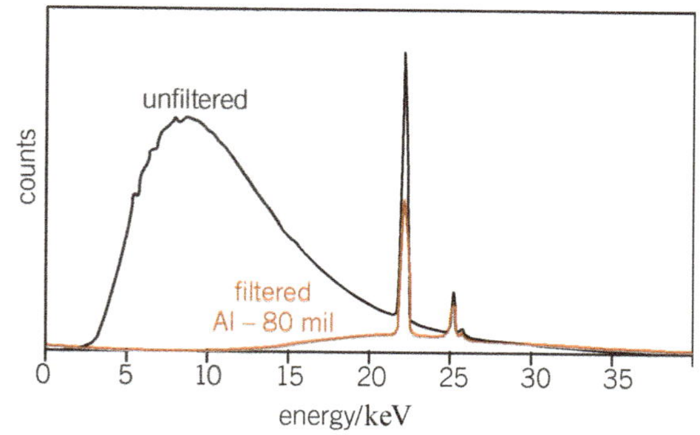

   (a) Describe the effect of the filtering for different photon energies. Give a reason for this.

   (b) Give an advantage in terms of the medical use of X-rays of filtering in this way.

   (c) Estimate the operating pd of the tube.

   (d) The spikes in the graph are the K_α and K_β lines. Calculate the wavelengths of these lines.

   (e) Sketch filtered and unfiltered graphs of the output of this tube at this pd with X-ray wavelength on the horizontal axis.

3. A website gives the following percentage penetration of various X-ray photon energies through 10 cm of soft tissue:

Energy / keV	50	60	70	90	100	120	130	150
% penetration	3.6	5.3	6.7	9.5	10.2	11.5	12.0	13.0

Use the information in the table:

   (a) to calculate the % of 100 keV photons which pass through (i) 20 cm, (ii) 5 cm and (iii) 15 cm of soft tissue

   (b) to estimate the attenuation coefficient of 80 keV X-rays in soft tissue.

4. A muscle with acoustic impedance $1.7 \times 10^6\ \text{kg m}^{-2}\ \text{s}^{-1}$ has an embedded cyst which contains fluid of density $1030\ \text{kg m}^{-3}$ in which sound travels at $1530\ \text{m s}^{-1}$. Calculate the percentage of an ultrasound wave which (a) is reflected at the first boundary and (b) which emerges from the other side of the cyst. The attenuation within the tissues may be ignored.

5. In a Doppler scan, the frequency shift in 2 MHz waves is 346 Hz for blood which is flowing at 30° to the direction of the waves. Taking the speed of sound in blood to be $1550\ \text{m s}^{-1}$, calculate the speed of the blood flow.

6. The main field of an MRI scanner has a flux density of 1.60 T. A secondary field has a gradient of $40\ \text{mT m}^{-1}$.

   (a) Calculate the Larmor frequency of hydrogen nuclei (protons) due to the main field.

   (b) Calculate the difference in Larmor frequency between two positions which are 5.00 mm apart in the direction of the secondary (not-uniform) field.

   (c) State the significance of the non-uniform field for locating a region in the body.

   (d) The scanner can achieve a resolution of 1 mm. What frequencies of radio waves must it be able to distinguish?

7. In a course of treatment for a brain tumour, a beam of radiation is directed at the tumour from different directions. Explain briefly:

   (a) How applying the radiation from different angles reduces the effect on the surrounding tissues.

   (b) How charged particle therapy (proton beam) reduces the effect on surrounding tissues. [Hint: protons are absorbed in a similar way to α particles.]

8. The thyroid gland absorbs iodine. Hence this element can be used both as a tracer and for radiotherapy (irradiating the thyroid from within). There are several radioactive isotopes of iodine with different decay modes:

$^{123}_{53}\mathrm{I}$ decays by γ emission with a half-life of 13 hours

$^{131}_{53}\mathrm{I}$ decays by β^- emission, with a half-life of 8.0 days, to an excited state of xenon (Xe^*) which rapidly decays by γ emission.

(a) Compare these two isotopes of iodine for use as a tracer and for therapy.
(b) Write decay equations for the two-stage decay of $^{131}_{53}\mathrm{I}$.

9. Ultrasound A-scans and ultrasound B-scans can both be used for investigation of the eye. Compare briefly the information which can be obtained by these two types of scan.

10. Ionising radiation, such as X rays and γ rays, causes genetic damage in cells which can lead to cancer. X- and γ- rays are also used in radiotherapy to treat cancer.

(a) Suggest a reason for this apparent contradiction.
(b) Explain how the damage caused by radiotherapy is kept as low as possible.
(c) People whose cancer has metastasised (i.e. several secondary cancers have formed in different parts of the body) are not usually treated with radiotherapy. Suggest why this is the case.

11. (a) The fraction of the ultrasound power reflected at the boundary between two materials, 1 and 2, with acoustic impedances Z_1 and Z_2 is given by:

$$\frac{(Z_1 - Z_2)^2}{(Z_1 + Z_2)^2}$$

In terms of the formula, explain why the fraction reflected is the same whether the incident wave is travelling in medium 1 or medium 2.

(b) The tibia (thigh bone) contains marrow which consists largely of fat. The bone is surrounded by muscle tissue. The acoustic impedances of the three tissues are given in the table:

Tissue	Muscle	Bone	Marrow
Acoustic impedance / 10^6 kg m^2 s^{-1}	1.70	5.40	1.35

(i) Calculate the fraction of ultrasound reflected at the boundary between muscle and bone.
(ii) The ultrasound which penetrates the bone is partially reflected at the bone/marrow boundary and again at the marrow/bone boundary. Ignoring multiple reflections calculate the fraction of the original power which emerges back into the muscle from reflections at these two boundaries. [Retain at least 3 sf in intermediate calculations.]

12. Following treatment for a cancer, a doctor suggests a whole body PET scan to detect possible secondary tumours. Briefly describe this process and explain why it detects tumours.

13. A rugby player attends hospital with a suspected broken collar bone. Why does the doctor use an X-ray examination rather than an MRI scan to investigate?

Option C: The physics of sports

This optional unit consists of basic physics concepts. The examiner can ask questions based upon these concepts in the context of any sport. For example, rotational dynamics can be assessed in the context of F1 cars or throwing the hammer. The selection of sports examples by the authors is entirely idiosyncratic and should not be taken as indicative of the actual contexts you will meet in a live examination.

C.1 The principle of moments

The moment of a force is a measure of its turning effect. The definition of the moment of a force about a point was covered in the AS textbook – see Sections 1.1.6 and 1.1.7. Those sections are concerned with bodies in equilibrium, and are applicable to many sporting contexts. In many sports, rotating objects are far from being in equilibrium.

(a) Stability and toppling

In spite of her best efforts, Wendy's Laser dinghy (Fig. C.1), is about to capsize (topple). This is a frequent occurrence in dinghy racing but very rare in F1 racing (or indeed in ocean yacht racing). We saw in Section 1.1.6(c) how to predict stability in situations where the only significant forces on an object are its own weight and the contact force with the ground. In sports this is rarely the case – toppling generally occurs when a force is applied from the side.

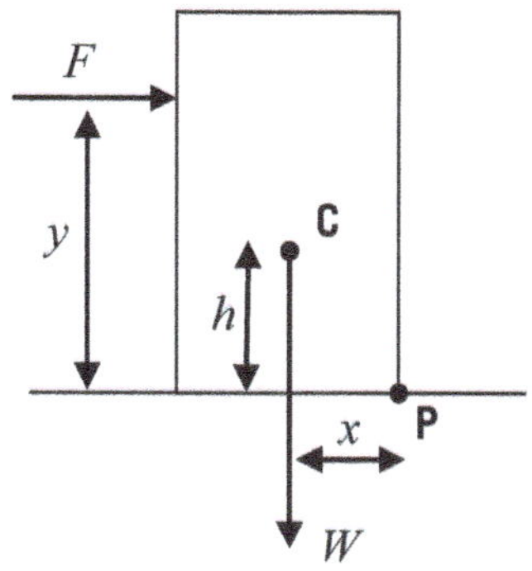

Fig. C.2 Will F topple the box?

Will the box in Fig. C.2 topple or not? We can use moments to find out.

If it topples, it will rotate clockwise about **P**. The resultant clockwise moment, M, is given by:

$$M = Fy - Wx$$

If $M > 0$, i.e. $F > \frac{Wx}{y}$

the box will topple. From this we can infer that the wider the box, the less likely it is to topple. So rugby players who are tackling or pushing will brace with one foot well in front of the other – to increase stability.

Knowledge check C1

How can you tell the box in Fig. C.2 is not of uniform composition?

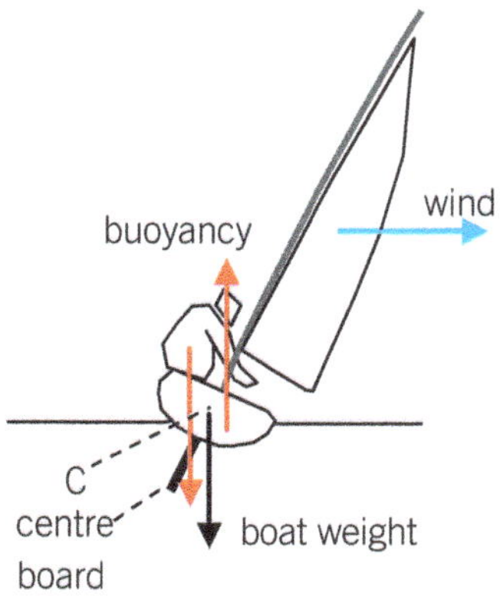

Fig. C.4 Forces on the dinghy

What about Wendy in Fig. C.1 and the rather more sedate mariner in Fig. C.3. What forces are involved?

The line drawing Fig. C.4 helps to clarify. The force of the wind acting on the sail exerts a **clockwise moment** about the centre of gravity, C. The buoyancy force (the resultant upward force of the water on the hull) and the weight of the dinghy pilot provide an anticlockwise moment.

Fig. C.3 A stable dinghy

But what if the wind blows a bit more strongly? The boat will heel over more, which will (a) reduce the wind area of the sail and (b) lower the height of the wind force above the centre of gravity, C – so stabilising the boat. Also the sailor might lean out a bit more: what effect will that have?

Notice two features of the design of the dinghy:

1. The shape of the hull. As the boat heels, the buoyancy force (more or less in the centre of the submerged part) shifts to the right.
2. The centre board, which is sometimes a heavy steel plate. Some dinghies also have ballast in the bottom of the hull. These have the effect of lowering the centre of gravity.

Knowledge check C2

Generally, sailing dinghies do not move sideways through the water. What force should be added to Fig. C.4?

C3 Knowledge check

If the dinghy heels over a bit more (e.g. due to a wave) explain why the resultant moment will cause it to return to its previous position (i.e. why is it stable).

The dinghy in Fig. C.3 has an additional stability feature. The tubes on either side of the *gunwale* are floatation devices – called *dogs*. If the dinghy heels so that one is partly submerged (as shown), the buoyancy force on it provides an additional restoring moment.

Generally, the lower the centre of gravity of an object and the wider the base, the more stable it is. The box in Fig. C.2 can rotate through $\tan^{-1}\frac{x}{h}$, where h is the height of the centre of gravity before it topples.

Rugby players who are going in for the tackle exploit this by keeping a low centre of gravity as well as bracing their feet apart in the front-rear direction.

The impact force, F, from the contact with the upright player is not far above the centre of gravity so has only a small moment. The player being tackled is much less stable – high centre of gravity – and can more easily be toppled.

Fig. C.5 Ideal tackling position

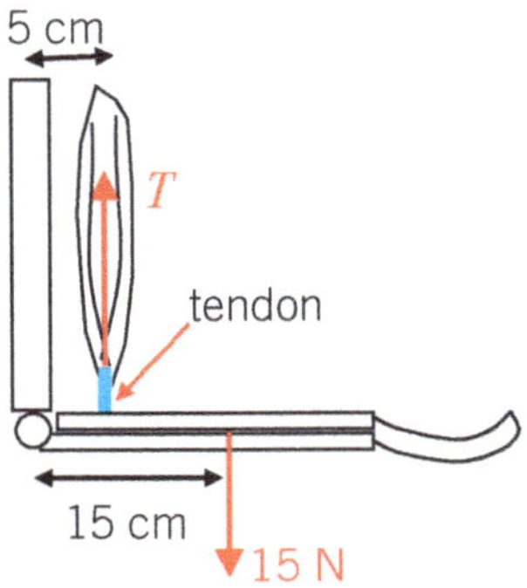

Fig. C.7 Tension in the distal biceps tendon

C4 Knowledge check

Calculate the tension in the biceps tendon if the hand in Fig. C.7 holds a 2 kg bag of potatoes (a distance of 35 cm from the elbow).

C5 Knowledge check

Explain why the tension in the biceps tendon doesn't change if the forearm in Fig. C.7 is 10° above (or below) the horizontal.

(b) Moments in the human skeleton-muscular system

Skeletal muscles (as opposed to heart and gut muscles) apply tensile forces to bones via tendons. Muscles never push – they only pull. They work in antagonistic (opposing) pairs, e.g. the biceps and triceps muscles of the upper arm.

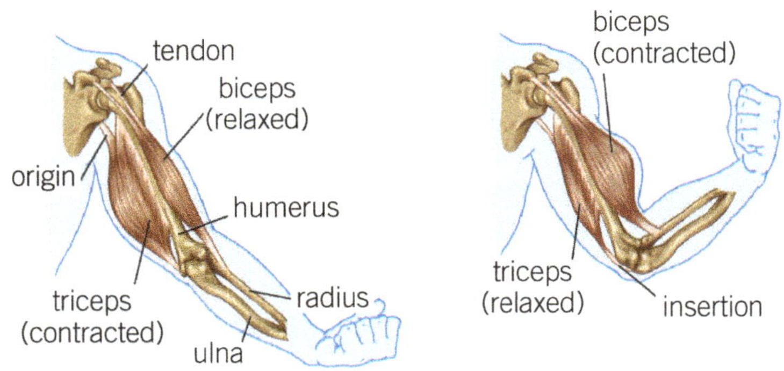

Fig. C.6 Muscles of the upper arm working in opposition

The arm is flexed by tightening the biceps, during which process the triceps is relaxed. Straightening the arm is achieved by the opposite process.

Example

Use Fig. C.7 to estimate the tension in the distal biceps tendon when the forearm is held in the horizontal position.

Answer

Taking moments about the elbow joint: $T \times 5\text{ cm} = 15\text{ N} \times 15\text{ cm}$

$$\therefore T = \frac{225\text{ N cm}}{5\text{ cm}} = 45\text{ N}$$

Another antagonistic pair of muscles is the quadriceps / hamstring pair in the thigh. During running, these muscles contract alternately and overtraining can lead to severe tendonitis (inflammation of the tendon) especially of the quadriceps.

Fig. C.8 Antagonistic muscle pair in the thigh

(c) More moments when sailing

One of the most surprising things about sailing is that you can sail against the wind. The simple diagram of a sailing boat, Fig. C.9, shows two of the horizontal forces on a boat with the wind coming from forward of the *beam* (sideways). The wind flowing over the sail produces a lift as shown. 'Lift' might seem strange, but we use the word *lift* to describe the force at right angles to a surface arising from the flow of air over the surface. This is the same as the lift by an aeroplane wing (see Section C7). The keel of the boat produces a balancing sideways force which prevents the boat being dragged to the side (though all boats make some *leeward* motion) and the resultant is a forwards force.

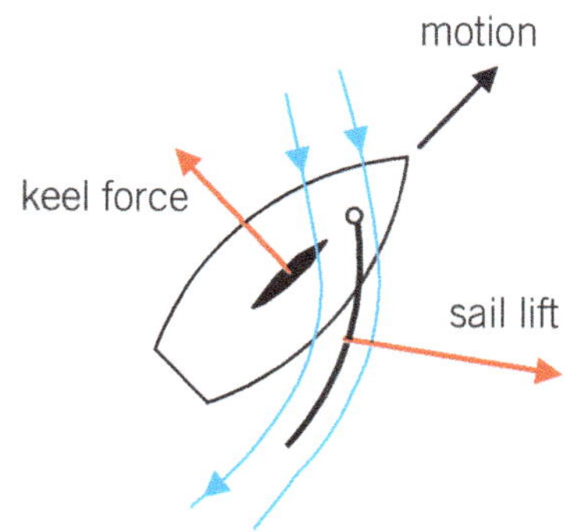

Fig. C.9 Heading into the wind

If we think about the moments of the forces about the centre of gravity of the boat, that cannot be the whole story. The keel force passes roughly through the centre but, as drawn, the sail lift passes aft of (behind) the centre and therefore has an anticlockwise moment – so the head of the boat would turn into the wind. One way of counteracting this is to use a second sail before the mast, a jib: Fig. C.10. This produces a clockwise moment, balancing the mainsail. It is also possible to steer away from the wind using the rudder, but this produces undesirable drag.

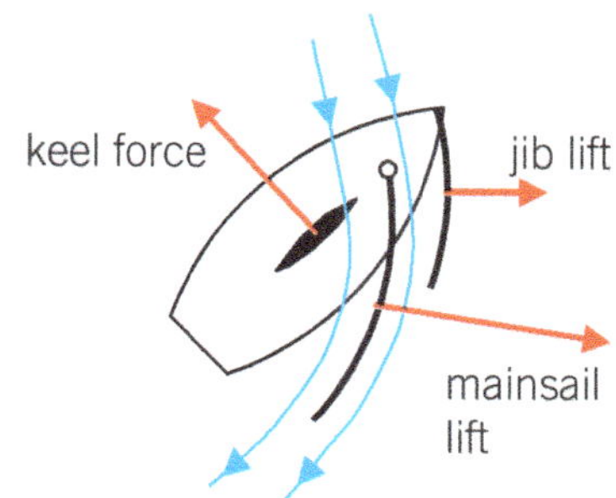

Fig. C.10 Close hauled with jib

C.2 Collisions

(a) Coefficient of restitution

When two elastic objects collide, their separation velocity is less than their collision velocity. To a good approximation the ratio of the magnitudes of their relative velocities after and before collision is a constant, which is called the **coefficient of restitution**, e. This is clarified in Fig. C.11.

Fig. C.11 Colliding spheres

Thus $$e = \frac{\text{relative velocity after collision}}{\text{relative velocity before collision}} = \left|\frac{v_2 - v_1}{u_2 - u_1}\right|$$

We can relate this to the relative bounce height of a ball, which is dropped on to the ground: see Fig. C.12.

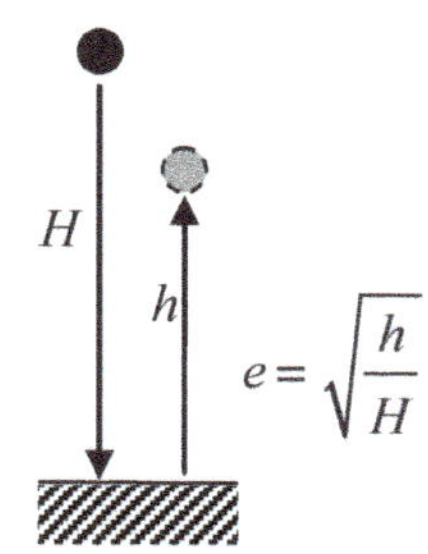

Fig. C.12 Bouncing ball

From an initial height, H, the speed of impact = $\sqrt{2gH}$.

If it bounces to height h the speed just after impact = $\sqrt{2gh}$.

Because the floor isn't moving we can write: $e = \frac{\sqrt{2gh}}{\sqrt{2gH}} = \sqrt{\frac{h}{H}}$.

Note: From this it is tempting to conclude that e^2 is the energy efficiency of the bounce, i.e. the fraction of the original kinetic energy which remains after the collision. This is true for an object bouncing off a stationary very massive object (e.g. the Earth) but it is not true in general (see Stretch & challenge).

Example

International Tennis Federation rules state that a tennis ball dropped from a height of 254 cm should bounce to between 135 cm and 147 cm when dropped on to a concrete floor.

Calculate the range of the coefficient of restitution specified.

Knowledge check C6

A basketball umpire throws the ball up to 3.0 m. If both players miss the ball, how high will it bounce?

[$e = 0.85$]

Stretch & challenge

A sphere of mass m_1 moving at speed u collides head on with a stationary sphere of mass m_2. Show that:

(a) the kinetic energy after the collision is given by

$$\frac{1}{2}m_1u^2\left[\frac{m_1 + e^2 m_2}{m_1 + m_2}\right]$$

(b) that this tends to $\frac{1}{2}m_1u^2$ as $m_2 \to \infty$.

Answer

$e_{min} = \sqrt{\frac{135}{254}} = 0.73\ (73\%)$ $e_{max} = \sqrt{\frac{147}{254}} = 0.76\ (76\%)$ [2 sf]

The rules for basketballs are slightly more complicated (see Q5 in Test yourself 4C]

C7 Knowledge check

Calculate the impact and rebound speed for a tennis ball in the drop test if the coefficient of restitution is 0.75.

(b) Forces in collisions

We can estimate the mean force exerted in a collision using N2, which is conveniently written

$$F\Delta t = mv - mu$$

Here, F is the mean force and Δt is the time over which the force is applied. The quantity $F\Delta t$ is called the impulse of the force, which as the equation shows is also equal to the change of momentum.

Consider a bouncing tennis ball.

Example

The diagram shows a tennis ball, of mass 58 g, in the interval between its first contact and last contact with the ground during a drop test. The middle shows the ball in mid-collision, when the centre of mass, **C**, is at its lowest point.

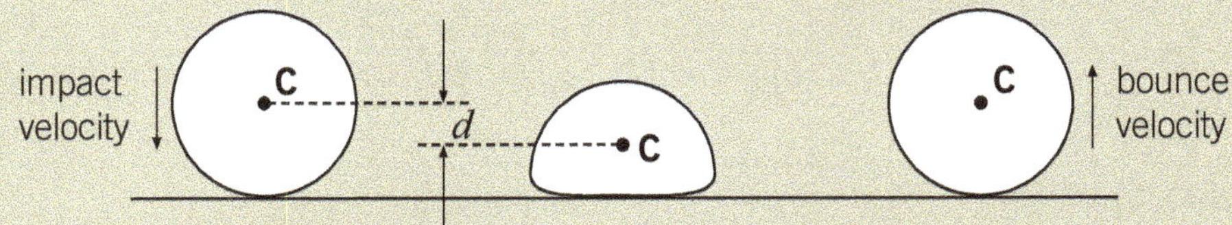

Assuming that the ball falls from a height of 254 cm, the distance d is 1.0 cm and that e is 0.75, estimate the mean force exerted by the tennis ball on the concrete floor.

Answer

Using the answer to Knowledge check C7, and the equation $F\Delta t = m(v - u)$ and taking the upward direction as positive:

$$F\Delta t = 0.058\ \text{kg}\left(5.295 - (-7.059)\right)\ \text{m s}^{-1} = 0.717\ \text{N s}$$

To estimate the duration of the collision, we'll assume that the deceleration to rest is uniform.

Hence the time to decelerate to rest $= \frac{0.010\ \text{m}}{\frac{1}{2}7.059\ \text{m s}^{-1}} = 2.83 \times 10^{-3}\ \text{s}$

Similarly we calculate the time to accelerate upwards from the ground.

Time to accelerate from rest $= \frac{0.010\ \text{m}}{\frac{1}{2}5.295\ \text{m s}^{-1}} = 3.78 \times 10^{-3}\ \text{s}$

The time Δt is the sum of these two times

$\therefore\ \Delta t = (2.83 + 3.78) \times 10^{-3}\ \text{s} = 6.61 \times 10^{-3}\ \text{s}$

$\therefore$ The mean force exerted on the ball by the ground, $F = \frac{0.717\ \text{N s}}{6.61 \times 10^{-3}\ \text{s}} = 110\ \text{N}$

Hence, by N3, the mean force exerted by the ball on the ground is -110 N, i.e. 110 N downwards.

C.3 Concepts and equations in rotational motion

Rotation is a feature of many sports. Before we analyse it we need to develop some concepts. To do this we recall that, in analysing linear motion we use the following quantities:

displacement velocity acceleration force

momentum kinetic energy mass time

With the exception of time, we shall look for rotational equivalents of all these quantities. We have met some of them in earlier parts of the A level Physics course. For the moment we'll consider only rigid bodies.

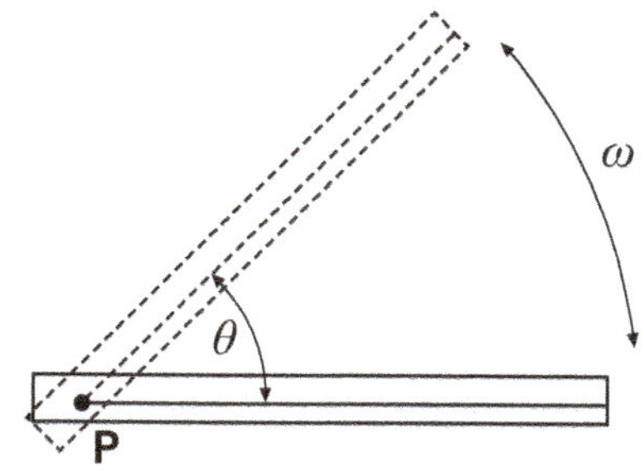

Fig. C.13 Rotating object

Consider the object in Fig. C.13 rotating about the point **P**. We can describe its orientation and angular motion by the angular position, θ , and angular velocity, ω. If ω is constant, it is defined by

$$\omega = \frac{\Delta\theta}{t} \quad \text{i.e.} \quad \omega = \frac{\theta_2 - \theta_1}{t}.$$

Similarly, we can define the angular acceleration, α, by analogy with linear acceleration by:

$$\alpha = \frac{\Delta\omega}{t} \quad \text{i.e.} \quad \alpha = \frac{\omega_2 - \omega_1}{t}$$

We can use linear equations to write equations of motion for rotation:

Linear equation	Condition	Rotational equation
$x = vt$	Constant velocity	$\theta = \omega t$
$v = u + at$	Constant acceleration	$\omega_2 = \omega_1 + \alpha t$
$x = \frac{u+v}{2}t$		$\theta = \frac{\omega_1 + \omega_2}{2}t$
$x = ut + \frac{1}{2}at^2$		$\theta = \omega_1 t + \frac{1}{2}\alpha t^2$
$v^2 = u^2 + 2ax$		$\omega_2 = \omega_1^2 + 2\alpha\theta$

Table C1 Linear and rotational equations

Study point

Don't bother learning the rotation equations in Table C1. Just know how to 'derive' them using the linear ones.

Looking again at Fig. C.13, we see that, even though the object has no linear (or *translational*) motion it must possess kinetic energy because all its particles are moving.

C.4 Rotational kinetic energy

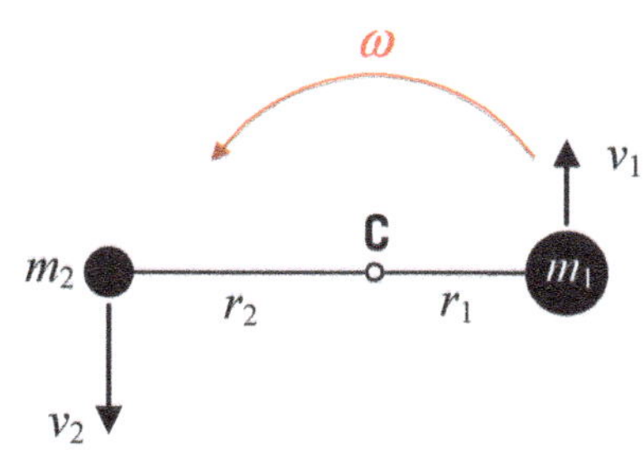

Fig. C.14 Rotating spheres

(a) Moment of inertia

In Fig. C.14, the spheres are connected by a rigid bar of negligible mass and are rotating about the point **C**. What is the kinetic energy of the system?

$$E_k = \tfrac{1}{2}m_1v_1^2 + \tfrac{1}{2}m_2v_2^2$$

Expressing this in terms of the angular speed, ω, by using the relationship, $v = r\omega$,

$$E_k = \tfrac{1}{2}m_1(r_1\omega)^2 + \tfrac{1}{2}m_2(r_2\omega)^2$$

This can be rewritten as: $E_k = \tfrac{1}{2}(m_1r_1^2 + m_2r_2^2)\omega^2$

Key terms

The **moment of inertia**, I, of an object consisting of n particles is defined by:

$$I = \sum_{i=1}^{n} m_i r_i^2$$

where m_i and r_i are the mass and distance from the axis of rotation of the ith particle.

Study point

Note the unit of moment of inertia: kg m^2.

Study point

Warning: Unlike mass, the moment of inertia of a body is not always the same. It depends upon the axis of rotation.

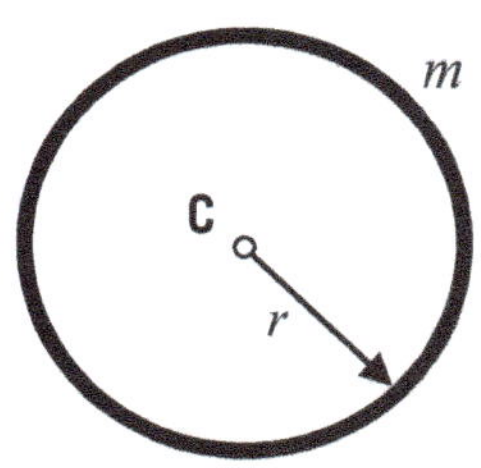

Fig. C.15 Cylindrical ring

Object	I
Uniform ring	mr^2
Uniform disc or cylinder	$\frac{1}{2}mr^2$
Solid sphere	$\frac{2}{5}mr^2$
Spherical shell	$\frac{2}{3}mr^2$
Uniform rod	$\frac{1}{12}ml^2$

Table C2 Some moments of inertia

Stretch & challenge

The barbell below is rotating about its centre and moving along. Show that its kinetic energy is given by:

$E_k = \frac{1}{2}mv^2 + \frac{1}{2}I\omega^2$

Hint: Calculate the speed of each sphere at the instant of the diagram.

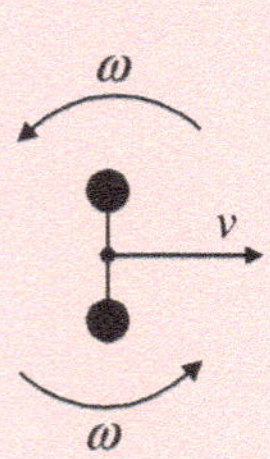

This is of the same form as the equation $E_k = \frac{1}{2}mv^2$ for linear motion, with ω replacing v and the quantity $m_1r_1^2 + m_2r_2^2$ replacing m. $m_1r_1^2 + m_2r_2^2$ is called the **moment of inertia** and has the symbol I. If there are more than two particles then we just add more terms to the sum and the rotational kinetic energy is given by:

$$E_{k\,rot} = \frac{1}{2}I\omega^2$$

The moment of inertia combines the mass of the object with the distance of its particles from the centre of rotation. The further out they are, the more they contribute to the moment of inertia.

Example

A speed regulator on a steam engine consists of two small metal spheres, each of mass 0.15 kg. Calculate the energy of rotation of the pair about the midpoint when they are rotating at 150 rpm and their centres are separated by a distance of 20 cm.

Answer

Distance of the centre of each sphere from midpoint = 10 cm = 0.10 m.

$\therefore$ Moment of inertia $= m_1r_1^2 + m_2r_2^2 = 0.15 \times 0.10^2 + 0.15 \times 0.10^2$
$= 0.003 \text{ kg m}^2$

Angular speed, $\omega = \dfrac{150 \times 2\pi}{60 \text{ s}} = 15.7 \text{ rad s}^{-1}$

$\therefore$ Kinetic energy $= \frac{1}{2}I\omega^2 = \frac{1}{2} \times 0.003 \text{ kg m}^2 \times (15.7 \text{ rad s}^{-1})^2 = 0.37 \text{ J}$

(b) Formulae for moment of inertia

Working out the moment of inertia of an extended solid object can be tricky, unless it has an easy symmetrical shape. Even then you need calculus. An exception is a thin uniform cylindrical ring (Fig. C.15).

This is easy because all the particles are the same distance, r, from the axis of rotation, **C**. Hence $I = mr^2$.

You won't need to remember formulae for moments of inertia – they will be given in the examination – but Table C2 has a few that you will come across. The formulae given are all for rotations about the centre of mass (centre of gravity). The axis is perpendicular to the plane of the disc and perpendicular to the uniform rod.

What if, say, a rod is rotating about its end (like a bowler's arm when delivering the ball)? Then the formula becomes $I = \frac{1}{3}ml^2$. Why is it bigger? This is left as a problem for you but, as a hint, consider the following diagram:

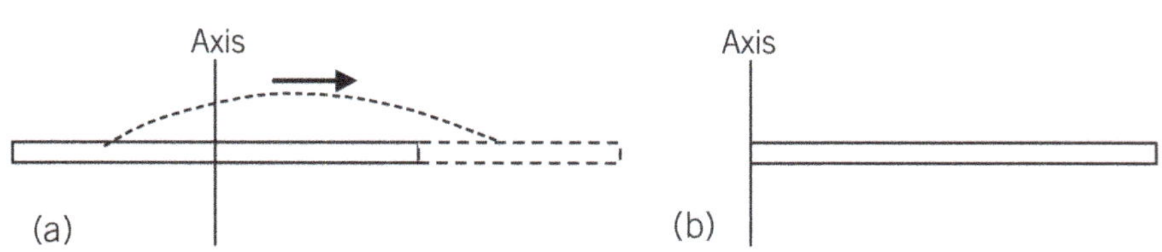

Fig. C.16 Why is the moment of inertia in (b) greater than in (a)?

(c) Rolling objects

What if an object is spinning and moving along? What is its total kinetic energy?

Conveniently, the answer is the sum of the translational and rotational KEs, i.e.

$$E_k = \tfrac{1}{2}mv^2 + \tfrac{1}{2}I\omega^2,$$

where v is the speed of the centre of gravity and ω the angular speed about the centre of mass (see Stretch & challenge).

For a rolling circular object, e.g. a bicycle wheel, the speed and angular speed are related by $v = r\omega$. This must be the case because the velocity of the rim is zero at the point of contact with the ground (Fig. C.17).

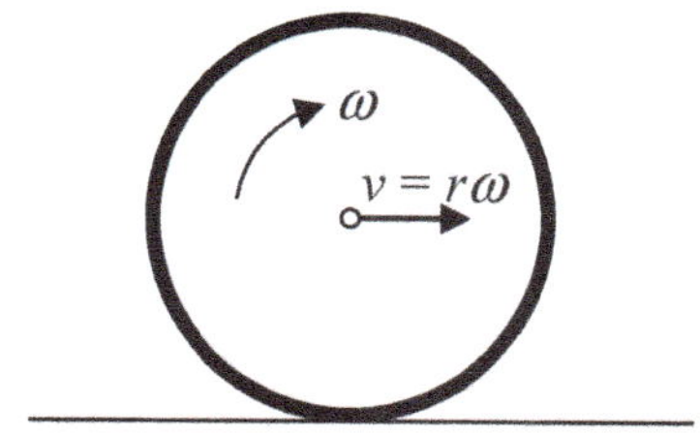

Fig. C.17 Rolling wheel

Example

Estimate the kinetic energy of the road wheel of a car which is rolling at $30\ \text{m s}^{-1}$, using the following estimated quantities:

Mass of wheel, $16\ \text{kg}$; radius, $30\ \text{cm}$; moment of inertia, $1\ \text{kg m}^2$

Answer

$$E_k = \tfrac{1}{2}mv^2 + \tfrac{1}{2}I\omega^2 = \tfrac{1}{2}16\ \text{kg} \times (30\ \text{m s}^{-1})^2 + \tfrac{1}{2}1\ \text{kg m}^2 \times \left(\frac{30\ \text{m s}^{-1}}{0.3\ \text{m}}\right)^2$$

$$= 7200 + 5000 = 12\,200\ \text{J}$$

Knowledge check C8

Show that the total kinetic energy of a rolling ring is mv^2.

Example

A *wood* (i.e. a lawn bowl) is launched by allowing it to roll down a slope of height 1.5 m. Approximating the wood as a uniform sphere, estimate its speed at launch.

Answer

Potential energy loss = $mg\Delta h = 1.5 \times 9.81m = 14.7m$

With a speed v, $E_k = \tfrac{1}{2}mv^2 + \tfrac{1}{2}I\omega^2$

But when the ball is rolling, $\omega = \frac{v}{r}$, and $I = \tfrac{2}{5}mr^2$

$\therefore\ E_k = \tfrac{1}{2}mv^2 + \tfrac{1}{2}\tfrac{2}{5}mr^2\left(\frac{v}{r}\right)^2 = \tfrac{7}{10}mv^2$

$\therefore\ \tfrac{7}{10}mv^2 = 14.7m,\ \therefore\ v = \sqrt{\frac{10 \times 14.7}{7}} = 4.6\ \text{m s}^{-1}$

Knowledge check C9

Calculate the kinetic energy of a uniform cylinder of mass $5\ \text{kg}$ which is rolling at a speed of $4\ \text{m s}^{-1}$.

C.5 Rotational dynamics

(a) Torque and angular acceleration

In order to produce an angular acceleration, with no translational acceleration, we need a couple – two equal and opposite forces with different lines of action (see Fig. C.18).

The turning effect of a couple is called its moment or **torque**, τ. The moment of a single force about a specified point depends upon the perpendicular distance from the point to the line of action of the force. But a little thought should convince you that the moment of the couple is $\tau = Fd$ about any point.

The effect of the couple is to cause an **angular acceleration**, α, given by

$$\tau = I\alpha$$

This equation can be taken as the definition of torque and is clearly analogous to $F = ma$.

Key terms

Torque, τ: The resultant torque, τ, on an object with moment of inertia I and angular acceleration α is equal to $\tau = I\alpha$.

UNIT: N m [= kg m^2 s^{-2}]

Angular acceleration, α; The rate of change of angular velocity

UNIT: rad s^{-2}

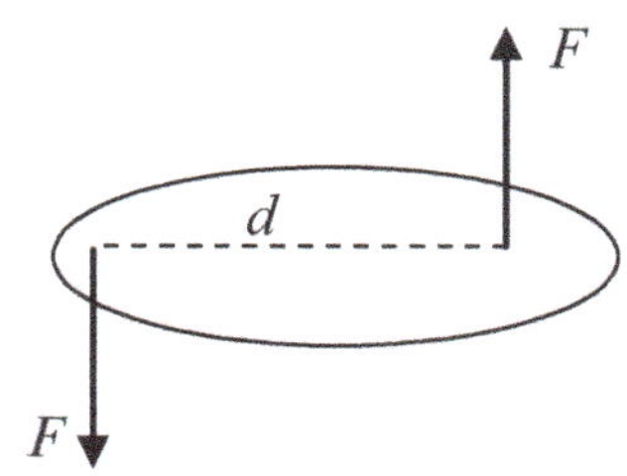

Fig. C.18 Couple

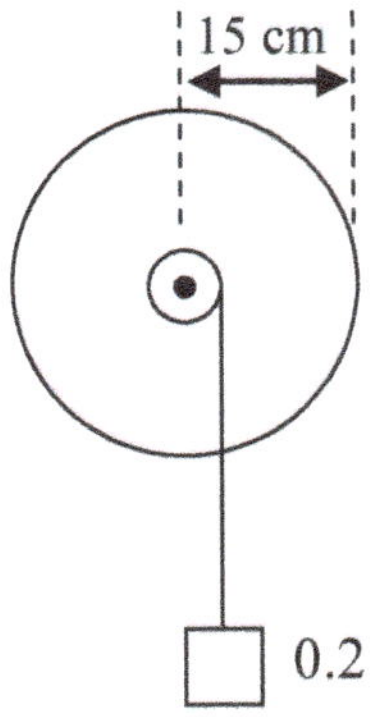

Fig. C.19 Wheel and axle

Study point

What are the approximations here (apart from the zero-friction mounting)?

- We are ignoring the moment of inertia of the axle (a minor point).
- The mass is accelerating, so there is a resultant force on it, so the tension in the cord is less than mg (but not by very much – see Knowledge check).

C10 Knowledge check

Use the calculated value of α in the example to find an approximate value of the acceleration of the 0.20 kg load and hence the percentage reduction in the tension as calculated in the example.

C11 Knowledge check

Give a unit for angular momentum.

C12 Knowledge check

The ice skater in the example starts with her arms outstretched and is rotating at 1 revolution per second. Use the estimates in the example to calculate (a) her angular momentum and (b) her rotational speed when she pulls her arms in.

Example

A smoothly mounted flywheel (a cylindrical disc) of mass 10 kg and radius 15 cm is accelerated by means of a 200 g mass attached to a cord which is wrapped around the 2 cm radius axle. Estimate the angular acceleration of the flywheel [see Study point].

Answer

Moment of inertia of the flywheel, $I = \frac{1}{2}mr^2 = 0.5 \times 10\text{ kg} \times (0.15\text{ m})^2 = 0.1125\text{ kg m}^2$.

Torque produced by applied weight, $\tau = Fd = mgd = 0.2\text{ kg} \times 9.81\text{ m s}^{-1} \times 0.02\text{ m} = 0.3924\text{ N m}$

$\therefore$ Angular acceleration $\alpha = \frac{\tau}{I} = \frac{0.3924\text{ N m}}{0.1125\text{ kg m}^2} = 3.5\text{ rad s}^{-2}$.

(b) Angular momentum

The **angular momentum**, L, of a body is defined by

$$L = I\omega \qquad \text{UNIT: kg m}^2\text{ s}^{-1}$$

It is the rotational equivalent of linear momentum, $p = mv$. The **law of conservation of angular momentum** states: In the absence of a resultant external torque, the angular momentum of a body or system of particles is constant. Its consequences are observed in many contexts, from pirouetting ballerinas to Olympic divers and Kepler's second law.

The ice-skater in Fig. C.20 starts herself spinning (a) with her arms outstretched. She pulls in her arms (b), reducing her moment of inertia.

Angular momentum is conserved, so $I_1\omega_1 = I_2\omega_2$.

But $I_2 < I_1$, so $\omega_2 > \omega_1$, i.e. the skater speeds up. The same principle holds for a diver going into a tuck. A ballerina can use this trick to speed up to 400 rpm during a leap! Calculating the moment of inertia of the human body is quite tricky because its shape and density are both non-uniform, so we'll need to make some approximations:

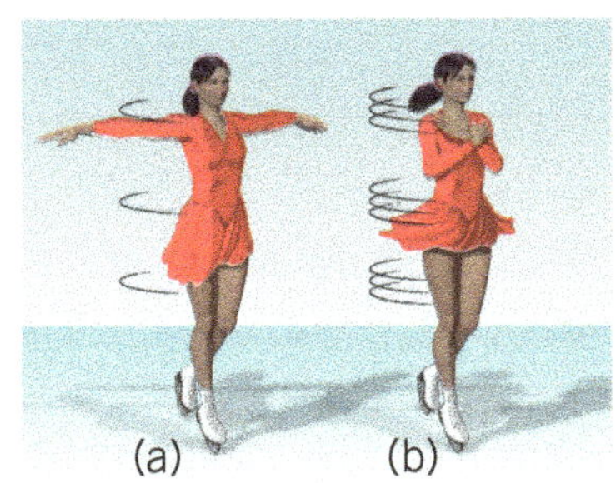

Fig. C.20 Spinning ice-skater

Example

Estimate the moment of inertia of an upright ice-skater (a) with her arms down, (b) with her arms outstretched. Model her outstretched arms by a single rod of total length 1.5 m and mass 6 kg. Model the rest of her body by a cylinder of mass 44 kg and radius 0.10 m.

Answer

(a) I for the cylinder $= \frac{1}{2} \times 44\text{ kg} \times (0.10\text{ m})^2 = 0.22\text{ kg m}^2$

What about her arms hanging by her sides? We could model these crudely as two concentrated 3 kg masses at 0.16 m from the rotation axis, so:

I for hanging arms $= 2 \times 3\text{ kg} \times (0.16\text{ m})^2 = 0.15\text{ kg m}^2$

So total I for skater $= 0.37\text{ kg m}^2$ [A safer estimate would be 0.4 kg m^2 (1 sf).]

(b) To I for the cylinder we need to add I for the rod.

I for rod $= \frac{1}{12} \times 6\text{ kg} \times (1.5\text{ m})^2 = 1.1\text{ kg m}^2$.

So total I for skater $= 0.22\text{ kg m}^2 + 1.1\text{ kg m}^2 = 1.3\text{ kg m}^2$

C.6 Sports involving projectiles

Many sports involve projectile motion: javelin, rugby, golf, football (all codes). The AS course included the analysis of the motion of objects which move under gravity, with the assumption that there were no other significant forces. This is not a very good assumption in the case of most sports. Air resistance and aerodynamic lift play a part in the flight of a javelin; a tennis player can cause a ball to swerve by applying spin, in a similar way to a footballer taking a corner; arrows and darts have flights whose function is to control the motion of the projectile through the air.

Fig. C.21 Jessica Ennis preparing to put the shot

(a) Shot putting

Perhaps the only projectile sport in which the air plays an insignificant part (other than allowing the athlete to stay alive) is shot putting. The mass of the shot is 4 kg (women) and 7.26 kg (men) and the speed of projection is around $12\ \mathrm{m\ s^{-1}}$. With these figures, air resistance is negligible.

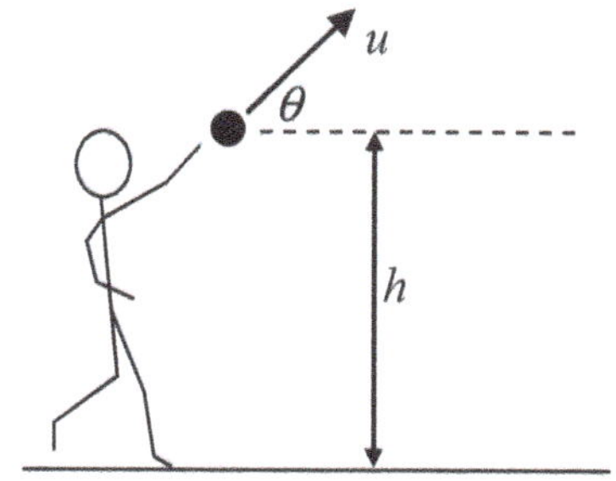

Fig. C.22 Projection from head height

If an object is projected from ground level with a speed u at an angle θ to the horizontal, its range, R is easily calculated, as follows:

Time in the air $= \dfrac{2u\sin\theta}{g}$, $\therefore R = u\cos\theta \times \dfrac{2u\sin\theta}{g} = \dfrac{u^2\sin 2\theta}{g}$.

Here we have used the trig identity: $\sin 2\theta = 2\sin\theta\cos\theta$. The maximum value of $\sin 2\theta$ is 1 (when $\theta = 45°$), so, with v at $12\ \mathrm{m\ s^{-1}}$, the maximum range should be $12^2/9.81 = 14.7\ \mathrm{m}$.

Stretch & challenge

Derive the formula for the range of a projectile released at height h, angle θ and velocity u.

Hint: Start by calculating the flight time.

But the world record (at the time of writing) is 23.12 m and most athletes seem to launch at about 35°, so what's going on? One clue is in Fig. C.22. The launch is obviously not from ground level but from height, h, which for most athletes is about 2 m. The range formula now is a little more complicated but a few lines of algebra give:

$$R = \frac{u^2\sin 2\theta}{2g}\left[1 + \sqrt{1 + \frac{2gh}{u^2\sin^2\theta}}\right].$$

The derivation is left as an exercise (see Stretch & challenge).

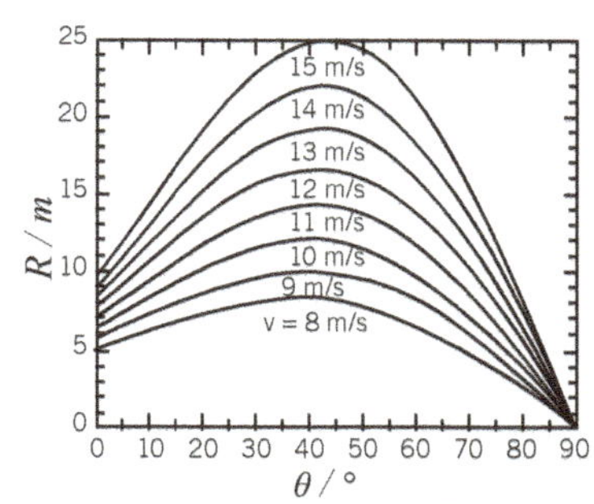

Fig. C.23 Variation of R with θ and u

Plotting R against θ for various release velocities is a nice spreadsheet exercise. The University Biomechanics of Athletics Research Group has produced Fig. C.23 which shows this variation. Notice that the peak range occurs at about 42° now – less than 45° but still greater than the observed 35° at events. The answer lies in the fact that the speed of projection that athletes can achieve is not constant – it is a function of the angle of projection. The Brunel group has taken measurements on athletes and the graph in Fig. C.24 is typical:

Why should the launch speed decrease with θ? For one thing, the shot athlete has to accelerate the shot against gravity for upward angles. For another, the human body finds it easier to throw horizontally. The effect of this variation is to reduce the optimum launch angle to about 35° for most athletes as observed in international events.

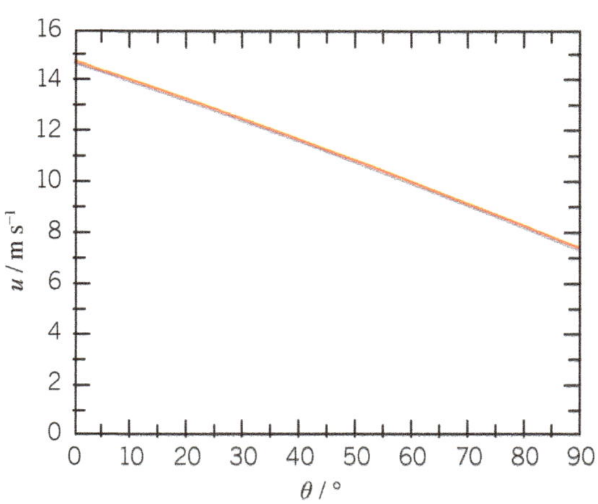

Fig. C.24 Variation of u with θ.

Knowledge check C13

Use the formula to calculate the range of a shot which is projected at 40° with a speed of $12\ \mathrm{m\ s^{-1}}$ from a height of 2.0 m and compare it with the appropriate graph in Fig. C.23.

Knowledge check C14

Consider the athlete whose results are as Fig. C.24. Calculate the ranges he should expect if launching the shot at (a) 45°, (b) 35°. [Take $h = 2.0$ m]

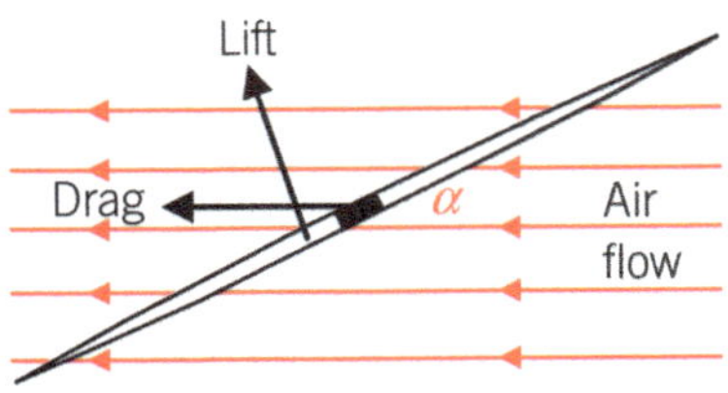

Fig. C.25 Javelin lift and drag

Study point

Fig. C.25 shows the motion of the air relative to the javelin. Notice that the centre of lift is behind the centre of mass, which is in the hand grip.

Key terms

The **angle of attack** is the angle between the air flow and the orientation of the javelin (or discus).

C15 Knowledge check

Account for aerodynamic drag and lift in terms of Newton's first and second laws of motion.

C16 Knowledge check

Explain why the departure from a parabolic curve is only noticeable for the second half. (Hint: see Fig. C.26.)

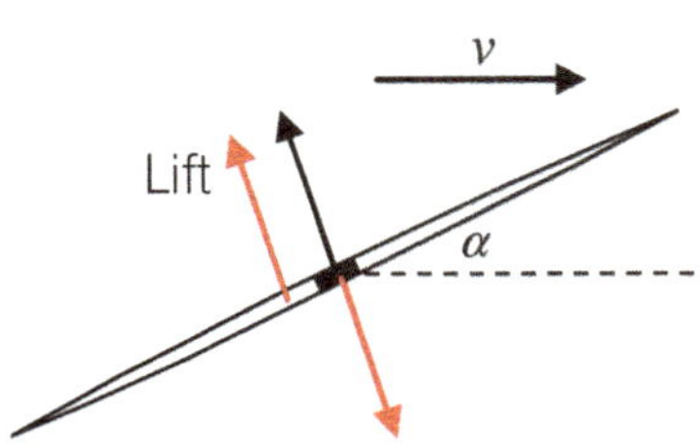

Fig. C.27 Turning effect of lift

C17 Knowledge check

Suppose the distance of the centre of lift from the centre of mass in the javelin in the example is 0.11 m.

(a) Find the mean value of the lift.

(b) A new design of javelin with a lift equal to that in part (a) has the centre of lift 0.22 m behind the centre of mass. Suggest why it is unlikely to be a successful design.

(b) Projectiles with non-gravitational forces

As we have already mentioned, most objects moving through the air experience other forces. In this section we'll focus on two of these, aerodynamic drag and lift (Fig. C.25), in the context of javelin flight. Fig. C.26 shows a typical flight path for a javelin.

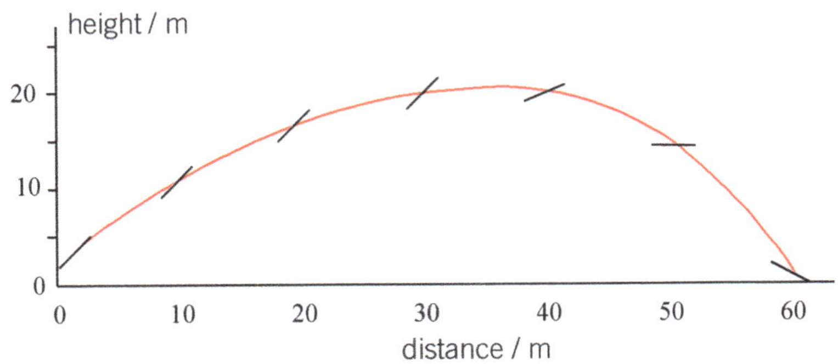

Fig. C.26 Javelin flight path and attitude

The flight path clearly increasingly diverges from a parabolic curve as the javelin travels. This is because of the joint effects of drag and lift. Drag arises mainly because the javelin has to do work against the excess pressure in front of it; lift arises because some of the air molecules are deflected downwards by the javelin. Both these forces increase with the **angle of attack** (Fig. C.25) and it happens that, for the javelin, they have roughly the same size (unlike the discus which has a lift-to-drag ratio of about 3).

As discussed in Section 1.3.5(c) of the AS textbook, drag is proportional to the square of the wind speed. It is given by $F_D = \frac{1}{2}\rho v^2 A C_D$ where ρ is the air density, A is the effective area of the object (which increases with the angle of attack, α) and C_D is a constant called the drag coefficient, which is roughly unity (data sheets give a value of 1.2–1.4). During the first half of the flight of the javelin, both the air resistance and the lift are very small (see Knowledge check C16) but, because of the orientation of the javelin, these become increasingly important towards the end of the flight.

Notice that the javelin changes its orientation through the flight.[1] In the absence of lift, this would not happen. Consider the javelin at the top of its flight. Because the flight has dipped down, the angle of attack is typically 30–40° at this point. Looking at Fig. C.25 again, the drag is through the centre of mass, as is the weight (by definition) so these two forces do not cause a rotation. The lift, on the other hand, does: because the centre of lift is behind the centre of mass so we can consider it to be a force through the CoM + a clockwise couple (Fig. C.27), which gives the javelin a clockwise angular acceleration, which increases as the angle of attack increases on the downward leg.

Example

A javelin is launched at 25 m s^{-1} at an angle of 45°, with an initial angle of attack of 0. The throw distance is 60 m and it makes an angle of 45° on impact. Estimate (a) the flight time and (b) the mean torque applied by the lift.

Javelin moment of inertia = 0.42 kg m^2.

Answer

(a) Initial horizontal component of velocity = 25 cos 45° = 17.7 m s^{-1}

Assuming constant velocity, time of flight = $\frac{60\text{ m}}{17.7\text{ m s}^{-1}} = 3.4\text{ s}$

(b) $\Delta\theta = 90° = 1.57$ rad. Using $\Delta\theta = \omega_1 t + \frac{1}{2}\alpha t^2$, with $\omega_1 = 0$

$\alpha = \frac{2\Delta\theta}{t^2} = \frac{2 \times 1.57\text{ rad}}{(3.4\text{ s})^2} = 0.27\text{ rad s}^{-2}$.

Then Torque, $\tau = I\alpha = 0.42\text{ kg m}^2 \times 0.27\text{ rad s}^{-2} = 0.11\text{ N m}$.

1 As a military weapon (in ancient Greece), it would have been pretty useless if the point didn't dip down towards the end of the flight.

C.7 Drag and lift in sports

These forces on objects which are moving through the air are important in ball games, athletics field sports, aviation and motor sports. The physics of this topic is far from easy and there are several misconceptions: this unit provides only a basic introduction.

Fig. C.28 Aerofoil showing turbulent flow

(a) Aerodynamic drag (air resistance)

The air flow around the aerofoil in Fig. C.28 is *turbulent*. The diagram below shows the same aerofoil with *laminar flow* (smooth flow) and introduces a few terms.

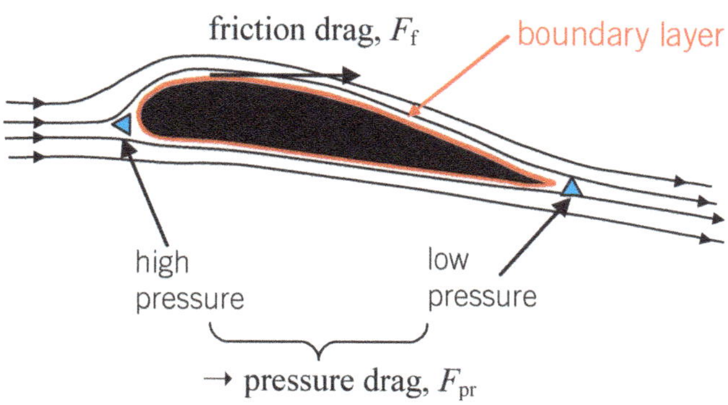

Fig. C.29 Aerofoil with laminar flow

>>> **Study point**

The aerofoil in Fig. C.28 has a much greater attack angle than that in Fig. C.29. Hence the air which separates at the front of the wing doesn't recombine smoothly and is turbulent. This effect happens at lower speeds with blunt ('non-aerodynamic') shapes and gives rise to a much greater pressure difference.

There are two main drag components:

- The *friction drag* is due to the interaction of the moving air with the *boundary layer* which is a thin layer of stationary air in contact with the object.
- The *pressure drag* is due to the difference in pressure between the front of the object and the back.

At the speeds usually encountered in sports, both these two components vary as v^2, the head-on area, A, of the object and the density, ρ, of the air, so the total drag force, F_D, can be expressed as $F_D = \frac{1}{2}\rho v^2 A C_D$. The constant C_D is called the drag coefficient, but unfortunately it isn't really constant. The variation of F_D with v for a football in Fig. C.30 shows that the air resistance is roughly proportional to v^2 up to $20\ m\ s^{-1}$, where the air flow is laminar. Then the flow becomes turbulent and the air resistance plummets with speed before starting to take off again. The lower speed region is dominated by the friction drag – at higher speeds the pressure drag dominates. This distinction controls the way that footballers curve the ball into the net (see later).

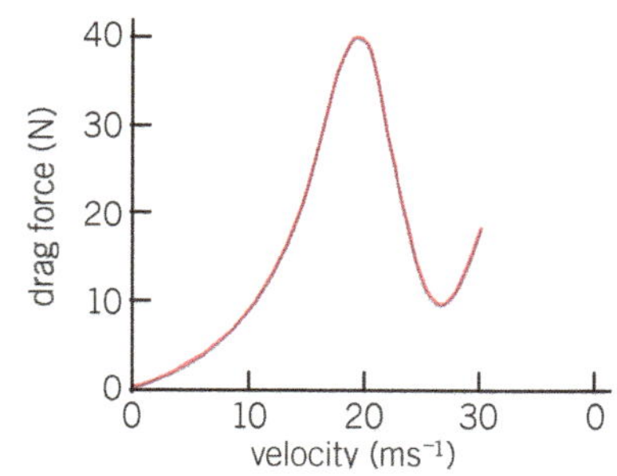

Fig. C.30 Variation of drag force with velocity for a football

Knowledge check C18

Does the low velocity portion of the drag graph in Fig. C.30 really show proportionality to v^2?

> **Example**
>
> Assuming that C_D in the drag equation, $F_D = \frac{1}{2}\rho v^2 A C_D$, is constant, how would you expect the fuel consumption of an F1 car to vary with v?
>
> **Answer**
>
> Assuming that the variation of fuel consumption is dominated by the air resistance, the power is proportional to v^3, because $P = Fv$. This suggests that the fuel consumed per unit time is proportional to v^3. But the distance travelled is proportional to v, so the fuel consumed per unit distance should be proportional to v^2 (see Study point).

>>> **Study point**

In practice, predicting fuel consumption is a lot more complicated: rolling resistance is roughly constant; the engine's efficiency varies with speed...... (so don't believe it!).

Question: How do we reduce the air resistance?

Answer: By making the object smoother.

Wrong! In the early twentieth century golfers found they could hit their old scuffed and cracked golf balls further than their nice new smooth ones. So what's going on?

Fig. C.31 Golf ball with dimples

It's all to do with the boundary layer. The dimples trap a turbulent layer of air next to the golf ball, which the air slides past smoothly and round the ball. This lamina flow separates much later from the ball leaving a smaller turbulent wake than the smooth golf ball. It is the pressure difference between the front and back which produces most of the drag. Because the low wake pressure acts over a smaller area in the dimpled ball, the drag is lower. This is summarised in Fig. C.32.

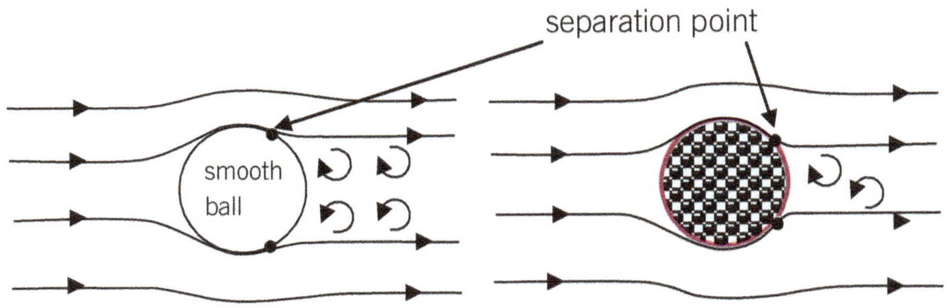

Fig. C.32 Dimpled golf balls have a narrow wake, so a lower drag

Study point

Note that all the diagrams of objects moving through the air are drawn as if the object is stationary and the air is flowing past (e.g. Fig. C.32). This is standard.

(b) Aerodynamic lift – explanation 1

The aerofoil in Fig. C.29 experiences an upwards force that we call lift. The spoilers on the back of sports cars (Fig. C.33) are there to create a downwards force to hold the back of the car on the road – this is also confusingly called lift. So how does it arise? Until recently the accepted explanation was the *Bernoulli effect* (pronounced burn-oo-ee). If a stationary gas at a pressure p_0 accelerates to a velocity v, the pressure falls to p given by:

$$p = p_0 - \tfrac{1}{2}\rho v^2.$$

An alternative way of writing the relationship between pressure and velocity is

$$p + \tfrac{1}{2}\rho v^2 = \text{constant}.$$

Bernouilli's equation tells us that, if the velocity of a flowing gas increases, its pressure decreases.

Fig. C.33 Aerodynamic spoiler

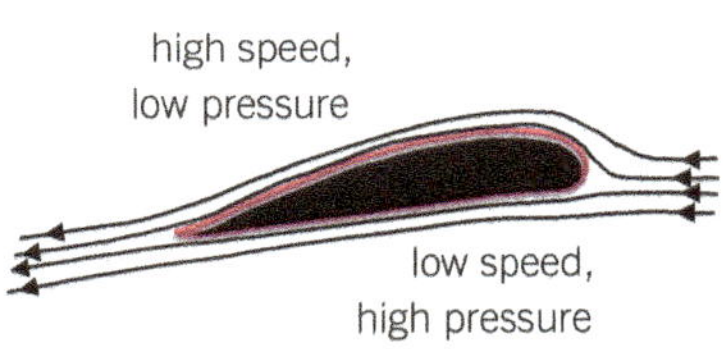

Fig. C.34 Bernoulli explanation for lift

Fig. C.35 Plane flying upside down

Applying this to the aerofoil, the argument runs: the air flowing over the upper curved surface of the wing has a longer path than that under the wing, so its speed is greater and its pressure lower. This difference in pressure produces the lift.

So how does the plane fly upside down?

C19 Knowledge check

The air flows at 160 m s^{-1} over the top of a wing of area 10 m^2. Underneath the speed is 140 m s^{-1}. Use Bernoulli's equation to calculate the lift.

ρ_{air} = 1.3 kg m^{-3}.

(c) Aerodynamic lift – explanation 2

The second explanation just uses Newton's 2nd and 3rd laws (N2 and N3). Look again at Fig. C.34. The initial velocity of the air is horizontal to the left; it leaves the wing downwards to the left. The wing pushes the air downwards, i.e. it has exerted a downwards force on the air (N2); the air exerts an equal and opposite force on the wing (N3), i.e. an upwards force.

In order to fly upside down, the pilot in Fig. C.35 has to angle the wings upwards against the design of the wing and plane, as shown in Fig. C.36. This takes more energy than flying with the curved side of the wing on top.

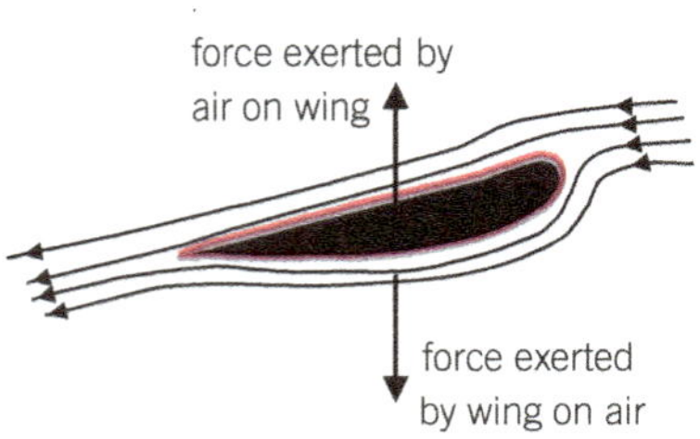

Fig. C.36 Flying upside down

(d) Bend it like Beckham – the Magnus effect[2]

Whether in tennis, football (e.g. Robert Carlos's legendary 2009 free kick – see Test yourself Q8) or golf, it is well known that a spinning ball will swerve away from the forward spinning side. Why is this? Fig. C.37 illustrates the situation for a spinning ball moving upwards on the page. As usual, the diagram is drawn reversed: the ball is stationary (but still spinning) and the air is moving downwards on the page. The ball is seen to veer to the left – why does it do this?

The air which flows over the left-hand side of the ball separates later than the air which flows round the right. This has the effect of bending the air flow to the right.

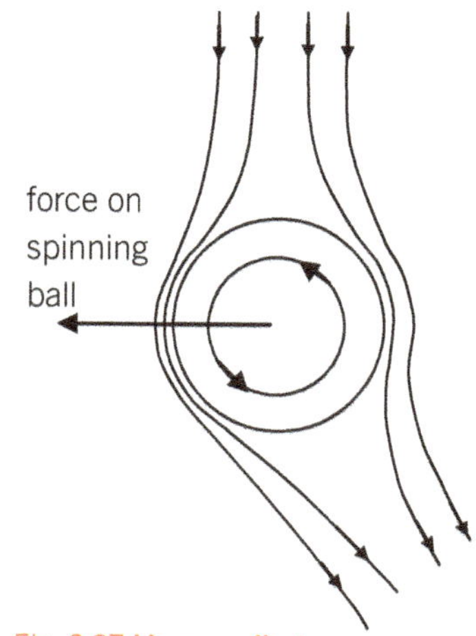

Fig. C.37 Magnus effect

Bernoulli explanation

Because the air is dragged round the left side of the ball (in Fig. C.37), it speeds ups and is moving faster than the air on the right. So, by Bernoulli, the pressure is lower on the left. The pressure difference makes the ball swerve to the left.

Newton's explanation

Newton is said to have explained the effect in this way when watching tennis in Trinity College nearly two centuries before Magnus described it. The air is given a rightward momentum as it is deflected so the ball acquires an equal and opposite momentum.

Study point

The explanation in terms of conservation of momentum is equivalent to applying N2 and N3: see AS textbook Section 1.3.4.

Which explanation is right?

The boring answer is 'both.' But it is generally accepted that the N2/3 (or conservation of momentum) effect is much greater than the Bernoulli effect (by an order of magnitude).

But how do you make the ball spin?

Fig. C.38 is an attempt to show the head of a tennis racket applying a slice to a tennis ball. The two forces arise because of the relative motion of the racket head and the ball. These forces will exist for a short time, Δt, and so each produces an impulse. What is their combined effect?

- The normal force passes through the centre of gravity and so just produces an impulse, $N\Delta t$, upwards and to the left and no rotation.
- As we saw in Section C5(a), we can think of the frictional force as being composed of a force, F, through the centre of mass and an anticlockwise couple, Fr, where r is the radius of the ball. So this produces an impulse $F\Delta t$ upwards and to the right and an anticlockwise angular impulse (see Study point) $Fr\Delta t$.

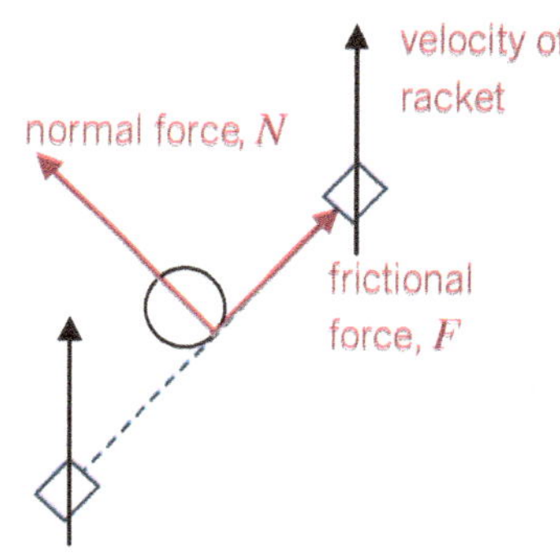

Fig. C.38 Slicing a tennis ball

Study point

Just as an **impulse** is equal to the change of momentum, $m(v - u)$ an *angular impulse* is equal to the change in angular momentum $I(\omega_2 - \omega_1)$.

The result of these is

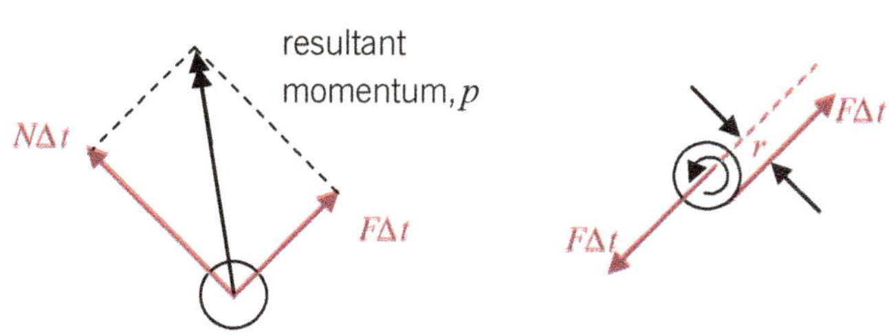

Fig. C.39 Linear and angular momentum produced by the slice

2 Named after the physicist Heinrich Gustav Magnus who described it in 1852.

Test yourself Option C

1. When sailing into the wind, the closer to upright a boat is, the faster it goes (for a particular angle to the wind). Explain this and also how it can be achieved by the sailor in Fig. C.4.

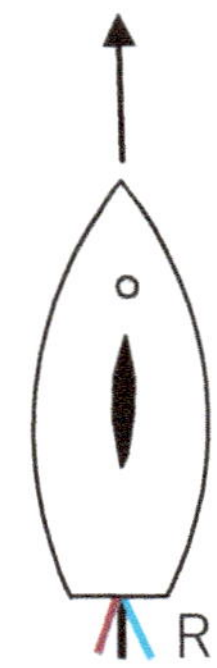

2. The diagram shows various directions that a rudder, **R**, can take in sailing. Copy Fig. C.9 and add the rudder in an appropriate direction to 'balance' the boat, i.e. stop it turning into the wind. Explain your answer and also explain why this will slow the boat.

3. Consider the dinghy in Fig. C.4. A less well-designed dinghy has a higher centre of gravity. It is (on the diagram) just where the mast disappears behind the boat stern. Explain briefly, using moments, why this boat is less stable.

4. It is difficult to get an accurate measure of bounce height, so a tennis club has decided to test its match balls by dropping from the regulation height and timing the interval between the first and second bounce, using a microphone and a data logger. [See the example in section C.2(a).]

   (a) Calculate the time intervals expected for the least and greatest legal bounce heights.
   (b) Suggest how accurately the timings need to be to detect a ball which bounces 1 cm lower or 1 cm higher than the allowed range.

5. The rules for basketball bounciness involve measuring to the top of the ball rather than the bottom. When dropped from a height of 6 feet (i.e. 72 inches) it should bounce to between 49 and 54 inches. Given that the circumference of a basketball is 30 inches, calculate the allowed range for the coefficient of restitution.

6. A sports aeroplane has two wings each of length 4.5 m. When flying horizontally they deflect the air within 1 m of the wing surface through an angle of 20° downwards. Calculate the lift achieved by the wings at an air speed of 60 m s^{-1}. Density of air = 1.3 kg m^{-3}.

7. A shot putter releases the shot from a height of 2.05 m, at a speed of 13.5 m s^{-1} and an angle of 35° to the horizontal.

   (a) Calculate the flight time of the shot and hence the distance achieved.
   (b) Investigate whether the athlete should consider changing the angle of release to 40° given that she can only release the shot at 13.0 m s^{-1} at this angle.

8. (a) Use Fig. C.32 and the answer to Knowledge check C17 to sketch a graph of the variation of the drag coefficient for a football with velocity.

   (b) Considering the Magnus force on a spinning object, one theory suggests than the angle of deflection of the air is linked positively to the drag coefficient – the greater the drag, the more spinning ball deflects the air.

   How does this theory account for the trajectory of Roberto Carlos's free kick for Brazil against France in 1997?

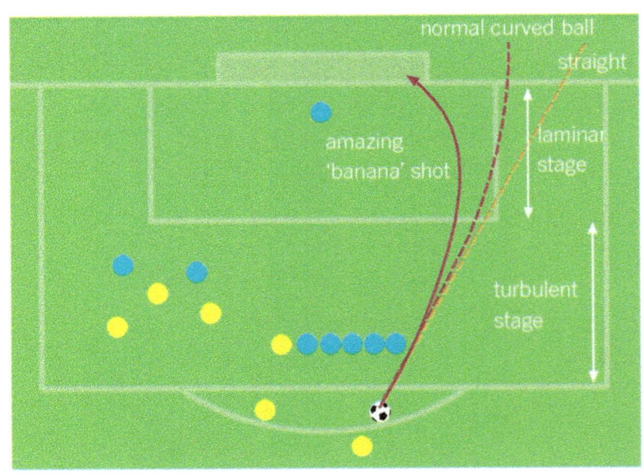

Option D: Energy and the environment

Unlike some of the other options, Option D contains little additional physics. The concepts which you have already covered, such as the radiation laws (Wien's, Stefan-Boltzmann, inverse square) are applied in the context of our need for energy and sustainability. Exceptions are Archimedes' principle and thermal conduction equations. Our starting point is the temperature of the Earth.

concentrations of greenhouse gases from 0 to 2005

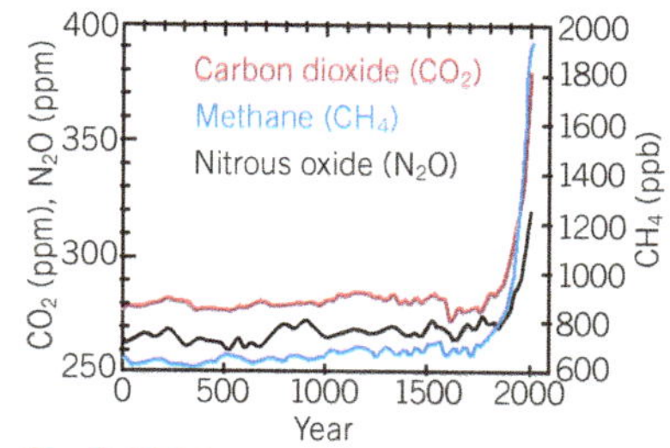

Fig. D.1 Rising greenhouse gas concentrations

D.1 The Earth's rising temperature

The vast majority of atmospheric scientists are convinced by the evidence that anthropogenic global warming (i.e. warming caused by humans) is a fact. The published data include 'hockey stick' graphs of the rising greenhouse gas concentrations and the variations of the Earth's temperature (Figs D.1 and D.2).

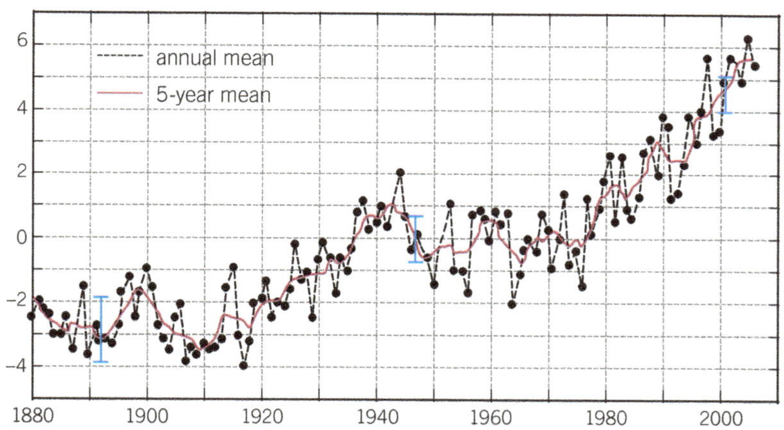

Fig. D.2 Global mean surface temperature anomaly (°C) 1880–2005

Before we examine the effect of greenhouse gases, we'll have a look at the radiative exchange between the Earth and space.

Study point

Note the different time scales on the two graphs – comparison graphs over the last two millennia are available. The word *anomaly* means the difference between the actual temperature and a baseline value. 1960 is often used, as in Fig. D.2.

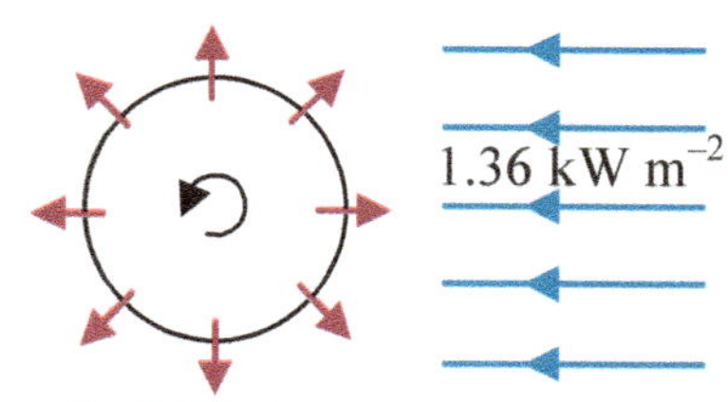

Fig. D.3 Radiation balance

(a) The Earth's temperature with no atmosphere

We'll start by looking at how the temperature of the Earth is established in a stable period. We know the atmosphere helps keep us warm. What would the Earth's mean temperature be without an atmosphere? As Fig. D.3 shows, there would be a balance between the power input from the Sun and the power emitted by the Earth. We receive 1.36 kW m^{-2} of radiation (mainly in the near UV, visible and near IR). This is the value of the **solar constant**, G.

∴ The total power received is given by, $P = \pi R^2 G$

Because the Earth is rotating, it will radiate from all around. If the Earth has a mean temperature, T, it radiates a power given by the Stefan-Boltzmann law

$$P = 4\pi R^2 \sigma T^4$$

Key terms

The **solar constant**, G, is the radiative power from the Sun per unit area incident on a surface at right angles to the solar radiation. Its value at the radius of the Earth is 1.36 kW m^{-2}.

Study point

The Earth rotates so the temperature of the day and night hemispheres will be similar.

D1 Knowledge check

The following data are mean values for Mars:

Surface temperature: 218 K

Orbital distance: 1.524 AU

[NB. inverse square law]

Albedo: 0.15 (meaning that Mars reflects 15% of the incoming solar radiation)

Explain whether these data are consistent with the model in Section D.1(a).

D2 Knowledge check

The Earth reflects 30% of the incoming solar radiation. Calculate its temperature using the model in Section D.1(a)

To calculate a theoretical value for the temperature of the Earth, we'll start off by making two basic assumptions:

- The power received and the power radiation are equal
- The Earth behaves as a black body.

With these assumptions: $4\pi R^2 \sigma T^4 = \pi R^2 G$

Simplifying and inserting known values gives:

$$T^4 = \frac{1360 \text{ W m}^{-2}}{4 \times 5.67 \times 10^{-8} \text{ W m}^{-2} \text{K}^{-4}}$$

which leads to a value of T of 278 K, i.e. 5 °C. Now try Knowledge check D1, which repeats this calculation for Mars.

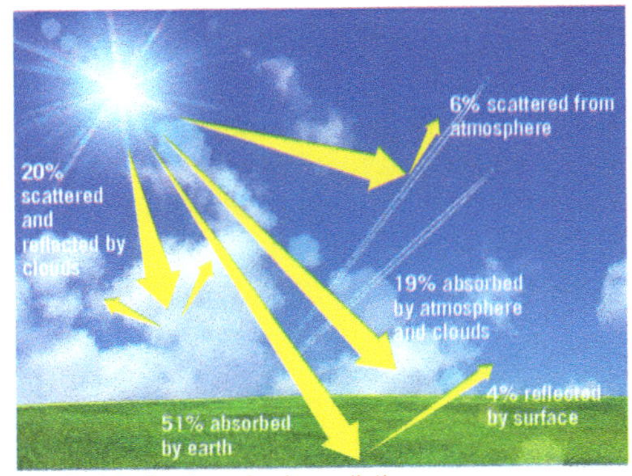

Fig. D.4 Reflection of solar radiation

The value of 5 °C doesn't look very far out from the observed mean temperature of the Earth of 15 °C. This sounds good for our theory but we have missed out an important fact:

The Earth reflects about 30% of the incident radiation back out into space.

This reflection occurs from the ground (especially the ice and snow near the poles), from the air and from the tops of clouds. See Fig. D.4 for a breakdown of the numbers. Why is this a problem for our theory? Because, if the Earth (and atmosphere) only absorb 70% of the incident solar radiation, they radiate less, which means their temperature is even lower. Our energy balance equation becomes:

$$4\pi R^2 \sigma T^4 = 0.7 \times \pi R^2 G$$

Now do Knowledge check D2 and show that the predicted mean temperature of the Earth comes out at a distinctly chilly −18 °C, a much worse answer and certainly way off the mark!

Note. Kirchhoff's law of radiation says that if the Earth only absorbs 70% of the incident radiation, the emissivity of the Earth should also be 70% of the black body value. However, the 30% albedo is only the figure for the visible and near IR and UV parts of the e-m spectrum, which are the main components of the incoming solar radiation. In the thermal infra-red the Earth behaves more like a perfect black body. This is similar to a white-painted domestic radiator – it behaves like a black body in the thermal infra-red.

(b) The effect of the atmosphere

It appears that Mars (Knowledge check D1) has a temperature that is in agreement with the calculation above. Why the difference? The biggest factor is the density of the Martian atmosphere at the surface, 0.020 kg m^{-3}, which is less than one sixtieth that of the Earth. We need to look at some of the properties of the Earth's atmosphere.

Composition (dry atmosphere): nitrogen (78.08%), oxygen (20.95%), argon (0.93%), carbon dioxide (0.04%) + trace gases, e.g. methane. Air also contains a variable amount of water vapour – typically 1% at sea level.

Transparency: Fig. D.5 has a wealth of information. Let's examine it in detail:

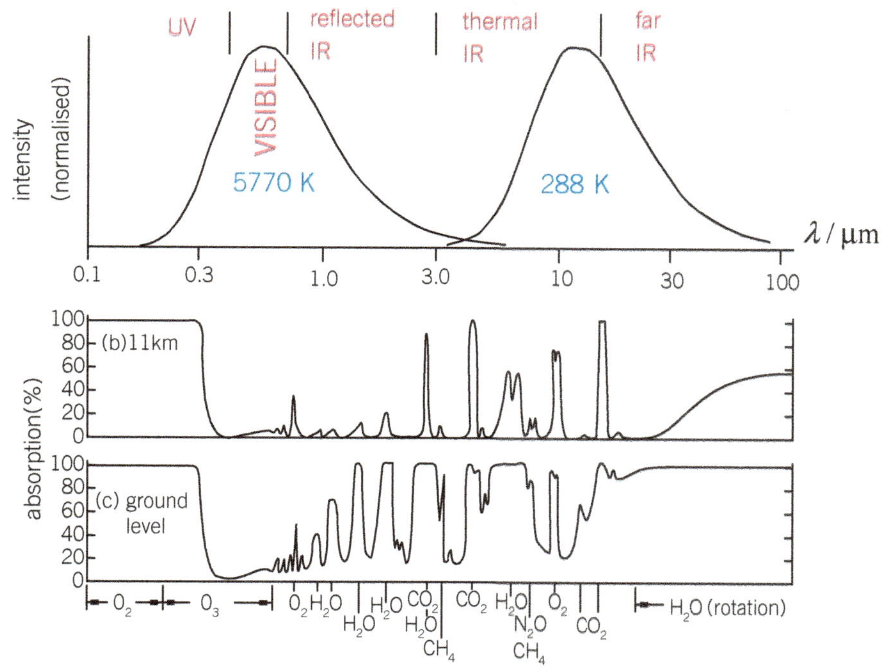

Fig. D.5 Absorption of radiation by atmospheric molecules

- The air at ground level is opaque to radiation of wavelength less than $0.3\ \mu\text{m}$ (in the UV) and longer than $20\ \mu\text{m}$ (in the far IR).
- Radiation from the Sun, which is mainly $0.3–2\ \mu\text{m}$, penetrates the atmosphere (with absorption in bands between 1 and $2\ \mu\text{m}$).
- Much of the radiation from the ground, mainly $5–50\ \mu\text{m}$, is absorbed by the atmosphere.
- Solar radiation which is reflected from the ground (near UV, visible and near IR) escapes largely into space.
- Various bands of wavelengths are wholly or partially absorbed by H_2O, CO_2, CH_4, O_3 and N_2O (see Study point).

A $6.5\ \mu\text{m}$ photon, for example, emitted from the ground is quite likely to be absorbed by a water molecule, promoting it to a higher vibrational energy level. The molecule can de-excite in two ways:

1 Collision with another molecule – increasing the translational and/or rotational kinetic energy of the molecules.

2 Re-radiation. This can occur in any direction, so roughly 50% of these photons will be reabsorbed by the ground.

The effect of these two processes is to raise the temperature of atmosphere and ground – the **greenhouse effect**. The equilibrium temperature is higher than would be the case without an atmosphere – but the rate of energy loss to space (emitted + reflected) is still equal to the net incoming power from the Sun.

It should be noted that the atmosphere also gets directly heated by the Sun: by absorption of solar radiation in one of the molecular bands between 1 and $3\ \mu\text{m}$. The heated atmosphere will lose this energy as above – some of it resulting in raised ground temperature and some escaping to space, reducing the magnitude of the greenhouse effect.

Knowledge check D3

Use Wien's law to confirm the position of the peak wavelengths of the solar and ground radiation in Fig. D.5.

Study point

The *partial* absorption by the polyatomic molecules is significant:

If the concentration of these gases increases so will the fraction of radiation absorbed.

Study point

Diatomic molecules don't absorb photons in the near UV, visible and IR regions of the e-m spectrum because the jumps in the rotational, vibrational or electron energy levels don't match the photon energies. Molecules with more than two atoms have additional vibrational energy modes (Fig. D.6) which absorb photons in the near and thermal infrared.

Fig. D.6 Vibrating CO_2 molecules

Knowledge check D4

N_2O absorbs radiation with a wavelength of $7.5\ \mu\text{m}$. Explain this in terms of the energy levels of an N_2O molecule.

Key terms

The **greenhouse effect** is the name given to the process of the warming of a planet by radiation from its atmosphere to a temperature above what it would be without an atmosphere.

D5 Knowledge check

The concentration (vertical) scale in Fig. D.7 is in ppm (parts per million). What would the labels be in %?

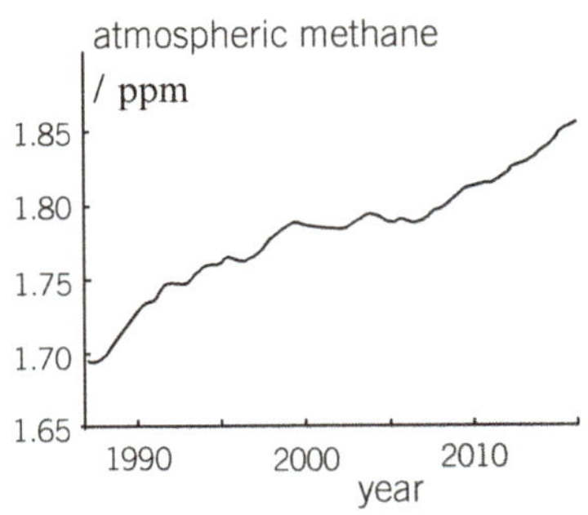

Fig. D.8 Rising methane levels

Study point

The 400 ppm level of CO_2 was recorded for the first time in Mauna Loa in May 2013.

Key terms

Archimedes' principle: A body wholly or partially immersed in a fluid, experiences an upthrust, U, equal to the weight of fluid displaced by the body.

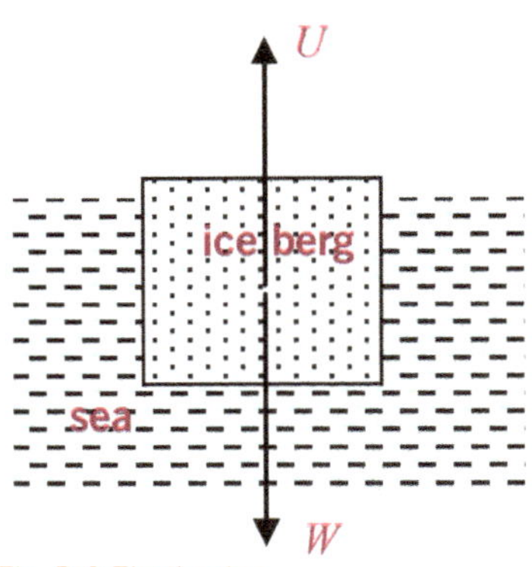

Fig. D.9 Floating ice

D6 Knowledge check

Calculate the volume of:

(a) an iceberg of mass 10^6 tonne

(b) the volume of sea water of mass 10^6 tonne.

(c) Anthropogenic global warming – with positive feedback

In spite of its low atmospheric concentration, carbon dioxide makes a major contribution to the greenhouse effect. Its natural concentration has varied widely over geological time, as shown in Fig. D.7. [Unit of concentration: ppm – parts per million]

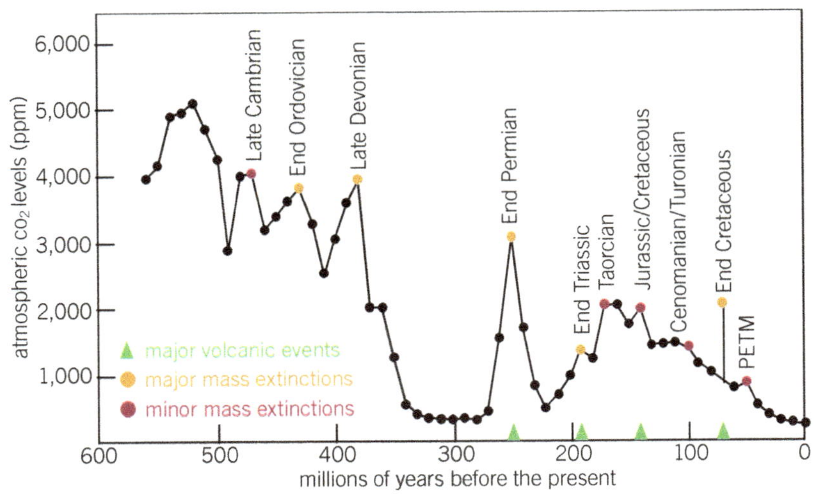

Fig. D.7 Variation of CO_2 concentration in the phanerozoic aeon

The pre-industrial CO_2 concentration was 280 ppm. Burning of fossil fuels has increased this to 400 ppm, which has increased the absorption of thermal infra-red radiation. A more potent greenhouse gas is methane. Increased intensive agriculture, especially animal rearing, has increased methane levels (Fig. D.8). The increased heating of the eath and atmosphere produced by human activity is referred to as **anthropogenic global warming**.

There are feedback mechanisms which amplify the warming effect:

- Melting ice cover reduces the fraction of radiation reflected back into space.
- Melting permafrost in Canada and Siberia releases methane.

Such feedback is called *positive feedback* because its effect is in the same direction as its cause.

(d) Melting ice and rising sea levels

The current rate of sea level rise is estimated as 2.9 ± 0.4 mm year^{-1} and appears to be accelerating at 0.013 mm year^{-2} from the 20th-century average of 1.7 mm year^{-1}. This appears to be due mainly to melting ice. However, the melting of floating ice (mainly in the Arctic ocean) does not contribute directly to this. Why not?

Archimedes' principle states, 'When a body is wholly or partially immersed in a fluid, it experiences an upthrust, U, which is equal to the weight of fluid displaced by the body.' Applying this to a floating object, e.g. the iceberg in Fig. D.9, this means that the weight, W, of the berg is equal to the weight of the sea water which would occupy the volume of the iceberg below the water line. Because ice is less dense than North Atlantic sea water (930 kg m^{-3} v 1030 kg m^{-3}) it floats with about 10% above the water line. Its volume when melted is only slightly more than that of the sea water, so it hardly causes the sea level to rise at all.

Example

An iceberg has a surface area of $1.0\ \text{km}^2$ and a height of $100\ \text{m}$. Calculate, (a) the mass of the iceberg and (b) the height of the iceberg above the water.

Answer

(a) Using the above values for density:

$$M_{\text{ice}} = \rho V = 930\ \text{kg m}^{-3} \times 1.0 \times 10^6\ \text{m}^2 \times 100\ \text{m}$$
$$= 9.30 \times 10^{10}\ \text{kg}$$

(b) By Archimedes' principle the iceberg displaces its own mass of sea water, which has a volume given by:

$$V = \frac{M_{\text{ice}}}{\rho_{\text{sea water}}} = \frac{9.30 \times 10^{10}\ \text{m}^3}{1030\ \text{kg m}^{-3}} = 9.03 \times 10^7\ \text{m}^3$$

∴ Fraction under water $= \dfrac{9.03 \times 10^7\ \text{m}^3}{(1.0 \times 10^3\ \text{m})^2 \times 100\ \text{m}} = 0.903.$

∴ Fraction above water $= 0.097$.

∴ Height above water $= 0.097 \times 100\ \text{m} = 9.7\ \text{m}$

Study point

When the iceberg in the example melts it will occupy a volume of $9.03 \times 10^7\ \text{m}^3$ once it has mixed with the seawater and attained the same density. There is a possible second order effect due to a drop in the salinity of the seawater.

D.2 Energy sources

This is a short review of **renewable** and **non-renewable** energy sources, which can be used for transport, electricity generation, heating and cooling. Most renewable energy sources are, directly or indirectly, solar energy. Apart from thermal and photovoltaic solar power, this includes wind, wave, hydroelectric and biomass. Geothermal energy relies on heat from within the Earth, the origin of which is the decay of radioisotopes within the core (supplemented by the release of latent heat from the growth of the inner core). Sections (a) – (c) are concerned directly with solar power.

Key terms

An energy source is regarded as **renewable** if it is replenished naturally in a human lifetime (some organisations use 50 years as the time scale). **Non-renewable** sources are regarded as effectively not capable of replenishment.

A **main sequence** star is one which generates most of its energy from the fusion of hydrogen to helium in its core.

(a) The origin of solar power

Without the input of radiation from the Sun, the Earth would be lifeless rock, with a surface temperature of a few tens of kelvin. The origin of this radiation is the energy released from the nuclear fusion of ^1H nuclei (protons) to ^4He nuclei:

- The *proton-proton chain* produces 98.3% of the ^4He nuclei. This is the dominant fusion route for **main sequence** stars of mass less than 1.3 times that of the Sun.
- More massive main sequence stars produce most of their energy by the CNO (carbon-nitrogen-oxygen) cycle. This accounts for only 1.7% of the energy output of the Sun.

The initial step of the proton-proton chain is the fusion of two protons to produce deuterium:

$$^1_1\text{H} + {}^1_1\text{H} \rightarrow {}^2_1\text{H} + {}^0_1\text{e}^+ + \nu_e$$

with the positron subsequently annihilating with an electron:

$$\text{e}^+ + \text{e}^- \rightarrow 2\gamma$$

We can tell from the change in quark flavour ($\text{u} \rightarrow \text{d}$) and the involvement of a neutrino that this is controlled by the weak interaction. As a result of this, even under the conditions of temperature and pressure, the lifetime of a proton at the centre of the Sun is of the order of 10^9 years. The next step is more rapid – the mean lifetime of the ^2H has been estimated to be 4 seconds!

Solar trivia

The temperature, pressure and density of the centre of the core are $15\ \text{MK}$, $26.5\ \text{PPa}$ and $150\ \text{g cm}^{-3}$ respectively.

99% of the power generation comes from the central $0.24R$ (1.4% of the volume) which also contains 40% of the mass.

The power production density is only $280\ \text{W m}^{-3}$, which is less than the metabolic production of a human being – but the Sun is rather larger.

Knowledge check D7

A website says that the second stage of the proton-proton chain is controlled by the strong interaction. Comment on this.

$$^2_1H + ^1_1H \rightarrow ^3_2H + ^0_1e^+ + \gamma$$

Most (83%) of the ^3He is converted to ^4He by the following reaction:

$$^3_2H + ^3_2He \rightarrow ^4_2He + 2^1_1H$$

We can summarise the proton-proton chain as:

$$4^1_1H + 2^{\ 0}_{-1}e^- \rightarrow ^4_2He.$$

D8 Knowledge check

Test the value of 26.73 MeV for the proton-proton chain using the following data:

$m_p = 1.007\ 276\ 47$ u

$m_{He4} = 4.001\ 506\ 47$ u

$m_e = 0.000\ 548\ 60$ u

$1\ u \equiv 931$ MeV

(b) The solar constant

The solar power received per unit area on a surface at right angles to its direction of travel (the **solar constant**) is $1360\ W\ m^{-2}$ at a distance of 1 AU from the Sun. The energy released inside the Sun in the production of a ^4He nucleus from four protons is 26.73 MeV (Knowledge check D8) of which 2.3% is carried away by neutrinos. The rest is energy of e-m radiation, initially γ in the core but emitted mainly as near UV, visible and IR radiation from the photosphere. The total power emitted as e-m radiation by the Sun is 3.83×10^{26} W, which we can relate to the solar constant using the inverse square law.

D9 Knowledge check

Jupiter's orbital distance is approximately 5 AU. Estimate the value of the solar constant at Jupiter.

Example

Use solar luminosity to calculate the solar constant, G. [1 AU = 1.50×10^{11} m]

Answer

Solar luminosity, $L_e = 3.83 \times 10^{26}$ W.

At the radius of the Earth's orbit, this radiation is spread over a sphere of surface area A, given by $4\pi r^2$.

$$\therefore G = \frac{L_e}{4\pi r^2} = \frac{3.83 \times 10^{26}\ W}{4\pi \times (1.50 \times 10^{11}\ m)^2} = 1350\ W\ m^{-2}$$

Note: This shows that the given data are reasonably consistent!

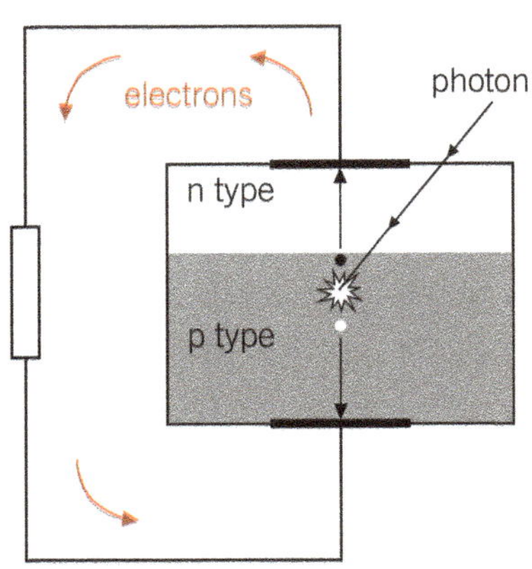

Fig. D.10 PV cell schematic

(c) Photovoltaic (PV) cells

A PV cell is a semiconductor (usually silicon) with n-type doping at one end and p-type doping at the other [see Section 4.4.4(b) on the Hall effect]. An incident photon with energy more than about 0.7 eV, liberates electrons and holes within the cell, which produce an electric current in the connected circuit. The details of this process are beyond this course.

The graph in Fig. D.11 is a typical current-voltage characteristic for a PV cell. It shows how the output current and voltage are related with constant lighting conditions. It is similar to the characteristic of batteries that we met in Unit 2 of the AS course – but notice two differences:

- The axes are the other way round – this is conventional for PV cells.
- The graph is not straight, showing that we cannot talk about PV cell as having an even approximately constant internal resistance.

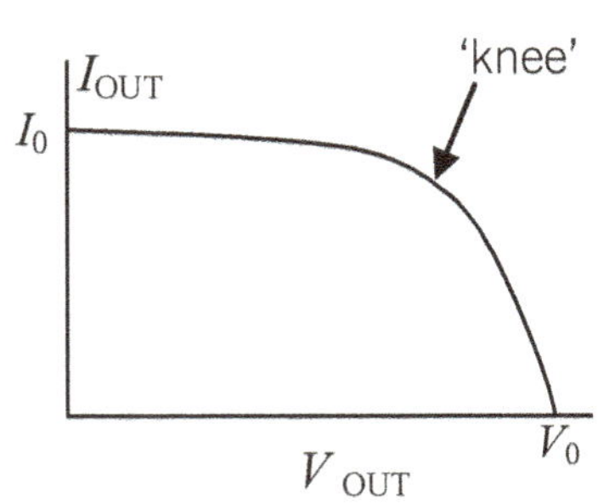

Fig. D.11 PV cell characteristic

Let's see what the characteristic is telling us. We can pick out the following points:

1. If we short out the terminals with a piece of (zero resistance) wire, the pd is zero and the current is a maximum, labelled I_0, which typically is a few mA. The power output, P_{OUT}, under these conditions is zero, because $P = IV$.

2. If we don't draw any current but just put a voltmeter across the terminals, we get the maximum pd, V_0, which is about 0.6 V (for a silicon PV cell). Here, too, $P_{OUT} = 0$.
3. If we put a load across the output, e.g. a variable resistor, and gradually increase its resistance from zero, the V_{OUT} initially increases significantly but the current, I_{OUT} decreases only slightly to start with, so P_{OUT} increases. Above a certain voltage (around the 'knee' on the characteristic) the situation is reversed with small changes in V_{OUT} and much larger changes in current.

To make a practical solar panel, the PV cells are typically arranged in a series / parallel combination. Fig. D.12 gives the principle of this. In this case the maximum pd available would be about $3V_0$ and the maximum current $2I_0$ (from the usual rules for combination of pd and current in series and parallel).

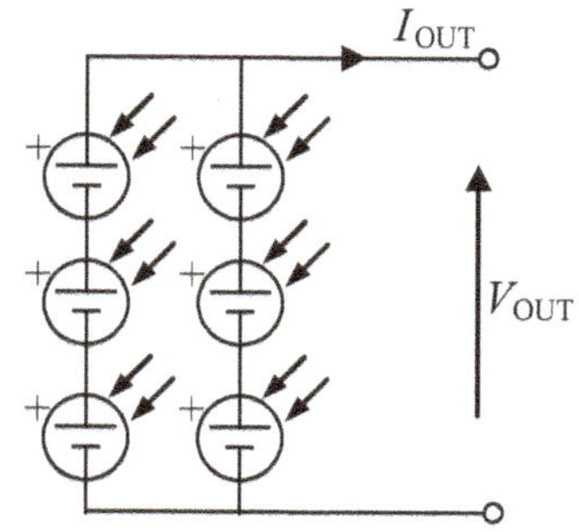

Fig. D.12 PV cells

The red graph in Fig. D.13 is the output characteristic of a PV module comprising a large number of PV cells. How is it used to maximise the output power? One thing to realise is that the output is not taken to a resistive load. Most PV installations feed power in to the National Grid, which is similar to charging up a rechargeable battery – the output voltage is fixed (and then turned into AC by a clever electronics device called an *inverter* that we don't need to worry about). Where on the graph should the control box select to give the maximum P_{OUT}? The dashed lines on the grid show values of I and V that give particular powers (20 W, 30 W ....). A V_{OUT} of about 7 V gives a P_{OUT} of 20 W; 11 V gives 30 W etc. You can confirm this using $P_{OUT} = I_{OUT}V_{OUT}$.

The point P_{max} on the characteristic is the author's estimate of the maximum power output, i.e. the maximum value of IV.

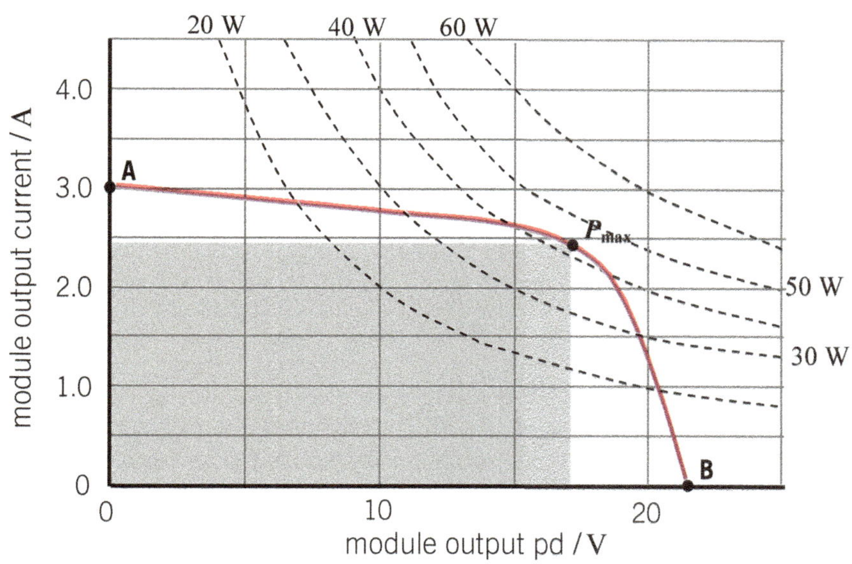

Fig. D.13 Output of a PV module

The power output is the product of pd and current: $P_{OUT} = V_{OUT} I_{OUT}$.

At **A** (short circuit), $V = 0$, so $P = 0$. Similarly $P = 0$ at **B**.

The grey rectangle is the author's estimate of the one under the graph with the maximum area, i.e. the maximum value of $V \times I$ (see Knowledge check D12).

Knowledge check D10

A 1 cm^2 silicon PV cell delivers 35 mA at 0.50 V when placed at right angles to a light beam of intensity 500 W m^{-2}. Calculate:

(a) the power output and
(b) the efficiency of the cell.

Study point

The dashed curves in Fig. D.13 are of $VI = 20$ W, 30 W .. 60 W. For example, the output is 20 W at (7.0 V, 2.86 A) and approximately (20 V, 1.0 A).

Knowledge check D11

Under the same lighting conditions as those for which the red graph in Fig. D.13 was obtained, each PV cell has I_0 and V_0 values of 30 mA and 0.60 V. Calculate the number of cells in the module and suggest how they are connected.

Knowledge check D12

Estimate P_{max} for the PV module in Fig. D.13.

Fig. D.14 Gwynt y Môr offshore windfarm

(d) Power from wind

The red graph in Fig. D.15 is a typical power curve for a wind turbine.

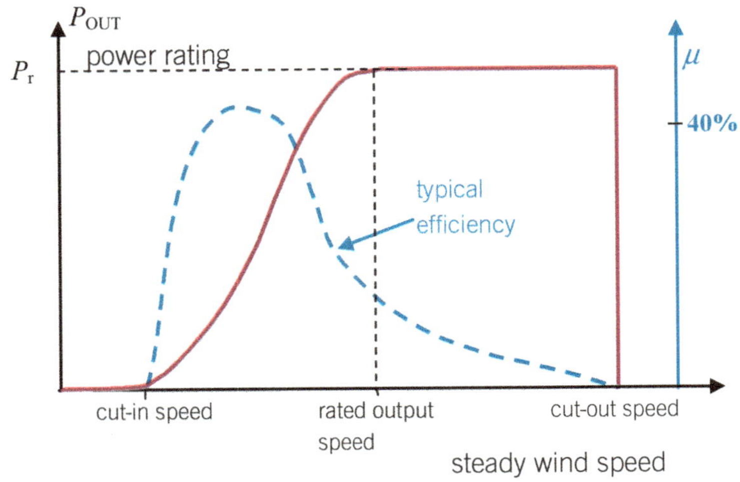

Fig. D.15 Typical characteristic of a wind turbine

Below the **cut-in speed** (typically 3–4 m s^{-1}) the turbine blades produce insufficient torque to rotate against friction. Above the **rated output speed** (typically about 12 m s^{-1}) the turbine is designed to extract a constant power from the wind, often by decreasing the angle that the blades make to the wind (see Option C.7). Turbines are designed to stop rotating and turn their blades edge on to the wind at the **cut-out speed** which is often ~25 m s^{-1}.

We can calculate the power available from a steady wind of speed v, to a turbine of area A, by identifying it with the kinetic energy of the incident air per unit time.

Volume of air incident per unit time = Av

∴ Mass of air incident per unit time = ρAv, where ρ = air density

∴ KE of air incident per unit time = $\frac{1}{2}mv^2 = \frac{1}{2}(\rho Av)v^2 = \frac{1}{2}\rho Av^3$

i.e. $P_{IN} = \frac{1}{2}\rho Av^3$

The efficiency, μ, is given by: $\mu = \frac{P_{OUT}}{P_{IN}} = \frac{P_{OUT}}{\frac{1}{2}\rho Av^3}$. It is usually expressed as a percentage.

The efficiency is also known as the *power coefficient*, c_P. The maximum value of efficiency is always less than 50% (see Fig. D.15). The main reason (apart from friction in the turbine) is that the kinetic energy of the air cannot be reduced to zero as it passes the blades. (If it did, it would block the following air!).

In order to estimate the practical power output of a wind turbine, several factors need to be taken into account in addition to its characteristic:

1 The effect of fluctuating wind speed. As shown in Fig. D.16, the general effect of fluctuations is to reduce the power output to below that of a steady wind. However, the output power for a range of mean speeds from below the cut-in speed upwards is actually greater than that of the equivalent steady speed; this is because of the disproportionate effects of the occasional high-speed gusts. For mean speeds slightly below the rated output speed, some of the gusts will take the speed above the rated output speed, causing safety measures to kick in and give an output below that of the steady speed value.

D13 Knowledge check

Assuming an efficiency of 42%, use the equation $P_{IN} = \frac{1}{2}\rho Av^3$ to estimate the power output of a turbine with 12 m rotor blades in a steady 10 m s^{-1} wind. [Take ρ = 1.2 kg m^{-3}]

Study point

The curve in Fig. D.16 is for a *steady* wind speed. The inertia of the turbine makes it slow to respond to rapidly fluctuating winds, so reducing the output.

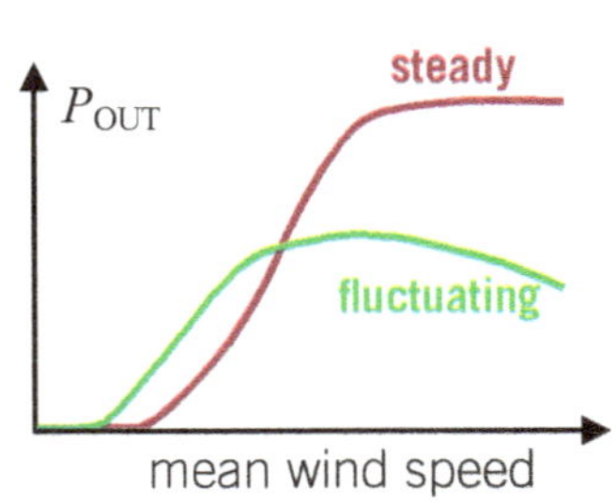

Fig. D.16 Effect of wind-speed variation

2 The daily variation of wind speed.

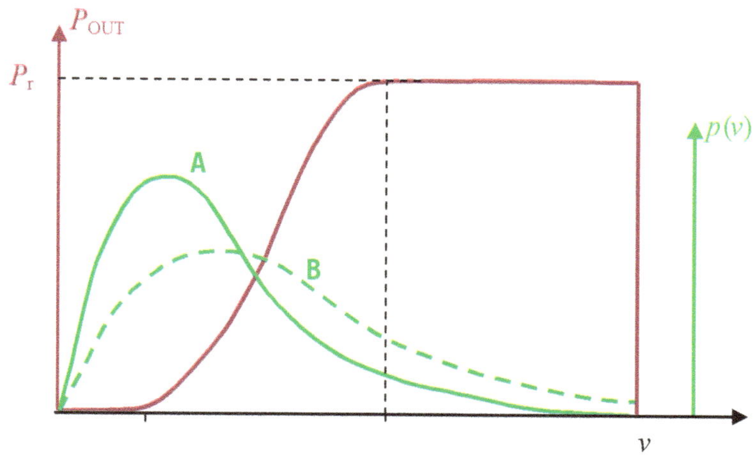

Fig. D.17 Wind-speed probability functions

Knowledge check D14

Explain briefly why the maximum value of the wind-speed probability function for location **B** in Fig. D.17 is lower than that for location **A**.

Windier places give a higher energy output. The green curves in Fig. D.17 are for two different locations: **B** is windier on the average than **A**. The green curves are the *wind-speed probability functions* for **A** and **B**. The area under both graphs is 1 (one), i.e. there is a 100% probability of the wind-speed being between 0 and ∞.

(e) Hydroelectric power

Tidal stream power stations operate on the same principle as wind turbines. Narrow tidal races, such as at St David's head in Pembrokeshire, provide ideal conditions. They have the advantages of tidal predictability and the much greater energy density of flowing water (because of its higher density).

Fig. D.18 Tidal stream generator in Strangford Lough in Northern Ireland

The release of the gravitational potential energy in water as it flows from an upper reservoir to a lower level is used to generate electricity by driving turbines. This principle is used in hydroelectric, pumped-storage and tidal-barrage power stations.

Consider the flow of water from the upper to the lower reservoir in Fig. D.19. Unless the pipe is very long compared to its diameter, we can ignore *pipe losses*, so the loss of potential energy per second is equal to the gain of kinetic energy. This means we can use the equation $\frac{1}{2}mv^2 = mgh$ to calculate the speed of flow, i.e. $v = \sqrt{2gh}$. The volume flow rate of the water is Av (where A is the cross-sectional area of the pipe) which allows us to calculate the rate of transfer of energy (see Knowledge check D15).

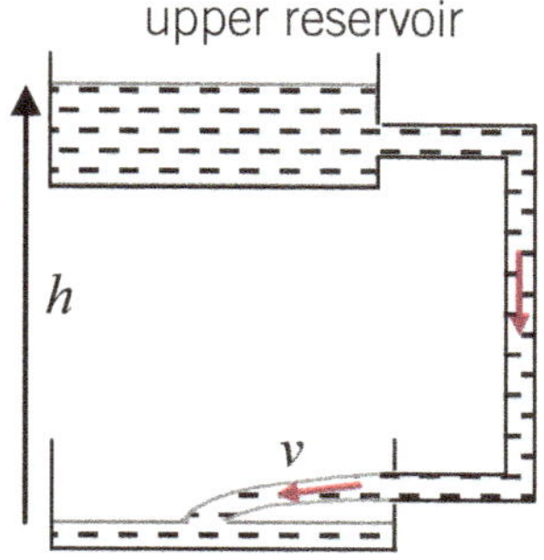

Fig. D.19 $E_p \rightarrow E_k$ in water

If we want to extract power, P_{OUT}, from the system, which is the point of a hydroelectric system, we need to build in a turbine, **T** (Fig. D.20). We can now model the energy transfer as:

GPE loss from upper reservoir	=	Work done on turbine	+	KE gained by water

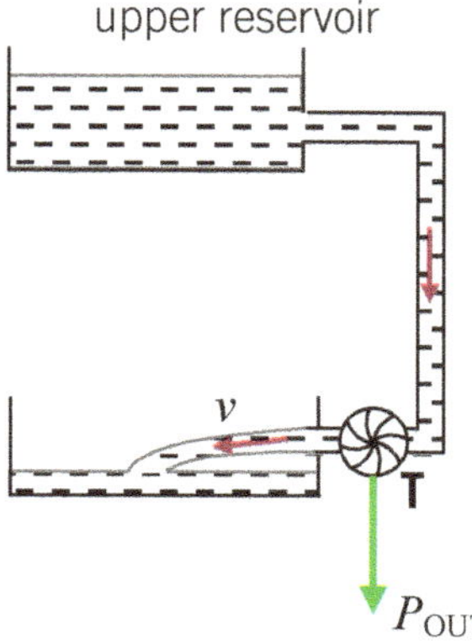

Fig. D.20 Power from a head of water

Now we see that the KE gained by the water must be less than it would have been without the turbine, i.e. we capture the GPE by allowing the water to flow more slowly and not gain as much KE. Using this model, then, if the volume flow rate is V then, in time Δt:

GPE loss = $\rho Vgh\Delta t$ and KE gain = $\frac{1}{2}(\rho V)\left(\frac{V}{A}\right)^2 \Delta t = \frac{1}{2}\frac{\rho V^3}{A^2}\Delta t$

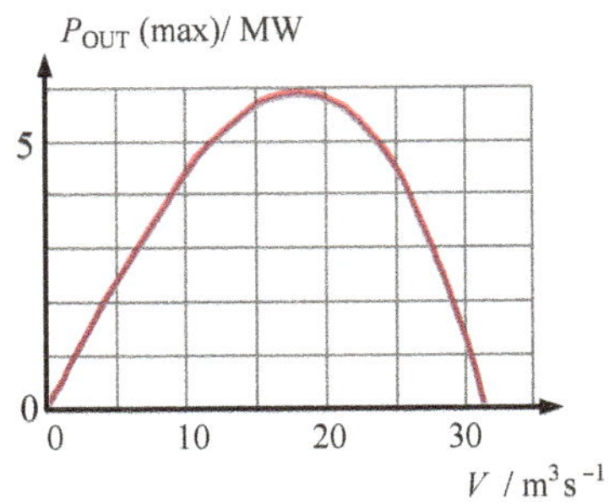

Fig. D.21 Model output power for a hydroelectric turbine

Note that, here the symbol A represents the cross-sectional area of the water tunnel at the turbine itself. So, the maximum output power possible (i.e. ignoring energy losses in the turbine, pipe losses, etc.) is given by:

$$P_{\text{OUT}} = \rho V g h - \tfrac{1}{2}\frac{\rho V^3}{A^2}$$

The graph in Fig. D.21 shows this output function for a turbine with a $50\ \text{m}$ head of water and a turbine cross-sectional area of $1\ \text{m}^2$.

And the maximum possible efficiency is found by dividing the power out by the power in, i.e. the loss in PE per second, $\rho V g h$, so Max. efficiency $= 1 - \frac{1}{2}\frac{V^2}{A^2 g h}$.

D15 Knowledge check

A reservoir drains through a $20\ \text{cm}$ diameter pipe to an outlet $20\ \text{m}$ below. Calculate:

(a) the speed of the water through the pipe,

(b) the volume flow rate,

(c) the mass flow rate, and

(d) the rate of energy transfer.

[$\rho_w = 1000\ \text{kg m}^{-3}$]

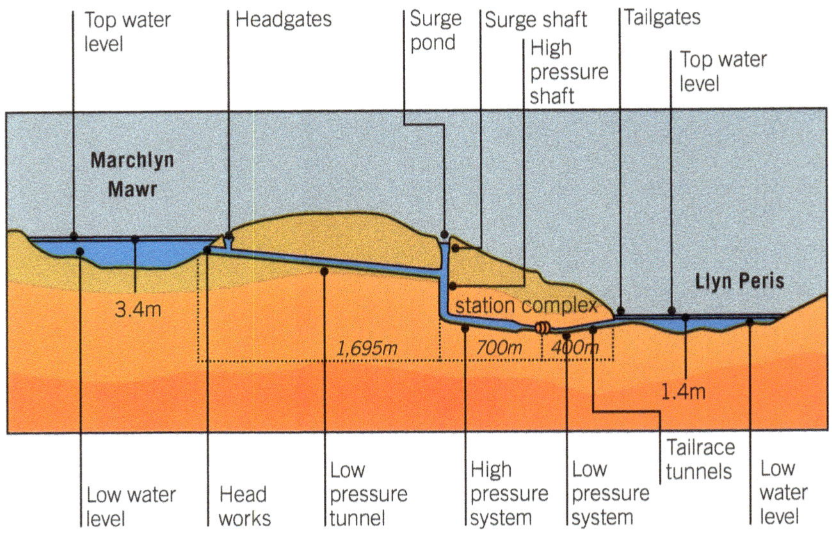

Fig. D.22 Dinorwig pumped-storage power station

D16 Knowledge check

Show that the equation for P_{OUT} is homogeneous.

Fig. D.22 is of a **pumped-storage** power station. Electrical energy itself cannot be stored and battery technology is not generally sufficient to allow it to be used for grid backup,[1] so pumping water from the lower to the upper reservoir provides a way of using power input to the grid (e.g. from wind and nuclear power providers) at times of low demand.

Stretch & challenge

Show that this model predicts that the maximum power output of a hydroelectric turbine occurs when $V = \sqrt{\frac{2ghA^2}{3}}$.

Show than the efficiency of the energy transfer at this volume flow rate is independent of h, A (and g).

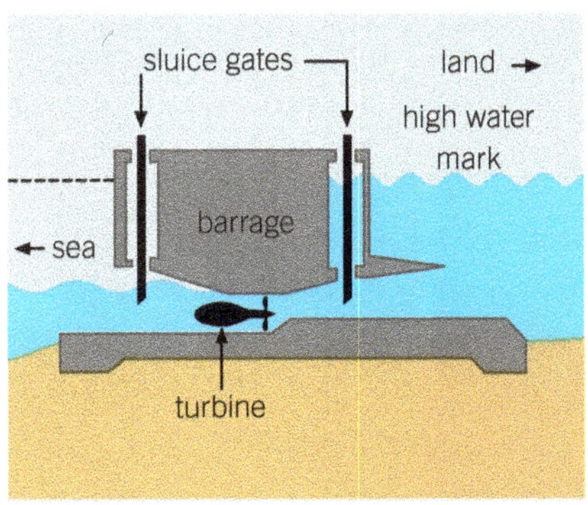

Fig. D.23 Tidal barrage principle

Tidal barrages have recently been proposed as hydroelectric energy sources. A barrage is constructed with inbuilt turbines as shown in Fig. D.23. As high tide approaches, the water level is higher outside the barrage than inside, so the sluice gates are opened allowing water to flow and its potential energy to be tapped.

When the water levels are equal the sluice gates are closed until the tide drops; they are then opened (as in Fig. D.23) and electricity is again generated from the water flow out past the turbines. As with tidal stream power stations, they have the advantage of predictability.

1 However, the Hornsdale Power Reserve in South Australia has a $194\ \text{MWh}$ lithium ion battery with a maximum power output of $150\ \text{MW}$, for a grid back-up.

Example

A tidal barrage has an area of $0.25\ \text{km}^2$. The maximum height difference across the barrage is about $6\ \text{m}$. Estimate the energy available in one outflow.
[Take $\rho_{\text{water}} = 10^3\ \text{kg m}^{-3}$]

Answer

Volume of water flowing = $0.25\ \text{km}^2 \times 6\ \text{m} = 1.5 \times 10^6\ \text{m}^3$.

$\therefore$ Mass of water flowing = $1.5 \times 10^9\ \text{kg}$

Assume mean height difference = $3\ \text{m}$

$\therefore$ Potential energy available = $mgh = 1.5 \times 10^9 \times 9.81 \times 3 = 4 \times 10^{10}\ \text{J}$ (1 sf)

Knowledge check D17

For the **Example**, estimate the mean power available. [Hint: there are two tides per day and generation occurs as the tide comes in **and** goes out.]

(f) Enriching uranium

Nearly all nuclear reactors in the world operate by the fission of ^{235}U. The principles of operation of nuclear fission power stations, including the need for moderator, control rods and the physics of the fission reactions themselves are not covered here. Natural uranium comprises 99.3% of ^{238}U which absorbs neutrons but does not fission. Most reactor designs require about 3%–5% ^{235}U so **enrichment** is required. Historically the separation was based on gaseous diffusion but the current technology uses gas centrifuges. This is a physical process based on the small mass difference between the two isotopes.

Fig. D.24 US gas centrifuge facility

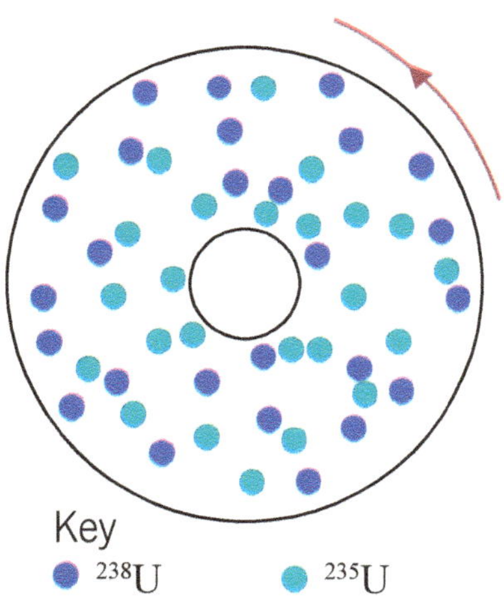

Fig. D.25 Gas centrifuge principle

The chemically identical isotopes are reacted with fluorine to produce the gaseous compound uranium hexafluoride, UF_6. The mixture of gases is fed into a spinning tube, which is shown schematically in Fig. D.25.

The heavier ^{238}U molecules are preferentially spun to the outside and the slightly enriched mixture (i.e. with a higher percentage ^{235}U) extracted along the central tube and fed to the next tube in a cascade. The outside gas with less ^{235}U is returned to the previous centrifuge. Each tube enriches the uranium by a factor of something like 1.2. The separation is this low because of the very small percentage difference in the masses of the molecules of the different isotopes.

Knowledge check D18

Calculate the percentage difference in the molecular masses of UF_6 containing the two different uranium isotopes.

Gas diffusion used the same starting point and was based upon rate of diffusion through a semi-permeable membrane being inversely proportional to the square root of the molecular mass (*Graham's law of diffusion*).

Knowledge check D19

The enrichment per stage in a gas centrifuge is a factor of 1.2.

(a) State the enrichment after 2 stages, 5 stages, n stages.

(b) Calculate the number of stages needed to achieve a U235 concentration of 5% from an initial feedstock concentration of 0.7%

(g) Breeding nuclear fuel

Although ^{238}U is not **fissile**, it does capture neutrons, giving rise to ^{239}U which decays in two stages by β^- emission to give the fissile plutonium 239, ^{239}Pu.

$$^{238}_{92}\text{U} + ^{1}_{0}\text{n} \rightarrow ^{239}_{92}\text{U} \xrightarrow[2.35\ \text{min}]{\beta^-} ^{239}_{93}\text{Np} \xrightarrow[2.3\ \text{day}]{\beta^-} ^{239}_{94}\text{Pu}$$

A smaller fraction of ^{238}U undergoes multiple neutron capture before decaying twice to give rise to the non-fissile ^{240}Pu and the fissile ^{241}Pu. Thus there is a gradual build-up of plutonium isotopes in fuel elements; it can be removed chemically at reprocessing. By the time the nuclear fuel element is removed for reprocessing, about half the fissile isotopes have undergone fission but half remain.

Key terms

A **fissile** nuclide is one that undergoes fission when it absorbs **thermal neutrons**, with the emission of enough neutrons to maintain a chain reaction.

Thermal neutrons are neutrons with a kinetic energy corresponding to temperatures of the order of 300 K.

Study point

The presence of ^{239}Pu and ^{241}Pu in spent nuclear fuel rods is of significance for nuclear non-proliferation because it is the fuel of choice in nuclear fission bombs and also in the primary stage of nuclear fusion bombs (*hydrogen bombs*).

Study point

In nuclear fusion, the **confinement time**, τ_E, is defined by:

$$\tau_E = \frac{W}{P_{loss}}$$

where W is the energy density and P the power loss per unit volume (e.g. by conduction and radiation) from the fusion fuel.

D20 Knowledge check

Scientists working on nuclear fusion often express temperature in the energy units of keV, by using the Boltzmann constant, e.g. in these units room temperature (~300 K) is $\frac{300 \text{ K} \times 1.38 \times 10^{-23} \text{ J K}^{-1}}{1.6 \times 10^{-16} \text{ J keV}^{-1}}$.

(a) Evaluate 300 K in keV.

(b) The temperature of the Sun's core is 15 MK. Express this in keV.

The fissile isotopes of plutonium can be incorporated into fuel rods together with ^{235}U in so-called MOX (*mixed oxide*) fuel. Most European PWRs (*pressurised water reactors*) operate with some MOX fuel rods; typically 30%. It has been proposed to use plutonium from nuclear weapon decommissioning in the production of MOX fuel, thus removing the existence of this highly dangerous material.

(h) The nuclear fusion triple product

In order to obtain fusion using, for example, ${}^2_1H + {}^3_1H \rightarrow {}^4_2He + {}^1_0n$ three conditions have to be satisfied:

1 A high enough temperature, T, to enable the reactant nuclei to approach close enough to overcome the Coulomb repulsion and allow fusion to occur; the strong nuclear interaction is short range so the nuclei have to approach within about 10^{-14} m; a typical temperature requirement is 100 MK (i.e. 10^8 K).

2 A high enough particle density, n, to allow a high enough collision rate between reactant nuclei; any individual collision of sufficient energy might not result in a fusion reaction, e.g. because the collision is slightly oblique.

3 A long enough **confinement time**, τ_E, which is a measure of how long the fuel maintains its internal energy; the hot fuel radiates away its energy by X-rays.

Each of these conditions needs to be achieved separately. The aim of fusion engineers is to maximise the product, $nT\tau_E$, which is called the *triple product*: the higher this product, the more likely it is for a sustainable nuclear fusion reaction to occur. It is suggested that you do Knowledge check D20 before reading the following **Example**.

Example

It has been estimated that the minimum value of the triple product for the deuterium-tritium reaction is 3×10^{21} keV s m^{-3}, where the temperature is expressed in keV. If a fusion reactor can attain a value of $n\tau_E$ of 3×10^{20} m^{-3} s, calculate the kelvin temperature required.

Answer

$$nT\tau_E = 3 \times 10^{20} \text{ m}^{-3} \text{ s} \times T = 3 \times 10^{21} \text{ keV m}^{-3} \text{ s}$$

$$\therefore T = \frac{3 \times 10^{21}}{3 \times 10^{20}} \text{ keV} = 10 \text{ keV} = \frac{10 \text{ keV} \times 1.6 \times 10^{-16} \text{ J keV}^{-1}}{1.38 \times 10^{-23} \text{ J K}^{-1}} = 116 \text{ MK}$$

D.3 Fuel cells

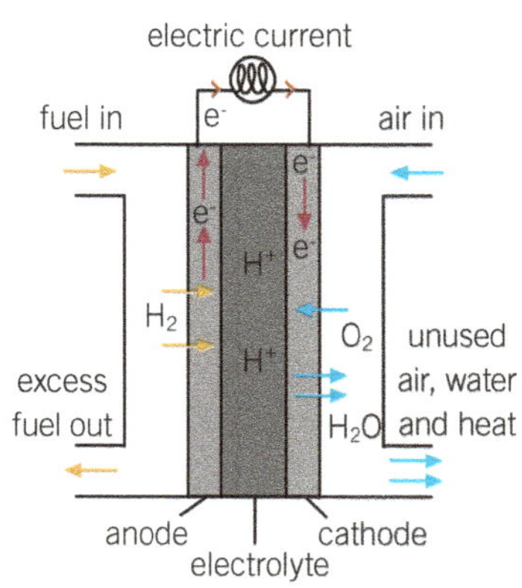

Fig. D.26 Principle of a proton-exchange fuel cell

The fuel cell is a battery with continuously replenished reactants. In the fuel cells which are proposed for transport use (cars, buses, goods vehicles) the fuel is hydrogen with the oxidant being oxygen. The principle of operation is:

- Hydrogen is drawn in to one side of the fuel cell, the anode.
- The hydrogen molecules are stripped of their electrons ('oxidised') at the anode by a catalyst, which is usually powdered platinum.
- The resulting hydrogen ions (i.e. protons, H^+) diffuse into the electrolyte through a non-conducting barrier, leaving the negatively charged electrons.
- The electrons travel along by electric wires into an external circuit, where they do work before entering the cathode.
- At the cathode, oxygen combines, under the influence of a catalyst (the *cathode catalyst*, often nickel or a nano-material) with the electrons and diffused protons from the anode to produce water, which is the waste product.

The half-reactions at the electrodes are:

Anode: $H_2 \rightarrow 2H^+ + 2e^-$

Cathode: $\frac{1}{2}O_2 + 2H^+ + 2e^- \rightarrow H_2O$

So the overall reaction is the familiar $2H_2 + O_2 \rightarrow 2\,H_2O$, but because the energy extraction is not via combustion and the release of heat, fuel cells are significantly more efficient than heat engines. Theoretical maximum efficiency is of the order of 80% and practical fuel cells regularly achieve 40% transfer to electrical energy.

The major advantage of the fuel cell is its lack of greenhouse gas emissions but this is only significant if the hydrogen can be produced without CO_2 emission, which requires electricity generated by non-fossil-fuel-based technologies (PV cells, wind, hydro...). Research cars using fuel cells currently obtain their hydrogen from hydrocarbons and so lead to carbon dioxide waste.

Knowledge check D21

A fuel cell produces $5\ kW$ of electrical output. Assuming a 40% efficiency, calculate the mass of hydrogen it consumes per hour.

Take the heat of combustion of molecular hydrogen H_2 to be $290\ kJ\ mol^{-1}$.

D.4 The conduction of heat

According to a 2020 report, the UK energy use in 2020 was $5.9\ EJ$. The domestic use accounts for 30% of this figure of which half is for space heating. We need to heat our homes because for most of the year we make the indoor temperature higher than the outside temperature. It is this temperature difference that drives heat loss, mainly by gas movement (convection) and conduction through the walls / windows. Indoor temperatures have risen markedly in the last 40 years, as Table D1 shows. This tends to increase the rate of heat loss and therefore requirement for heating, with the consequent greater use of fossil fuels and increased greenhouse emissions. However, increasing insulation standard and the move to low-energy light bulbs, TVs, etc., means that domestic energy use has actually fallen by 10% in the last 20 years.

Knowledge check D22

Express the annual domestic space heating energy use in standard form.

Location	Winter temperature / °C	
	1979	2008
Living room	18.3	21.3
Hall	15.8	21.1
Bedroom	15.2	21.0

Table D1 Mean British winter room temperatures

(a) The conduction equation

The conduction of heat through good conductors can be investigated using the set-up which is shown schematically in Fig. D.27.

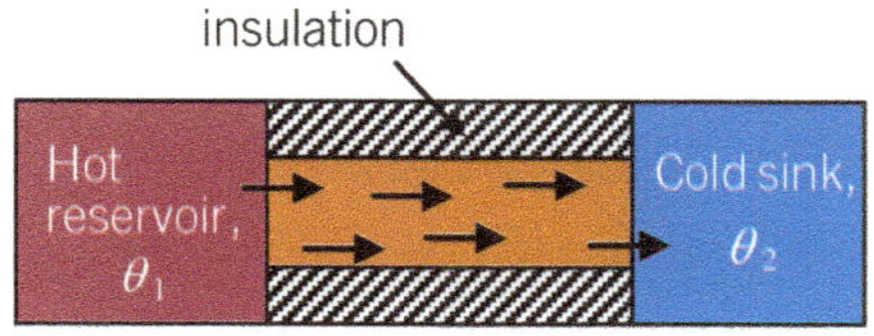

Fig. D.27 Measuring conductivity

A constant temperature gradient is established between the ends of a specimen of material and the flow of heat measured. One way of maintaining the temperatures θ_1 and θ_2 constant is to use a jet of steam in the hot reservoir and a stream of cold water in the sink. The heat flow is measured by the small temperature rise in the water flow in the sink.

The results of such experiments are the conduction equation:

$$\frac{\Delta Q}{\Delta t} = -AK\frac{\Delta \theta}{\Delta x} \text{ (see Study point)}$$

where A and x are the cross-sectional area and distance in the direction of heat flow. The constant K is the (coefficient of) thermal conductivity and is characteristic of the material and has the unit $W\ m^{-1}\ K^{-1}$. The values of K in this unit range from 430 (silver) and 400 (copper) through glass (1–2) and brick (0.1–0.7) to rock wool (0.05) and air (0.025). The range is not nearly as great as the range of electrical conductivities.

Study point

Note that the use of ΔQ rather than Q for heat flow is not consistent with the first law of thermodynamics equation. However, in this context it is conventional.

In order to measure the thermal conductivities of the insulators, a different experimental arrangement is needed. The specimen is usually a thin disc (large value of A and small value of Δx) rather than a long bar.

Knowledge check D23

Show that $W\ m^{-1}\ K^{-1}$ is the basic SI unit of thermal conductivity.

(b) Insulation and heat loss

Consider the house wall in Fig. D.28. It was built in the days before cavity wall insulation. Using the symbols on Fig. D.28, the rate of heat loss through an area, A, of wall is given by:

$$\frac{\Delta Q}{\Delta t} = AK_b\frac{\theta_1 - \theta_2}{d_b}$$

where K_b is the thermal conductivity of the brickwork.

What is the effect of putting on a layer of insulation, such as expanded polystyrene on the interior wall (Fig. D.29)? The temperature between the insulation and the bricks must now be in between θ_1 and θ_2 so $\Delta\theta$ for the wall is now less, so the rate of heat loss is less.

We can calculate the temperature at the join by noticing that the heat flow through the bricks and insulation must be the same.

Hence: $AK_b\frac{\theta - \theta_2}{d_b} = AK_i\frac{\theta_1 - \theta}{d_i}$

We can solve this equation for θ and then use the heat flow equation for either the bricks or the insulation (it doesn't matter which).

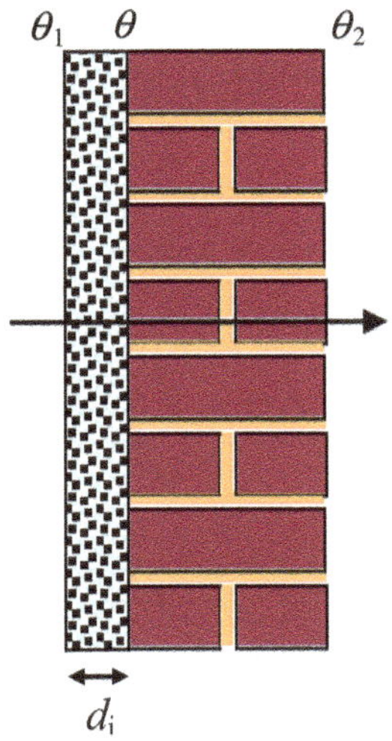

Fig. D.29 Insulated wall

The next **Example** uses this way of working to calculate the heat loss through an insulated brick wall. **Warning**: the answer we get is too high by a large factor for reasons we'll discuss immediately afterwards.

Example

A 0.110 m thick brick wall ($K_b = 0.5\ \text{W m}^{-1}\ \text{K}^{-1}$) has a 0.050 m layer of expanded polystyrene ($K_b = 0.033\ \text{W m}^{-1}\ \text{K}^{-1}$) added as insulation. Calculate the rate of heat loss per unit area for the insulated wall when the temperatures of the inside and outside surfaces are $24\ °\text{C}$ and $5\ °\text{C}$ respectively.

Answer

First find the temperature, θ, of the insulation-wall boundary:

Equating heat loss through each layer: $A \times 0.5\frac{\theta - 5}{0.110} = A \times 0.033\ \frac{24 - \theta}{0.050}$

Solve this to give $\theta = 7.41\ °\text{C}$ – this is left as an exercise.

$\therefore$ heat loss through brick $= 0.5\ \text{W m}^{-1}\ \text{K}^{-1}\ \frac{(7.41 - 5.00)\ \text{K}}{0.110\ \text{m}} = 11\ \text{W m}^{-2}$

Discussion: As stated above, the real heat loss is usually much lower than that calculated in the Example. With an interior temperature of $24\ °\text{C}$ and an exterior temperature of $5\ °\text{C}$, the actual temperature difference across the composite wall will be much less than $19\ °\text{C}$. This is because of the existence of a thin 'skin' of stationary air in contact with both the interior and exterior surfaces – see Fig. D.30. Air such a good insulator – hence the thermal properties of woollen clothes – that most of the temperature drop occurs across these skins, rather than the wall.

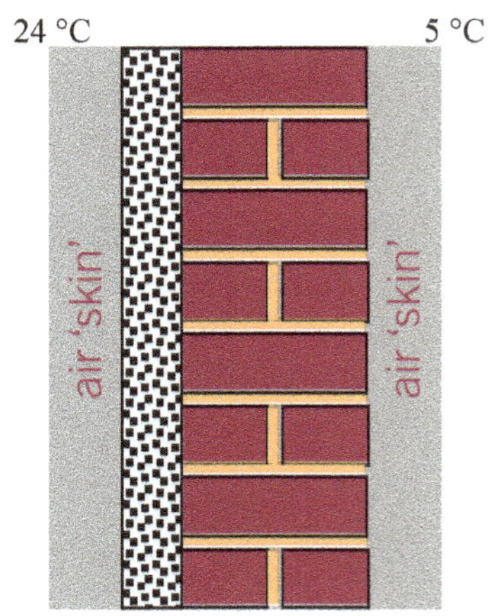

Fig. D.30 The insulating effect of air

D24 Knowledge check

Use the conduction equation to explain the use of a thin disc when determining K for poor conductors.

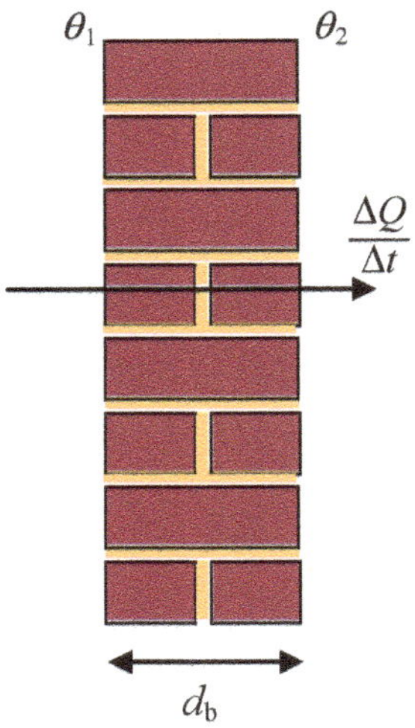

Fig. D.28 Uninsulated wall

Study point

Don't learn the final equation for multi-layer calculations. Just apply the basic conduction equation to the two layers as in the main text.

D25 Knowledge check

Calculate the rate of heat loss per unit area for the uninsulated wall under the same conditions.

Study point

Temperature differences have the same value in °C as in K:

$24\ °\text{C} - 5\ °\text{C} = 297\ \text{K} - 278\ \text{K}$

Hence there is no need to change to K for conduction calculations.

The interior surface of the wall will be several degrees cooler than the room. The exterior surface will be warmer than the outside air temperature (though this effect is reduced on windy days). Because $\Delta\theta$ across the wall is less, the rate of heat conduction will also be less.

This 'insulating skin' effect is taken into account in the determination of U values (see below).

(c) U values

Building regulations for insulation purposes are framed in terms of the **U values** of building elements, such as walls, doors and windows, rather than the thermal conductivities of the materials themselves. Because of the problems of weather conditions, U values are specified under standardised conditions: usually 24 °C (297 K) temperature difference, under wind-free conditions at 50% humidity.

The U value of a structure is defined by: $\frac{\Delta Q}{\Delta t} = UA\Delta\theta$

Notice that we have dropped the minus sign (the direction of the heat transfer is by inspection).

Key terms

The **U value** of a building element (e.g. a wall) is the rate of heat transfer per unit area per unit temperature difference.

UNIT: $\text{W m}^{-2}\text{ K}^{-1}$

Example

The U value of a brick wall (no cavity) is $2.0\text{ W m}^{-2}\text{ K}^{-1}$. Calculate the rate of heat loss per unit area with a temperature difference of 15 °C.

Answer

Rate of heat loss = $UA\Delta\theta$.

∴ Heat loss per unit area = $U\Delta\theta = 2.0 \times 15 = 30\text{ W m}^{-2}$.

Now compare this answer with one calculated using the thermal conductivity of the bricks, out of which the wall is constructed (KC D26)

Knowledge check D26

The bricks which make up the wall in the example have a thermal conductivity of $0.60\text{ W m}^{-2}\text{ K}^{-1}$. The thickness of the wall is 110 mm. Show that the rate of heat loss calculated using these figures is about 2½× that determined using the U value.

Compound structures – combining U values

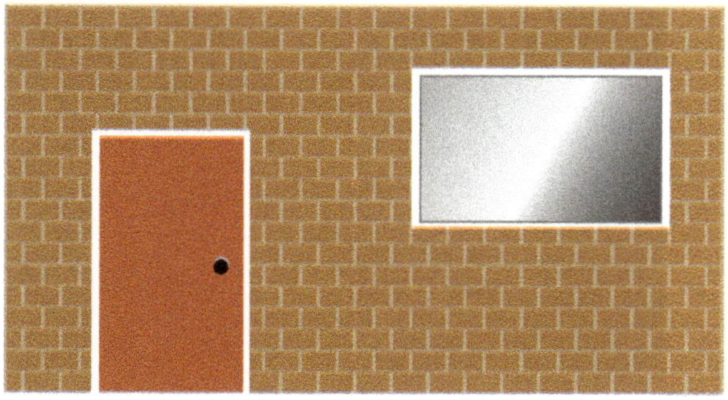

Fig. D.31 Wall + door + window

Calculating the heat loss from (say) a wall containing a door and window is straightforward: The total heat loss is the sum of the losses through the wall, the door and the window, i.e.

$$\frac{\Delta Q}{\Delta t} = (U_{\text{wall}}A_{\text{wall}} + U_{\text{door}}A_{\text{door}} + U_{\text{window}}A_{\text{window}})\Delta\theta$$

The U value of a multi-layered structure, e.g. the insulated wall in Fig. D.29, can be estimated using a similar approach to the one in the previous example. The next example shows how this can be done. However, the method has a problem: an individual layer has a U value which is affected by the insulating skin of air; when the layers are combined, there is no air skin in between them. So the result of the calculation will underestimate the U value.

Example

An exterior brick wall has a U value of $2.0\ \text{W m}^{-2}\ \text{K}^{-1}$. The interior temperature is $20\ ^{\circ}\text{C}$ and the outside is $12\ ^{\circ}\text{C}$. A layer of insulation with a U value of $0.4\ \text{W m}^{-2}\ \text{K}^{-1}$ is added to the interior of the wall. Estimate:

(a) the rate of heat loss per unit area and (b) the new U value of the wall.

Answer

(a) Let the temperature of the interface be θ. The rates of heat loss through the two layers are the same.

$\therefore$ Heat loss per unit area $= 0.4 \times (20 - \theta) = 2.0 \times (\theta - 12)$

Dividing by 0.4 and multiplying out: $20 - \theta = 5\theta - 60$

Giving $\theta = \dfrac{80}{6} = 13.3\ ^{\circ}\text{C}$

$\therefore$ Heat loss through the insulation $= 0.4 \times (20 - 13.3) = 2.67\ \text{W m}^{-2}$

(b) $\therefore 2.67\ \text{W m}^{-2} = U_{\text{wall}} \times 8\ \text{K}$

$\therefore U_{\text{wall}} = \dfrac{2.67}{8} = 0.33\ \text{W m}^{-2}\ \text{K}^{-1}$

Stretch & challenge

For a layered structure:

(a) show that thermal resistances add

(b) use this result to confirm the U value of the insulated wall in the Example.

Building engineers define the thermal resistance, R, of a structure as the reciprocal of its U value, i.e. $R = U^{-1}$. The advantage of this is that, for a structure with several layers, the thermal resistances add to give the total thermal resistance [like electrical resistances in series]. See Stretch & challenge.

Test yourself Option D

1. A student measures the wavelength of the peak of the solar spectrum to be $493\ \text{nm}$. Use Wien's constant [$2.90 \times 10^{-3}\ \text{m K}$] to calculate the temperature of the Sun's photosphere.

2. A website gives the diameter and temperature of the photosphere of the Sun as 1.392 million km and $5778\ \text{K}$ respectively. The mean Sun–Earth distance is $1.496 \times 10^{8}\ \text{km}$. Use these data to calculate: (a) the power output of the Sun, and (b) the Solar constant.

3. A student drops a piece of ice into a measuring cylinder containing $200\ \text{cm}^3$ of water. The water level rises to a reading of $240\ \text{cm}^3$.

   (a) State the volume of ice which is submerged.
   (b) Calculate the total volume of the ice.
   (c) State and explain what happens to the water level in the measuring cylinder when the ice melts.
   (d) Briefly state the relevance of this experiment to our understanding of the effect of melting ice on rising sea level [$\rho_{\text{ice}} = 917\ \text{kg m}^{-3}$; $\rho_{\text{water}} = 1000\ \text{kg m}^{-3}$]

4. The dwarf planet, Ceres, has a mean orbital radius of $2.77\ \text{AU}$.

   (a) Calculate the ratio of the intensity of sunlight at Ceres to that at the orbit of the Earth.
   (b) Estimate the mean temperature of the surface of Ceres given that it possesses no atmosphere.

5. A tidal flow turbine of diameter $5.0\ \text{m}$ is placed in Ramsey Sound where the peak tidal current in $3\ \text{m s}^{-1}$. Estimate the peak power available given a power coefficient of 0.4.

6. The low power turbine at Cwm Rheidol power station uses a water head of $7.5\ \text{m}$. The flow speed is approximately $6\ \text{m s}^{-1}$ through a turbine with a cross-sectional area of $4\ \text{m}^2$. Estimate:

   (a) the mass flow rate
   (b) the loss in gravitational potential energy per second
   (c) the gain in kinetic energy per second of the water
   (d) the power generated, assuming a turbine and generator combined efficiency of 80%
   (e) the overall efficiency achieved.

7. Thorium-232 has been successfully used in a breeder reactor. It is present in the Earth's crust in much larger amounts than any uranium isotope. Its half-life is 1.41×10^{10} years. Not itself fissile, it absorbs thermal neutrons, with an energy of around $0.025\ \text{eV}$, giving an excited nuclide which decays by γ emission followed by two β^- decays to a fissile isotope of uranium. [Consult a Periodic Table if necessary.]

   (a) Use the Boltzmann constant to justify the figure of $0.025\ \text{eV}$ for thermal neutrons.
   (b) Write reactions for the three decays mentioned above.
   (c) Explain why a thorium-232 reactor requires the presence of a fissile nuclide such as ^{235}U.
   (d) Why is ^{232}Th more prevalent that ^{235}U or ^{238}U in the Earth's crust?

8. The nuclear material in a nuclear reactor consists initially of 5% of ^{235}U and 95% of the non-fissile ^{238}U. Up to 30% of the energy produced in a reactor comes from reactions other than the fission of ^{235}U. Briefly explain this.

9. A $2.0\ \text{m} \times 1.0\ \text{m}$, singled-glazed window is made of glass. The coefficient of thermal conductivity of the glass out of which a window is made in $1.0\ \text{W m}^{-1}\ \text{K}^{-1}$. The thickness of the glass is $0.60\ \text{cm}$.

   (a) Calculate the rate of heat flow through the window if a temperature difference of $20\ {}^\circ\text{C}$ is maintained between its faces.
   (b) The U value of the window is $4.8\ \text{W K}^{-1}$. Use this value to calculate the rate of heat loss on a calm day from a room with this window when the inside and outside temperatures are $25\ {}^\circ\text{C}$ and $5\ {}^\circ\text{C}$ respectively.
   (c) Explain the (huge) difference in the rates of heat loss in the answers to parts (a) and (b).
   (d) Calculate the actual temperature difference between the inside and outside faces of the glass window pane in part (b).
   (e) A double-glazed unit is made from two $4\ \text{mm}$ sheets of this glass separated by a ($12\ \text{mm}$) layer of air. The U value of the unit is $2.8\ \text{W m}^{-2}\ \text{K}^{-1}$. Assuming the U value of the glass panes is inversely proportional to their thickness, estimate the U value of the air layer.

10. A new design of solar panel consisting of a number of series combinations of 90 PV cells connected in parallel. The graph shows the output characteristic of the panel illuminated at $90°$ by solar radiation of intensity 560 W m^{-2}.

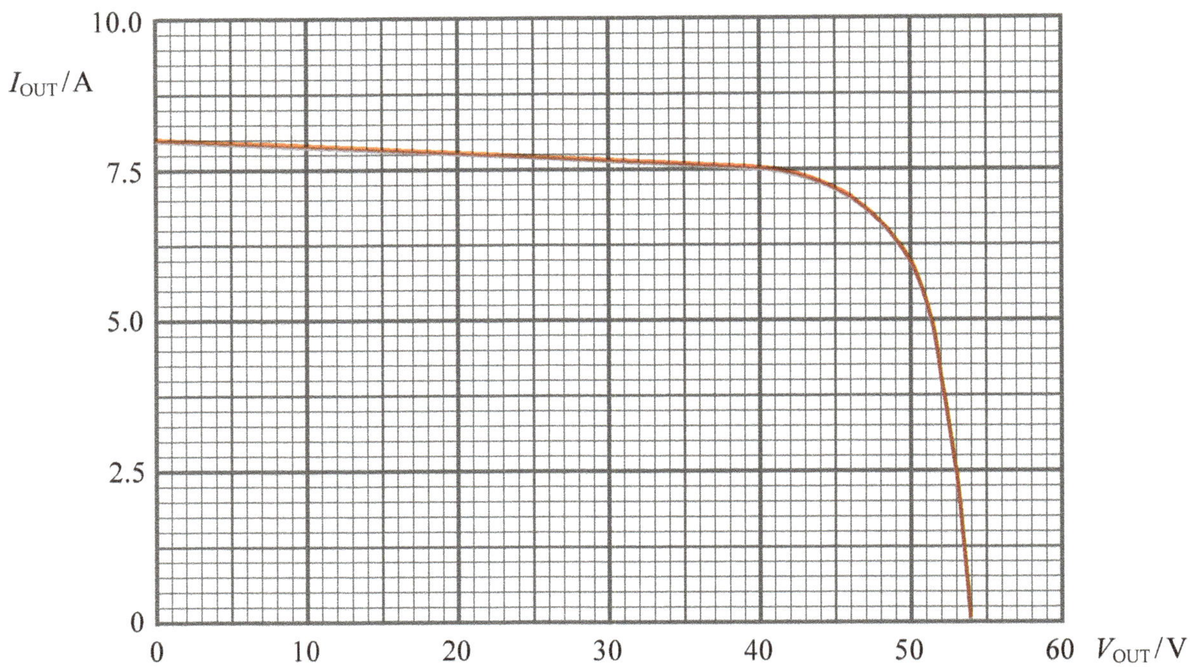

(a) The maximum current produced by each solar cell is 40 mA. Calculate:
    (i) the number of parallel combinations of solar cells
    (ii) the maximum output pd of each solar cell.
(b) Use the graph to calculate the output power, P_{out}, of the solar panel when operating at an output voltage, $V_{out} = 20 \text{ V}$.
(c) Using a series of calculated values, draw a graph of P_{out} against V_{out}.
(d) Use the graph to determine the maximum power output of the solar panel and state the operating values of V_{out} and I_{out}.
(e) Calculate the area of the solar panel, assuming an operational efficiency of 30%.

11. A tidal lagoon (a form of tidal barrage) has been discussed as a means of providing renewable energy. Among objections to it is the fact that the output varies with the tide – so there are times when no electricity is generated. The following table gives the times of high tides in various locations relative to Swansea.

Location	Fishguard	Aberystwyth	Porthmadog	Holyhead	Bangor
High tide difference (min)	+65	+100	+130	+265	+310

Discuss how this information alters the case for tidal lagoon construction.

Unit 4 equations

Equation	Description
$C = \frac{Q}{V}$	C = capacitance; Q = charge on either plate of capacitor
$C = \frac{\varepsilon_0 A}{d}$	Parallel plate capacitor with air between the plates A = area of each plate; d = plate separation
$E = \frac{V}{d}$	Uniform electric field: E = field strength, V = pd d = distance in the direction of the field.
$U = \frac{1}{2}QV$	Energy stored on by a capacitor
$Q = Q_0(1 - e^{-t/RC})$	Charging equation for a capacitor; R = resistance
$Q = Q_0 e^{-t/RC}$	Discharging equation for a capacitor
$F = \frac{1}{4\pi\varepsilon_0}\frac{Q_1Q_2}{r^2}$	Electrostatic force between two point charges; r = separation
$\text{PE} = \frac{1}{4\pi\varepsilon_0}\frac{Q_1Q_2}{r}$	Electric potential energy of two point charges
$E = \frac{1}{4\pi\varepsilon_0}\frac{Q}{r^2}$	Electric field due to a point charge; r = distance from charge
$V_E = \frac{1}{4\pi\varepsilon_0}\frac{Q}{r}$	Electric potential due to a point charge
$W = q\Delta V_E$	Work done in moving a charge in an electric field
$F = G\frac{M_1M_2}{r^2}$	Gravitational force between two point masses
$\text{PE} = -G\frac{M_1M_2}{r}$	Gravitational potential energy of two point masses
$g = \frac{GM}{r^2}$	Gravitational field strength due to a point mass
$V_g = -G\frac{M}{r}$	Gravitational potential due to a point mass
$W = m\Delta V_g$	Work done in moving a mass in a gravitational field
$\frac{\Delta\lambda}{\lambda} = \frac{v}{c}$	Doppler shift; $\Delta\lambda$ = change in wavelength; v = radial velocity
$v = H_0 D$	v = recession velocity; H_0 = Hubble constant ; D = distance
$\rho_c = \frac{3H_0^2}{8\pi G}$	ρ_c = critical density
$r_1 = \frac{M_2}{M_1 + M_2}d$	r_1 = distance of centre of mass from M_1
$T = 2\pi\sqrt{\frac{d^3}{G(M_1 + M_2)}}$	Period of orbit for two bodies orbiting about their centre of mass
$F = BI\ell\sin\theta$	Force on a current-carrying conductor in a magnetic field, B
$F = Bqv\sin\theta$	Force on a moving charge, q, in a magnetic field
$B = \frac{\mu_0 I}{2\pi a}$; $B = \mu_0 nI$	Magnetic field due to a straight wire; in a long solenoid n = number of turns per unit length
$\Phi = AB\cos\theta$	Flux in an surface of area A; θ = angle between field and normal
Flux linkage = $N\Phi$	N = number of turns

Because of the requirement to set synoptic questions, the WJEC Data Booklet for Unit 4 includes the equations for AS Units 1 and 2, and A2 Unit 3. The equations on this page are the ones specific to the core content of Unit 4. Those on the next page relate to the Unit 4 Options.

The descriptions are not given in the Data Booklet – just the symbol equations.

Unit 4 equations (Options)

	Equation	Description
Option A	Flux linkage = $BAN\cos\omega t$	Flux linkage in a spinning coil in a magnetic field
	$V = \omega BAN\sin\omega t$	Output pd of by a spinning coil in a magnetic field
	$V_{\text{rms}} = \frac{V_0}{\sqrt{2}}$; $I_{\text{rms}} = \frac{I_0}{\sqrt{2}}$	rms values for sinusoidally varying pd and current I_0 and V_0 are the peak values
	$V_{\text{rms}} = \frac{\omega BAN}{\sqrt{2}}$	rms pd for a spinning coil in a magnetic field
	$X_L = \omega L$; $X_C = \frac{1}{\omega C}$	Reactance of an inductor (X_L) and a capacitor (X_C)
	$Z = \sqrt{X^2 + R^2}$	Relationship between impedance (Z), reactance and resistance
	$Q = \frac{V_L}{V_R}\left(= \frac{V_C}{V_R}\right) = \frac{\omega_0 L}{R}$	Q factor in terms of the pds at resonance and in terms of the resonance frequency, inductance and resistance
Option B	$I = I_0 e^{-\mu x}$	Decay of intensity, I, of X-rays in a medium
	$Z = c\rho$	Z = acoustic impedance; c = speed of sound; ρ = density
	$\frac{\Delta f}{f_0} = \frac{2v}{c}\cos\theta$	Doppler shift applied to ultrasound scans
	$f = 42.6 \times 10^6 B$	Larmor frequency for protons in a magnetic field
	$H = DW_R$	H = equivalent dose; D = absorbed dose; W_R = radiation weighting factor
	$E = HW_T$	E = effective dose; W_T = tissue weighting factor
Option C	$Ft = mv - mu$	Impulse (Ft) = change in momentum
	$e = \frac{\text{relative speed after collison}}{\text{relative speed before collision}}$	Definition of coefficient of restitution
	$e = \sqrt{\frac{h}{H}}$	h = bounce height; H = drop height
	$I = \frac{2}{5}mr^2$; $I = \frac{2}{3}mr^2$	Moments of inertia for a solid ball and a hollow ball
	$\alpha = \frac{\omega_2 - \omega_1}{t}$	α = angular acceleration
	$\tau = I\alpha$	τ = torque
	$L = I\omega$	L = angular momentum
	$\text{KE} = \frac{1}{2}I\omega^2$	Kinetic energy of rotation
	$p = p_0 - \frac{1}{2}\rho v^2$	Bernoulli's equation
	$F_D = \frac{1}{2}\rho v^2 AC_D$	Drag force equation; C_D = drag coefficient
Option D	$I = \frac{P}{A}$	I = intensity of radiation; P = power; A = area at right angles
	$P = \frac{1}{2}A\rho v^3$	Power available from a moving fluid (air or water)
	$\frac{\Delta Q}{\Delta t} = -AK\frac{\Delta\theta}{\Delta x}$	Thermal conduction equation; K = thermal conductivity
	$P = UA\Delta\theta$	Rate of heat loss (P) in terms of U value

Unit 4

1 **(a)** **(i)** State Kepler's 1st and 2nd laws. **[3]**

(ii) Kepler's 3rd law can be derived from Newton's gravitational law and the equation for centripetal motion. Show that, for any object in a circular orbit about the Earth:

$$T^2 = \frac{4\pi^2}{GM_E} r^3$$

where T = the period of orbit, r = the radius of orbit, G = the gravitational constant and M_E = the mass of the Earth. **[3]**

(b) The radius of the geostationary orbit above the Earth's equator is $42\,000$ km and the Earth–Moon distance is $380\,000$ km. Use Kepler's 3rd law to calculate the period of the Moon's orbit. **[3]**

(c) Estimate the minimum possible orbital period for a satellite orbiting the Earth, stating any assumption that you make (the radius of the Earth is 6370 km). **[3]**

(Total 12 marks)

[*WJEC A level Physics 2019 Unit 4 Q3*]

2 **(a)** The current in a long wire is 290 A and is too great to be measured by an ammeter. Explain how you could use a Hall probe, calibrated in tesla (T), to determine this current. **[3]**

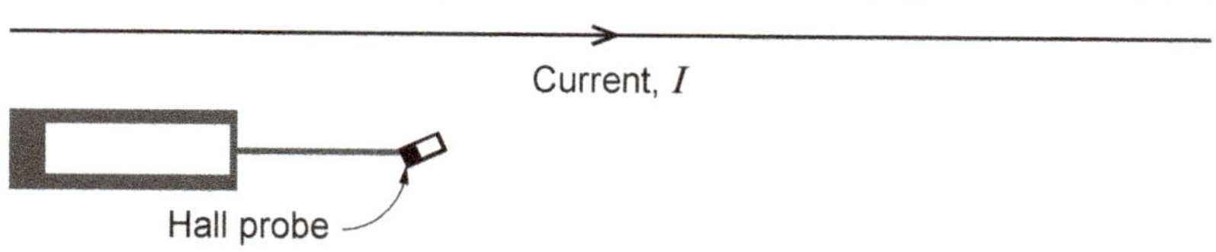

(b) Another long wire carrying the same large current is placed parallel to the original wire as shown. Calculate the force per unit length on each wire also stating the direction of the force on each wire. **[4]**

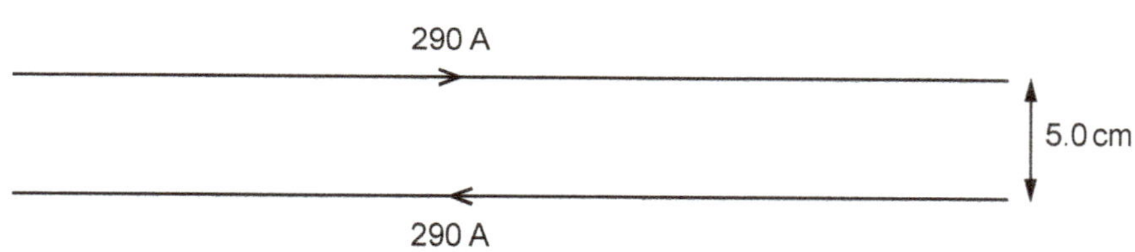

(c) **(i)** Tirion claims that an electron halfway between the wires, travelling at a speed of 2.9×10^7 m s^{-1} parallel to the wires will perform perfect circular motion between the wires. Determine, using a suitable calculation, whether or not Tirion's claim is correct. The magnetic flux density halfway between the wires is 4.64 mT. **[4]**

(ii) Sketch the motion of the electron. **[2]**

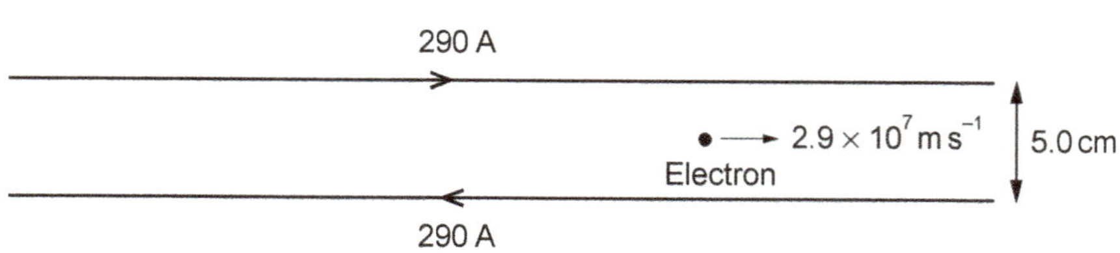

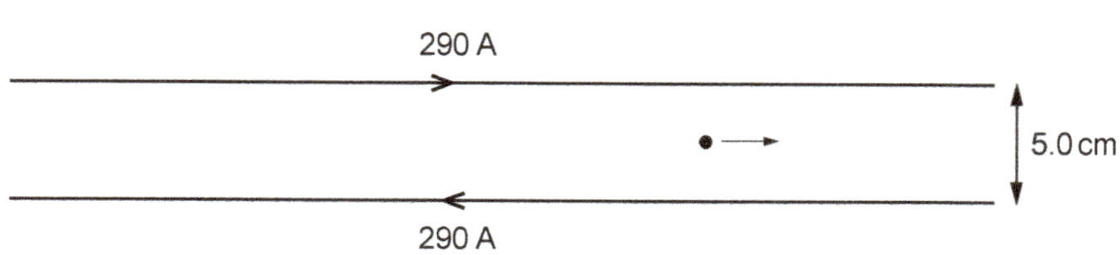

(d) Suppose that an electron travels in a region with a magnetic field and an electric field due to two parallel metal plates as shown below. Deduce whether or not the electron continues with constant velocity ($B = 4.64$ mT). **[5]**

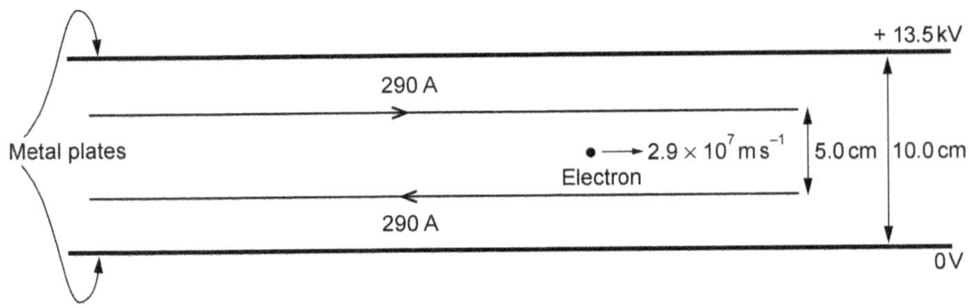

(Total 18 marks)

[*WJEC A level Physics 2019 Unit 4 Q4*]

3. (a) State the laws of Faraday and Lenz for electromagnetic induction. **[2]**

(b) Two strong bar magnets are dropped through two copper pipes (P and Q). Pipe P has a slit running along its length but Q is complete. When the magnet is dropped through pipe P it accelerates almost uniformly but the magnet dropped through pipe Q quickly reaches a very low terminal velocity. Explain these observations. **[6 QER]**

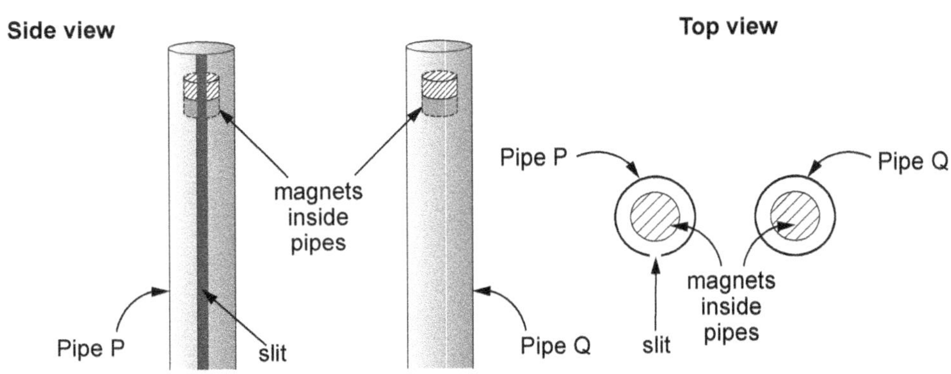

(c) By applying the principle of conservation of energy, calculate the temperature increase of pipe Q after the magnet has fallen at constant speed. **[4]**

- mass of magnet = 0.300 kg,
- cross-sectional area of the copper walls of pipe Q = 7.85×10^{-6} m^2 (see diagram)
- density of copper = 8960 kg m^{-3}
- specific heat capacity of copper = 385 J K^{-1} kg^{-1}.

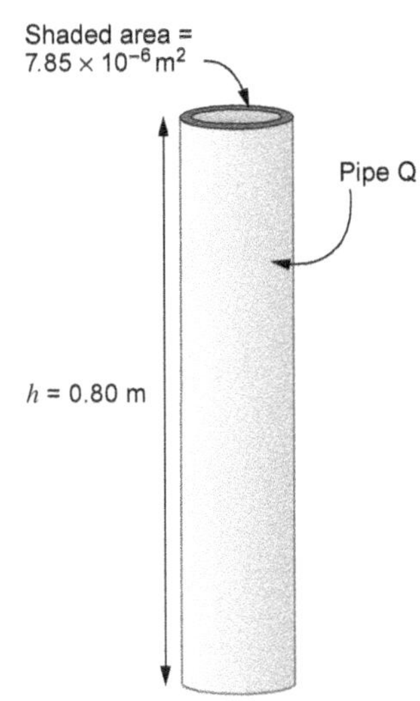

(Total 12 marks)

[*WJEC A level Physics 2019 Unit 4 Q5*]

Optional topics

Option A – Alternating currents

1 **(a)** The following RCL circuit is constructed.

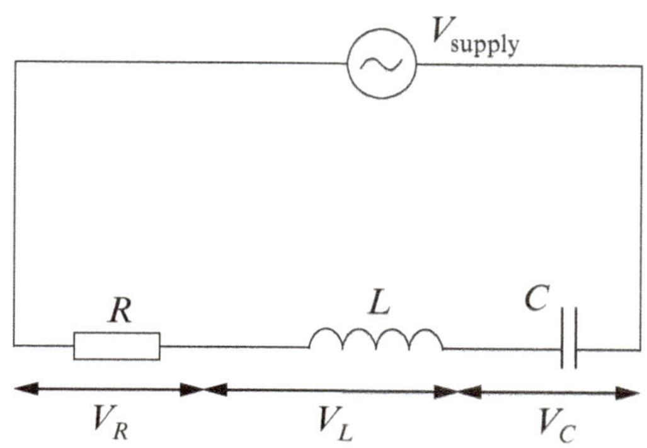

For this RCL circuit, describe the relationships between the rms pds V_{supply}, V_R, V_L and V_C:

(i) when the circuit is not in resonance; **[3]**

(ii) when the circuit is in resonance. **[2]**

(b) The following RCL circuit is **at resonance**. The values of R and C are provided along with their rms pd values.

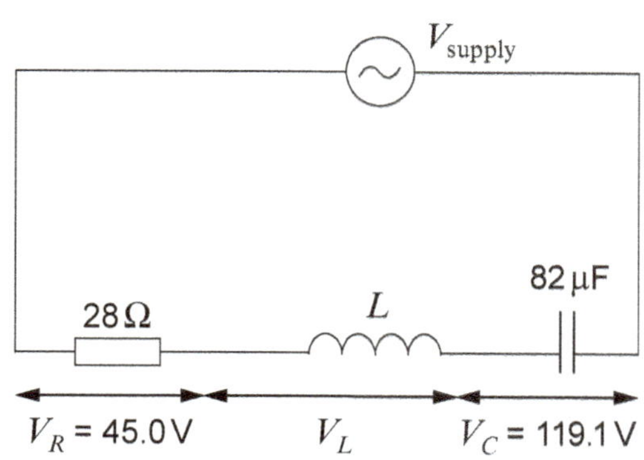

Calculate:

(i) the Q factor of the circuit; **[1]**

(ii) the rms current; **[1]**

(iii) the frequency of the power supply; **[2]**

(iv) the inductance of the inductor. **[2]**

(v) Sketch a graph of the instantaneous pd across the capacitor on the grid provided. (The instantaneous pd across the resistor is shown.) **[2]**

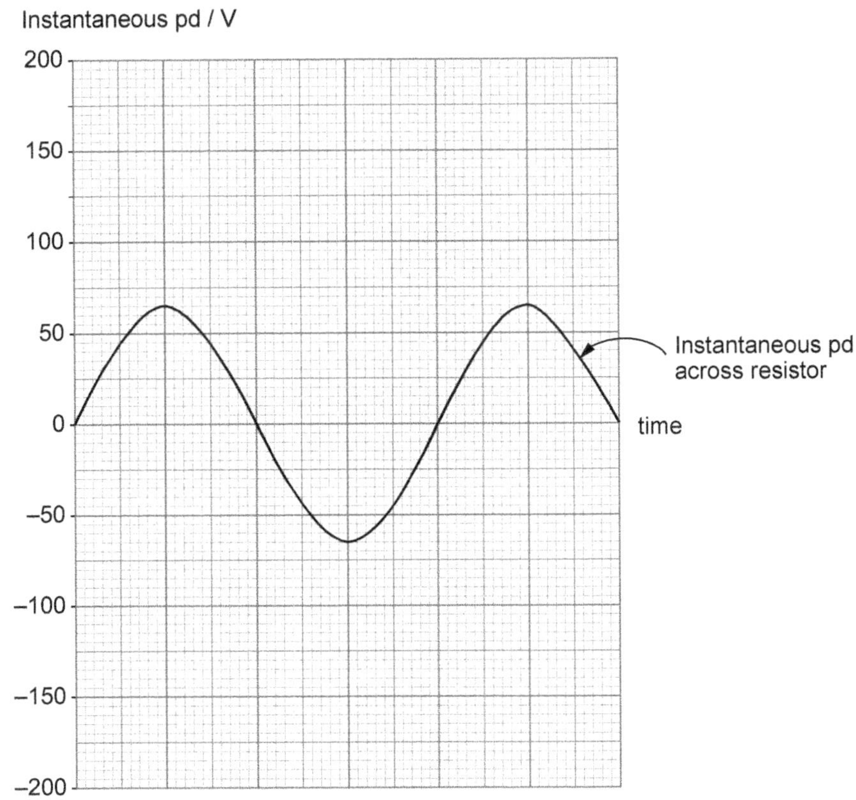

(vi) Without further calculation, explain why the current decreases when the frequency of the power supply is increased. **[2]**

(c) Sam claims that the output pd (V_{OUT}) in the following circuit is greater than 6.0 V when the frequency is greater than 20 kHz, but less than 6.0 V when below 20 kHz. Investigate whether or not Sam is correct. **[5]**

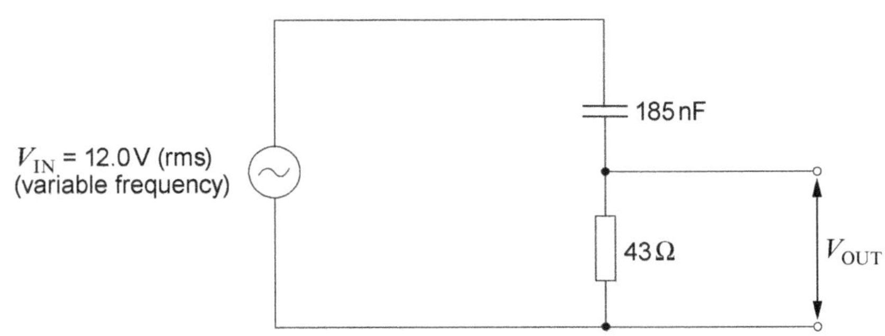

(Total 20 marks)

[*WJEC A level Physics 2019 Unit 4 Q6*]

Option B – Medical physics

2 **(a)** **(i)** Discuss how the properties of X-rays make them suitable for medical imaging. **[3]**

(ii) An X-ray machine has an operating pd of $18\ \mathrm{kV}$ and a current of $12\ \mathrm{mA}$.
If only 0.5% of the power is converted into X-rays find:

I. the velocity with which the electrons strike the target; **[2]**

II. the power of the emitted X-rays. **[2]**

(b) **(i)** The intensity, I, of an ultrasound beam decreases with the thickness, x, of a material according to the equation:

$$I = I_0 e^{-\mu x}$$

The half-value thickness, $x_{1/2}$ is the thickness of the material that reduces the intensity of an incident beam by 50%. Show that:

$$\mu x_{1/2} = \ln 2$$

[2]

(ii) The half-value thickness of muscle for ultrasound of frequency $1.0\ \mathrm{MHz}$ is $2.7\ \mathrm{cm}$. Determine the thickness of muscle required to reduce the intensity to 70% of its original value. **[3]**

(c) A patient has a suspected heart murmur possibly caused by a problem with their aortic valve. You have the choice of the following forms of medical imaging available:

MRI scan **ultrasound B-scan** **fluoroscopy** **CT scan** **X-ray**

Evaluate the effectiveness of each type of imaging in confirming the diagnosis. **[5]**

(d) **(i)** When discussing radiation exposure, medical scientists will mention the *absorbed dose, D, and the equivalent dose, H*. Explain the difference between these two terms. **[2]**

(ii) During treatment for a cancerous tumour using gamma radiation, a patient's lungs received an equivalent dose of $4\ \mathrm{mSv}$. If the weighting factor of lung tissue is 0.12 calculate the effective dose. **[1]**

(Total 20 marks)

[*WJEC A level Physics 2019 Unit 4 Q7*]

Option C – The physics of sports

3 This question is about the physics of the motion of an ice hockey puck, which is a hard rubber disc of mass $0.17\ \text{kg}$ and diameter 76 mm.

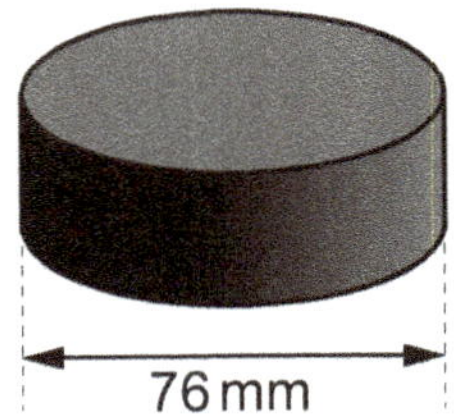

(a) When the disc is at room temperature, the coefficient of restitution between the puck and ice is 0.55.

(i) Explain what is meant by the statement 'the coefficient of restitution is 0.55'. **[2]**

(ii) When a puck is cooled from room temperature to $0\ °\text{C}$ its coefficient of restitution is reduced by 30%. Calculate the bounce height of a puck at $0\ °\text{C}$ when it is dropped from an initial height of $0.50\ \text{m}$ on to ice. **[4]**

(b) The image shows an ice hockey player taking a shot at goal.

(i) When striking the puck, the player changes its speed from $3\ \text{m s}^{-1}$ to $34\ \text{m s}^{-1}$ without changing its direction. The puck remains in contact with the hockey stick for $25\ \text{ms}$. Calculate the mean force exerted by the hockey stick on the puck. **[2]**

(ii) The player aims for the top corner of the goal, which is at a height of $1.2\ \text{m}$. The initial angle of the puck's motion is $8°$ to the horizontal and its speed is $34\ \text{m s}^{-1}$. Determine whether or not the puck ever exceeds the height of the goal.

Ignore the effect of air resistance. **[3]**

(iii) The spin rate of the puck at its maximum height is 14 revolutions per second. Calculate its **rotational** kinetic energy.

The moment of inertia of the puck is given by: $I = \dfrac{mr^2}{2}$ **[3]**

(iv) Wayne thinks that the answer to part (b)(iii) is actually the total kinetic energy of the puck at the maximum height. Determine whether Wayne is correct. **[2]**

(v) During the shot at goal, the puck is moving to the right. The diagram below shows the velocity of the air relative to the puck. During its flight, the velocity of the air above the puck is greater than that below it, creating lift.

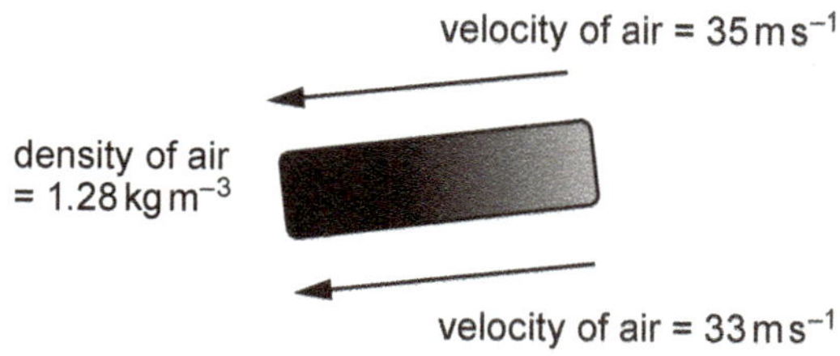

Use the Bernoulli equation to calculate the lift force on the puck and show that this is small compared with its weight. **[4]**

(Total 20 marks)

[*WJEC A level Physics 2019 Unit 4 Q8*]

Option D – Energy and the environment

4 (a) (i) State Archimedes' principle. **[1]**

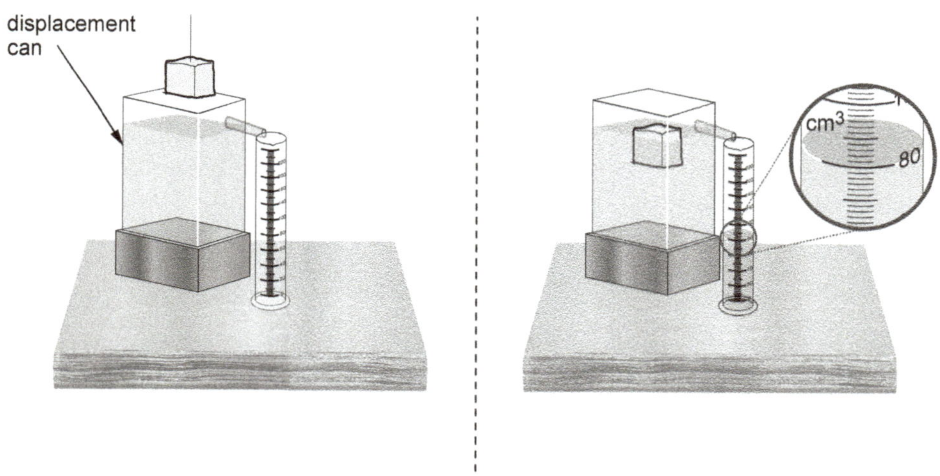

(ii) A student lowers a block of ice into a displacement can containing salt water. As shown in the diagram, $80\ \text{cm}^3$ of displaced salt water is collected in the measuring cylinder.

Calculate the mass of salt water displaced taking the density of salt water, $\rho_{\text{salt water}}$, to be $1030\ \text{kg m}^{-3}$. $[1\ \text{cm}^3 = 1 \times 10^{-6}\ \text{m}^3]$ **[2]**

(iii) The density of ice, ρ_{ice} , is $920\ \text{kg m}^{-3}$. Use this and the Archimedes' principle to determine the volume of the block of ice above the surface. **[3]**

(iv) The Greenland ice sheet is a vast body of ice covering roughly 80% of the surface of Greenland. It is estimated that Greenland's melting ice is releasing $2.2 \times 10^{11}\ \text{m}^3$ of water into the surrounding ocean every year.

I. If the surface area of oceans on Earth is $3.6 \times 10^{14}\ \text{m}^2$ calculate the rise in water levels per year from Greenland ice melting. **[1]**

II. Explain why the melting of the ice sheet would have a greater effect on global sea levels than the melting of the same mass of icebergs. **[2]**

III. It is believed that melting ice sheets may lead to further increases in global temperatures. Suggest a reason for this. **[1]**

(b) **(i)** Use an appropriate equation to show that the unit of the coefficient of thermal conductivity, K, is $\text{W m}^{-1}\text{ K}^{-1}$. **[2]**

(ii) The recommended thickness of loft insulation has changed over time. In 1985, a 100 mm layer of fibre glass loft insulation ($K = 0.041\text{ W m}^{-1}\text{ K}^{-1}$) was used to cover an area of 72 m^2 in the roof space of a house. During winter, the air temperature just above the insulation was $5\,°\text{C}$ and the temperature of the surface supporting the insulation was $20\,°\text{C}$. Calculate the rate of heat flow through the insulation. **[2]**

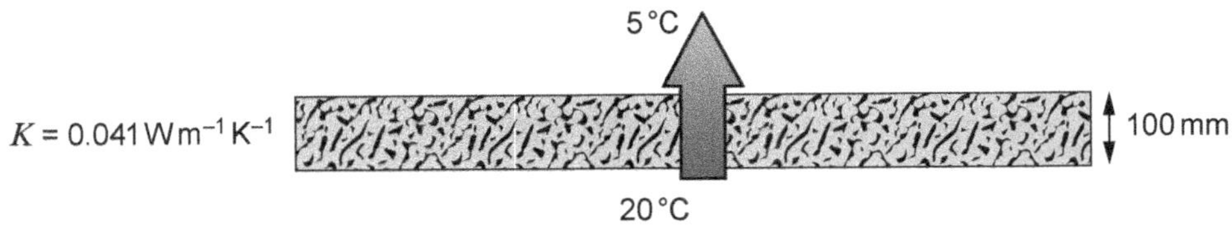

(iii) A present-day loft insulation manufacturer recommends a loft insulation thickness of **270 mm**. They suggest the house could achieve this modern standard by adding **170 mm** of cellulose loft insulation ($K = 0.035\text{ W m}^{-1}\text{ K}^{-1}$) and that it would reduce the rate at which energy is transferred by more than $60\,\%$.

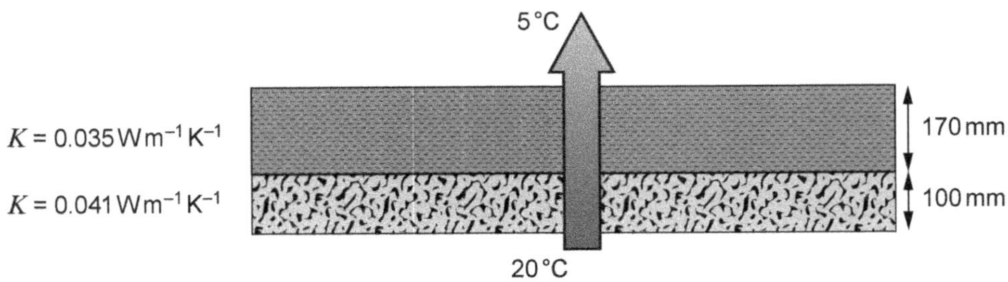

Again, taking the air temperature just above the insulation to be $5\,°\text{C}$ and the temperature of the surface supporting the insulation as $20\,°\text{C}$, investigate whether their recommendation is correct. **[4]**

(iv) Starting from the lower surface at $20\,°\text{C}$, sketch a graph of temperature against distance through the loft insulation on the axis provided. **[2]**

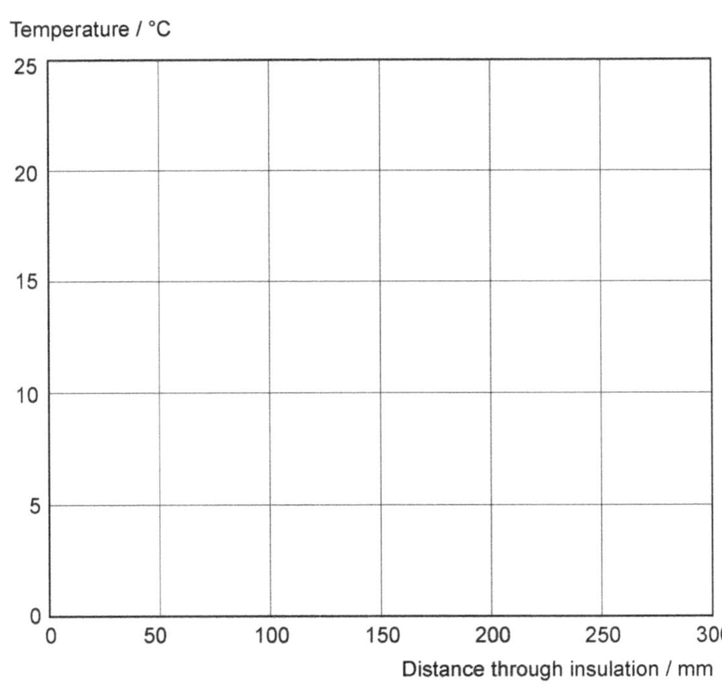

(Total 20 marks)

[*WJEC A level Physics 2019 Unit 4 Q9*]

Mathematical and data-handling skills

The content of Units 3 and 4 of the A level specification requires more advanced analytical skills than that of Units 1 and 2. There are two main areas of increased mathematical demand, which are developed in this chapter to be applied in the relevant sections of Units 3 and 4:

- The study of rotations and oscillations in Unit 3 and the optional topic of AC electricity requires a deeper understanding of trigonometrical functions.
- Exponential and logarithm functions are essential to the description of growth and decay, such as in the study of damped oscillations, radioactive emissions and the absorption of γ radiation. These two functions are also used in analysing data which involves power law and exponential decay.

Most of the mathematical skills and data-handling needed for A Level physics are the same as those for AS and are covered in the Year 1 & AS textbook. That book was written with A Level as well as AS students in mind, so the demands of the exercises and examination-style questions are at an appropriate level. However, there are some additional skills specified for A Level and some of the contexts in the non-AS section of the course require a deeper understanding of the skills that were covered at AS.

Most of the analytical work on orbits in Sections 4.2 and 4.3 is in terms of circles, but some understanding of ellipses is required. Section M2 gives a reference for a few terms.

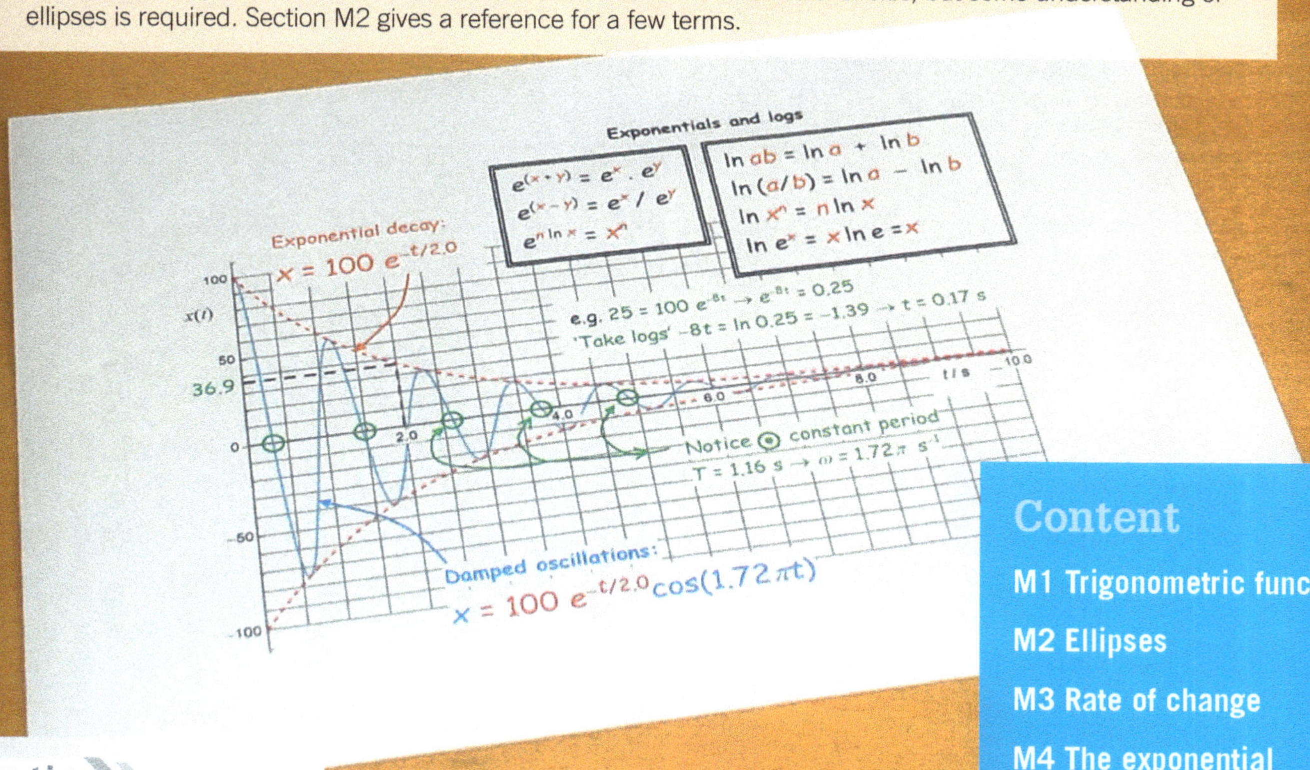

Top tip

As with the Year 1 book, this chapter is not a substitute for a maths textbook. For practice in maths techniques you should use a maths textbook, e.g. *Maths for A Level Physics* (Kelly & Wood).

Content

M1 Trigonometric functions

The trigonometric sine and cosine are used to model the behaviour of oscillating systems. It is important to know how to sketch and interpret them as well as to understand the relationships between them.

Maths tip

For oscillation calculations, your calculator should be set in radian mode.

Key terms

The **angular frequency**, ω, of an oscillation is $2\pi f$ where f is the frequency.

UNIT: s^{-1} or rad s^{-1}.

M1.1 The form of sine and cosine functions

Simple harmonic oscillating motions and alternating currents (and voltages) can be expressed in the form:

$$f(t) = A\cos(\omega t + \varepsilon) \qquad \text{and} \qquad f(t) = A\sin(\omega t + \phi).$$

Here $f(t)$ is the displacement, velocity, acceleration, current or pd expressed as a function of time, A is the amplitude of the oscillating quantity, ω the **angular frequency** and ε (and ϕ) are constant phase angles. [You should assume that angles ε and ϕ are expressed in radians unless otherwise told.]

We recall from Section 4.4.4 of the Year 1 & AS textbook, that the extreme values of the sine and cosine functions are ±1 and the function repeats every 2π radians change in ωt. Fig. M.1 shows these functions with ε and ϕ both zero.

M1 Knowledge check

Use Fig. M.1 to give three values of t for which:

(a) $A\cos\omega t = A$

(b) $A\cos\omega t = 0$

(c) $A\sin\omega t = A$

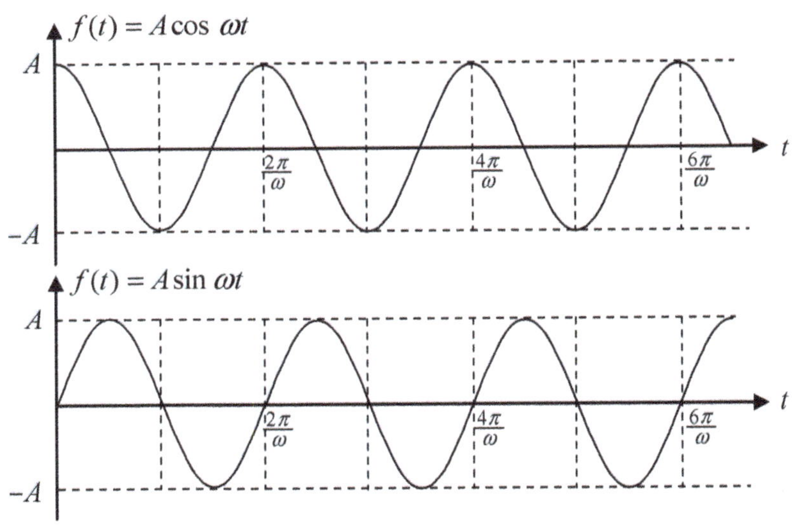

Fig. M.1 Sine and cosine functions with phase constant zero

M2 Knowledge check

The displacement, x, of an oscillating object is given by $x = A\cos\omega t$, where A is 12 cm and ω is 31.4 rad s^{-1}.

(a) State the amplitude of the oscillation.

(b) Calculate the frequency of the oscillation.

(c) Calculate the first two times after $t = 0$, for which $x = 0$.

What if the phase constant, ε in $f(t) = A\cos(\omega t + \varepsilon)$ is not zero? If $\varepsilon > 0$, the function $f(t) = A\cos(\omega t + \varepsilon)$ is the same as $f(t) = A\cos\omega t$ but shifted earlier in time by $\Delta t = \varepsilon/\omega$ (see Study point). A negative phase constant shifts the graph to the right. Alternative expressions for Δt are:

$$\Delta t = \frac{\varepsilon}{\omega} = \frac{\varepsilon}{2\pi f} = \frac{\varepsilon T}{2\pi}$$

Similarly, $f(t) = A\sin(\omega t + \phi)$ is the same as $f(t) = A\sin\omega t$, just shifted earlier in time by $\Delta t = \phi/\omega$.

Study point

$A\cos(\omega t + \varepsilon)$ is shifted **earlier** than $A\cos\omega t$ (if ε is positive) because a smaller value of t is now needed to give the same value of the argument of the cosine.

M3 Knowledge check

An alternating current, I, with I in mA, and t in s, varies as $I = 0.1\sin(1000t + 0.5)$

(a) State the angular frequency, ω.

(b) Calculate the frequency, f.

(c) Calculate the period, T.

(d) Calculate the times between $t = -T$ and $t = T$ for which $I = 0$.

M1.2 The relationship between the sine and cosine functions

(a) Phase

The shape of the cosine function is the same as that of the sine function; it is just shifted earlier in phase by $\frac{\pi}{2}$ (i.e. $\frac{1}{4}$ of a cycle), that is $T/4$ earlier in time. So we can write:

$$\cos \omega t = \sin (\omega t + \tfrac{\pi}{2}) \qquad \text{or, equivalently,} \qquad \sin \omega t = \cos (\omega t - \tfrac{\pi}{2})$$

This means that we can always write any sine function as a cosine function instead, with the appropriate phase angle.

We can go further than this. If we take a sine or cosine function and shift it along in either direction by $T/2$ in time, i.e. by π radians, we get the negative of the same function. That means

$$\cos (\omega t \pm \pi) = - \cos \omega t \qquad \text{and} \qquad \sin (\omega t \pm \pi) = - \sin \omega t$$

This gives us further flexibility in writing sines or cosines.

(b) Gradients

By applying calculus techniques, which are beyond the scope of this book, we can show that the gradient of a sine curve is a cosine curve and vice versa. More precisely:

$$\text{gradient of } \sin t = \cos t \qquad \text{and} \qquad \text{gradient of } \cos t = - \sin t$$

Multiplying t by ω in $\sin \omega t$ and $\cos \omega t$ squashes the oscillations horizontally by the factor ω so increases the gradient by the same factor. Multiplying by A, as in $A \sin \omega t$ and $A\cos \omega t$, stretches the oscillations vertically by the factor A, so also increases the gradient by the same factor. Hence:

$$\text{gradient of } A \sin \omega t = A\omega \cos \omega t$$

and

$$\text{gradient of } A \cos \omega t = -A\omega \sin \omega t$$

In studying simple harmonic motion we usually write the displacement, x, as:

$$x = A \cos \omega t, \qquad \text{or, with phase constant} \qquad x = A \cos (\omega t + \varepsilon)$$

The velocity is the gradient of the displacement–time graph, so

$$v = -A\omega \sin \omega t \quad \text{or} \quad v = -A\omega \sin (\omega t + \varepsilon)$$

And acceleration is the gradient of the velocity–time graph, so

$$a = -A\omega^2 \cos \omega t \quad \text{or} \quad a = -A\omega^2 \cos (\omega t + \varepsilon)$$

Example

The displacement (in cm) of a body of mass 0.002 kg is given by

$$x = 2.5 \cos (600t + 0.8),$$

Find the maximum magnitude of the resultant force on the body and the first time (> 0) when this occurs.

Knowledge check M4

Write the function for the alternating current in KC M3 as a cosine function.

Knowledge check M5

Two students come up with the following answers for an oscillation:

$x = 10 \cos (100t - 1.0)$

$x = 10 \sin (100t + 0.57)$

Can they both be right?

Stretch & challenge

Estimate the gradient of $\sin t$, at $t = 0$ by considering the interval -0.1 rad to $+0.1$ rad. Repeat this for smaller and smaller intervals, e.g.

-0.01 rad to $+0.01$ rad;

-0.001 rad to $+0.001$ rad...

How does this support the suggestion that the gradient of $\sin \theta$ is $\cos \theta$. Try repeating for other central values, e.g. $\theta = 0.5$ rad.

Study point

If $x = A\cos \omega t$ and $a = -A\omega^2 \cos \omega t$

then $a = -\omega^2 x$.

So the acceleration is directly proportional to the displacement and in the opposite direction. This kind of motion (SHM) arises when an object is displaced slightly from an equilibrium position.

Study point

It helps to be familiar with the values of θ for which trig functions are 0 and ±1. For example,

$\cos\theta = 1$ when θ is $\dots -2\pi, 0, 2\pi, 4\pi \dots$

$\cos\theta = -1$ when θ is $\dots -\pi, \pi, 3\pi, 5\pi \dots$

So $\cos\theta = \pm 1$ when $\theta = n\pi$

$\cos\theta = 0$ when θ is $\dots -\frac{\pi}{2}, \frac{\pi}{2}, \frac{3\pi}{2}, \frac{5\pi}{2} \dots$

Answer

From above, the acceleration $a = -2.5 \times 600^2 \cos(600t + 0.8)$ cm s^{-2}

$\therefore a_{max} = 2.5 \times 600^2 = 900\,000$ cm s^{-2} = 9000 m s^{-2}

$\therefore$ By N2: $F_{max} = m\,a_{max} = 0.002$ kg × 9000 m s^{-2} = 18 N

This occurs when $\cos(600t + 0.8) = \pm 1$, i.e. when $600t + 0.8 = n\pi$, where $n = 0, \pm 1$ (see Study point).

$\therefore t = \frac{n\pi - 0.8}{600}$, so the smallest $t > 0$ is when $n = 1$,

i.e. $t = \frac{\pi - 0.8}{600}$ s = 3.9 ms.

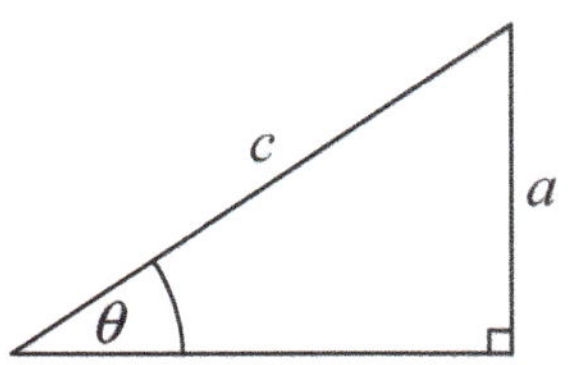

Fig. M.2 Right-angled triangle

(c) Pythagoras' theorem in terms of trig functions

Consider the angle θ in the right-angled triangle in Fig.M.2

Pythagoras' theorem: $a^2 + b^2 = c^2$

Divide by c^2: $\frac{a^2}{c^2} + \frac{b^2}{c^2} = 1$

But $\frac{a}{c} = \sin\theta$ and $\frac{b}{c} = \cos\theta$, so $\sin^2\theta + \cos^2\theta = 1$

A common application for this equation in simple harmonic motion is the relationship between velocity and displacement:

If the phase constant is 0: $x = A\cos\omega t$ and $v = -A\omega\sin\omega t$

$\sin\omega t = \pm\sqrt{1 - \cos^2\omega t}$ $\quad\therefore v = \pm A\omega\sqrt{1 - \frac{x^2}{A^2}}$

which can be simplified to $v = \pm\omega\sqrt{A^2 - x^2}$

M1.3 Graphs of $\sin^2\theta$ and $\cos^2\theta$

A body undergoing simple harmonic motion has a velocity which varies with time as a sinusoid (i.e. a sine or cosine function). Its kinetic energy therefore varies as the square of this. The potential energy is also the square of a sinusoid (it is the area under the F–x graph which is $\frac{1}{2}kx^2$). The graphs of $\sin^2\theta$ and $\cos^2\theta$ are sinusoidal in shape themselves. The graphs in Fig. M.3 show how $\cos^2\theta$ is related to $\cos\theta$.

Study point

Notice that the period of the $\cos^2\theta$ graph is half that of the $\cos\theta$ graph; $\cos^2\theta$ is never less than 0 (why not?). The mean value of $\cos^2\theta$ is 0.5.

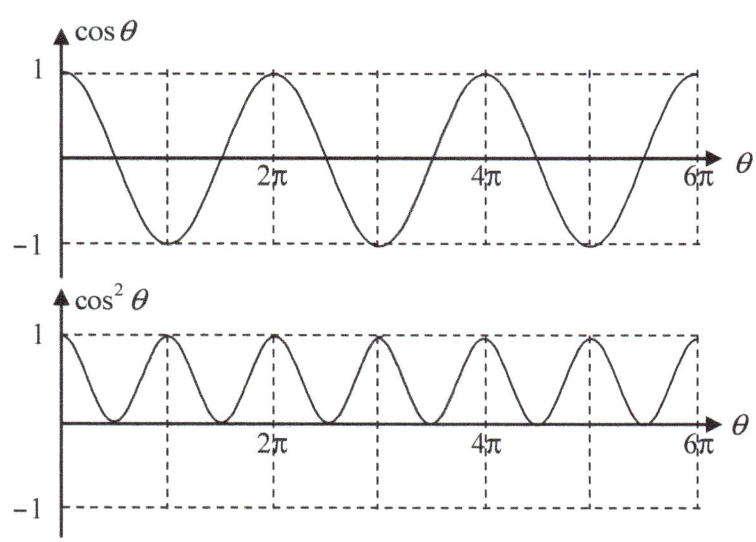

Fig. M.3 The relationship between $\cos\theta$ and $\cos^2\theta$

M6 Knowledge check

Sketch a graph of $y = 2\sin\pi t$ for t between 0 and 6 seconds. Sketch $y = 4\sin^2\pi t$.

Arrange the graphs like Fig. M.3 to show the connection between them.

M2 Ellipses

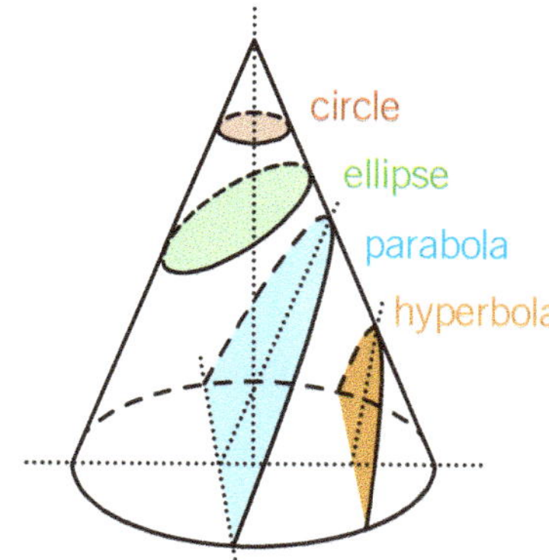

Fig. M.4 Conic sections

Ellipses belong to a family of curves called conic sections because they can all be produced by cutting circular cones in different ways – see Fig. M.4. They are important because orbits around a single object are always conic sections. This includes the orbits of planets, minor planets and comets about the Sun and satellites (moons) about planets. Objects from outside the Solar System move past the Sun in hyperbolic paths.

Fig. M.5 represents an object, P in an elliptical orbit about the Sun.

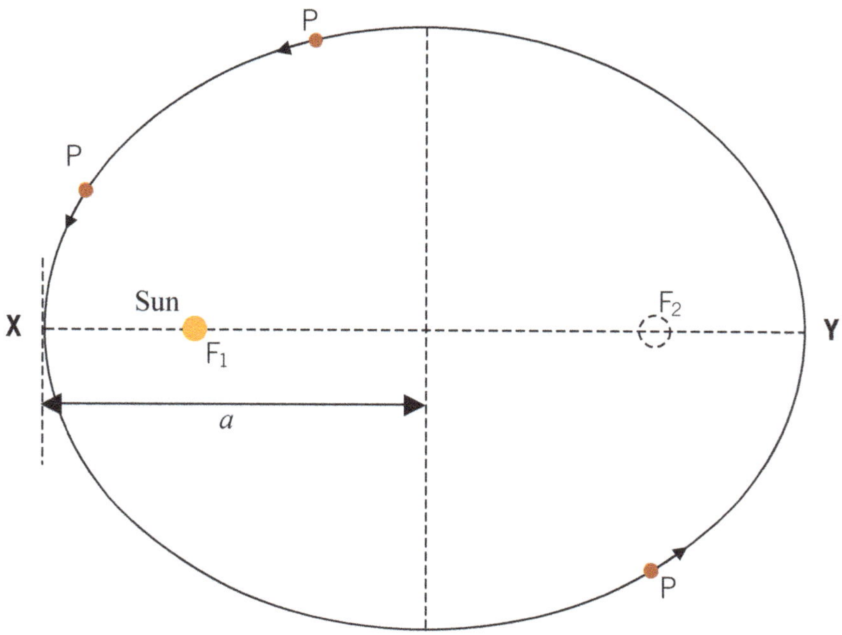

Fig. M.5 Elliptical orbit about the Sun

The line XY is called the major axis of the ellipse: it is the longest straight line that can be drawn within the ellipse. Kepler's third law refers to the length of the semi-major axis, which is labelled a.

Note that the orbits of the planets are almost circular. For example, the near and far points of the Earth in its orbit are 147 and 152 million kilometres from the Sun respectively.

Study point

The period, T, and the total energy, E, of a body of mass m in orbit around a body of mass M are given by

$T^2 = \frac{4\pi^2 a^3}{GM}$ and

$E = -\frac{1}{2}\frac{GMm}{a}$

where a is the length of the semi-major axis of the orbit.

M3 The exponential function

This function is often referred to as the *growth function*. There are many physical situations in which a variable gets larger or smaller (usually smaller) at a rate which is proportional to the value of that variable.

Study point

Notice that both these examples are for decrease rather than growth. There **are** examples of exponential growth but only for a short time, e.g. a runaway nuclear explosion. The approximate Moore's law in computing is also an example.

Examples

- Radioactive decay: the activity of a sample of a radioactive nuclide (i.e. the number of decays per second) is proportional to the number, N, of undecayed nuclei in the sample. The activity is the rate of *decrease* of N, so

  $\frac{dN}{dt} = -\lambda N$, where λ is the *decay* constant.
- Capacitor discharge: the rate of *decrease* of charge, Q, on a capacitor is proportional to the charge on the capacitor,

  $\frac{dQ}{dt} = -\frac{Q}{RC}$ where C is the capacitance.

M3.1 Rate of change

In Section M1, we wrote 'the gradient of' several times and we now consider how to write 'the gradient of' mathematically.

The mean rate of change of the variable x between points **A** and **B** in Fig. M.6 is given by the gradient of the **chord** **AB**.

Hence: Mean rate of change between **A** and **B** is $\frac{\Delta x}{\Delta t}$.

The *instantaneous* rate of change of x at any point, e.g. **A**, is the gradient of the tangent at that point and is written $\frac{dx}{dt}$. Usually (as in Fig. M.6) this rate of change varies with time, i.e. it is itself a function of time. Note that x can be any variable. It could be displacement, current, velocity, activity of a radioactive material..... So in these cases the rate of change would be written $\frac{dx}{dt}$, $\frac{dI}{dt}$, $\frac{dv}{dt}$ and $\frac{dA}{dt}$ respectively.

so, if $x = A \cos \omega t$, $v = \frac{dx}{dt} = -A\omega \sin \omega t$, etc.

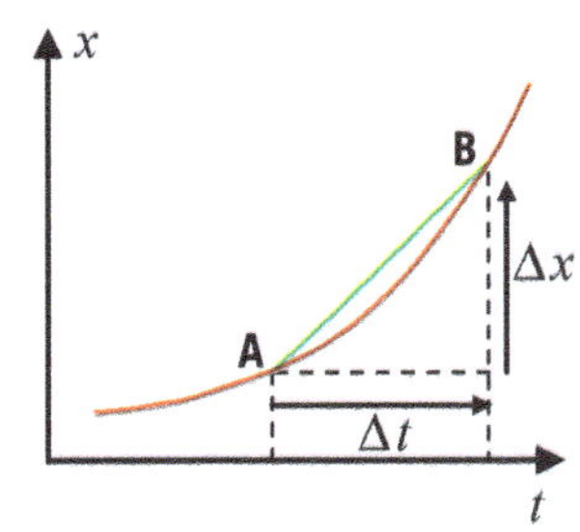

Fig. M.6 Rate of change

M3.2 The exponential growth function

The red line in Fig. M.7 is the graph of the function $y = e^x$ where e is a constant, sometimes called Euler's constant (after the Swiss mathematician) but usually just referred to as e. The value of e is 2.718 28 (to 6 sf); like π, it is an irrational number. There is a special button on your calculator to calculate e^x; for reasons that will become clear later, it often shares the button with $\ln x$.

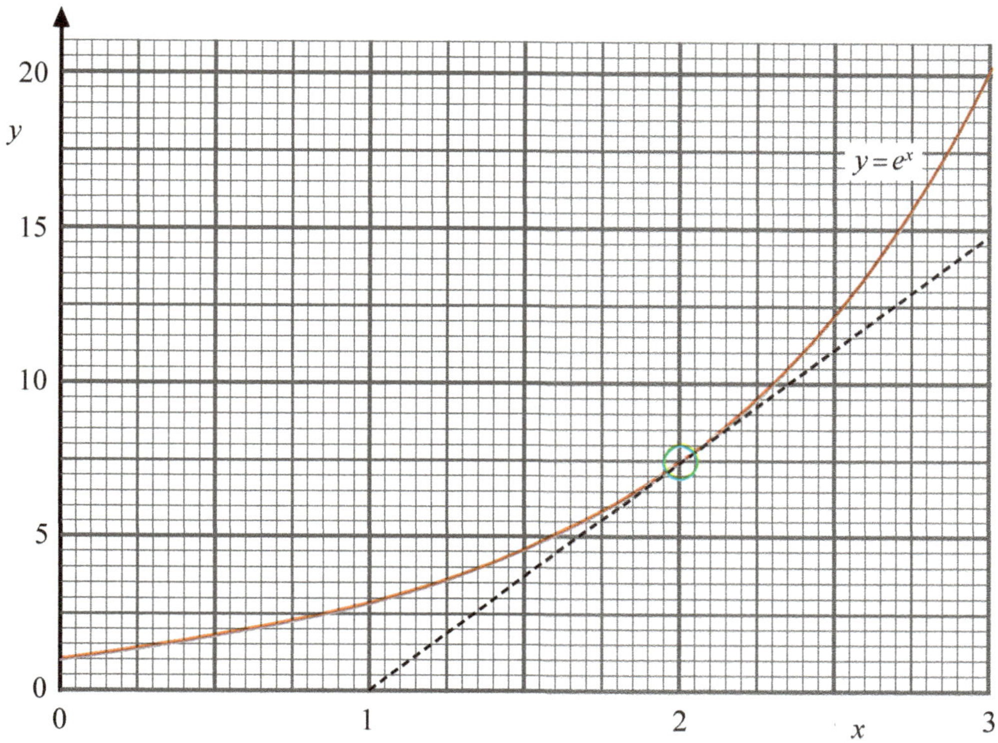

Fig. M.7 Exponential growth function $y = e^x$

To see what is special about the function $y = e^x$, have a look at the point in the green circle, with $x = 2$:

- The value of the function, i.e. $y = e^x = 7.39$ (to 3sf). Check this on your calculator.
- The dotted line is the tangent at $x = 2$.
- The gradient of the tangent, which you should also check, is $\frac{14.8}{2.0} = 7.4$, by eye.
- Hence, to within the uncertainty of reading the graph, $\frac{dy}{dx} = y$.

This is not an accident. Use the graph to check: Find the gradient of the tangent at $x = 2.5$. It should be equal to $e^{2.5}$. [NB: you should find that the tangent crosses the x axis at 1.5.]

So we have a function whose rate of growth is equal to its value. Why is this useful? Because there are quite a few growth situations where the rate of growth of a variable is proportional to the size of the variable, e.g. an uncontrolled fission reaction in ^{235}U, the number of yeast cells in a sugar solution. *Proportional* to, not *equal* to. We can easily fix that: Whilst we are doing so, we'll change the variables to x and t, because we are usually interested in how things grow over time, and write:

$$x = Ae^{\lambda t},$$

where A and λ are positive constants.

Maths tip

Calculators have an e^x key. (It is often the 'Inverse' or 'Shift' key plus the 'ln' key.) Pressing this key with an entry of 1 should give 2.71828 … . Try it.

Knowledge check M7

(a) Use your calculator to find $e^0, e^1, e^{1.5}, e^2, e^3, e^{4.605}, e^5, e^{10}, e^{100}$

(b) What is the relationship between e^5 and e^{10}?

(c) What is the relationship between e^{10} and e^{100}?

Knowledge check M8

(a) Use your calculator to find $e^{-1}, e^{-1.5}, e^{-2}, e^{-3}, e^{-4.605}, e^{-5}, e^{-10}, e^{-100}$

(b) What is the relationship between e^5 and e^{-5}?

(c) What is the relationship between e^{-10} and e^{-100}?

In the same way the gradient of $A\sin\omega t$ is $A\omega\cos\omega t$, the gradient of $x = Ae^{\lambda t}$ is $A\lambda e^{\lambda t}$, i.e. it is $\lambda \times$ the value of the function. The constant A is the value of the function at $t = 0$.

The important feature of the exponential growth function, $x = Ae^{\lambda t}$, is that the time interval, Δt, for the functions to double (or triple, or $\times 2.5$, $\times 10^2$, etc.) is a constant. To see what this means, try KC M10.

M9 Knowledge check

The number of bacteria in a colony increases from 100 to 10 000 in 4 hours. Assuming exponential growth, calculate:

(a) the number of bacteria in 8 hours

(b) [slightly tricky] the number of bacteria in 10 hours

(c) how long it would take for the number to reach 10^{10}.

M3.3 The exponential decay function

We'll now modify our exponential function to reflect the fact that A level Physics situations which use this function involve decay (radioactivity, capacitor discharge, damped oscillations) rather than growth. We modify the function by simply introducing a '–' sign into the function, e.g. in radioactive decay, if the initial number of radioactive nuclei is N_0 we have:

$$N = N_0 e^{-\lambda t}$$

The activity, $A = \lambda N$ (see Study point) so we can also write:

$$A = A_0 e^{-\lambda t}$$

Study point

For radioactive decay:

$A = -\frac{dN}{dt} = -[-\lambda N_0 e^{-\lambda t}]$,

because the activity is the rate of loss of radioactive nuclei.

So $A = A_0 e^{-\lambda t}$,

where $A_0 = \lambda N_0$

For an example, Fig. M.8 is a decay curve for a sample of iodine-123, which is used as an imaging tracer to investigate the thyroid gland.

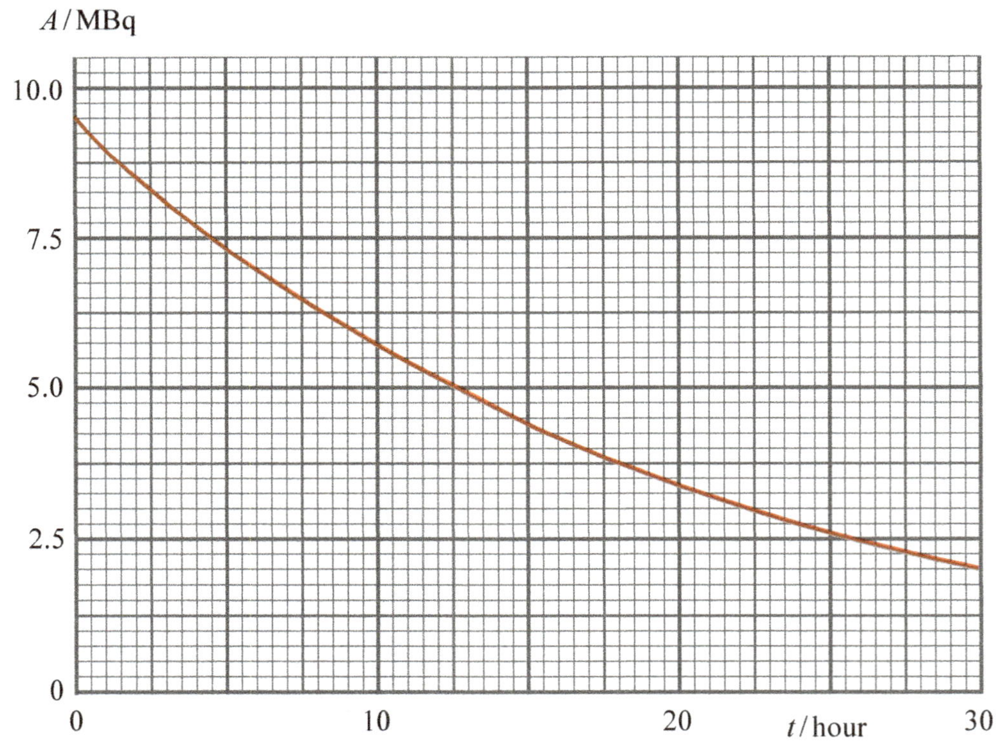

Fig. M.8 Exponential decay of iodine-123 $x = e^{-\lambda t}$

M10 Knowledge check

Use Fig. M.7 to determine the increase in x that doubles the value of e^x.

[Hint: For accuracy it is better to ask, for example, what Δx multiplies e^x by 16 (= 2^4) and divide that by 4.]

M11 Knowledge check

Use Fig. M.7 to determine the value of Δx that increases the value of e^x by a factor of 2.7. What is special about 2.7?

The characteristic feature of the exponential decay function is that the time taken for the function to fall by any given factor is always the same. In radioactivity, we often refer to the *half-life*, which is the time taken to fall by half. You should be able to use Fig. M.8 to show that the half-life of ^{123}I is a little more than 13 hours.

λ and the decay curve

The decay constant, λ, is related to the speed of decay. In a time, t, such that $\lambda t = 1$, the value of $e^{-\lambda t}$ falls to $e^{-1} \sim 0.37$ of its original value. The decay constant has units of time^{-1} (usually s^{-1}, but it could be hour^{-1}, year^{-1}, etc.) and is related to the half-life, $t_{\frac{1}{2}}$, by:

$$t_{\frac{1}{2}} = \frac{\ln 2}{\lambda} = \frac{0.69}{\lambda}.$$

We'll meet ln, the logarithm function, in the next section.

Knowledge check M12

Use Fig. M.8 to determine the decay constant, λ, for the radioactive decay of ^{123}I. Express your answer (a) in day^{-1}, (b) in s^{-1}.

Capacitor discharging

If a capacitor discharges through a resistor, (see Fig. M.9), Q, I and V, all decay with time in the same way:

$$Q = Q_0 e^{-\lambda t} \qquad I = I_0 e^{-\lambda t} \qquad V = V_0 e^{-\lambda t} \qquad \text{where} \qquad \lambda = \frac{1}{RC}$$

Why is this? We have three relevant equations:

$$I = \frac{V}{R} \qquad V = \frac{Q}{C} \qquad \text{and importantly} \qquad I = \frac{dQ}{dt}$$

which can be combined to give:

$$\frac{dQ}{dt} = -\frac{1}{RC}Q$$

Hence, $Q = Q_0 e^{-\lambda t}$ and the other two equations follow.

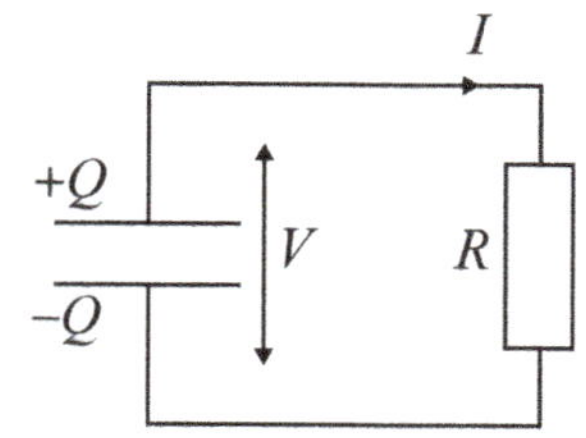

Fig. M.9 Capacitor discharge

Example

A capacitor discharges through a 2.2 kΩ resistor.

A student measures the time taken for the pd across the capacitor to drop from 10.0 V to 3.7 V to be 12.0 s. Calculate the capacitance, C of the capacitor.

Answer

To drop to 37% of the original pd, $\frac{1}{RC}t = 1$

$\therefore C = \frac{12.0 \text{ s}}{2.2 \times 10^3 \ \Omega} = 5.5 \text{ mF}$

Note: The figure of 37% in the example was a special case. To deal with other values we need logarithms, which is the topic of the next section.

Study point

The decay constant, λ, is commonly used for radioactive decay. For a discharging capacitor, the time constant, τ, is commonly referred to instead, defined as RC. The decay equations are written as:

$Q = Q_0 e^{-t/RC}$; $I = I_0 e^{-t/RC}$; $V = V_0 e^{-t/RC}$

The time constant is the time taken to decay to e^{-1} (≈ 0.37 or 37%) of the original value.

Knowledge check M13

For the capacitor decay in the **Example**:

(a) State the time constant of the decay (see Study point).

(b) What will be the pd after a **further** 24 s?

M4 Logarithms

Study point

We normally write $\ln(x)$ without the brackets, i.e. $\ln x$, unless there would be confusion.

We write the log of $a + b$ as $\ln(a + b)$ but $\ln 2a$ would be taken to mean $\ln(2a)$ just as $\sin 2x$ means $\sin(2x)$.

M4.1 Definition of the 'natural' logarithm function

Just as $\sin^{-1}$ is the inverse function to the sine function, meaning that $\sin^{-1}(\sin\theta) = \theta$ or $\sin(\sin^{-1}\theta) = \theta$, the *natural logarithm* function, ln, (or *logarithm to the base e*), is the inverse function to the exponential function, e^x: i.e.

$$\ln(e^x) = x \qquad \text{and} \qquad e^{\ln x} = x.$$

Remember these defining equations. Because $\ln x$ is the inverse function of e^x it is often on the same calculator key, with one of the functions needing the inverse key. Finding the logarithm of a number is often referred to as, 'taking logs'.

M14 Knowledge check

Evaluate:

(a) $\ln 2$,

(b) $\ln e^2$, i.e. $\ln(e^2)$,

(c) $e^{\ln 5}$,

(d) $2 \ln 10$,

(e) $3e^{2\ln 12}$.

You are (for the moment) allowed to use your calculator for (a) and (e). If you must, use your calculators for the others then try and explain the answer!

Example

From KC M8 we know that $e^{4.605} = 100$. Write this in terms of the logarithm function.

Answer

'Taking logs' of both sides of the equation:

$$\ln(e^{4.605}) = \ln 100$$

But from the definition of the logarithm function, $\ln(e^{4.605}) = 4.605$

$\therefore \ln 100 = 4.605$

Study point

Check the examples in properties 1–5 of Section M4.2. They do work. Make sure you know how the author was sure they would, e.g. $125 = 5^3$

M4.2 Properties of logs

You need to know the following properties of logs:

1	$\ln ab = \ln a + \ln b$	e.g. $2 \times 5 = 10$, so	$\ln 10 = \ln 2 + \ln 5$
2	$\ln \frac{a}{b} = \ln a - \ln b$	e.g.	$\ln 6 = \ln 30 - \ln 5$
3	$\ln \frac{1}{x} = -\ln x$	e.g.	$\ln 5 = -\ln 0.2$
4	$\ln x^n = n \ln x$	e.g	$\ln 125 = 3 \ln 5$
5	$\ln(\sqrt[n]{x}) = \frac{1}{n}\ln x$	e.g.	$\ln 100 = \frac{\ln 1\,000\,000}{3}$

M15 Knowledge check

Given that $\ln x = 5$,

(a) Write down $\ln x^2$.

(b) Use your calculator to find x.

Example

Work out the answer to KC M14 (e) without using a calculator.

Answer

$2 \ln 12 = \ln 12^2 = \ln 144$ — Rule 1 (or 4)

$e^{\ln 144} = 144$ — Definition of ln

$\therefore 3e^{2\ln 12} = 3 \times 144 = 432$

M4.3 Logs to the base 10

The logarithm function can be, and often is, defined using 10 as the base. The log to the base 10 of x is written $\log_{10}x$ or just $\log x$. The definitions are
$\log_{10}(10^x) = x$ and $10^{\log x} = x$.

Notice that we have left the subscript $_{10}$ off the second one. It is optional. The properties 1–5 from Section M.4.2 apply equally to the log function as to the ln, e.g.
$\log ab = \log a + \log b$.

M4.4 Using logs to find unknown exponents

If an unknown quantity is the exponent in an equation, we solve the equation by taking logs, as the following two examples show.

Example 1

Find x if $2^x = 35$

Answer

Taking logs gives: $\ln 2^x = \ln 35$

But $\ln 2^x = x \ln 2$ $\quad \therefore x \ln 2 = \ln 35$

$\therefore x = \dfrac{\ln 35}{\ln 2} = 5.13$ (to 3 sf)

Knowledge check M16

You can also solve Example 1 by taking logs to the base 10. Do this and show that you get the same answer.

Example 2

The intensity, I, of a parallel beam of X-rays decreases with distance, x, into a material according to the equation $I = I_0 e^{-\lambda x}$, where λ is a constant called the attenuation coefficient. Find λ if the intensity after 8.0 cm is 10% of the initial value.

Answer

At $x = 8.0$ cm, $I = 0.1I_0$. $\therefore 0.1 = e^{-\lambda \times 8.0\,\text{cm}}$

Taking logs: $\ln 0.1 = \ln(e^{-\lambda \times 8.0\,\text{cm}})$

So from the definition of the ln function: $\ln 0.1 = -\lambda \times 8.0$ cm.

$\therefore \lambda = \dfrac{\ln 0.1}{-8.0\text{ cm}} = \dfrac{-2.30}{-8.0\text{ cm}} = 0.29\text{ cm}^{-1}$ (to 2 sf)

Knowledge check M17

Find n if $15^n = 2.5$

Knowledge check M18

The pd, V, across a capacitor falls with time t according to the equation $V = V_0 e^{-\lambda t}$, where V_0 is the initial pd. Show how we can use logs to make λ the subject of the equation.

M5 Using logs to linearise functions

M5.1 Graphs of exponential functions

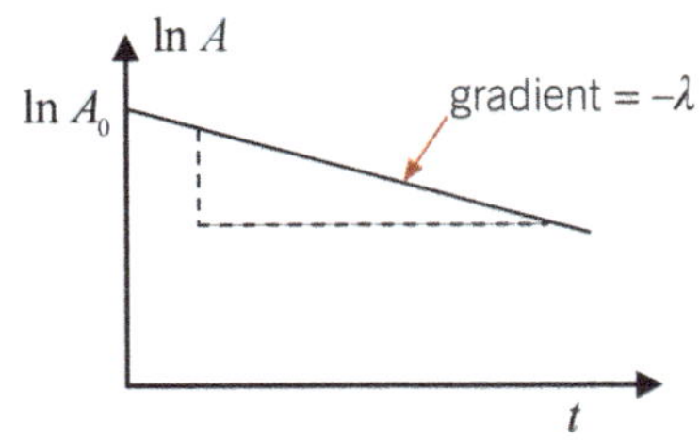

Fig. M.10 Semi-log graph

Logs can be used to transform a decay curve into a linear graph, which is: (a) easier to draw, (b) enables the suspected exponential form of the decay to be verified and (c) allows the decay constant to be determined.

We start from the suggested decay equation, e.g. for the activity, A, of a radioisotope: $A = A_0e^{-\lambda t}$.

We take (natural) logs of this equation: $\ln A = \ln (A_0e^{-\lambda t})$

Applying Rule 1 from Section M4.2 $\ln A = \ln A_0 + \ln e^{-\lambda t}$

Applying the definition of natural logs $\ln A = \ln A_0 - \lambda t$

Comparing this with a linear equation $y = mx + c$

The red arrows indicate the variables and the blue arrows the constants in the equations. Hence, if the suggested equation is correct, a graph of $\ln A$ against t will be a straight line of gradient $-\lambda$ with an intercept of $\ln A_0$ on the $\ln A$ axis.

Key terms

A graph of the logarithm of a variable against a linear variable is called a **semi-log graph** or a **log-linear graph**.

A graph of the logarithm of one variable against the logarithm of another variable is called a **log-log graph**.

M5.2 Graphs of power functions

Sometimes we suspect that two variables, say x and y are connected by a power function,

$$y = Ax^n$$

but we don't know what value n should have, possibly because we have no working theory. Taking logs can be used to find n.

Taking logs of the putative equation $\ln y = \ln A + n \ln x$

Comparing this with a linear equation $y = mx + c$

In the same way as for the **semi-log graph**, we see that, if the suggested equation is correct, a graph of $\ln y$ against $\ln x$ will be a straight line. Its gradient will be n. The intercept on the $\ln y$ axis (i.e. the value of $\ln y$ when $\ln x = 0$) is $\ln A$. [Note: you can also use log10 instead of ln, when solving this kind of equation.]

Note: Often, when plotting a **log-log graph**, the data points are a long way from the origin. This means that the equation of the line needs to be calculated using one of the techniques in Section 4.5.4 of the AS textbook. See also Question 10 in the Test yourself exercise on the next page.

M19 Knowledge check

Suppose two variables, x and y, are connected by the equation

$$y = 10x^3.$$

What will be the gradient of a graph of $\ln y$ against $\ln x$? What will be the intercept on the $\ln y$ axis?

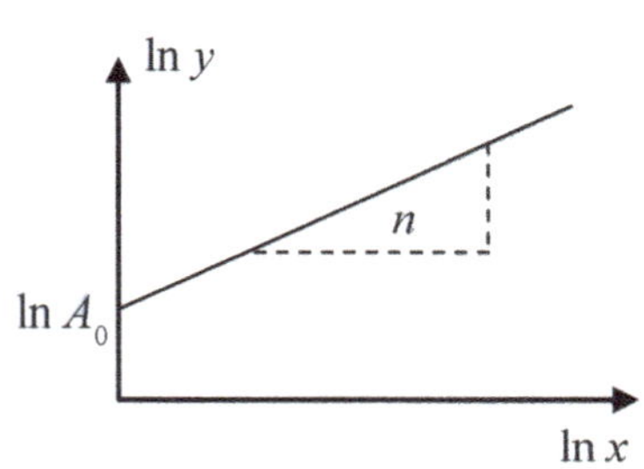

Fig. M.11 log-log graph

Test yourself M

1. Express the following functions as functions of the form $\pm\sin \omega t$ or $\pm\cos \omega t$.

   (a) $\sin(\omega t - \frac{\pi}{2})$ (b) $\cos(\omega t + \frac{3\pi}{2})$ (c) $\cos(\omega t + 2\pi)$ (d) $\sin(\omega t - \frac{3\pi}{2})$ (e) $\sin(\omega t - \pi)$

2. Express the following functions in the form $\cos(\omega t \pm \varepsilon)$.

   (a) $-\cos \omega t$ (b) $-\sin \omega t$ (c) $\sin \omega t$ (d) $-\cos(\omega t - \frac{\pi}{2})$ (e) $\sin(\omega t - \frac{\pi}{2})$

3. A particle oscillates with its displacement, x, varying with time (in seconds) by $x = 5 \cos 100\pi$, with t in s. Calculate the times, between -20 ms and $+20$ ms for which (a) $x = 0$, (b) $x = 5$, (c) $x = 2.5$ and (d) $x = -2.5$.

4. Give the times for the same displacements as in Q3 if the particle now oscillates with time (in s) according to the function $x = 5 \cos(100\pi t + \frac{\pi}{2})$.

5. An alternating electric current varies with time, in seconds, as $I = 3.5\cos(800\pi t - \frac{\pi}{4})$.

   (a) Calculate the frequency, f, and the period, T.
   (b) Calculate the value of I at (i) $t = 0$, (ii) $t = 0.625$ ms and (iii) $t = 1.25$ ms.
   (c) Calculate the times between $t = 0$ and $t = 5$ ms when the current is instantaneously zero.
   (d) Calculate the times between $t = 0$ and $t = 5$ ms when $I = +3.5$ A.

6. Use your calculator to find the values of (a) $e^{3.5}$, (b) $4\,e^{-2.5}$ (c) $\ln 25$, (d) $\frac{6}{7 \ln 5}$, (e) $2.5\ln(e^{2.5} - 6)$.

7. Use the definition and rules of natural logs to simplify and evaluate the following expressions:
   (a) $\ln(e^{-1})$, (b) $5\ln(e^{3})$, (c) $e^{\ln 6}$, (d) $e^{3\ln 6}$, (e) $4e^{(\ln 2 + \ln 3)}$.

8. Solve the following equations for x:
   (a) $25 = 10\ln x$, (b) $0.6 = 8\ln(2x - 3)$, (c) $6 = 4\ln\left(\frac{1}{1-x}\right)$, (d) $80 = 500e^{-0.2t}$, (e) $4 \times 10^{-6} = 25e^{-0.001t}$.

9. The pd, V, across a capacitor falls with time t, in s, according to the equation $V = V_0 e^{-\lambda t}$, where V_0 is 12.0 V and λ is $0.005\ \text{s}^{-1}$. Calculate (a) the pd after 5 minutes and (b) the time taken for the pd to fall to 1.0 V.

10. The illuminances (light-sensor readings), E_v, at a fixed distance from a filament lamp, for different supply voltages, V, are as follows:

V / V	6.0	8.1	9.4	10.1	11.1	12.1	12.9	13.4
E_v / lux	20	58	99	127	172	221	274	307

    The background illuminance, i.e. with the lamp off, is 2 lux. It is suspected that the E_v from the lamp depends on V by a relationship of the form $E_v = kV^n$ for some unknown values of k and n. By plotting a log-log graph, show that this is approximately correct and use your graph to determine values of n and k.

11. A capacitor discharges through a 4.7 kΩ resistor. In 10 s the current, I, decreases from 8.0 mA to 1.5 mA.

    (a) Determine the gradient of a graph of $\ln I$ against time.
    (b) Calculate the capacitance of the capacitor given that the decay equation is $I = I_0 e^{-t/RC}$.

Knowledge check answers

Unit 3

3.1

3.1.1 1.3 rad s^{-1}

3.1.2 (a) 0.052 cm

(b) Shorter wavelength

3.1.3 (a) 1145 N

(b) 877 N

3.1.4 Circumference = $2\pi r$

$\therefore \text{AB} = \frac{1}{4} 2\pi \times 5 = \frac{5\pi}{2}$

$\therefore \text{Time} = \frac{\frac{5\pi}{2}}{20} = \frac{\pi}{8}$ s

3.1.5 For $\pm\frac{\pi}{6}$ rad, $\Delta t = \frac{\pi}{12}$ s

$\Delta v = 2 \times 20 \sin\frac{\pi}{6} = 20$ m s^{-1}

$\rightarrow \langle a \rangle = \frac{20}{\frac{\pi}{12}} = 76.4$ m s^{-2}

3.1.6 80 m s^{-2}

3.1.7 $v_{max} = 6.3$ m s^{-1}

3.1.8 Because the required centripetal force and the maximum grip are both proportional to the mass

3.1.9 1.4 m s^{-1}

3.1.10 $g = 0.0027$ m s^{-2}

3.1.11 $r\omega^2 = 385 \times 106 \times \left(\frac{2\pi}{27.3 \times 86400}\right)^2$

$= 0.0027(3)$ m s^{-2}

Very close!

3.1.12 $r\omega^2 = 6731 \times 103 \times \left(\frac{2\pi}{86400}\right)^2$

$= 0.0356$ m s^{-2}

3.2

3.2.1 (a) $\Delta F_A = 4$ N

(b) $\Delta F_B = -6$ N

(c) $F_{res} = 10$ N (to the left)

3.2.2

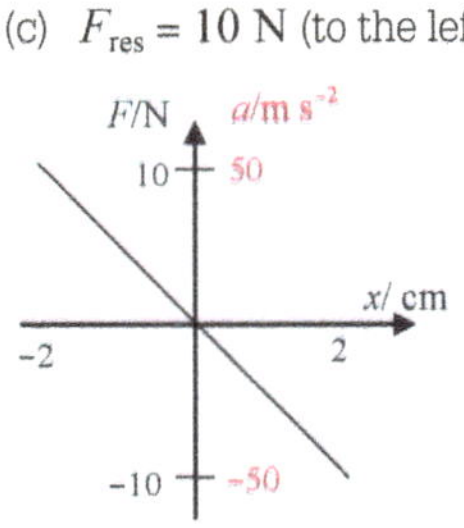

3.2.3 $[k] = \text{N m}^{-1} = \text{kg m s}^{-2} \text{ m}^{-1} = \text{kg s}^{-2}$

$[m]$ = kg

$\left[\sqrt{\frac{k}{m}}\right] = \sqrt{\left[\frac{k}{m}\right]} = \sqrt{\frac{\text{kg s}^{-2}}{\text{kg}}} = \sqrt{\text{s}^{-2}} = \text{s}^{-1}$

3.2.4 (a) 0, -2π, 0, 2π, 0 [in m s^{-1}]

(b) $-4\pi^2$, 0, $4\pi^2$, 0, $-4\pi^2$ [in m s^{-2}]

(c) 2.0 s

3.2.5 (a) (i) 12, 26, 40 [all in μs]

(ii) 5, 33

(iii) 19, 47

(b) 5, 19, 33 47

(c) 12, 26, 40

3.2.6 $v_{max} = 450$ m s^{-1}

$a_{max} = 1.0 \times 10^8$ m s^{-2}

3.2.7 The 'answers' which can be ruled out are: $\varepsilon = 2.020$ and -4.264

3.2.8 x / cm = 0.2 cos (220(t / ms) – 1.12)

3.2.9

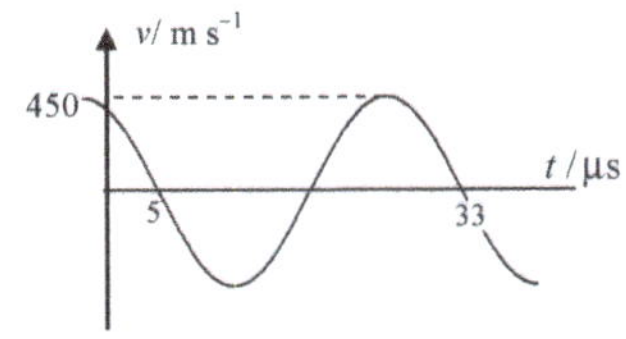

3.2.10 $\omega = 10$ s^{-1}, $f = 1.59$ Hz, $T = 0.628$ s

3.2.11 (a) 0.039 m

(b) 2.5 Hz

3.2.12 0.99 m

3.2.13 Tension greatest at bottom because upward centripetal force greatest

∴ Length greatest at the bottom

3.2.14 Immediately and after each ½ cycle after release. $F_{max} = 1.3$ N

3.2.15 (a) $E_{total} = 250$ J

(b) $E_p(0) = 250$ J, $E_k(0) = 0$

(c) $E_p\left(\frac{\pi}{200}\right) = 0$, $E_k\left(\frac{\pi}{200}\right) = 250$ J

3.2.16 (a) $1000\pi = 3100$ s^{-1} (2 sf)

(b) 300 s^{-1} (2 sf)

3.2.17 (a) 0.70

(b) Measure velocity using the gradient where $x = 0$ a whole number of cycles apart. then calculate the ratio of v^2

3.2.18 The amplitude should come down to (every 2.0 ms) 7.3 cm, 5.3 cm, 3.9 cm, 2.8 cm, 2.1 cm

3.2.19 Initial gradient = 0

3.2.20 Over-damping – takes a long time to close

Under-damping – continues to swing or hits the stop with a bang

3.2.21 1.3 m s^{-1} and 3.0 m s^{-2}

3.2.22 Oven frequency ~ 0.000 02 × absorption peaks

3.2.23 $T^2 = \frac{4\pi^2}{g} l + \frac{4\pi^2}{g} r$ where r = radius

Gradient the same, intercept = $\frac{4\pi^2}{g} r$

3.3

3.3.1 (a) 1.99×10^{-14} J = 124 keV

(b) 1.32×10^{-18} J = 8.2 eV

3.3.2 1.17×10^{-26} kg

3.3.3 4.0 cm

3.3.4 6.97×10^{-26} kg [propene, C_6H_6]

3.3.5 (a) (i) $M(\text{O}) = 16$ g mol^{-1} $= 0.016$ kg mol^{-1}

(ii) $M(\text{O}_2) = 32$ g mol^{-1} $= 0.032$ kg mol^{-1}

(b) 32

(c) 32 u

(d) 5.3×10^{-26} kg

3.3.6 $R = \frac{pV}{nT}$, ∴ for homogeneity,

$[R] = \frac{\text{N m}^{-2}\text{ m}^3}{\text{mol} \times \text{K}}$

$= \text{N m mol}^{-1}\text{ K}^{-1}$

$= \text{J mol}^{-1}\text{ K}^{-1}$

3.3.7 (a) 1 mol occupies 18 cm^3

∴ Volume per molecule

$= \frac{18 \text{ cm}^3 \text{ mol}^{-1}}{6.02 \times 10^{23} \text{mol}^{-1}}$

$= 2.99 \times 10^{-23}$ m^3

$\therefore \frac{4}{3}\pi\left(\frac{d}{2}\right)3 = 2.99 \times 10^{-23}$ cm^3

Leading to $d = 390$ pm

(b) ~ 5 nm

3.3.8 (a) 3.5 mm

(b) 5.5×10^{-19} m

3.3.9 (a) Higher speeds make a proportionately greater contribution to Σv^2

(b)

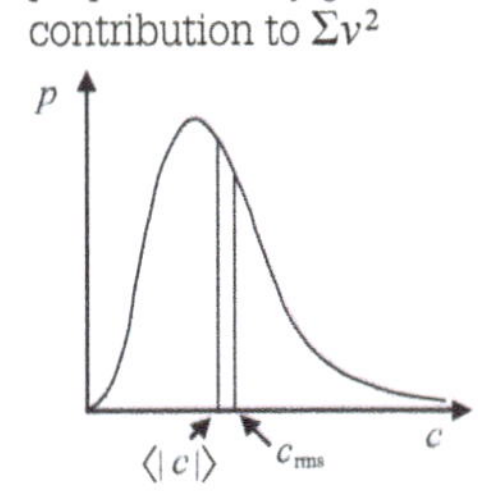

3.3.10 (a) 80.8 mol

(b) 4.86×10^{25} molecules

3.3.11 $k = \frac{R}{N} = \frac{8.31 \text{ J mol}^{-1}\text{ K}^{-1}}{6.02 \times 10^{23}\text{ mol}^{-1}} = 1.38 \times 10^{-23}$ JK^{-1}

3.4

3.4.1 (a) 11.2 kJ

(b) 3.4 MJ

3.4.2 1.025×10^5 Pa

3.4.3 (a) 0.25 m^3

(b) 75 kJ

(c) $Q = 125$ kJ

3.4.4 Equal. Same amount of gas at the same temperature – both monatomic.

3.4.5 $U_{He} = 38U_{Ar}$

3.4.6 (a) $(pV)_C = 4.8 \times 10^5$ J, $(pV)_B = 3.2 \times 10^5$ J, $(pV)_A = 4.0 \times 10^5$ J

$pV \propto T$ hence the result

(b) 0.8

3.4.7 1.28 kg

3.4.8 Electrical work output = VIt

Heat output = I^2rt

Change in internal energy = $-EIt$

3.4.9 (a) $\Delta U > 0$

(b) $Q > 0$

(c) $W < 0$

(d) $\Delta U < 0$

(e) $\Delta U > 0$

3.4.10 $\Delta U = 2490$ J, $W = 0$, $Q = 2490$ J

3.4.11 $W_{out} = 84$ kJ (to 2 sf); $W_{in} = 42$ kJ

∴ Net work = (84 – 42) kJ = 42 kJ

3.4.12 $T_A = T_B = 600$ K

$T_C = T_D = 300$ K

3.4.13 $U_A = U_B = 90$ kJ

$U_C = U_D = 45$ kJ

3.4.14 $\Delta U = 45$ kJ; $W = 84$ kJ

$\therefore Q = \Delta U + W = 129$ kJ

3.4.15 67%

3.4.16 –264 ± 14 °C

3.5

3.5.1 ${}^{235}_{92}U \rightarrow {}^{231}_{90}Th + {}^{4}_{2}He$

3.5.2 ${}^{106}_{52}Te \rightarrow {}^{102}_{50}Sn + {}^{4}_{2}He$

3.5.3 (a) Baryon number (1)
Charge (0)
Lepton number (0)

(b) Neutrino involvement: only feels the weak interaction

Change of quark flavour: d → u

3.5.4 Nucleus with atomic number Z–1 and mass number A; positron ${}^{0}_{1}e$; electron neutrino ${}^{0}_{0}\nu$ [The positron subsequently annihilates with an electron to produce to γ photons.]

3.5.5 (a) 1 MeV

(b) 1.8×10^{-30} kg

(c) 0.0011 u

3.5.6 (a) A unchanged; Z → Z – 1

(b) $A \rightarrow A - 1$; Z → Z – 1

3.5.7 ${}^{11}_{7}N \rightarrow {}^{10}_{6}C + {}^{1}_{1}H$

3.5.8 (a) 50 000

(b) There are more atoms per μm in a solid than in a gas

3.5.9 (a) 0.34

(b) $(0.34)^5 = 0.0045$

3.5.10 Because it loses more energy in passing through any atom

3.5.11 Concentration in aluminium is 0.4 times that in lead.

3.5.12 24.7 (3 sf)

3.5.13 1 – 11: 24.3

2 – 12: 24.0

3 – 13: 24.8

4 – 14: 24.9

5 – 15: 24.5

6 – 16: 23.6

7 – 17: 23.6

8 – 18: 23.8

9 – 19: 24.8

10 – 20: 25.1

Similar spread

3.5.14 The β^+ tracks would have a similar spread of curvatures to the β^- but be oppositely directed (same direction as α)

3.5.15 3.65×10^{10} Bq [36.5 GBq]

Note: Nowadays the curie is defined as 37 GBq

3.5.16 $5.0 \times 10^{-9}\ s^{-1}$

3.5.17 (a) 1.39×10^8 s = 4.39 years

(b) 4.1 GBq [4.1×10^9 Bq]

3.5.18 $T_{400} - T_{200} = 50$ s

From $C_{140} = 100$ s and using

$C = C_0 2^{-x} \rightarrow 50.5$ s

From $C_{140} = 100$ s and using

$C = C_0 e^{-\lambda t} \rightarrow 50.6$ s

3.5.19 (a) $\lambda = 0.086\ day^{-1} = 9.9 \times 10^{-7}\ s^{-1}$

(b) Using 0.086 day^{-1}

$A = 3 \times 10^6 e^{-0.086 \times 21} = 497$ kBq (~500)

3.5.20 $10 \times 10^{-3} = 5 \times 10^6 \times 2^{-n} \rightarrow n = 8.97$

→ 225 min

3.5.21 830, 690, 160

3.5.22 $\left(\frac{11}{12}\right)^n = \frac{1}{2}, \therefore n\ln\left(\frac{11}{12}\right) = \ln 2 \rightarrow n = 7.967$

3.5.23 $[\varepsilon]$ = cm, $[k]$ = $cm^2\ s^{-1}$

3.5.24 Gradient = $\frac{1}{\sqrt{k}}$, intercept = $\frac{\varepsilon}{\sqrt{k}}$

$\therefore \varepsilon = -\frac{\text{gradient}}{\text{intercept}}$

3.6

3.6.1 44.01 g

3.6.2 2.5×10^{-12} kg

3.6.3 (a) 4.29×10^{-12} J

(b) 27 MeV

~ 2 million times as much

3.6.4 (a) Mass loss = 1.0 kg × 0.71%

$= 7.1 \times 10^{-3}$ kg

∴ Energy release = mc^2

$= 7.1 \times 10^{-3} \times (3.0 \times 10^8)^2$

$= 6.4 \times 10^{14}$ J

(b) 6000 (1 sf)

3.6.5 Figures give 931.49 MeV

3.6.6 Mass of $6{}^1H + 6n = 12.09894$ u

∴ Mass deficit = 0.09894 u

∴ BE = 0.09894 × 931 MeV

= 92.1 MeV

3.6.7 (a) Mass deficit = 0.09894 u

$= 1.64 \times 10^{-28}$ kg

$\therefore BE = 1.64 \times 10^{-28}\ kg \times (3.0 \times 10^8\ m\ s^{-1})^2$

$= 1.49 \times 10^{-11}$ J

(b) 1 MeV = 1.60×10^{-13} J

∴ BE = 92.4 MeV

3.6.8 (a) 263 MeV

(b) 8.5 MeV nuc^{-1}

3.6.9 (a) ${}^{2}_{1}H$, ${}^{6}_{3}Li$, ${}^{11}_{5}B$, ${}^{21}_{10}Ne$

(b) Only one particle in the nucleus

3.6.10 4.2×10^{-13} J, 2.6 MeV

3.6.11 (a) $\lambda = 0.00122\ s^{-1}$,

$A = \frac{1 \times 10^{-6}}{27} \times 6.02 \times 10^{23} \times 0.00122$

$= 2.72 \times 10^{13}$ Bq

$\therefore P = 2.72 \times 10^{13} \times 4.2 \times 10^{-13}$ W

= 11.4 W

(b) $P = 230$ μW

3.6.12 (a) ${}^{180}_{74}W \rightarrow {}^{176}_{72}W + {}^{4}_{2}He$

(b) $\Delta m = 179.946\,70 - 175.941\,51 - 4.002\,604$

= 0.002 586 u

$\therefore E = 0.002\,586 \times 931$ MeV

= 2.41 MeV

$= 3.9 \times 10^{-13}$ J

3.6.13 If $r = kA^{1/3}$, then mass = $\frac{4}{3}\pi k^3 A\rho$

$\therefore \rho = \frac{3 \times 1.66 \times 10^{-27} A}{4\pi \times (1.1 \times 10^{-15})^3 A}$

$= 3.0 \times 10^{17}\ kg\ m^{-3}$

3.6.14 Mass of 92 ${}^{1}_{1}H$ atoms = 92.719 9 u

Mass of 143 neutrons = 144.239 095

∴ Total = 236.958 995

∴ Mass deficit = 1.915 065 u

= 1783 MeV

∴ Binding energy per nucleon = $\frac{1783}{235}$

= 7.6 MeV nuc^{-1}

3.6.15 (a) Mass of a nucleon ~ 1.67×10^{-27} kg

∴ Nucleons in 1 kg ~ 6.0×10^{26}

(b) Energy release = $6.0 \times 10^{26} \times 0.9$ MeV

$= 5.4 \times 10^{26}$ MeV

$= 8.6 \times 10^{13}$ J

3.6.16 4.78 MeV

3.6.17 $r_{H2} = 1.1\sqrt[3]{2} = 1.39$ fm

$r_{H3} = 1.1\sqrt[3]{3} = 1.59$ fm

∴ Total separation = 1.39 + 1.59 fm ~ 3 fm

Unit 4

4.1

4.1.1 Because otherwise the backs of the conductor would be in contact

4.1.2 F = $kg^{-1}\ m^{-2}\ s^4\ A^2$

4.1.3 68 nF = 0.068 μF

4.1.4 30 V. Check the polarity

4.1.5 0.025 m^2

4.1.6 (a) 55 μF, (b) 13.2 μF

4.1.7 (a) Q = 180 nC; V = 8.2 V
(b) Q = 320 nC; V = 6.8 V

4.1.8 (a) Parabola
(b) 0.48 μF = 480 nF
(c) Probably not – could be a film capacitor

4.1.9 ~ 4 minutes

4.1.10 (a) 33.0 s
(b) 15.5 μC
(c) 46 s

4.1.11 (a) 10 V
(b) 6 MΩ

4.1.12 (a) Gradient = −0.85 μC s^{-1}
(b) Current [0.85 μA discharge current at 2.0 s]
(c) 0.084 μA [allow ±0.02]

4.1.13 (a) 0.71 s
(b) 0.35 s
(c) Charge decay time = 2 × energy decay time

4.1.14 Transferred as internal energy in the connecting wires

4.1.15 (a) 3.3 s
(b) 7.8 V

4.1.16 1.5 MV m^{-1} downwards

4.1.17 2.7 μN

4.1.18 The acceleration is constant, so the mean velocity is half the final velocity

4.1.19 3.34×10^{-7} s; 3.67 m

4.1.20 (a) 2920 V
(b) 2.5×10^{-9} s
(c) 6.4×10^{15} m s^{-2}
(d) 1640 V
(e) 1.6×10^{7} m s^{-1}
(f) 3.6×10^{7} m s^{-1} at 0.464 rad (26.6°)

4.1.21 (a) In each discharge: Energy supplied
$= \frac{1}{2}C(V^2 - (1.00\ \text{V})^2)$
∴ Total energy supplied
$= 5 \times \frac{1}{2}C(V^2 - (1.00\ \text{V})^2)$
Energy to raise temp of coil
$= mc(\theta_2 - \theta_1)$
Hence
$= 5 \times \frac{1}{2}C(V^2 - (1.00\ \text{V})^2)$
(b) 3.5 F

4.2

4.2.1 $\text{F m}^{-1} = \text{C V}^{-1}\text{ m}^{-1}$
$= \text{C J}^{-1}\text{ C m}^{-1}$
$= \text{C (N m)}^{-1}\text{ C m}^{-1}$
$= \text{C}^2\text{ N}^{-1}\text{ m}^{-2}$
Other methods are available

4.2.2 (a) 8.2×10^{-8} N
(b) 2.65×10^{-11} m

4.2.3 22 pC

4.2.4 (a) Gradient = – 5.0 V m^{-1}
(b) 5.0 V m^{-1} – in agreement

4.2.5 (a) Assume it behaves as a complete sphere
$Q = -1.1 \times 10^{-6}$ C
(b) 670 kV m^{-1} towards the centre

4.2.6 0.021 m

4.2.7 0 (zero) – because the potential is 0 at B

4.2.8 (a $E_x = -24.4$ V m^{-1}
(b) $E_y = -19.2$ V m^{-1} downwards
(c) $E = 31$ V m^{-1} at 218° anticlockwise from x

4.2.9 $W = 1.2 \times 10^{-13}$ J
Positive as C is closer to B than A

4.2.10 864 V m^{-1} horizontally to the right

4.2.11 ΔV = 108 V

4.2.12 Electric field strength is greater

4.2.13 1.2×10^{36}

4.2.14 8.3×10^{-8} N

4.2.15 3.7 N kg^{-1}

4.2.16 17

4.2.17 7.5×10^{30} kg

4.2.18 Gradient = 0.016
Value read off at ~ 0.015 (close)

4.2.19 1370 m s^{-1}

4.2.20 $\frac{x}{y} = 32.36$, $x = 7.55 \times 10^{11}$ m,
$y = 2.3 \times 10^{10}$ m

4.2.21 From the ratio of distances to the neutral point, $\frac{M_1}{M_2} = 9$

4.3

4.3.1 $\frac{E_P(\text{P})}{E_P(\text{A})} = \frac{9.3}{0.65} \sim 14$

4.3.2 (a) $E_P = -7.1 \times 10^{21}$ J
(b) $E_k = 5.9 \times 10^{21}$ J
(c) $E_{Tot} = -1.3 \times 10^{21}$ J

4.3.3 $v \propto \frac{1}{\sqrt{r}}$

4.3.4 0.39 AU

4.3.5 Change of velocity in the same direction as the applied force

4.3.6 8.6×10^{25} kg

4.3.7 -1.28×10^{21} J (in agreement)

4.3.8 2130 m s^{-1}

4.3.9 (a) gradient = $\sqrt{GM_\odot}$, intercept = 0
(b) gradient = $GM_\odot$, intercept = 0
(c) gradient = $-\frac{1}{2}$, intercept = $\frac{1}{2}\ln GM_\odot$

4.3.10 (a) 28 500 ly
(b) 8.7 kpc

4.3.11 Applying $v^2 = \frac{GM}{r}$ gives 157 km s^{-1}
i.e. 160 km s^{-1} (2 sf)

4.3.12 $\lambda_{\text{measured}} = 656.939 \pm 0.002$ nm

4.3.13 Subst for r_2 in $m_1r_1 = m_2r_2$
$\rightarrow m_1r_1 = m_2(d - r_1)$
Rearranging: $(m_1 + m_2)r_1 = m_2d$
$\therefore r_1 = \frac{m_2d}{m_1 + m_2}$ QED

4.3.14 4900 km from centre of the Earth

4.3.15 (a) 2.0 million km
(b) 9.6×10^{29} kg and 6.4×10^{29} kg

4.3.16 Ignoring 15° → 3.5% error.
Uncertainty in velocity = 7%

4.3.17 (a) $r^3 = \frac{GMT^2}{4\pi^2} \rightarrow 1.69 \times 10^{30}$ m^3
$\therefore r = 1.19 \times 10^{10}$ m
And
$v = \frac{2\pi \times 1.19 \times 10^{10}\ \text{m}}{1.0 \times 10^6} = 74.7$ km s^{-1}
(b) $v_S = 102$ m s^{-1}
(c) $m_C = 1.36 \times 10^{27}\text{kg} = \frac{1}{740} m_s$

4.3.18 (a) 1.19×10^{10} m
(b) $v_s = 102$ m s^{-1}
$r_s = 16.2 \times 10^6$ km
(c) 1.36×10^{27} kg

4.3.19 $m_A = 1.09 \times 10^{31}$ kg
$m_B = 1.74 \times 10^{31}$ kg

4.3.20 Duration = 47 000 s
0.008% dimming
Assumption: exactly in plane of orbit.

4.3.21 67.9 km s^{-1} Mpc^{-1}
$= \frac{67.9 \times 10^3\ \text{m s}^{-1}}{3.085\ 68 \times 10^{22}\ \text{m}}$
$= 2.20 \times 10^{-18}$ s

4.3.22 (a) 6.9×10^{24} m (b) 220 Mpc

4.3.23 $t_H = 4.55 \times 10^{17}$ s
$= \frac{4.55 \times 10^{17}\ \text{s}}{3.16 \times 10^7\ \text{s year}^{-1}}$
$= 1.44 \times 10^{10}$ years

4.3.24 The wavelength has doubled so the universe has doubled in size, i.e. the light has been on its way for 7×10^9 years. Hence the current distance of the source is 14×10^9 years.

4.4

4.4.1 Like poles repel: the plotting compass N

pole, which indicates the direction of the field line, points away from the N pole of the magnet

4.4.2 Field lines must get closer together across region 2

4.4.3 Upwards ↑; to the right →

4.4.4 0.042 T

4.4.5 (a) 0 (zero)

(b) 2.4 mN upwards

(c) 4.8 mN upwards

(d) 3.4 mN downwards

4.4.6 160 A

4.4.7 $r = \dfrac{mv}{Bq} = \dfrac{\sqrt{2E_k m}}{Be}$

$= \dfrac{\sqrt{2eVm}}{Be} = \sqrt{\dfrac{2Vm}{B^2 e}}$

Inserting the data

$\rightarrow r = 0.070$ m = 70 mm

This is within the error, so consistent.

Note that there are other ways of showing this, e.g. by calculating the electron speed from its KE, then substituting.

4.4.8 After n gaps, $E_k = nqV$

$\therefore \frac{1}{2}mv^2 = nqV$

$\therefore v = \sqrt{\dfrac{2nqV}{m}} = w\sqrt{n}$

4.4.9 (a) 4.57 MHz

(b) 0.65 m

4.4.10 $A = 24$

4.4.11 (a) Accelerates parallel to the field

(b) Remains at rest

4.4.12 1.1×10^6 m s^{-1}

4.4.13 In silicon electrons move much faster for the same current, so the magnetic deflection force is much greater

4.4.14 e.g. Mains AC circuit or discharging capacitor

4.4.15 Fields due to currents on opposite sides of the coil reinforce at P but oppose at Q

4.4.16 (a) 28 μT at 045°

(b) 14 μN at 090°

4.4.17 5.3 A

4.4.18 4.7 μWb

4.4.19 Slightly smaller – same flux over a larger area because field lines spread out

4.4.20 1.1 μN downwards (↓)

4.4.21 Forces due to the magnetic fields on opposite sides of the tube are attractive, thus pulling the tube inwards

4.4.22 (a) 8.2 mV

(b) Corrected V_H (0.5 A) = 41.6 mV

Corrected V_H(1.0 A) = 82.8 mV

$\therefore V_H(1.0) \sim 2V_H(0.5)$, so proportional

Using 83 mV for 1 A → 30 V T^{-1}

4.4.23 gradient $= \dfrac{\mu_0 I}{2\pi}$

4.4.24 They are (a) horizontal and (b) equal and opposite

4.4.25 $B = 21 \pm 2$ mT [accept 21.0 ± 1.6 mT]

4.5

4.5.1 (a) PQ and SR – not cutting flux

QR – outside the field

(b) Down the page, P⟶S

(c) Up the page, S⟶Q

4.5.2 From Faraday's law:

$[\mathscr{E}] = \left[\dfrac{\Delta N\Phi}{\Delta t}\right] = \dfrac{[B]\,[A]}{[t]}$

$\therefore [B] = \text{V s m}^{-2}$

$\therefore [Blv] = \text{V s m}^{-2}\text{ m m s}^{-1} = \text{V}$

$= [\mathscr{E}]$ QED

4.5.3 22 mT

4.5.4 (a) 0.94 V

(b) There would be an equal and opposite emf induced in the connecting wires

4.5.5 $\langle \mathscr{E} \rangle = 0.99$ μV

4.5.6 1.03 mV

4.5.7 (a) $P = B\ell vI$

(b) $F_{motor} = BI\ell$

$\therefore P = F_{motor} \times v = BI\ell v = B\ell vI$ QED

4.5.8 Into page

4.5.9 (a) The flux is decreasing but the rate of change of flux magnitude is increasing. So $|\mathscr{E}|$ is increasing

(b) Force on side XY is up the page to oppose change. So by FLHR current is out of page

Opposite for WZ

4.5.10 (a) 8.0 V

(b) Faraday's law [of electromagnetic induction]: when the magnetic flux linking a circuit changes an emf is induced in the circuit proportional to the rate of change of flux linkage.

4.5.11 Eddy currents are induced in the copper pipe which produce a magnetic field which (by Lenz's law) opposes the relative motion of the magnet and pipe

4.5.12 $[J] = \text{A m}^{-2}$; $[\rho] = \Omega\text{ m} = \text{V A}^{-1}\text{ m}$

$\therefore [J^2\rho] = (\text{A m}^{-2})^2\text{ V A}^{-1}\text{ m}$

$= \text{V A m}^{-3}$

$= \text{W m}^{-3}$

$= [\Pi]$

∴ Homogeneous

4.5.13 30 ms

4.5.14 It increases at the rate of 1.5 A s^{-1}

4A

A1 30 Wb-turn

A2 Peak voltage = 470 V

V = 470 cos (100πt + ε)

A3

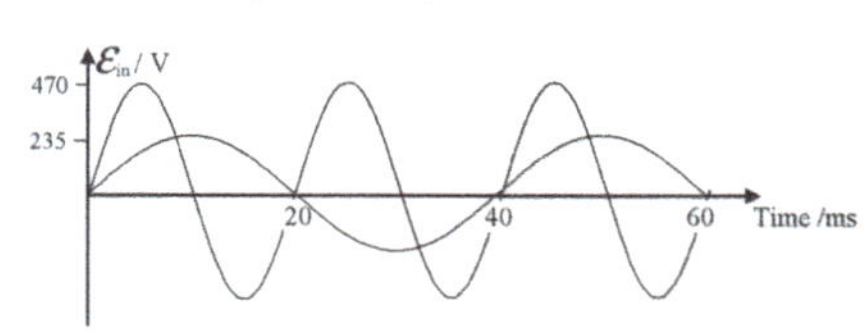

A4 12 ms

A5 Any 3 of …$-4\pi, -2\pi, 0, 2\pi, 4\pi$ …

A6 $V = 17\sin(100\pi t - 1.1\pi)$ or

$17\sin(100\pi t + 0.9\pi)$

A7 $V_{rms} = \dfrac{17}{\sqrt{2}} = 12.0$ V

A8 (a)

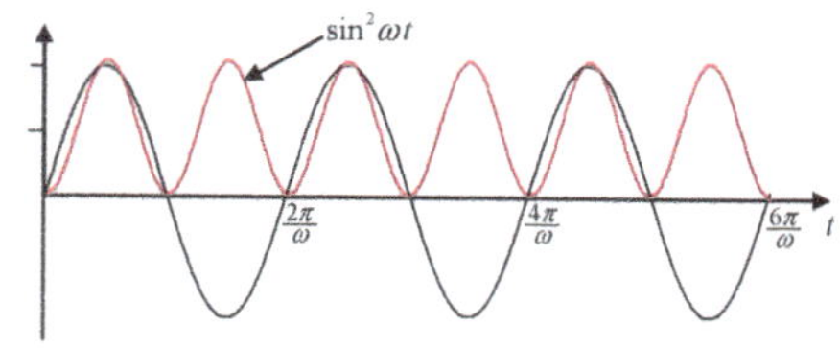

(b) The function $\sin^2\omega t$ is sinusoidal, oscillating between 0 and 1. By symmetry, its mean value is $\frac{1}{2}$.

A9 (a) 9.4 V

(b) 1.88 W

A10 (a) 75 Ω

(b) $f = 250$ Hz; $\omega = 500\pi$ s^{-1} (1570 s^{-1})

(c) Vrms = 10.6 V; $I_{rms} = 0.141$ A

(d) 1.5 W

A11 (a) 3 W, (b) 3 W, (c) 0.

A12

A13 (a) 20 V, 1.5 mA, 14.1 V, 1.06 mA, 100 μs, 10 kHz, $2 \times 10^4\pi$ s^{-1}

(b) 1.2 nF

(c) $\omega CV_{rms} = 2 \times 10^4\pi \times 1.2 \times 10^{-6} \times 14.1$

$= 1.06\text{ A} = I_{rms}$

A14 (a) 1.0 V, 10.0 mA, 0.707 V, 7.07 mA, 2 ms, 500 Hz, 1000π s^{-1}

(b) 31.8 mH

A15 $[\frac{1}{2}\varepsilon_0 E^2]$ $= \text{F m}^{-1}(\text{V m}^{-1})^2$

$= \text{C V}^{-1}\text{ m}^{-1}\text{ V}^2\text{ m}^{-2}$

$= \text{C V m}^{-3}$

$= \text{J m}^{-3}$ = [energy density]

$\left[\dfrac{B^2}{2\mu_0}\right]$ $= (\text{V s m}^{-2})^2 \times (\text{H m}^{-1})^{-1}$

$= \text{V}^2\text{ s}^2\text{ m}^{-4}(\text{A V}^{-1}\text{ s}^{-1})\text{ m}$

$= (\text{V A s})\text{ m}^{-3}$

$= \text{J m}^{-3}$ = [energy density]

A16 40 μF

A17 (b) $I < 0$ decreasing; $V > 0$ decreasing

(c) $I < 0$ increasing; $V < 0$ decreasing

(d) $I > 0$ increasing; $V < 0$ increasing

A18 0 (zero)

A19 (a) $V = V_R$

(b) $V = V_C$

A20 141 Ω

A21 $\frac{\pi}{4}$ (45°)

A22 160 nF = 0.16 μF

A23 $Z = \sqrt{5^2 + 12^2} = 13\ \Omega$

$\therefore V_0 = 13\ I_0 = 5.2\ \text{V}$

$\theta = \tan^{-1}\frac{X_L}{R} = \tan^{-1}\frac{12}{5} = 1.18\ \text{rad}$

$V = 5.2\cos(\omega t + 1.18)$

A24 (a) $40\ \Omega$

(b) $50\ \Omega$

(c) $0.24\ \text{A}$

(d) $0.93\ \text{rad}$

A25

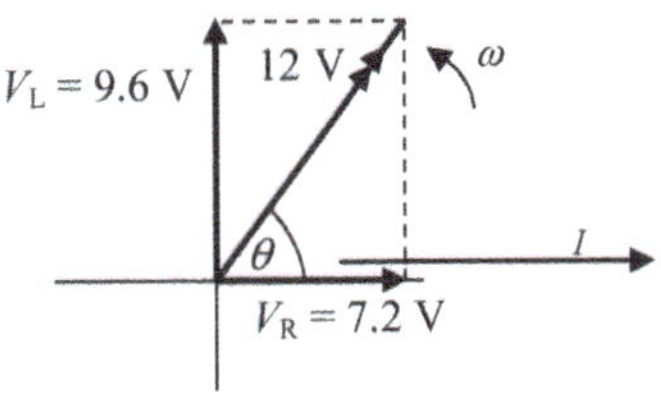

A26 (a) $22.4\ \Omega$

(b) $1.11\ \text{rad}$

A27 $\frac{\omega_0 L}{R} = \frac{I\omega_0 L}{IR} = \frac{V_L}{V_R} = \frac{V_L}{V_R}$

[because $V_L = V_C$ at resonance]

$\omega_0 = \frac{1}{\sqrt{LC}}\ \therefore\ \frac{\omega_0 L}{R} = \frac{1}{\sqrt{LC}}\frac{L}{R} = \frac{1}{R}\sqrt{\frac{L}{C}}$

A28 $V_0 = \sqrt{2}V_{rms}$ – these are not time variable

But instantaneous values of V are time variable and out of phase

A29 $Q = 3.2$

A30 Peak current would be greater (0.6 A); curve much sharper

A31 $20\ \text{mV div}^{-1}$

A32 $5\ \mu\text{s div}^{-1}$

A33 $f = 80\ \text{kHz}$; $V_0 = 48\ \text{mV}$

A34 $V_0 = 1.4\ \text{V}$; $T = 3.1\ \text{ms}$

A35 $V_{rms} = 1.0\ \text{V}$; $f = 320\ \text{Hz}$

A36 $V_{rms} = 9.2\ \text{V}$; $I_{rms} = 0.21\ \text{mA}$, $X_C = 44\ \text{k}\Omega$, $C = 36\ \text{nF}$

4B

B1 (a) $E = \frac{hc}{\lambda}$, $\therefore \lambda = \frac{hc}{E}$,

$= \frac{6.63 \times 10^{-34} \times 3 \times 10^8}{1.610^{-14}}$

$= 1.24 \times 10^{-11}\ \text{m} \sim 10\ \text{pm}$

(b) 100 pm, 1 nm

B2 1 kW

B3 (a) 31 pm

(b) Must be greater than 10.2 kV

B4 (a) 36%

(b) $0.26\ \text{cm}^{-1}$

B5 This would need whole body irradiation which would cause too much damage to healthy tissue

B6 Fraction = 1.1×10^{-6} (0.00011%)

B7 1.52 MHz

B8 9.9×10^{-5} s

B9 2.5 cm

B10 Because of the difference in the acoustic impedance between blood fluid and air

B11 130 Hz

B12 (a) 2.5×10^{-26} J

(b) $0.16\ \mu\text{eV}$ (1.6×10^{-7} eV)

B13 6.2×10^{-21} J, i.e. 250 000 ×

B14 15 MHz

B15

6 tissues × 0.12	=	0.72
1 tissue × 0.08	=	0.08
4 tissues × 0.04	=	0.16
4 tissues × 0.01	=	0.04
Total		1.00

B16 0.76 mSv

B17 (a) Electron anti-neutrino $\overline{\nu_e}$

(b) ${}^{99}_{43}\text{Tc}_m \rightarrow {}^{99}_{43}\text{Tc} + \gamma$

(c) 1.5×10^{-4} u

B18 ${}^{124}_{53}\text{Te} + {}^{1}_{1}\text{H} \rightarrow {}^{125}_{53}\text{I} \rightarrow {}^{123}_{53}\text{I} + 2{}^{1}_{0}\text{n}$

${}^{123}_{53}\text{I} + {}^{0}_{-1}\text{e} \rightarrow {}^{123}_{52}\text{Te}_m + \nu_e$

${}^{123}_{52}\text{Te}_m \rightarrow {}^{123}_{52}\text{Te} + \gamma$

B19 A photon has momentum $\frac{h}{\lambda}$. The e^+ and e^- will have little momentum at the point of annihilation, hence, to conserve momentum, two photons are emitted having equal energy (and hence wavelength) in opposite directions

4C

C1 The C of G is below the midpoint

C2 Horizontal force to left on boat due to water resistance

C3 The force of the wind on the sail will decrease, so its clockwise moment will decrease.

C4 182 N

C5 Because there will be a common factor of $\cos 10°$ in all the terms in the moments equation

C6 2.17 m

C7 Impact speed = $7.06\ \text{m s}^{-1}$

Rebound speed = $5.29\ \text{m s}^{-1}$

C8 $E_{k\ trans} = \frac{1}{2}mv^2$

$E_{k\ rot} = \frac{1}{2}I\omega^2 = \frac{1}{2}(mr^2)\left(\frac{v}{r}\right)^2 = \frac{1}{2}mv^2$

$\therefore E_{k\ tot} = \frac{1}{2}mv^2 + \frac{1}{2}mv^2 = mv^2$ QED

C9 60 J

C10 acceleration = $0.007\ \text{m s}^{-2}$

Reduction in tension = 0.07%

C11 $\text{kg m}^2\ \text{s}^{-1}$ (or N m s)

C12 (a) $13.8\ \text{kg m}^2\ \text{s}^{-1}$

(b) $16.7\ \text{rad s}^{-1} = 2.7\ \text{rev. s}^{-1}$

C13 Equation gives 16.5 m; so does the figure

C14 (a) 14.3 m [Allow yourself ±0.2]

(b) 16.4 m (speed wins over angle) [Allow ±0.5]

C15 Drag: [javelin] exerts a forward force on air molecules [N2]

∴ air exerts an equal backward force on [javelin] [N3]

Lift: [wing] exerts a downward force on air molecules [N2]

∴ air exerts an equal upward for on [wing] [N3]

C16 'Lift' is backwards for the second half of the flight.

C17 (a) Lift = 1.0 N

(b) For the same lift, the torque will be double so the javelin point rotates downwards too soon

C18 Reasonably so: the drag at $20\ \text{m s}^{-1}$ is ~4× that at $10\ \text{m s}^{-1}$

C19 Lift = 39 kN

4D

D1 With 85% energy absorption (albedo = 0.15) the model leads to a temperature of 218 K, which is in good agreement

D2 $T^4 = \frac{0.7 \times 1360}{4 \times 5.67 \times 10^{-8}}\ \text{K}^4$

Leads to $T = 255\ \text{K} = -18\ °\text{C}$

D3 Sun: Using 5800 K → λmax = 0.5 μm ✓

Ground: Using 15 °C (288 K) → 10 μm ✓

D4 Energy of 7.5 μm photon = 0.17 eV, so vibrational energy levels in N_2O are 0.17 eV apart

D5 0.1%, 0.2% 0.6%

D6 (a) $1.08 \times 10^6\ \text{m}^3$

(b) $0.97 \times 10^6\ \text{m}^3$

D7 Presence of γ photon suggests e-m interaction

D8 Data give: $\Delta m = -0.002\ 669\ 661$ u

Using 931.5 MeV / u → 26.73 MeV

D9 ~ $54\ \text{W m}^{-2}$

D10 (a) $P_{out} = 17.5\ \text{mW}$

(b) $P_{in} = 50\ \text{mW}$, ∴ efficiency = 0.35 (35%)

D11 100 parallel branches of ~35 cells in series, ∴ ~3500 cells

D12 42 W

D13 110 kW

D14 Total area under each graph is 1.0 by definition; location B has a greater area under the graph at higher speeds so must have a lower maximum probability

D15 (a) $19.8\ \text{m s}^{-1}$

(b) $0.62\ \text{m}^3\ \text{s}^{-1}$

(c) $620\ \text{kg s}^{-1}$

(d) 120 kW

D16 $[\rho V] = \text{kg s}^{-1}$

$\therefore [\text{RHS}] = \text{kg s}^{-1} \times \text{N kg}^{-1} \times \text{m} - \dfrac{(\text{kg s}^{-1}) \times \text{m}^6\ \text{s}^{-1}}{\text{m}^4}$

$= \text{N m s}^{-1} - \text{N m s}^{-1}$

$= \text{J s}^{-1} - \text{J s}^{-1}$

$= \text{W} - \text{W}$

$= [\text{LHS}]$ so homogeneous

D17 Total energy available in 24 hours $= 4 \times 4.4 \times 10^{10}$ J

$\therefore$ Mean power = 2 MW (to 1 sf)

D18 0.9%

D19 (a) 1.4(4), 2.5, $(1.2)^n$

(b) 10.8 (11 to the next whole number)

D20 (a) 2.6×10^{-5} keV

(b) 1.3 keV

D21 0.31 kg

D22 9.8×10^{17} J

D23 $[K] = \dfrac{[\Delta Q][\Delta x]}{[\Delta t][A][\Delta \theta]}$

$= \text{J m (s m}^2\ \text{K)}^{-1}$

$= \text{J s–1 m}^{-1}\ \text{K}^{-1}$

$= \text{W m}^{-1}\ \text{K}^{-1}$

D24 If K is very small, A needs to be large and Δx small to get a measurable ΔQ

D25 86 W m^{-2}

D26 Rate of heat loss $= K\dfrac{\Delta \theta}{\Delta x}$

$= 0.6 \times \dfrac{15}{0.11}$

$= 82\ \text{W m}^{-2}$

Maths

M1 (a) $t = 0, \pm\dfrac{2\pi}{\omega}, \pm\dfrac{4\pi}{\omega}, \ldots$

(b) $t = \pm\dfrac{\pi}{2\omega}, \pm\dfrac{3\pi}{2\omega}, \pm\dfrac{5\pi}{2\omega}, \ldots$

(c) $t = -\dfrac{3\pi}{2\omega}, \dfrac{\pi}{2\omega}, \dfrac{5\pi}{2\omega}, \ldots$

M2 (a) 12 cm

(b) 5 Hz

(c) 0.05 s, 0.15 s

M3 (a) 1000 s^{-1}

(b) 159 Hz

(c) 6.28 ms

(d) −3.6 ms, −0.5 ms, 2.6 ms, 5.8 ms

M4 $I = 0.1 \cos\left(1000t + 0.5 - \frac{\pi}{2}\right) = 0.1 \cos(1000t + 5.212 \text{ rad})$

M5 $\cos\theta = \sin\left(\theta + \frac{\pi}{2}\right)$

$\therefore \cos(100t - 1.0) = \sin\left(100t - 1.0 + \frac{\pi}{2}\right)$

$= \sin(100t + 0.57)$

$\therefore$ Yes!

M6

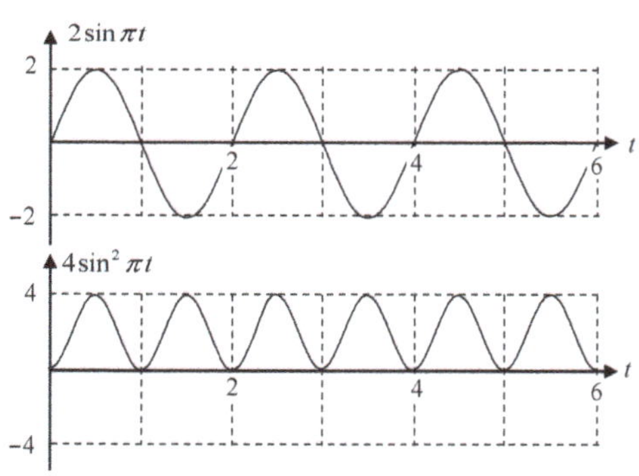

M7 (a) 1, 2.7, 4.5, 7.4, 20, 100, 148, 22 000, 2.7×10^{43}

(b) $\sqrt{e^{10}} = e^5$ or $(e^5)^2 = e^{10}$

(c) $(e^{10})^{10} = e^{100}$

M8 (a) 0.37, 0.22, 0.14, 0.050, 0.010, 0.0067, 4.5×10^{-5}, 3.7×10^{-44}

(b) $e^{-5} = \dfrac{1}{e^5}$

(c) $e^{-100} = (e^{-10})^{10}$

M9 (a) 10^6

(b) 10^7

(c) 16 hours (from the start)

M10 0.69

M11 1

$e = 2.7$ (to 2 sf)

M12 Decay constant = 0.0526 hour^{-1}

$\therefore$ (a) 1.26 day^{-1}, (b) 1.46×10^{-5} s^{-1}

M13 (a) 12.0 s

(b) 0.51 V (using 37% in 1 time constant)

M14 (a) 0.69

(b) 1.39

(c) 5.0

(d) 4.61

(e) 432

M15 (a) 10

(b) 148

M16 $\log_{10}2^x = \log_{10}35$

$\therefore x = \dfrac{\log_{10}35}{\log_{10}2} = 5.13$

M17 $n = 0.34$ (to 2 sf)

M18 Rearranging: $e^{\lambda t} = \dfrac{V_0}{V}$

Taking logs: $\lambda t = \ln\left(\dfrac{V_0}{V}\right)$

$\therefore \lambda = \dfrac{1}{t}\ln\left(\dfrac{V_0}{V}\right)$

Alternative forms:

$\lambda = -\dfrac{1}{t}\ln\left(\dfrac{V}{V_0}\right), \lambda = \dfrac{1}{t}(\ln V_0 - \ln V)$

M19 Gradient = 3; intercept = ln 10

Test yourself answers

Unit 3

3.1

1 (a) 754 rad s^{-1} (b) 25000 m s^{-2}

2 (a) 5000 N (b) 26.8 m s^{-1}

3 (a) The direction of the velocity is always changing [because of the gravitational force of the star].
(b) Towards the centre of the orbit [i.e. the star].
(c) Towards the centre of the orbit.
(d) If a body is moving in a circle (at a constant speed) the resultant force on it is towards the centre of the circle.

4 20 rad s^{-1}

5 (a) (i) The slotted masses are in equilibrium so the resultant force on them is zero, i.e. $T - mg = 0$.
(ii) $T\cos\theta$ is the horizontal component of the force exerted by the string on the small object. This provides the centripetal force for the circular motion.
(iii) The small object is not accelerating vertically. Hence the sum of the vertical components of the forces on it is zero. $T\sin\theta$ is the vertical component of the force of the string on the object.
(b) 1.13 kg
(c) From equation 1: $T = 1.13 \times 9.81$ N,
So from equation 3: $\theta = \sin^{-1}\left(\frac{0.12 \times 9.81}{1.13 \times 9.81}\right) = 6.09°$
$\therefore \cos\theta = 0.994$, so the approximation is good.
(d) 1.14 kg, $\theta = 6.05°$ so the same to 2 sf.

6 (a) (i) Provides the centripetal force, so radially inwards.
(ii) 2.7 N
(b) 7.1 N radially outwards

7 Anwen is correct that the speed is constant. The gravitational potential energy is constant so the kinetic energy must be constant too. The direction of the velocity is always changing so there must be a resultant force on it. Rhodri should remember that the gravitational force is at right angles to the direction of motion of Venus; hence it does no work on Venus and so the kinetic energy must stay the same.

8 (a) 1.72 m s^{-1}
(b) $E_k = 0.059$ J; $E_p = 0.235$ J
(c) 3.84 m s^{-1}

9 (a) Vertically upwards
(b) Vertically downwards
(c) $L_1 - mg$
(d) $L_2 + mg$
(e) The gravitational potential energy, E_p, at **A** is less than that at **B**. If there are no energy losses, $E_k + E_p$ is constant. Hence the kinetic energy, E_k, is greater at **A** than **B**.
(f) The centripetal force is less at **B** than **A** because the speed is less. Hence:
$L_2 + mg < L_1 - mg$
So $L_2 < L_1 - 2mg$ and so $L_2 < L_1$

10 At **A** the ball is, momentarily, at rest but accelerating towards the lowest point in the basin. Hence, the resultant force is tangentially to the right. Similarly at **C** the resultant force is tangentially to the left. At **B** the speed is constant so there is zero tangential component of force. Because of the circular motion the resultant force is towards the centre, **X**.

3.2

1 Force oscillations are those which occur when a sinusoidally varying 'driving' force is applied to an oscillatory system, causing it to oscillate with the frequency of the applied force. Free oscillations are those which occur when an oscillatory system is displaced from its equilibrium position and released.

2 The direction of the acceleration is towards the fixed point.

3 (a) Damping, mainly from air resistance in this case.
(b) Hysteresis.

4 (a)

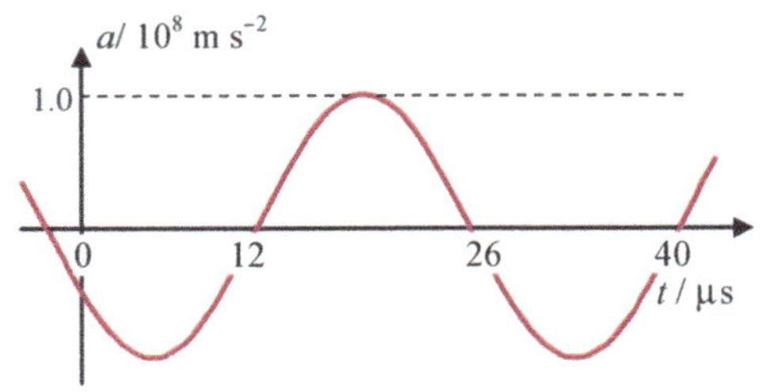

(b) $x(0) = 0.87$ mm (2 s.f.)
$v(0) = 400$ m s^{-1}
$a(0) = -4.4 \times 10^7$ m s^{-2}
(c) ±300 m s^{-1} (2 sf); at any displacement, the motion can be towards or away from the equilibrium point

5 (a) x / cm = 2.39 cos $\left(\frac{4}{3}\pi t - \frac{\pi}{2}\right)$
(b) −2.1 cm, −5.0 cm s^{-1}, 36 cm s^{-2}

6 (a) 47 mm s^{-1}, 0.30 m s^{-2}
(b) x / mm = 7.5 cos $(2\pi t + 0.84)$

7 (a) (i) A and C (ii) B
(b) (i) 0.10 rad (ii) 12 cm
(c) $T = 2.20$ s, $\omega = 2.86$ rad s^{-1}
(d) $v_{max} = 0.34$ m s^{-1}; $a_{max} = 0.98$ m s^{-2}
(e)

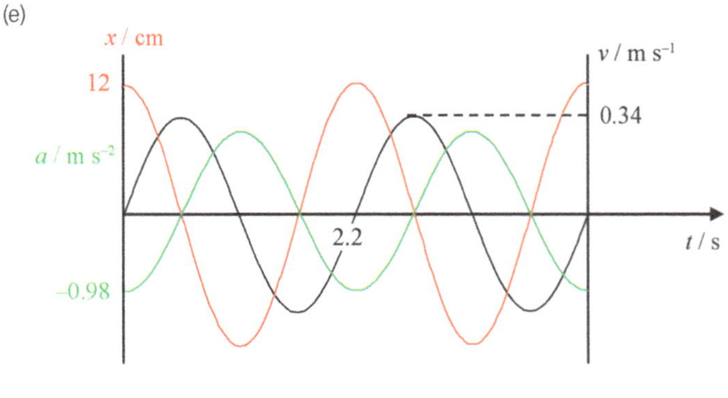

8 (a) $T = \pi(\sqrt{2} + 1)\sqrt{\frac{l}{2g}}$

(b)

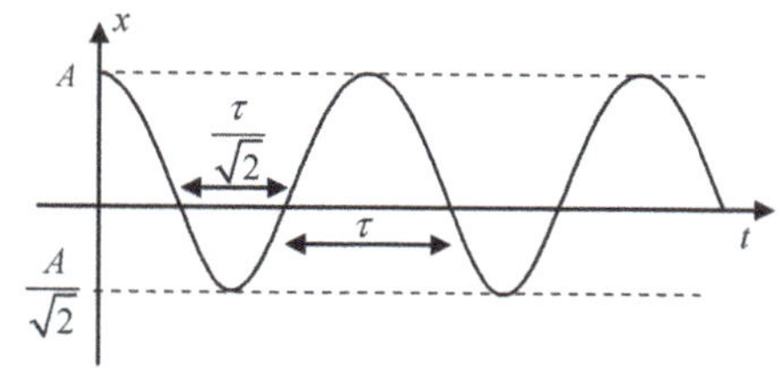

9

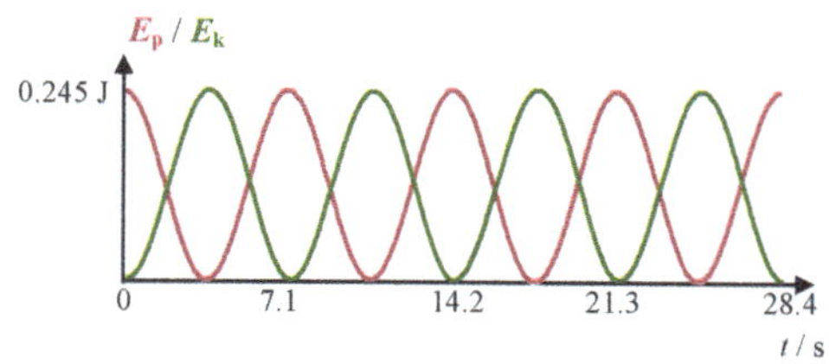

10 (a) 4.5 cm (b) 3.4 cm (c) 2.0 cm

11 (a) x / cm = $20e^{-0.555t}\cos 4\pi t$

(b)

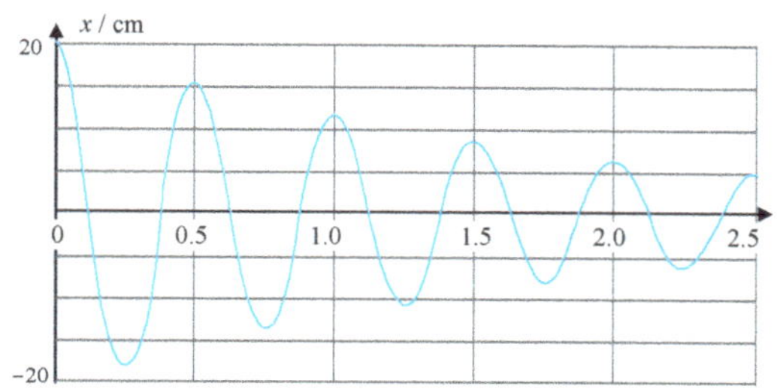

(c) 3.75 cm

12 Results give $g = 9.83 \pm 0.05$ m s^{-2}

The range includes 9.81 m s^{-2}, so consistent

13 There should be no uncertainty in the number oscillations. The only uncertainties therefore will be those associated with starting and stopping the stopwatch. The absolute uncertainty in the timing will be the same whatever the timing period, so the longer the timing period, the lower the percentage uncertainty. Hence, Iestyn's method is better.

14 Ratio of successive amplitudes are all approximately 0.8 (range, 0.74 – 0.85) so approximately exponential [alternatively, a graph of ln A against t is a straight line]

$\lambda \sim 0.0069$ s^{-1}

15 It is difficult to release the pendulum without giving it a small nudge, so it is better to allow it to swing before starting the timing. The ends of the swing are not good places to time between because the pendulum is moving slowly and changing direction. If there is a vertical fiducial mark, it is easy to use it to measure the times and possible to anticipate when the pendulum will pass it. Hence Eleri's method is superior!

16 (a) (i) A straight line graph of y against x has the equation, $y = mx + c$, where m and c are constants. If the line passes through the origin, $c = 0$. The expected relationship can be rearranged to give $\ell = \frac{g}{4\pi^2}T^2$. This is of the same form as $y = mx + c$, with $y = \ell$, $m = \frac{g}{4\pi^2}$, $x = T^2$ and $c = 0$.

(ii) In the equation $y = mx + c$, the gradient is m.

Hence $g = 4\pi^2 \times$ gradient.

(b) At 40.0 cm, $T^2 = (1.49 \pm 0.10)$ s^2;
at 80.0 cm, $T^2 = (3.24 \pm 0.14)$ s^2

(c) (i), (ii)

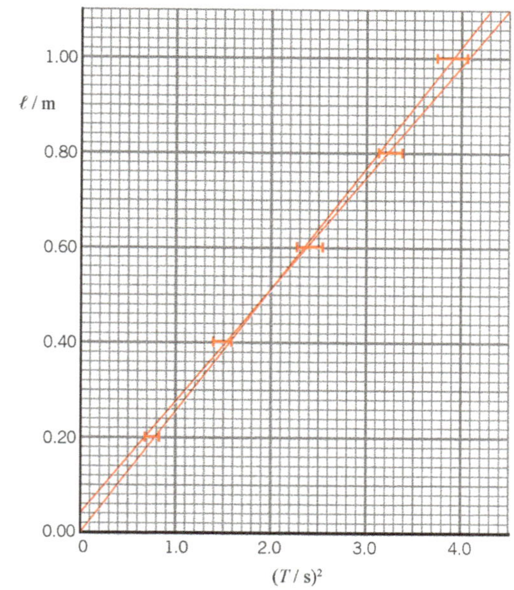

(iii) The error bars are consistent with a straight-line relationship and the origin lies within the max/min gradient lines. Hence consistent.

(iv) Max gradient $= \frac{1.024 - 0.000}{4.00 - 0.04} = 0.259$ m s^{-2}

Min gradient $= \frac{1.000 - 0.038}{4.05 - 0.00} = 0.238$ m s^{-2}

∴ Gradient = 0.245 (± 8.6%) m s^{-2}

(d) $4\pi \times$ gradient $\rightarrow g = (9.7 \pm 0.8)$ m s^{-2} so the result is consistent with the accepted value.

(e) The true values of ℓ were all smaller than those plotted, by the radius of the bob. Hence the graph should be slightly lower, which would affect the intercept but not the gradient.

3.3

1 p = pressure / Pa = N m^{-2}; V = volume / m^3; n = amount of substance / mol

R = ideal gas constant / J mol^{-1} K^{-1}; T = temperature / K

N = number of molecules (no unit); k = Boltzmann constant / J K^{-1}

2 An ideal gas is one which obeys the ideal gas equation ($pV = nRT$) exactly.

3 (a) N is the number of molecules and m is the molecular mass.

(b) Nm is the total mass of the gas molecules.

(c) The product, number of moles × molar mass, is also equal to the total mass of the gas molecules.

4 (a) 18

(b) (i) 18 u, (ii) 2.99×10^{-26} kg

(c) Water vapour (H_2O)

5 3.1×10^{24}

6 $n = 1070$ mol, $N = 6.4 \times 10^{26}$

7 2.23×10^{19} Na^+ ions

8 (a) $M_r = 352.02$

(b) $M = 352.02$ g mol^{-1} = 0.352 02 kg mol^{-1}

(c) 20 mol

9 (a) $498\ \text{m s}^{-1}$
(b) 42.3 mol
(c) $28.95\ \text{g mol}^{-1}$, c.f. $28.94\ \text{g mol}^{-1}$ from composition data

10 $R = 8.30\ \text{J kg}^{-1}\ \text{K}^{-1}$

11 (a) (i) $290\ \text{m s}^{-1}$ (ii) $91\,500\ \text{m}^2\ \text{s}^{-2}$ (iii) $302\ \text{m s}^{-1}$
(b) 117 K

12 (a) Speed ratios – $1 : \frac{1}{\sqrt{2}} : \frac{1}{2} : \frac{1}{4} : \frac{1}{4\sqrt{2}}$
(b) Speeds in m s^{-1}, 1110, 787, 557, 278, 197 respectively

13 (a) 1×10^{44} molecules
(b) It is indeed reasonable.

14. (a) 1.21 MJ
(b) (i) 6.0×10^{-21} J (ii) 330

15. At 40 °C, the relationship clearly departs from that expected for an ideal gas, especially around 8 MPa and $0.15\ \text{m}^3$ where the graph has a kink in it. At 50 °C, the shape of the graph corresponds qualitatively to a pV = constant relationship. However, the value of pV changes from 960 J, at 12 MPa, to 2300 J at 4 MPa, so it is not in quantitative agreement.

3.4

1 (a) 720 kJ
(b) At a higher temperature there are fewer molecules, each with a greater mean KE; these effects cancel
(c) (i) 288 K, (ii) $1340\ \text{m s}^{-1}$

2 (a) 374 kJ
(b) $484\ \text{m s}^{-1}$
(c) The molecules possess 250 kJ of rotational kinetic energy

3 (a) $U_A = 3.6$ MJ, $U_B = 2.4$ MJ
(b) $T_A = 578$ K, $T_B = 385$ K
(c) (i) 481.5 K
(ii) $p_A = 1.00 \times 10^6$ Pa, $p_B = 5.00 \times 10^6$ Pa
(iii) 608 kJ from A to B
(d) $p = 667$ Pa, $T = 482$ K

4

	ΔU	Q	W
(a)	> 0	$= 0$	< 0
(b)	> 0	> 0	> 0
(c)	> 0	> 0	~ 0
(d)	$= 0$	> 0	> 0
(e)	> 0	> 0	~ 0
(f)	$= 0$	$= 0$	$= 0$
(g)	$= 0$	$= 0$	$= 0$

5 (a) 170 kJ
(b) 0.075 (7.5%)
(c) $Q = 2.26$ MJ, $\Delta U = 2.09$ MJ, $W = 0.17$ MJ

6 (a) 76 kJ
(b) $Q = 76$ kJ, $\Delta U = 76$ kJ, $W = 0$

7 (a) 0.2, 0.4 and 2.0 respectively
(b) 0.2, 0.4 and 2.0 respectively
(c) & (d)

	ΔU / kJ	Q / kJ	W / kJ
A→B	−12	−20	−8
B→C	3	3	0
C→D	24	40	16
D→A	−15	−15	0
Total	0	8	8

(e) 3.4 mol

8 1.2 MW

9 On a plot of p / bar against θ / °C the steepest line has a gradient 0.003 78 and passes though (0, 0.947). The least steep line has a gradient = 0.003 41 and passes through (0, 0.970). These figures give a value of absolute zero of (−268 ± 17) °C or (−270 ± 20) °C.

10. Mean rate of temperature rise = 0.162 °C^{-1} (from the gradient of a graph of temperature against time).
This leads to $c_{\text{water}} = 4100\ \text{J kg}^{-1}\,°\text{C}^{-1}$ (to 2 sf)
Did you spot the irrelevant information? [The mains voltage.]

3.5

1 (a) Gradient = −0.0133, intercept = 6.306
(b) $\lambda = 0.0133\ \text{s}^{-1}$, $T = 52$ s

2 (a) $^{192}_{78}\text{Pt} \rightarrow ^{188}_{76}\text{Os} + ^{4}_{2}\text{He}$
(b) $^{181}_{72}\text{Hf} \rightarrow ^{181}_{73}\text{Ta} + ^{0}_{-1}\text{e} + \overline{\nu_e}$
(c) $^{48}_{23}\text{V} \rightarrow ^{48}_{22}\text{Ti} + ^{0}_{1}\text{e} + \nu_e$
(d) $^{59}_{28}\text{Ni} + ^{0}_{-1}\text{e} \rightarrow ^{59}_{27}\text{Co} + \nu_e$
(e) $^{77}_{32}\text{Ge} \rightarrow ^{77}_{33}\text{As} + ^{0}_{-1}\text{e} + \overline{\nu_e}$
(f) $^{65}_{30}\text{Zn} \rightarrow ^{65}_{29}\text{Cu} + ^{0}_{1}\text{e} + \nu_e$
(g) $^{65}_{30}\text{Zn} + ^{0}_{-1}\text{e} \rightarrow ^{65}_{29}\text{Cu} + \nu_e$
(h) $^{239}_{92}\text{U} \rightarrow ^{239}_{93}\text{Np} + ^{0}_{-1}\text{e} + \overline{\nu_e}$
$^{239}_{93}\text{Np} \rightarrow ^{239}_{94}\text{Pu} + ^{0}_{-1}\text{e} + \overline{\nu_e}$
(i) $^{17}_{7}\text{N} \rightarrow ^{16}_{8}\text{O} + ^{1}_{0}\text{n} + ^{0}_{-1}\text{e} + \overline{\nu_e}$

3 For the isotopes with a neutron excess, ^{23}Ne & ^{24}Ne, there is a vacant proton state with a low enough energy level for the transformation n → p + e− + $\overline{\text{ve}}$ to occur
For the isotopes with a neutron deficit, ^{18}Ne & ^{19}Ne, there is a vacant neutron state with a low enough energy level for the transformation p → n + e+ + ve to occur
For ^{20}Ne, ^{21}Ne and ^{22}Ne neither of these is the case, so the nucleus is stable

4 (a) The fractional uncertainty decreases with the total count; a longer count time is used for the lower count rates to increase the total count
(b)

Distance / cm	10	15	20	25	30	50	70
Corrected count, C	729	256	129	89	65.7	22.0	10.5
$1/\sqrt{C}$ /10^{-3}	37	62.5	88	106	123	213	309

(c) $k = 51700\ \text{cpm cm}^2 = 5.2\ \text{cpm m}^2$
$\varepsilon = -1.2$ cm

5 (a) $\lambda_{234} = 8.8 \times 10^{-14}\ s^{-1}$, $\lambda_{235} = 3.1 \times 10^{-17}\ s^{-1}$, $\lambda_{238} = 4.9 \times 10^{-18}\ s^{-1}$

(b) ^{234}U: 63.5%, ^{235}U: 1.6%, ^{238}U: 34.9%

(c) 35 MBq

6 Most α particles absorbed by the material; The kinetic energy of the alpha particles is lost in collisions with atoms.

7 (a) Number of half-lives ~ 10^4

$\therefore$ Fraction remaining $= 2^{-10000} = (2^{-10})^{1000} = (10^{-3})^{1000}$
$= 10^{-3000}$

10^{3000} is (vastly) more than the number of baryons in the universe (estimated as 10^{72}), hence there is nothing left

(b) $^{238}_{92}U \xrightarrow{\alpha} {}^{234}_{90}Th \xrightarrow{\beta^-} {}^{234}_{91}Pa \xrightarrow{\beta^-} {}^{234}_{92}U$

8 (a) A can only change by 0 or −4; no combination of 0s and 4s can produce a change of −3

(b) ^{238}U: 94.4%, ^{235}U: 5.6%

(c) 6×10^9 years

9 (a) Argon is an unreactive gas – unlikely to bond to other atoms in the rock, hence any argon would have escaped from the original (liquid) magma

(b) $^{40}_{19}K \rightarrow {}^{40}_{18}Ar + {}^{0}_{1}e + \nu_e$

$^{40}_{19}K + {}^{0}_{-1}e \rightarrow {}^{40}_{18}Ar + \nu_e$

$^{40}_{19}K \rightarrow {}^{40}_{20}Ca + {}^{0}_{-1}e + \overline{\nu_e}$

(c) As the rock ages, the number of potassium atoms decreases and the number of argon atoms increases. Hence the ratio of argon to potassium increases, assuming that the argon remains trapped.

(d) $5.33 \times 10^{-10}\ year^{-1} = 1.69 \times 10^{-17}\ s^{-1}$

(e)

Time / Myr	0	100	200	300	400	500	600
^{40}K	100	94.8	89.9	85.2	80.8	76.6	72.6
^{40}Ar	0	0.52	1.01	1.48	1.92	2.34	2.74
Ar/K	0	0.006	0.011	0.017	0.024	0.031	0.038

(f) 340 million years, this is the least possible age.

If some argon escaped the true age would be higher

(g) (i) Number of ^{40}K atoms, $N = N_0 e^{-\lambda t}$

Number of decayed ^{40}K atoms $= N_0(1 - e^{-\lambda t})$

$\therefore$ Number of ^{40}Ar atoms $= 0.1 \times N_0(1 - e^{-\lambda t})$

$\therefore R = \dfrac{^{40}Ar}{^{40}K} = \dfrac{0.1 \times N_0(1-e^{-\lambda t})}{N_0 e^{-\lambda t}} = \dfrac{1-e^{-\lambda t}}{10e^{-\lambda t}}$

(ii) From (i), $10Re^{-\lambda t} = 1 - e^{-\lambda t}$

Dividing by $e^{-\lambda t}$ and rearranging $\rightarrow e^{\lambda t} = 1 + 10R$

$\therefore$ (Definition of ln) $\lambda t = \ln(1 + 10R)$, hence the result.

(iii) $t = \dfrac{1}{5.33 \times 10^{-10}\ year^{-1}} \ln(1 + 0.20)$

$= 340 \times 10^6$ year (to 2 sf) ✓

10 Using $\pm\sqrt{N}$ as the uncertainty, over 10 minutes:

Background = 250 ± 16 counts

No absorber = 570 ± 24 counts,
so source = 320 ± 40 counts

With absorber = 525 ± 23 counts,
so source now = 275 ± 39 counts

The uncertainty ranges overlap considerably so it is not safe to conclude that the source is an alpha emitter

11 Likely to be β

12 (a) $13.5\ s^{-1}$ (b) $10.1\ s^{-1}$

3.6

1 (a) 3.0×10^{16} J

(b) 0.33 kg

2 (a) 3.6×10^{26} W

(b) 1.6×10^{17} W

3 (a) $^{7}_{4}Be + {}^{0}_{-1}e \rightarrow {}^{7}_{3}Li + \nu_e$

(b) Weak: change of quark flavour (u→d), neutrino involvement

(c) 0.86 MeV; mainly carried by neutrino because it has equal (and opposite) momentum to the daughter nucleus and is much lighter

4 (a) $^{241}_{95}Am \rightarrow {}^{237}_{93}Np + {}^{4}_{2}He$

(b) 5.50 MeV = 8.8×10^{-13} J

(c) $2.6 \times 10^5\ m\ s^{-1}$

(d) 1.2×10^8 Bq

5 (a) β^- gives ^{40}Ca: $^{40}_{19}K \rightarrow {}^{40}_{20}Ca + {}^{0}_{-1}e + \overline{\nu_e}$

β^+ and K capture $\rightarrow {}^{40}Ar$ (see Test yourself 3.5, Q9)

(b) 1.3 MeV

6 (a) $^{239}_{92}U \rightarrow {}^{239}_{93}Np + {}^{0}_{-1}e + \overline{\nu_e}$

$^{239}_{93}Np \rightarrow {}^{239}_{94}Pu + {}^{0}_{-1}e + \overline{\nu_e}$

(b) U→Np: 1.26 MeV

Np→Pu: 0.72 MeV

(c) Some energy released as KE of the neutrinos, which are not absorbed in the reactor

7 7.3 MeV = 1.2×10^{-12} J

8 (a) 168 MeV = 2.7×10^{-11} J

(b) From the Cs: 11 MeV = 1.8×10^{-12} J

From the Rb: 13.7 MeV = 2.2×10^{-12} J

(c) 7.9×10^{13} J (including the subsequent decays)

6.9×10^{13} J from the fission only

9 (a) Step 1: 1.44 MeV (including the e^+e^- annihilation)

Step 2: 5.49 MeV

Step 3: 12.85 MeV

(b) [Note: Total release = 2 × (step 1 + step 2) + step 3]

6.3×10^{11} J

10 (a) $Q_{Sn} = 8.0$ aC, $Q_{Mo} = 6.7$ aC

(b) $r_{Sn} = 5.5$ fm, $r_{Mo} = 5.2$ fm

(c) 4.5×10^{-11} J (~280 MeV)

(d) Mo: $1.2 \times 10^7\ m\ s^{-1}$

Sn: $1.0 \times 10^7\ m\ s^{-1}$

11 Both graphs are useful: the ln graph for the value of the exponent, the linear graph for k. For just the heavier nuclides (A~ 120+) r is proportional to the cube root of A. The relationship departs from this for the lighter nuclides. Again for just the heavier nuclides, the data return a value of 0.9 fm, slightly lower than 1.1 fm.

Unit 4

4.1

1 (a) 1.8×10^{-7} F = 180 nF
(b) 8.9×10^{-5} C = 89 μC
(c) 0.022 J = 22 mJ
(d) 5.0×10^{6} V m^{-1} = 5.0 MV m^{-1}

2 2.8 μF (ignoring dielectric effect)

3 (a) $\frac{2}{3}C$, (b) $3C$

4 (a) $V_R = V_S$, $V_C = V_{2C} = 0$, $I_0 = \frac{V_S}{R}$
(b) (i) $\frac{V_S}{2R}$, (ii) $\frac{V_S}{2}$, (iii) (L→R) $\frac{1}{3}CV_S, -\frac{1}{3}CV_S, \frac{1}{3}CV_S, -\frac{1}{3}CV_S$
(iv) $V_C = \frac{1}{3}V_S$, $V_{2C} = \frac{1}{6}V_S$,
(v) $U_C = \frac{1}{18}CV_S^2$, $U_{2C} = \frac{1}{36}CV_S^2$
(vi) $P_R = \frac{V_S^2}{4R}$
(c) (i) 0, (ii) 0, (iii) (L→R) $\frac{2}{3}CV_S, -\frac{2}{3}CV_S, \frac{2}{3}CV_S, -\frac{2}{3}CV_S$
(iv) $V_C = \frac{2}{3}V_S$, $V_{2C} = \frac{1}{3}V_S$
(v) $U_C = \frac{2}{9}CV_S^2$, $U_{2C} = \frac{1}{9}CV_S^2$
(vi) $P_R = 0$

5 (All in μF). Individual capacitors: 10, 22, 33
Series combinations: 5.7, 6.9, 7.7 and 13.2
Parallel combinations: 32, 43, 55, 65
Parallel pair in series with single: 8.5, 14.6, 16.2
Series pair in parallel with single: 23.2, 29.7, 39.9

6 11 mF

7 112 μF

8 (a) 2.2 mC, 11 mJ
(b) Q_{220} = 1.51 mC, Q_{100} = 0.69 mC
(c) 7.6 mJ

9 Capacitance $\rightarrow \frac{1}{2}C$ using $C = \frac{\varepsilon_0 A}{d}$
pd → $2V$ because Q constant and C halved [or electric intensity, E, stays the same and double the separation]
$Q \rightarrow Q$ because there is nowhere else for the charge to go
$U \rightarrow 2U$, from any 2 of C, V and Q

10 The pd is the pd across the battery, hence $V \rightarrow V$
As in Q9, $C \rightarrow \frac{1}{2}C$
$Q \rightarrow \frac{1}{2}Q$ because C halved and V constant
$U \rightarrow \frac{1}{2}U$ from any two of C, V and Q

11 (a) 56.4 mC
(b) I / mA = $120e^{-2.13t}$
(c) Area under graph = 56.4 mC

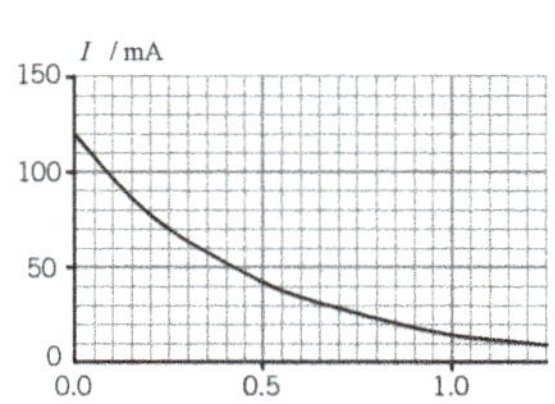

(d) U_0 = 0.34 J
(e) U / mJ = $340e^{-4.26t}$
(f) P / W = $1.44e^{-4.26t}$
(g) Area under graph = 0.34 J

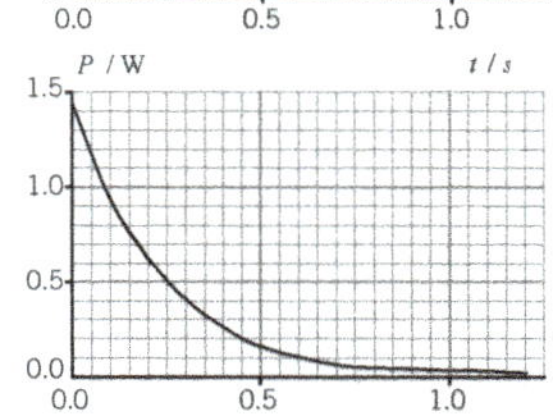

12. (a) 450 N C^{-1} downwards (b) 6.8 V

13. 12 electrons lost.

14. (a) 2.3×10^{7} m s^{-1}
(b) Increasing the separation decreases the field strength and therefore the force.

15. (a) At P, the y component of velocity = 3.0×10^{7} m s^{-1}
$a_y = \frac{F}{m} = \frac{eE}{m} = \frac{1.60 \times 10^{-19} \times 12\,000}{9.11 \times 10^{-31}}$ m s^{-2} = 2.11×10^{15} m s^{-2}
Using $v = u + at$ with $u = 0 \rightarrow t = \frac{3.0 \times 10^{7}}{2.11 \times 10^{15}}$ s = 14.2 ns
(b) 0.42 m
(c) 0.21 m

4.2

1

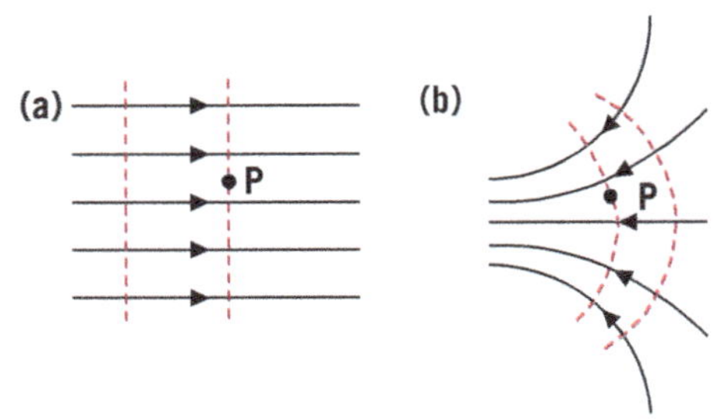

2 (a) X is at the higher potential; work needs to be done to move a positive charge from Y to X
(b) 15 kV
(c) 150 mJ
d) 1.70×10^{6} m s^{-1}
(e) 1.23×10^{6} m s^{-1}

3 (a) Uniform distribution on outer surface of the metal
(b) V = 108 kV, E = 1.08 MV m^{-1}
(c) (i) 54 kV, 270 kV m^{-1}
(ii) 108 kV, 0

4 (a) Repelled by like charge on the sphere
(b) 1080 m s^{-2}
(c) 14.6 m s^{-1}

5 (a) − 29.6 J
(b) −20 μC: 254 N at 45° up to the right
top 10 μC: 152 N at 12.1° to the left of vertical
right 10 μC: 152 N at 12.1° below horizontal to the left

6 (a) 254 km s^{-2}
(b) Accelerates along line of symmetry, reaches maximum speed at midpoint of the two 10 μC charges, decelerates to rest at opposite corner of square; accelerates back and slows down to rest at initial point, etc.
(c) 173 m s^{-1}

7 (a) 1.49×10^{22} kg, 3580 kg m^{-3}
(b) 1.00 m s^{-2}
(c) -1.00×10^{6} J kg^{-1}
(d) -1.00×10^{8} J

8 (a) -2.00×10^{5} J kg^{-1}, 0.04 N kg^{-1}
(b) -5.00×10^{5} J kg^{-1}, 0.25 N kg^{-1}

9 (a) 1.00×10^{6} J
(b) 9.1×10^{5} J
(c) 9%

10 (a) 0.100 N kg^{-1}

(b) (i) $-10\,000 \text{ J kg}^{-1}$

(ii) $-60\,000 \text{ J kg}^{-1}$

(c) The gravitational potential due to the satellite alone is the same at all points on its surface, i.e. -10 kJ kg^{-1}. The potential due to the dwarf planet varies from $-50.25 \text{ kJ kg}^{-1}$ at the closest point on the moon to $-49.75 \text{ kJ kg}^{-1}$ at the furthest point.

Hence the total gravitational potential varies as shown in the diagram.

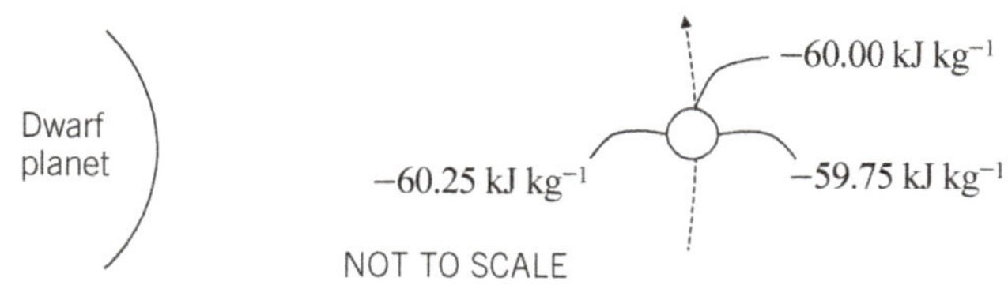

11 (a) 0.0025 m s^{-2}

(b) 6.5 days

4.3

1 0.63 (or 1.59)

2 3.375 (or 0.296)

3 (a) $M_J = 1.90 \times 10^{27}$ kg

(b) $T = 16.9$ days

4 (a) First three points all very close – last one a long way away

(b) gradient = 1.5, intercept = $\ln\frac{2\pi}{\sqrt{GM}} = \frac{1}{2}\ln\frac{4\pi^2}{GM}$

(c) (i) Kepler's third law states that $T^2 \propto r^3$,

- so (taking the square root), $T \propto r^{3/2}$, hence T against $r^{3/2}$ is a straight line through the origin
- now taking the cube root $T^{1/3} \propto r^{1/2}$, hence $T^{1/3}$ against $r^{1/2}$ is a straight line through the origin
- and finally squaring $T^{2/3} \propto r$, hence $T^{2/3}$ against r is a straight line through the origin.

(ii) The values of orbital radius are roughly evenly spaced, so a plot of $T^{2/3} \propto r$ will have a uniform spread of points. [Try them]

5 (a) $a = 1.43 \times 10^{14}$ m

$M = 7.0 \times 10^{36} \text{ kg} = 3.5 \times 10^6\, M_\odot$ – (so close)

(b) $p = 0.024 \rightarrow \Delta M = \pm\, 0.8\, M_\odot$

(c) $v = 0.023c$

6 (a) $d = 1.88 \times 10^9$ m; $r_{1.4} = 8.09 \times 10^8$ m; $r_{0.6} = 1.88 \times 10^9$ m

(b) $v_{1.4} = 93.9 \text{ km s}^{-1}$; $v_{0.6} = 219 \text{ km s}^{-1}$

(c) $\Delta\lambda_{1.4} = 0.205$ nm; $\Delta\lambda_{0.6} = 0.479$ nm

4.4

1 (a) ↓ (b) → (c) ↓ (d) → (e) ↑

2 (a) 0.0125 N 20° above horizontal North

(b) 0.0117 N; 0.0043 N

3

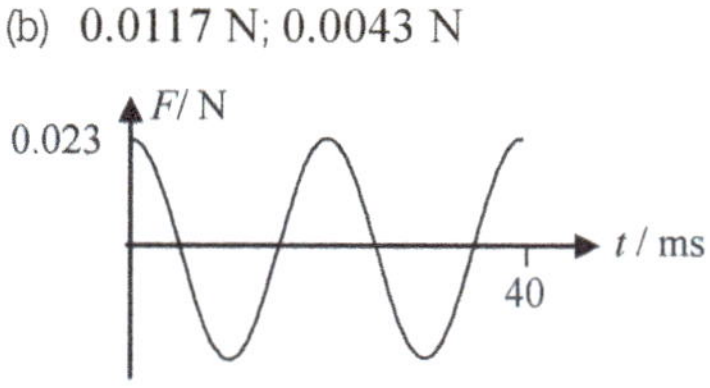

4 (a) 1.5 N m

(b) 0.75 N m

5 (a) 1.71×10^{-23} N s

(b) 0.00213 T

(c) 60 MHz

6 (a) 6.08×10^{-5} T

(b) 930 Hz

7 (a) 7.62 MHz

(b) 4.0 MeV (6.4×10^{-13} J)

(c) 0.578 m

(d) Energy > m_ec^2, so need for relativistic correction

8 (a) $29\,000 \text{ m s}^{-1}$

(b) $3.80 \times 10^{-26} \text{ kg} = 22.9 \text{ u}. \therefore \text{Na}^+$

9 (a) Into the diagram

(b) $v_X = \sqrt{\frac{2eV}{m_X}}$ and $v_Y = \sqrt{\frac{2eV}{m_Y}}$

(c) $r_X = \sqrt{\frac{2Vm_X}{B^2e}}$ and $r_Y = \sqrt{\frac{2Vm_Y}{B^2e}}$

(d) (i) 120 V (ii) 0.39 cm

10 (a) 3380 Ω

(b) 16.9 V

(c) 0.313 V T^{-1}

(d) 0.34 V

4.5

1 The flux linking the coil is decreasing

2 (a) Decreasing

(b) Decreasing

(c) To the right – the direction opposes the decrease in flux

(d) Clockwise viewed from the left

3 (a) The force is to the left – this opposes the increasing separation of the magnet and loop

(b) Towards the viewer at the top

(c) At the bottom, the current is away from the viewer, so clockwise viewed from the left

4 The flux would be increasing to the left, so the flux due to the induced current is to the right as in 2

The force on the loop must be to the right, so (by FLHR) the current is towards the viewer at the top, as in 3

5 (a) The flux changes slower then more quickly; when the middle of the magnet is passing though the plane of the coil, the rate of change of flux is zero; after this the direction of the change of flux is opposite

(b) The magnet is travelling more quickly so the rate of change of flux is more rapid

(c) $\mathscr{E}_{\text{in}} = -\frac{d(N\Phi)}{dt}$

Area under graph = $\int \mathscr{E}_{\text{in}} dt = -\Delta(N\Phi)$

The flux linkage due to the magnet at the beginning and the end is very close to zero, so the total change in flux linkage must be close to zero. Hence the +ve and −ve areas between the graph and the axis must be the same.

6 (a) 0.85 V

(b) There is no return path for any current

7 (a) 1 Wb-turn, (b) 0.5 Wb-turn, (c) 0

8 8.0 V

9 (a) Long sides sweep out a cylinder of radius 0.20 m in 0.50 s.

$\therefore$ Speed $= \dfrac{2\pi \times 0.20\ \text{m}}{0.50\ \text{s}} = 2.51\ \text{m s}^{-1}$

(b) $90° / \dfrac{\pi}{2}$

(c) $\ell = 100 \times (0.50\ \text{m} + 0.50\ \text{m}) = 100\ \text{m}$

$\therefore\ \mathscr{E} = 0.050\ \text{T} \times 100\ \text{m} \times 2.5\ \text{m s}^{-1}$

(d) The short sides do not cut the flux.

10 (a)

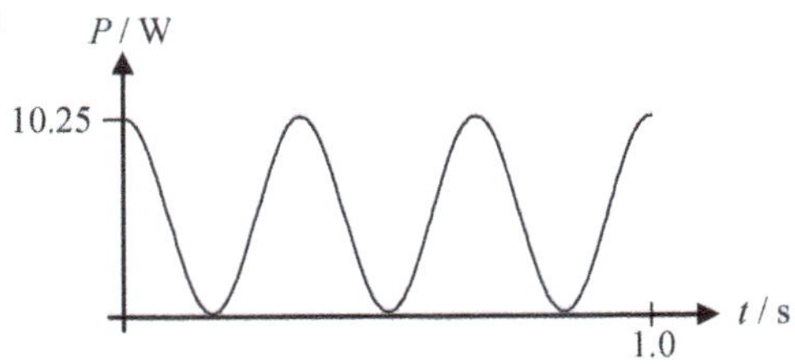

(b) Because the $P(t)$ function is sinusoidal with maximum and minimum values of 10.25 W and 0 respectively, the mean power is 5.1 W (to 2 sf)

11 (a) $f = 40$ Hz, $\omega = 251$ rad s^{-1}

(b) $B = 0.24$ T

(c) $T = 12.5$ ms, $V_0 = 12.0$ V

12. (a) (i) 0.05 s, 0.15 s, 0.25 s and 0.35 s. When the plane is at right angles to the field, the field lines are not being cut [i.e. the rate of change of flux linkage is zero].

(ii) The field lines are being cut at a maximum rate, giving the greatest magnitude of the induced emf, so the times are 0, 0.10 s, 0.20 s, 0.30 s and 0.40 s.

(b)

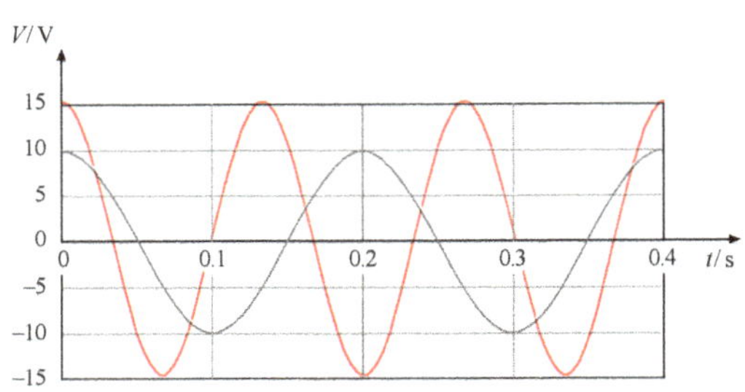

Because the frequency is 1.5×, the maximum rate of cutting flux is 1.5×, so by Faraday's law the peak voltage will be 1.5× as much (15 V = 1.5 × 10 V). The period is now 0.13 s because it is two thirds the original.

4A

1. For the coil $\Phi = BA\cos\omega t$

$\mathscr{E}_{\text{in}} = -\dfrac{\text{d}(N\Phi)}{\text{d}t} = BAN\omega\sin\omega t$

So $\mathscr{E}_{\text{in}} \propto B, A, N$ and ω

2. (a) 78.5 s^{-1}; 3.5 V

(b)

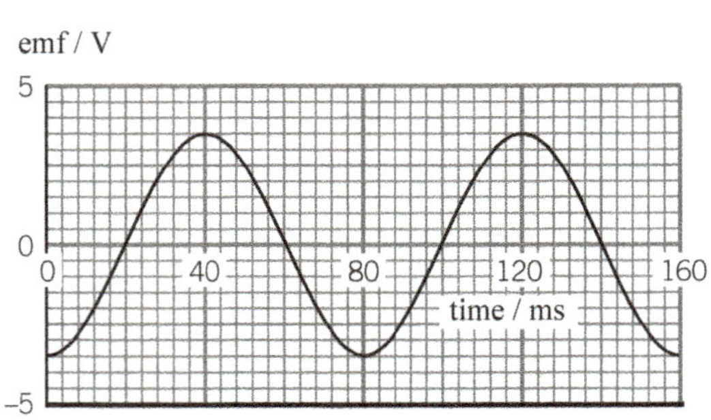

3 (a) (i) 0.90 A, (ii) 8.1 W

(b) $V = 12.7\cos 100\pi t = 12.7\cos 314t$

(c) $I = 1.27\cos 100\pi t = 1.27\cos 314t$

(d)

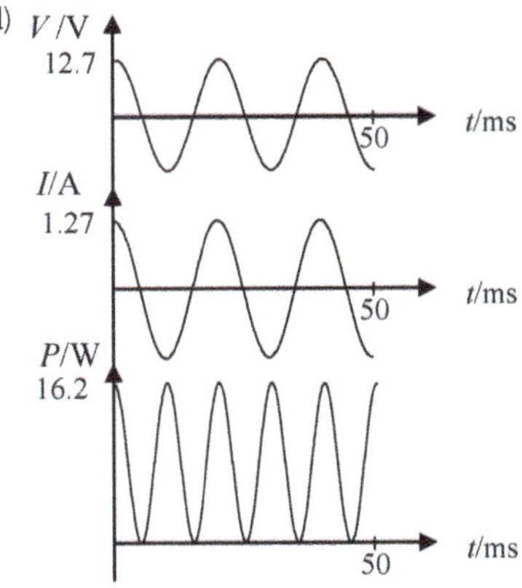

4 1 μF cap: $I_{\text{rms}} = 0.075$ A

100 mH ind: $I_{\text{rms}} = 0.019$ A

5 B = resistor (current and voltage in phase). $R = 1.4$ kΩ

The frequency is 31.25 Hz (needed for A and C)

A = capacitor (current leads the voltage). $C = 5.1$ μC

C = inductor (voltage leads the current). $L = 6.1$ F (rather large)

6 (a) 11.2 V; 5.6 Ω

(b) Resistor: $R = 2.5\ \Omega$, $X = 0$

Capacitor: $R = 0$, $X = 10\ \Omega$

Inductor: $R = 0$, $X = 5\ \Omega$

$R = 2.5\ \Omega$; $C = 320\ \mu$F; $L = 16$ mH

(c) (i) $\varepsilon = 1.11$ rad (current leads)

(ii) 10 W

7 $X_C = 5\ \Omega$; $X_L = 10\ \Omega$.

All other quantities remain the same except that the voltage phasor now leads (by 1.11 rad)

8. Capacitor reactance $Z = \dfrac{1}{400\pi \times 12 \times 10^{-6}} = 66.3\ \Omega$

$\therefore$ Total impedance $Z = \sqrt{47^2 + 66.3^2} = 81.3\ \Omega$

$\therefore\ I_{\text{rms}} = \dfrac{I_0}{\sqrt{2}} = \dfrac{24}{\sqrt{2} \times 81.3} = 0.21$ A

$\varepsilon = 55°$ (current leads)

9. (a) (i) 11.0 kHz

(ii) $V_A = 5.09$ V; $V_B = 1.41$ mV (rms)

(iii) 0.25π rad; 45°

(b) If the claim were correct, the voltage, V_A, would lead the voltage, V_B, across the resistor (which is in phase with the current). This is true. But $V_B << V_A$. So, assuming the inductor has negligible resistance, the resistance of the resistor would be very much less than the reactance of the inductor and hence the phase difference should be close to $\frac{1}{2}\pi$, instead of $\frac{1}{4}\pi$. So the inductor cannot have zero resistance.

[A 0.1 H inductor would have a reactance of 6 800 Ω. With these traces, the resistance of the inductor would also need to be 6 800 Ω and the resistance of the resistor would be 2.8 Ω.]

10. (a) 2.53 nF; 62.8 Ω

(b) (i) 1.59 W (ii) 0.77 W

(c) The power at 95 kHz is close to half the maximum power [0.77 W rather than 0.80 W]. 95 kHz is $0.95f_0$ so the same power would be dissipated at $1/0.95f_0$ which is approximately 105 kHz, giving a half-power bandwidth of ~10 kHz. So there is reasonable agreement.

4B

1 (a) $E_{154} = 8.07$ keV; $E_{139} = 8.94$ keV

(b) More than 8.94 kV

2 (a) Filtering preferentially cuts out low energy photons because they are less penetrating

(b) Low energy photons have carcinogenic properties without being useful for imaging

(c) between 35 kV and 40 kV

(d) $\lambda_{22\,\text{keV}} = 57$ pm; $\lambda_{25.5\,\text{keV}} = 49$ pm

(e)

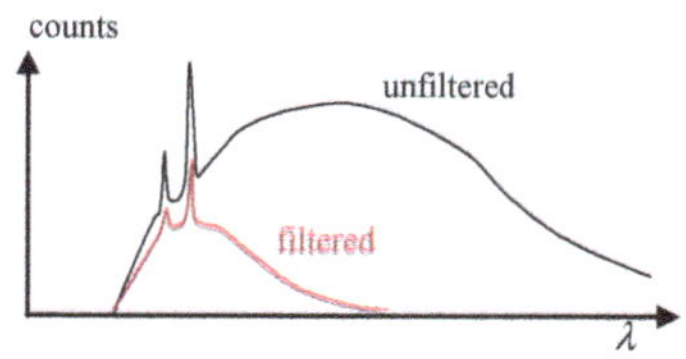

3 (a) (i) 1.0%, (ii) 31.9%, (c) 3.3%

(b) Assuming 8% penetration at 80 keV, $\mu = 0.25$ cm^{-1}

4 (a) Fraction reflected = 0.0014 = 0.14%

(b) 99.7%

5 0.16 m s^{-1}

6 (a) 68.2 MHz

(b) 8520 Hz

(c) Different locations emit different Larmor frequencies and can therefore be located

(d) 1.7 kHz

7 (a) All the X-rays pass through the tumour; other places receive only a fraction

(b) Most of the energy from a proton beam is discharged at the end of the track. If the correct energy is selected (depending on the depth of the tumour) nearly all the energy is released in the tumour, minimising the effect on tissues before and behind the tumour

8 (a) ^{123}I – useful as a tracer in conjunction with a gamma camera; the γ-rays are not absorbed by the body; its short half-life minimises the long-term effects

^{131}I – can be used for radiotherapy, the β particles being absorbed by the thyroid (including the tumour); it can also be used as a tracer but is less safe because of the β emissions and the longer half-life

(b) ${}^{131}_{53}\text{I} \rightarrow {}^{131}_{54}\text{Xe}^* + {}^{0}_{-1}\text{e} + \overline{\nu_e}$

${}^{131}_{54}\text{Xe}^* \rightarrow {}^{131}_{54}\text{Xe} + \gamma$

9. **A**-scans do not produce images – they just reveal the depth of the various structures in the eye, e.g. front and back of the lens, back of the eyeball.

B-scans produce images of the structure of the eye.

10. (a) The cancer is a current, life-limiting threat. Cancers will not definitely be produced by the treatment and any that are will only be a danger in the long term, so that a period of cancer-free life is possible.

(b) The radiotherapy beams are directed at the cancer from several directions, and/or the patient rotates about the position of the cancer. This limits the dose of radiation received at tissues extraneous to the cancer. The received dose is also calculated to achieve a maximum effect whilst keeping effects to other tissues as small as possible.

(c) With several cancers, beams of radiation would need to be aimed at several different places in the body, increasing the whole body dose. There is an increased risk to benefit ratio.

11. (a) $(Z_2 - Z_1)^2 = (Z_1 - Z_2)^2$ and $(Z_2 + Z_1)^2 = (Z_1 + Z_2)^2$

(b) (i) 0.27 (27%)

(ii) From first bone/marrow boundary: 0.19

From second bone/marrow boundary: 0.08

12. A solution is injected, which contains a biologically active molecule (e.g. glucose) tagged with a positron emitter. Cancers are actively growing so concentrate these molecules. Emitted positrons annihilate with electrons in the body giving rise to γ rays which are detected in a γ camera. Concentrations of emission sites reveal the location of tumours.

13. Although both techniques produce images of sufficient detail, a standard X ray is much quicker and cheaper to administer. The slight risk associated with the received radiation dose from the X ray is negligible.

4C

1 A more upright boat presents a larger sail area to the wind. If it heels at angle θ the effective sail area is A cos θ. The sailor in Fig. C.4 should lean further out, thus increasing the anti-clockwise moment of his/her weight.

2 The rudder diverts the water to the starboard (right). The water exerts an opposite force (N3) on the rudder and hence provides a clockwise moment to the boat.

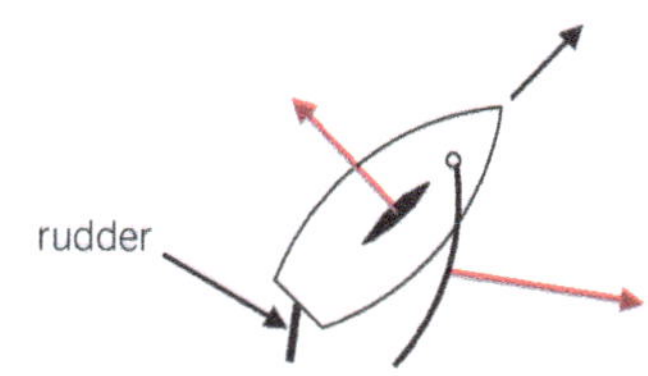

The rudder also exerts a forward force on the water which exerts an equal and opposite force on the rudder.

3 The boat is stable because of the anticlockwise moment exerted by its weight. If the CoG is higher this moment will be less for any particular angle of heel.

4 (a) [Ignoring bounce duration] $t_{max} = 1.095$ s, $t_{min} = 1.049$ s

(b) 1 cm outside range gives a 4 ms difference

5 $e_{max} = 0.84(4)$, $e_{min} = 0.79(5)$

6 29 kN

7 (a) Flight time = 1.81 s; distance = 20.0 m

(b) The flight time is greater (1.92 s) but because the horizontal velocity is less, so is the distance (19.1 m), so this is not advisable

8 (a)

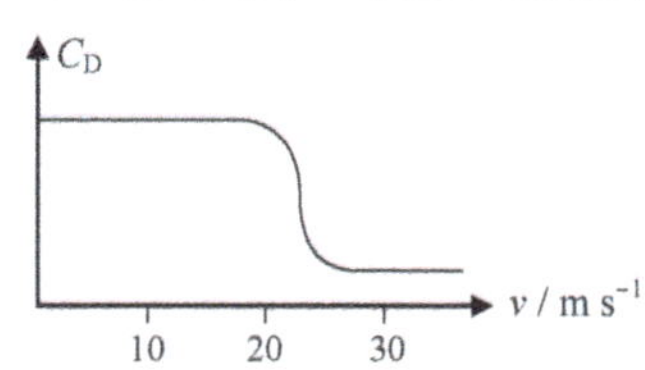

(b) Initially, the speed of the ball is high (~ 30 m s^{-1}) giving a low drag force and therefore a small deflection. As the ball slows down (past the 'wall') its drag coefficient becomes much greater, increasing the drag force and therefore the Magnus force. This, in conjunction with the lower speed, gives a greater curvature to the path.

4D

1 5880 K

2 (a) 3.847×10^{26} W

(b) 1.368 kW m^{-2}

3 (a) 40 cm^3

(b) 43.6 cm^3

(c) No change; the 43.6 cm^3 of ice melts down into 40 cm^3 of water

(d) Only melting land-bound ice raises sea level significantly; melting floating ice (e.g. Arctic ice cap) has only a only a small effect, due to the difference in densities of the sea water and the (fresh) meltwater.

4 (a) 0.130

(b) Assuming zero reflection → 167 K; the actual albedo is 0.09, which leads to 163 K

5 106 kW

6 (a) 24 000 kg s^{-1}

(b) 1.77 MW

(c) 430 kW

(d) 1.1 MW

(e) 60%

7 (a) At 300 K, $kT = 4.14 \times 10^{-21}$ J = 0.026 eV

(b) ${}^{233}_{90}\text{Th}^* \rightarrow {}^{233}_{90}\text{Th} + \gamma$

${}^{233}_{90}\text{Th} \rightarrow {}^{233}_{91}\text{Pa} + {}^{0}_{-1}\text{e} + \overline{\nu_e}$

${}^{233}_{91}\text{Pa} \rightarrow {}^{233}_{92}\text{U} + {}^{0}_{-1}\text{e} + \overline{\nu_e}$

(c) To provide the source of neutrons for the first step in the reaction chain

(d) Because its half-life is longer than ^{235}U (7×10^8 years) or ^{238}U (4.2×10^9 years)

8 When ^{238}U absorbs a neutron it decays in two stages to ^{239}Pu which is fissile; the fission of ^{239}Pu provides the additional energy

9 (a) 6.7 kW

(b) 192 W

(c) The temperature of the inner surface of the glass is much less than 25 °C and that of the outer surface much greater than 5 °C; a 'skin' of still air on either side of the pane adds to the insulation

(d) 0.58 °C

(e) 12.6 W m^{-2} K^{-1}
[the assumption is not a very good one]

10 (a) (i) 200 (ii) 0.60 V

(b) 155 W

(c)

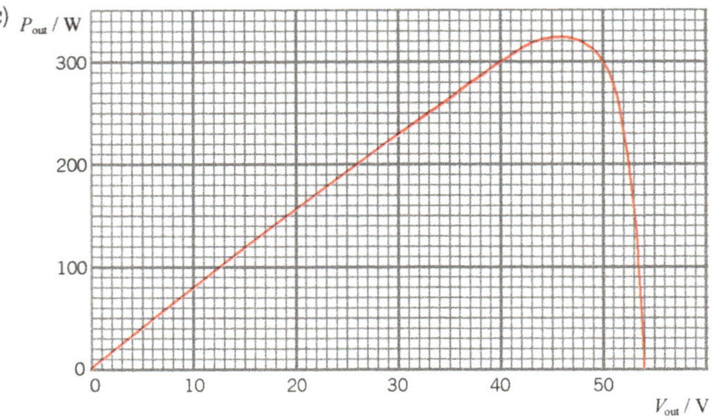

(d) P_{max} = 322 W (ish), at 46 V and 7.0 A

(e) 1.9 m^2.

11. In any location, there are 4 peak generating times per day – around high and low water – separated by about 6 hours. The tide time differences with Swansea are ~1hr, 2 hrs, 4 hrs and 5 hrs. So, if tidal lagoons were constructed at all these locations, there might be a roughly constant energy generation – increasing the benefit. Other factors to be taken into account though are the variation of tidal range with location, the availability of suitable sites and the impact on wildlife.

M

1 (a) $-\cos \omega t$

(b) $\sin \omega t$

(c) $\cos \omega t$

(d) $\cos \omega t$

(e) $-\sin \omega t$

2 (a) $\cos(\omega t \pm \pi)$

(b) $\cos(\omega t + \frac{\pi}{2})$

(c) $\cos(\omega t - \frac{\pi}{2})$

(d) $\cos(\omega t - \frac{\pi}{2})$

(e) $\cos(\omega t \pm \pi)$

3 (a) ± 15 ms, ± 5 ms

(b) ± 20 ms, 0 ms

(c) ± 3.3 ms, ±16.7 ms

(d) ± 6.7 ms, ±13.3 ms

4 (a) –20 ms, –10 ms, 0, 10 ms, 20 ms

(b) –5 ms, 15 ms

(c) –8.3 ms, –1.7 ms, 11.7 ms, 18.3 ms

(d) –18.3 ms, –11.7 ms, 1.7 ms, 8.3 ms

5 (a) f = 400 Hz, T = 2.5 ms

(b) (i) 2.5 A, (ii) 2.5 A, (iii) –2.5A

(c) 0.9 ms, 2.2 ms, 3.4 ms and 4.7 ms

(d) 0.3 ms and 2.8 ms

6 (a) 33, (b) 0.33 (c) 3.2 (d) 0.53 (e) 4.55

7 (a) –1 (b) 15 (c) 6 (d) 216 (e) 24

8 (a) $x = 12.2$

(b) $x = 2.04$

(c) $x = 0.777$

(d) $t = 9.16$

(e) $t = 1.56 \times 10^4$

9 (a) 2.7 V (b) 497 s

10 The graph of ln (corrected Ev) against ln V is a good straight line
n = gradient = 3.6; $k = e^{\text{intercept}} = 0.03$

11 (a) gradient = –0.167 [s^{-1}]

(b) C = 1.3 mF

Exam practice answers

WJEC bears no responsibility for the example answers to questions taken from its question papers which are contained in this publication.

Unit 3

1 (a) The radian is the angle between two radius lines of a circle for which the arc length is equal to the radius length.

(b) (i) Period, $T = \frac{60 \text{ s}}{36.0} = 1.67 \text{ s}$

(ii) Speed $= \frac{36.0 \times 2x \times 2.80 \text{ m}}{60.0 \text{ s}} = 10.6 \text{ m s}^{-1}$

(c) N is the only force with a horizontal component so must provide the centripetal force.

$\therefore F = \frac{mv^2}{r} = \frac{66.2 \times (10.6)^2}{2.8} = 2634 \text{ N} \sim 2600 \text{ N}$

There is no vertical acceleration,

so $F = W = mg = 66.2 \times 9.81 = 649 \text{ N} \sim 650 \text{ N}$

(d) (i) To prevent sliding $F_{max} \geq mg$, $\therefore \mu N \geq mg$,

$\therefore \mu = \frac{mg}{N} = \frac{649 \text{ N}}{2634 \text{ N}} = 0.247 \sim 0.25$

(ii) As the angular velocity falls, so does the centripetal force (N) because $N = mr\omega^2$. Hence the maximum frictional force, ($= \mu mr\omega^2$) falls and the floor must be raised when this is becomes equal to the weight (mg).

To support the weight, $\mu mr\omega^2 \geq mg$, so

$\omega \geq \sqrt{\frac{g}{\mu r}} = \sqrt{\frac{g}{0.45 \times 2.80}} = 2.79 \text{ rad s}^{-1}$

So if ω falls below this value, the floor will be needed for support.

2 (a) a = acceleration

ω = angular frequency

x = displacement

(b) (i) $a_{max} = A\omega^2 = 0.012 \text{ m} \times \left(\frac{2x}{0.40 \text{ s}}\right)^2 = 2.96 \text{ m s}^{-2}$

(ii)

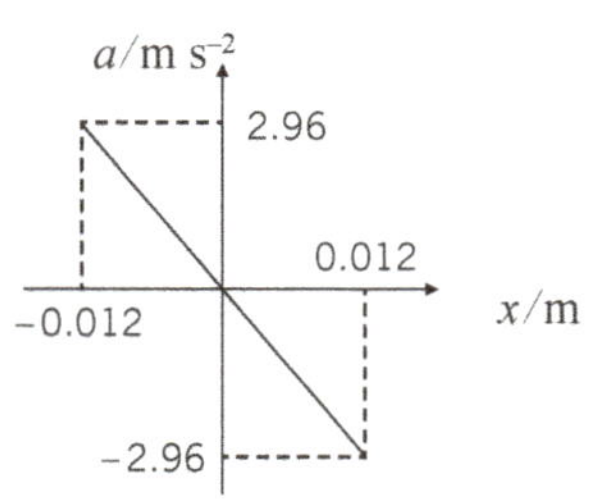

(c) $x = 0.012 \cos\left(15.7t + \frac{3\pi}{2}\right)$ or $x = 0.012 \cos\left(15.7t + \left[-\frac{\pi}{2}\right]\right)$

(d) (i) Note that there is no single correct answer to this question. Here are two possible answers:

In a microwave oven, the electric field vibrations in the microwaves force the water molecules (in the food) into vibrations, the energy transfer leading to the food's being heated.

In a violin, the vibrations of the strings are passed through the bridge and into the wood of the violin's body making it vibrate, which in turn causes vibrations in the surrounding air, resulting in (pleasurable) sound waves.

(ii) Here again different examples are possible, all involving resonance:

If a group of people walk in step over a bridge, so that the period of their steps happens to match one of the natural frequencies of oscillation of the bridge, resonance occurs and uncomfortable large-amplitude oscillations of the bridge are produced, arising from the impact of the footsteps (as in the London Millennium Bridge).

3 [QER]

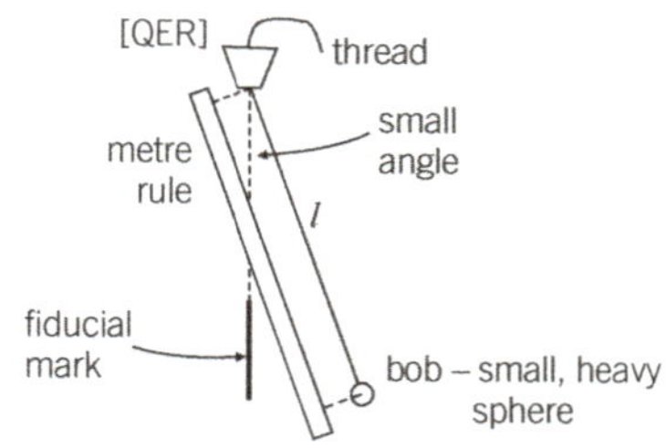

Procedure: The apparatus is set up as shown. The length l is measured from the base of the cork to the centre of mass of the bob, using the metre rule. The bob is pulled a small angle (~5°) to the side and released. The period, T, is determined by measuring the time taken for 20 oscillations, (measured from the time the thread passes in front of the fiducial mark) and dividing by 20. The measurement is repeated to check for miscounts [the timing error is negligible because anticipation is possible]. The procedure is repeated for at least 5 different values of l between 20 cm and 100 cm.

Analysis: The relationship between T and l is $T^2 = 4\pi^2 \frac{l}{g}$, so a graph is plotted of T^2 against l and the gradient, m, determined.

Hence, g is determined from $g = \frac{4x^2}{m}$.

4 (a) (i) $pV = nRT$,

so $n = \frac{pV}{RT} = \frac{5.0 \times 10^5 \text{ Pa} \times 8.5 \times 10^{-3} \text{ m}^3}{8.31 \text{ J mol K}^{-1} \times 285 \text{ K}} = 1.79 \text{ mol}$

(ii) $N = nN_A = 1.79 \times 6.02 \times 10^{23} = 1.08 \times 10^{24}$

(iii) Mass of oxygen $= 32 \times 10^{-3} \text{ kg mol}^{-1} \times 1.79 \text{ mol} = 0.0574 \text{ kg}$

$p = \frac{1}{3}\rho\overline{c^2}$

so $c_{rms} = \sqrt{\frac{3p}{\rho}} = \sqrt{\frac{3pV}{M}} = \sqrt{\frac{3 \times 5.0 \times 10^5 \times 8.5 \times 10^{-3}}{0.0574}}$

$= 471 \text{ m s}^{-1}$

(iv) $F = pA = 5.0 \times 10^5 \text{ Pa} \times 0.040 \text{ m}^2 = 20 \text{ kN}$

(b) (i) $p_2 = p_1 \frac{V_2}{V_1} = 5.0 \times 10^5 \text{ Pa} \times \frac{8.5(\times 10^{-3})}{10.2(\times 10^{-3})} = 4.17 \times 10^5 \text{ Pa}$

(ii) The temperature is constant, so the internal energy is unchanged. Hence the heat input into the gas must equal the work output (First law of thermodynamics), i.e. heat input is 773 J.

(iii)

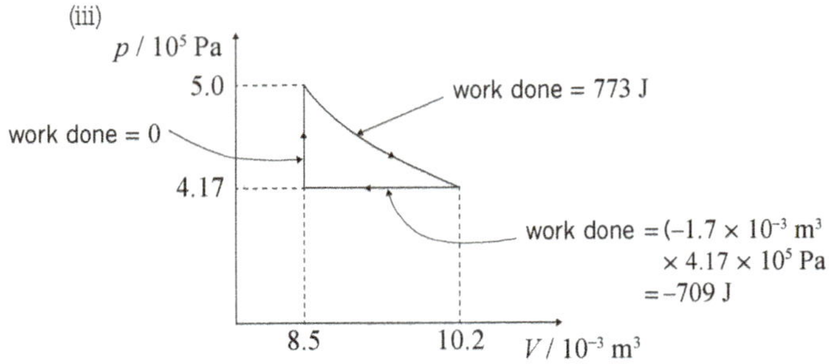

Total work done by gas, W, = 773 J – 709 J = 64 J

The internal energy of the gas is unchanged at the beginning and end of the cycle, i.e. ($\Delta U = 0$). $\Delta U = Q - W$, so the heat input, Q, to the gas is 64 J and the student is correct.

5 (a) (i) ${}^{228}_{90}\text{Th} \rightarrow {}^{224}_{88}\text{Ra} + {}^{4}_{2}\alpha$

(ii) ${}^{90}_{38}\text{Sr} \rightarrow {}^{90}_{39}\text{Y} + {}^{0}_{-1}\beta$

(b) Mass deficit = 38 × (1.007 276 + 0.000 549) +

52 × 1.008 664 − 89.907 738

= 0.840 14 u

∴ Binding energy = 0.840 14 × 931 MeV = 782 MeV

∴ BE per nucleon = $\frac{782}{90}$ = 8.69 MeV nuc^{-1}

(c) (i) $\frac{1}{4}$ (0.25 or 25%)

(ii) I. p(undecayed after 1 throw) = 1 – 0.25 = 0.75

∴ p(undecayed after n throws) = $(0.75)^n$

II. After 10 throws, expected number

= 564 × $(0.75)^{10}$ = 31.8 (3 sf)

[Note: The 'expected' number does not have to be an integer. Neither 31 nor 32 is the expected number]

III. From the results, the number remaining after 10 throws is 35, which is close to 31.8 (within expected random variation). The individual throws mainly give results which are close to 0.75 – with just one being 0.9, which is quite a long way off, but that is to be expected in a random process.

Unit 4

1 (a) (i) 1 A planet moves in an elliptical orbit with the Sun at one focus.

2 The line joining the planet to the Sun sweeps out equal areas in equal times.

(ii) Equating gravitational force on a body of mass m to mass ◊ centripetal acceleration:

$$\frac{GM_E m}{r^2} = mr\omega^2 = mr\frac{pV}{RT} = \left(\frac{2\pi}{T}\right)^2$$

Re-arranging: $T^2 = \frac{4\pi^2}{GM_E} r^3$

(b) $\left(\frac{T^2}{r^3}\right)_{\text{moon}} = \left(\frac{T^2}{r^3}\right)_{\text{geostat}}$ so $T_{\text{moon}} = \left(\frac{r_{\text{moon}}}{r_{\text{geostat}}}\right)^{\frac{3}{2}} T_{\text{geostat}}$

so $T_{\text{moon}} = \left(\frac{380\,000}{42\,000}\right)^{\frac{3}{2}} \times 1.00 \text{ day} = 27 \text{ day}$

(c) If satellite could orbit just above ground level (that is neglecting effects of atmosphere),

$r_E\omega^2 = g_{\text{surface}}$ so $r_E\left(\frac{2\pi}{T}\right)^2 = g_{\text{surface}}$

so $T = 2\pi\sqrt{\frac{r_E}{g_{\text{surface}}}} = \sqrt{\frac{6.37 \times 10^6}{9.81}}$ s = 5065 s = 84 minutes

2 (a) Orientate the probe to measure fields directed into paper on diagram. Perpendicular distance, r, between centre of probe and centre-line of wire should be a few cm. Measure r with a ruler and note the Hall probe reading, B. Calculate the current, I, using

$$I = \frac{2\pi rB}{\mu_0}$$

(b) The wires repel, that is the force on the top wire is up the page and the force on the bottom wire is down the page.

Force on either wire, $F = BI\ell = \frac{\mu_0 I}{2\pi a} I\ell$

So $\frac{F}{\ell} = \frac{\mu_0 I^2}{2\pi a} = \frac{4\pi \times 10^{-7} \times 290^2}{2\pi \times 0.050}$ N m^{-1} = 0.34 N m^{-1}

(c) (i) Flux density is directed into page.

Initial radius, R, of path given by $\frac{mv^2}{R} = Bev$

So $R = \frac{mv}{Be} = \frac{9.11 \times 10^{-31} \times 2.9 \times 10^7}{4.64 \times 10^{-3} \times 1.60 \times 10^{-19}}$ m = 3.6 cm

The claim is incorrect as the electron will be deflected down the page and out of the central region between the wires, where the field is fairly uniform, into a region of increasing flux density.

(ii)

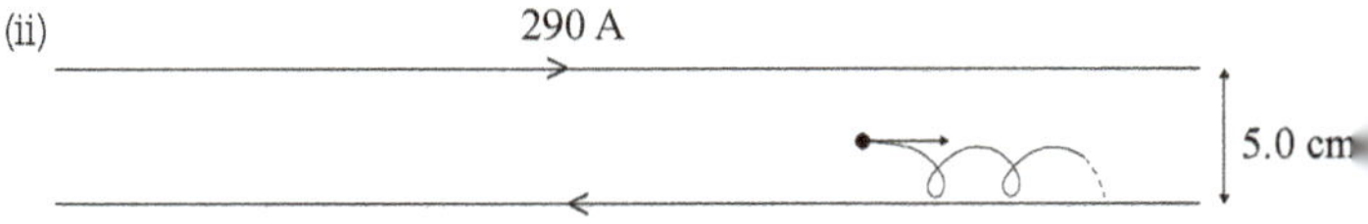

(d) Magnetic force on electron

= Bev = 4.64 × 10^{-3} × 1.60 × 10^{-19} × 2.7 × 10^7 N

= 2.15 × 10^{-14} N down the page

Electric force on electron = eE

= 1.60 × 10^{-19} × $\frac{13\,500}{0.10}$ N = 2.16 × 10^{-14} N up the page

Therefore the resultant force on the electron is almost zero and it will continue with almost constant velocity.

3 (a) Faraday's Law. When the magnetic flux linking a circuit changes, an emf is induced in the circuit. The emf is proportional to the rate of change of flux linkage.

Lenz's Law. The direction of any current resulting from an induced emf is such as to oppose the change in flux linkage that is causing the current.

(b) Initially the magnets accelerate downwards because of the downward gravitational forces on them. Their motion causes a changing magnetic flux linked with circumferences of the pipes and inducing circumferential emfs. In pipe **P** these emfs drive currents around the pipe's circumference, setting up magnetic fields inside the pipe. These fields exert a force on the magnet, which, by Lenz's law, is upwards. At a certain, low speed the upward force balances the downward gravitational force and there is no further acceleration. In pipe **Q** the slit prevents the emf from driving a current, so there is no upward force (except air resistance). The magnet therefore accelerates at an almost constant rate.

(c) Loss of gravitational PE of magnet = gain in internal energy of tube

So $m_{magnet}gh = m_{pipe}c_{copper}\Delta T = Ah\rho_{pipe}c_{copper}\Delta T$

in which A is the shaded area.

Re-arranging:

$$\Delta T = \frac{m_{magnet}g}{A\rho_{copper}c_{copper}} = \frac{0.300 \times 9.81}{7.85 \times 10^{-6} \times 8960 \times 385}\text{ K} = 0.11\text{ K}$$

Optional topics

Option A

1 (a)

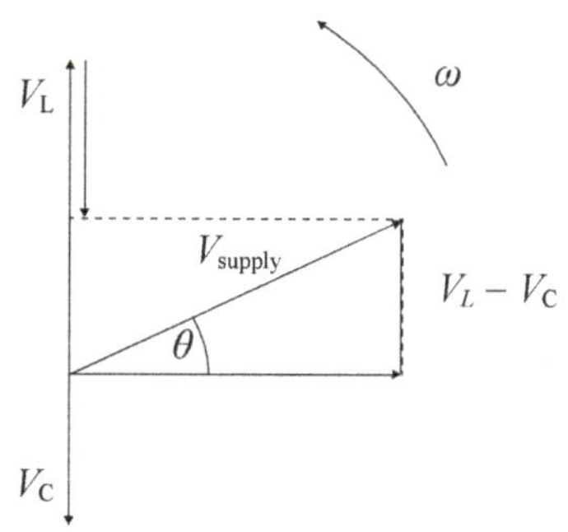

(i) V_L leads V_R by $\frac{\pi}{2}$, and V_C lags V_R by $\frac{\pi}{2}$.

The angle θ is given by $\tan\theta = \frac{V_L - V_C}{V_R}$.

Rms values are related by $V_{supply} = \sqrt{(V_L - V_C)^2 + V_R^2}$

(ii) At resonance, $V_L = V_C$ so $\tan\theta = 0$, that is V_{supply} is in phase with V_R, and $V_{supply} = V_R$

(b) (i) $Q = \frac{V_L}{V_R} = \frac{V_C}{V_R} = \frac{119.1\text{ V}}{45.0\text{ V}} = 2.65$

(ii) $I = \frac{V_R}{R} = \frac{45.0\text{ V}}{28\ \Omega} = 1.61\text{ A}$

(iii) $\frac{1}{2\pi fC} = \frac{V_C}{I}$ so $f = \frac{I}{2\pi CV_C}$

$$= \frac{1.61}{2\pi \times 82 \times 10^{-6} \times 119.1}\text{ Hz} = 26.2\text{ Hz}$$

(iv) $2\pi fL = \frac{V_L}{I}$ so $L = \frac{V_L}{2\pi fI}$

$$= \frac{119.1}{2\pi \times 26.2 \times 1.61}\text{ H} = 4.52\text{ mH}$$

(v)

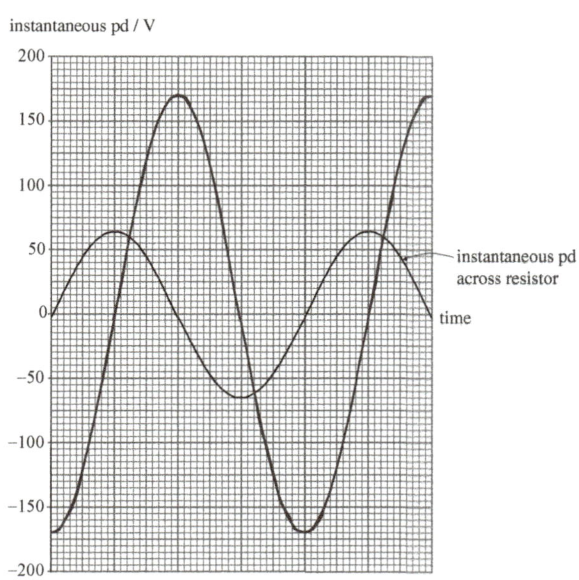

(vi) The reactance of the inductor increases and that of the capacitor decreases so that $X_L - X_C$ is no longer zero and the circuit impedance is greater than R, so the current decreases.

(c) At 20 kHz, $X_C = \frac{1}{2\pi \times 20\,000 \times 185 \times 10^{-9}} = 43\ \Omega$, so

$$V_{out} = 12.0\text{ V} \times \frac{43}{43\sqrt{2}} = 8.5\text{ V}$$

At a higher frequency than 20 kHz, X_C is smaller so V_{out} is larger than 8.5 V; at a lower frequency, X_C is larger so V_{out} is smaller. With 8.5 V instead of 6.0 V Sam would have been right. He added X_C to R instead of using $Z = \sqrt{X_C^2 + R^2}$

Option B

2 (a) (i) X-rays can be produced with quite simple apparatus. X-ray images can be conveniently produced on photographic film or on fluorescent screens. Different tissues, including abnormal tissue, attenuate X-rays to different degrees, and can be distinguished on images formed by transmitted X-rays. Bone attenuates X-rays strongly and so bones and fractures show up particularly well. Soft tissue can also be imaged – in great detail using X-ray CT scanning.

(ii) $\frac{1}{2}mv^2 = eV$

so $v = \sqrt{\frac{2eV}{m}} = \sqrt{\frac{2 \times 1.60 \times 10^{-19} \times 18\,000}{9.11 \times 10^{-31}}}\text{ m s}^{-1}$

$= 8.0 \times 10^7\text{ m s}^{-1}$

(iii) $P = 0.005 \times 18\text{ kV} \times 12\text{ mA} = 1.1\text{ W}$

(b) (i) If $x = x_{1/2}$, then $\frac{I}{I_0} = \frac{1}{2}$, so $\frac{1}{2} = e^{-\mu x_{1/2}}$

Therefore $\ln\frac{1}{2} = -\mu x_{1/2}$ that is $\mu x_{1/2} = \ln 2$

(ii) $\mu = \frac{\ln 2}{2.7\text{ cm}}$ so $\ln 0.70 = -\frac{\ln 2}{2.7\text{ cm}}x$ giving $x = 1.4\text{ cm}$

(c) An MRI scan would show the heart in detail, but cheaper forms of imaging usually suffice. An ultrasound B scan shows the heart in some detail if a contrast agent is injected. The scan can be done in conjunction with Doppler measurements of blood flow speeds. X-ray fluoroscopy with a (different) contrast medium and an image intensifier produces real-time images of a beating heart. CT scans can also show the heart in detail. Conventional X-rays imaging does not reveal enough detail because X-rays penetrate all soft tissue easily.

(d) (i) The absorbed dose, D, is the energy absorbed per unit mass of tissue. The equivalent dose, H is DW_R, in which W_R is a number without units that represents the amount of damage done by a particular sort of radiation relative to that done by X-rays.

(ii) Equivalent dose = 4 mSv × 0.12 = 0.48 mSv.

Option C

3 (a) (i) $\left|\dfrac{\text{relative velocity of separation (of puck and ice)}}{\text{relative velocity of approach}}\right| = 0.55$

(ii) Coefficient, e, at room temperature $= 0.55 \times 0.70 = 0.385$.

$$e = \left|\frac{v_{\text{separation}}}{v_{\text{approach}}}\right| = \sqrt{\frac{\text{rise height}}{\text{drop height}}}$$

so rise height = drop height × $e^2 = 0.50\text{ m} \times 0.385^2 = 0.074\text{ m}$

(b) (i) Mean force $= \dfrac{\Delta(mv)}{\Delta t} = \dfrac{0.17 \times (34 - 3)}{0.025}\text{N} = 210\text{ N}$ (2 sf)

(ii) $u_{\text{vert}} = 34 \times \cos 82° = 4.73\text{ m s}^{-1}$

So greatest height reached $= \dfrac{0 - 4.73^2}{-2 \times 9.81}\text{m} = 1.14\text{ m}$

Height of goal not exceeded.

(iii) Rotational KE $= \dfrac{1}{2}I\omega^2 = \dfrac{1}{2} \times \dfrac{1}{2}mr^2 \times (2\pi f)^2$

$= 0.17 \times 0.038^2 \times \pi^2 \times 14^2\text{ J}$

$= 0.47\text{ J}$

(iv) The puck is also moving horizontally and so has translational kinetic energy as well as its rotational kinetic energy. So Wayne is incorrect.

(v) Pressure below – pressure above $= \dfrac{1}{2}\,1.28(35^2 - 33^2)\text{ Pa}$

$= 87.0\text{ Pa}$

So net upwards force $= 87.0 \times \pi \times 0.038^2\text{ N} = 0.39\text{ N}$

Whereas the puck's weight is $0.17 \times 9.81\text{ N} = 1.7\text{ N}$

Option D

4 (a) (i) The net upward force on a body partly or wholly immersed in a fluid is equal to the weight of fluid that it displaces.

(ii) Mass of salt water displaced $= \rho V = 1030 \times 80 \times 10^{-6}$

$= 0.082(4)\text{ kg}$

(iii) So mass of ice cube $= 0.082(4)\text{ kg}$

and volume of cube $= \dfrac{m}{\rho} = \dfrac{0.0824}{920}\text{ m}^3$

$= 8.96 \times 10^{-5}\text{ m}^3 = 89.6\text{ cm}^3$

So volume of cube above surface $= 89.6 - 80.0 = 9.6\text{ cm}^3$

(iv) I $\Delta h = \dfrac{\Delta\,\text{volume}}{\text{area}} = \dfrac{2.2 \times 10^{11}}{3.6 \times 10^{14}}\text{m}^3 = 0.61\text{ mm}$

II All the water from the melting ice sheet goes into the ocean, raising its level. The submerged part of an iceberg has *already* raised the ocean level. Melting raises the levels only a little because the volume of melt water is greater than the submerged volume of the berg only in the ratio density of sea water/density of fresh water.

III Ice sheets reflect an appreciable fraction of the Sun's radiation. The land exposed by melting would not reflect as much.

(b) (i) $\dfrac{\Delta Q}{\Delta t} = -KA\dfrac{\Delta\theta}{\Delta x}$

So $K = \dfrac{\Delta Q}{\Delta t} \div A\dfrac{\Delta\theta}{\Delta x}$

so $[K] = \text{J s}^{-1}(\text{m}^2\text{ K m}^{-1})^{-1} = \text{W m}^{-1}\text{ K}^{-1}$

(ii) $\dfrac{\Delta Q}{\Delta t} = -0.041 \times 72 \times \dfrac{5 - 20}{0.10}\text{ W} = 440\text{ W}$ (2 sf)

(iii) Let θ be the temperature of the interface between old and new,

Then $\left[\dfrac{\Delta Q}{\Delta t} =\right] -0.041 \times \cancel{72} \times \dfrac{\theta - 20}{0.10} = -0.035 \times \cancel{72} \times \dfrac{5 - \theta}{0.17}$

So $8.2 - 0.41\,\theta = 0.026\,\theta - 1.03$

so $0.616\,\theta = 9.23$ so $\theta = 15.0\,°\text{C}$

So $\dfrac{\Delta Q}{\Delta t} = -0.041 \times 72 \times \dfrac{15 - 20}{0.10}\text{W} = 150\text{ W}$ (2 sf)

The heat loss rate is reduced by $\frac{2}{3}$, so the recommendation is correct

(iv)

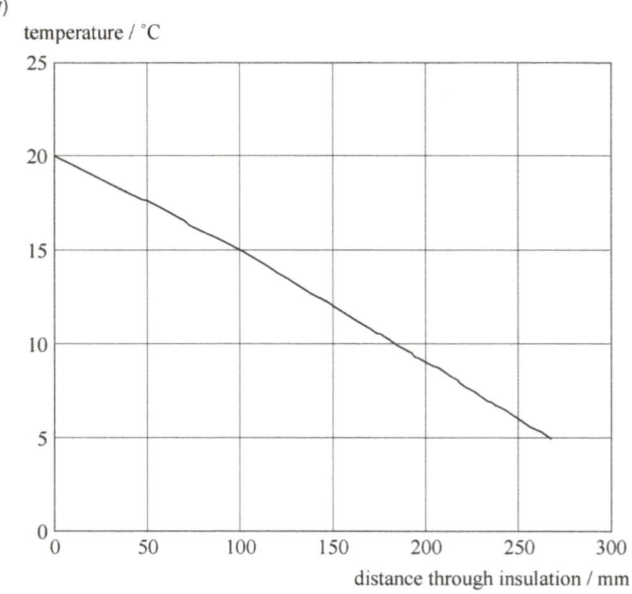

Glossary

Absorbed dose (D) The energy of the radiation absorbed per unit mass of tissue (units are Gy).

Acoustic impedance (Z) The acoustic impedance of a material is defined by $Z = c\rho$ where ρ is its density and c the speed of sound in the material.

Activity (A) The activity, A, of a radioactive source is the number of radioactive decays per unit time (units are Becquerel (Bq) $\equiv$ s^{-1}).

Acute effects (radiation poisoning) – illness and death due to a sudden large dose of radiation.

Alpha radiation (α) Fast moving particles consisting of two protons and two neutrons (or helium nuclei) ejected from certain radioactive nuclei.

Amplitude (A) of an oscillation The maximum value of the object's displacement from its equilibrium position (units are m).

Angle of attack The angle between the air flow and the orientation of the javelin (or discus).

Angular acceleration (α) The rate of change of angular velocity (units are rad s^{-2}).

Angular frequency (ω) The quantity, ω, is called the angular frequency of the oscillation (units are s^{-1} [or rad s^{-1}]).

Angular momentum (L) The product of the moment of inertia and the angular velocity (units are N m s).

Angular velocity (ω) The rate of change of angular position (units are rad s^{-1}).

Archimedes principle A body wholly or partly immersed in a fluid experiences an upthrust equal to the weight of fluid displaced by the object.

Avogadro constant (N_A) The Avogadro constant is the number of particles per mole (units are mol^{-1}).

Background radiation In the context of laboratory investigations into radioactive sources, background radiation is the nuclear radiation which is detected in the absence of the source under investigation.

Beta radiation (β) High energy electrons (β^-) or positrons (β^+) ejected from certain radioactive nuclei.

Binding energy The binding energy of a nucleus is the energy required to disassemble the nucleus into its constituent protons and neutrons (units are J or MeV).

Capacitance (C) The constant ratio - charge on either plate/pd between plates. That is

$$C = \frac{Q}{V}$$

(Units are $C\ V^{-1}$ = farad (F))

Capacitor Two conductors (so-called 'plates') separated by an insulator.

Closed system A closed system is one which matter cannot enter or leave (so an isolated system is also closed).

Coefficient of restitution The ratio of the relative speed of a pair of colliding bodies after to collision to the relative speed before collision.

Conservation of mass/energy Energy cannot be lost or gained, only transferred from one form to another. We can measure the energy in a body by multiplying its mass by c^2.

Contrast medium A substance introduced into a part of the body (e.g. gut, blood vessels) to give a greater contrast with surrounding tissues.

Coulomb's law The force between point charges, Q_1 and Q_2, in a vacuum and separated by distance r is given by

$$F = \frac{Q_1 Q_2}{4\pi\varepsilon_0 r^2}$$

in which ε_0 is the permittivity of free space
$= 8.85 \times 10^{-12}$ F m^{-1}

[NB $\frac{1}{4\pi\varepsilon_0} = 9.0 \times 10^9$ F^{-1} m (2 sf)]

Coupling medium Gel or oil used to exclude air between the skin and the ultrasound probe, reducing the mismatch in Z and hence allowing greater transmission of ultrasound.

Critically damped system A critically damped system returns to equilibrium as rapidly as possible without oscillating.

Damping The dying away, due to resistive forces, of the amplitude of 'free oscillations'.

Decay constant (λ) The fraction of the atoms of a radioactive nuclide which decay per unit time.

Dielectric Insulator between the plates of a capacitor, which increases the capacitance of the capacitor.

Direction of a magnetic field The direction of a magnetic field in a certain place is the direction in which the North pole of a small, freely pivoted magnet tends to point.

Doppler effect Change in frequency of waves due to the relative motion of wave source, sensor, medium or reflector

Eddy currents Currents that take closed loops in conducting material in bulk. They arise from electromagnetic induction.

Effective dose (E) The equivalent dose multiplied by a tissue weighting factor (units are Sv).

Electric field line A line whose direction at each point along it is the direction of the electric field at that point.

Electric field strength (E) At a point, P, electric field strength is defined as

$$E = \frac{[\text{force on test charge, } q\text{, at P}]}{q}$$

It is a vector whose direction is that of the force on a positive test charge
(units are N C^{-1} = V m^{-1}).

Electrical potential (V) Electric potential at a point is the work done per unit charge in bringing a positive charge from infinity to that point (units are V or J C^{-1}).

Electrical potential energy The electrical potential energy of a charge is the work it can do because of its position in an electric field.

Electromagnetic damping This is the slowing of motion due to forces experienced by induced currents, in accordance with Lenz's law.

Enrichment of uranium The process of increasing the ratio of the fissile U-235 to the non-fissile U-239 in a sample of uranium.

Equipotential surface In an electric or gravitational field, an imaginary surface on which all points are at the same potential.

Equivalent dose (H) The absorbed dose multiplied by a radiation weighting factor (units are Sv).

Exoplanet A planet which orbits a star other than the Sun.

Faraday's law When the magnetic flux, Φ, through a circuit changes, an emf is induced in the circuit, proportional to the rate of change of flux linkage. For a circuit with N identical turns

$$E = \frac{-\Delta(N\Phi)}{\Delta t}$$

$N\Phi$ is the flux linkage. If the circuit is a coil of N turns, with the same flux, Φ, through each, the emf is the same as if there were N times the flux through a single turn!

First law of thermodynamics The increase, ΔU, in the internal energy of a system is $\Delta U = Q - W$, in which Q is the heat entering the system and W is the work done by the system.

Fissile nuclide A fissile nuclide is one that undergoes fission when it absorbs thermal neutrons, with the emission of enough neutrons to maintain a chain reaction.

Fleming's left hand motor rule Hold the left hand so that:

- the First finger points straight out from the palm and in the direction of the magnetic Field,
- the seCond finger points in the direction of the conventional Current
- the Thumb is at right angles to both these fingers and gives the direction of the force (or Thrust) on the wire.

Fleming's right hand generator rule Hold the right hand so that:

- the First finger points straight out from the palm and in the direction of the magnetic Field,
- the ThuMb points in the direction of Motion of the wire,
- the seCond finger is held in towards the palm...
  The seCond finger points in the direction of the conventional Current in the conductor (and so in the emf direction).

Forced oscillation A forced oscillation occurs when a sinusoidal driving force is applied to an oscillatory system, causing it to oscillate with the frequency of the applied force.

Free oscillations These occur when an oscillatory system is displaced and released

Frequency (f) The frequency, f, is the number of revolutions or oscillations per unit time (units are $\text{Hz} \equiv \text{s}^{-1}$).

Fuel cell A device which uses chemical energy from fuel to provide electrical energy directly.

Gamma (γ) radiation High energy photons ejected from the nuclei of radioactive atoms.

Gravitational field line A line whose direction at each point along it is the direction of the gravitational field at that point.

Gravitational field strength (g) The gravitational field strength at a point is defined as

$$g = \frac{F}{m}$$

that is

$$g = \frac{\text{Force on test mass } (m)}{m}$$

at that point. It is a vector.

Gravitational potential (V_g) Gravitational potential at a point is the work done per unit mass in bringing a mass from infinity to that point (units are J kg^{-1}).

Gravitational potential energy The gravitational potential energy of a test mass, m, at a point P in a gravitational field is the work done by the field on m if m goes from P to infinity.

Greenhouse effect The name given to the process of the warming of a planet by radiation from its atmosphere to a temperature above what it would be without an atmosphere.

Half-life The time taken for the number of nuclei of a radioactive nuclide (or the activity) to halve.

Heat (Q) The energy flow from a region at higher temperature to a region at lower temperature, due to the temperature difference

Heat engine A heat engine is a system which converts thermal energy into mechanical work in a cyclic process.

Hubble time The Hubble time is an estimate of the age of the universe and is defined as the reciprocal of the Hubble constant.

Hubble's law Hubble's law states that the recession velocity, v, of deepspace objects is proportional to their distance.

Ideal gas An ideal gas is one which strictly obeys the equation $pV = nRT$.

Impedance (Z) The ratio

$$\frac{V}{I}$$

for a component (or circuit), for any phase relationship between current and pd, where V and I are the rms pd and current [or peak pd and current] respectively of sinusoidal a.c.

Induced fission The splitting of a nucleus caused by the absorption of a neutron.

Inductance (L) The inductance of an inductor is defined by the equation:

$$V = \frac{L\Delta I}{\Delta t}$$

or, in calculus,

$$V = \frac{L\mathrm{d}I}{\mathrm{d}t}.$$

The unit is the henry (H).

Internal energy of a system The internal energy of a system is the sum of the potential and kinetic energies of its particles.

Isolated system A system which does not allow energy or matter to enter or leave is called an isolated system.

Kepler's first law of planetary motion (K1) Each planet moves in an ellipse with the Sun at one focus.

Kepler's second law of planetary motion (K2) The radius vector (the line joining a planet to the centre of the Sun) sweeps out equal areas in equal times.

Kepler's third law of planetary motion (K3) The square of the period of a planet's orbit is proportional to the cube of its semi-major axis.

Larmor frequency The photon frequency which resonates with the energy level difference between the magnetic moment states of a nucleus in a magnetic field.

Lenz's law The direction of an induced emf is such that its effects oppose the change producing it.

Log-log graph A graph of the logarithm of one variable against the logarithm of another variable.

Long-term effects (or stochastic effects) cancers, cataracts, development problems (e.g. foetal).

Magnetic field We say that a magnetic field exists in a region in which a moving electric charge experiences a force, but a stationary charge does not.

Magnetic field line (or line of magnetic flux) A line whose direction at every point is the direction of the magnetic field at that point.

Magnetic flux The magnetic flux, Φ, through an area, A is defined by $\Phi = BA \cos\theta$, in which θ is the angle between the normal to the area and the flux density. It is a scalar.

Magnetic flux density (B) A vector of magnitude

$$B = \frac{F_{\text{max}}}{I\ell},$$

where F_{max} is the force on a short wire of length ℓ, carrying current I, angled so as to maximise the force on it. The direction of the vector is given by Fleming's left hand rule, knowing the force and current directions.

Main sequence A main sequence star is one which generates most of its energy from the fusion of hydrogen to helium in its core.

Mean square speed The quantity $\overline{c^2}$ in equation [3] is called the mean square speed. It is the mean value of the squares of the molecular speeds [squaring it makes it a scalar quantity].

Mole (mol) The SI unit of amount of substance. It is the amount containing as many particles (e.g. molecules) as there are atoms in 12 g (exactly) of carbon-12.

Moment of inertia (I) The moment of inertia of an object consisting of n particles is defined by:

$$I = \Sigma m_i r_i^2$$

where m_i and r_i are the mass and distance from the axis of rotation of the ith particle.

Natural frequency The frequency of 'free oscillations'.

Newton's law of gravitation Every body with mass attracts every other body with mass. For two 'point masses', m_1 and m_2, separated by distance r, the force is given by

$$F = \frac{Gm_1m_2}{r^2},$$

in which G is a constant called *Newton's gravitational constant.* $G = 6.67 \times 10^{-11}\ \text{N m}^2\ \text{kg}^{-2}$

Nonrenewable energy source An energy source regarded as effectively not capable of replenishment.

North pole (or North-seeking pole) of a magnet The end that tends to point roughly geographically North if only the Earth's magnetic field is present.

Nuclear binding energy curve A plot of the binding energy per nucleon against nucleon number for known nuclides.

Nuclear decay (or radioactive decay) Nuclear decay (or radioactive decay) is the process by which the nucleus of an atom loses energy by emitting (nuclear) radiation.

Nuclear fission The process of splitting of a heavy nucleus into two large fragments with the release of energy.

Nuclear fusion The process of the merging of two light nuclei with the emission of energy.

Nuclear radiation Nuclear radiation is the emission of energy in the form of waves or particles from the atomic nucleus. The term is also used for the emitted waves or particles themselves.

Nuclide A distinct variety of atom or nucleus, indicated by a specific symbol, ${}^{A}_{Z}\text{X}$.

Period (T) The period, T, of circular motion is the time taken for one complete revolution.

Phase of the oscillation The angle $\omega t + \varepsilon$ in the equation $x = A\cos(\omega t + \varepsilon)$ is called the phase of the oscillation. The angle ε is called the phase constant.

Phasor A vector rotating at the frequency of the alternating current, which can represent the current or voltage in magnitude and phase.

Potential The potential, V, at a point P in an electric field is the work done by the field per unit charge on a test charge as it goes from P to infinity. UNIT: J C^{-1} = volt (V)

Potential difference The potential difference, ΔV, between two points, A and B, in an electric field is the work done per unit charge (W/q) by the field on a test charge, q, going from A to B. If W/q is positive, A is at the higher potential. UNITS: J C^{-1} = volt (V)

Potential energy The potential energy of a test charge, q, at a point P in an electric field is the work done by the field on q, as q goes from P to infinity.

Q factor (Q) A measure of the sharpness of the resonance curve - the larger the Q factor, the sharper the resonance curve.

Radial velocity The radial velocity of an object (e.g. a star) is the component of its velocity relative to an observer on the Earth, in the direction from the Earth to the object.

Radian The angle subtended at the centre of a circle by an arc equal in length to the radius.

Range The range of a nuclear radiation is the distance it will travel before being absorbed. It depends upon the nature and energy of the radiation and the nature of the medium through which it is travelling.

Reactance (X) The ratio $\frac{V}{I}$ for a component (or circuit) for which the current and pd are $\frac{\pi}{2}$ (90°) out of phase, where V and I are the rms pd and current [or peak pd and current] of sinusoidal a.c.

Relaxation time (T) The time for the relaxation emission is the time for the intensity of the radiation to drop to $\frac{1}{e}$ of its original intensity.

Renewable energy source An energy source is regarded as renewable if it is replenished naturally in a human lifetime (some organisations use 50 years as the time scale).

Resistance (R) The ratio of $\frac{V}{I}$ for a component (or circuit) for which the current and pd are in phase, where V and I are the rms pd and current [or peak pd and current] of sinusoidal a.c.

Resonance frequency The frequency of the applied force (or pd) for which the amplitude of the oscillations (or current) is a maximum.

Right hand grip rule Grasp the conductor in your right hand with your thumb sticking out in the direction of the current. The fingers will curl round in the direction of the magnetic field.

RMS pd, V_{rms} the square root of the mean value of the pd (taken over a whole number of cycles), i.e. $V_{\text{rms}} = \sqrt{\langle V^2 \rangle}$ Similarly: $I_{\text{rms}} = \sqrt{\langle I^2 \rangle}$

Root mean square speed (or rms speed) Taking the square root, gives $\sqrt{\overline{c^2}}$ which is called the root mean square speed (or rms speed) which is written c_{rms}.

Root-mean-square (rms) value The square root of the mean of the square value (e.g. of molecular speed or alternating currents and voltage)

Self-induction The induction of an emf in a coil (or circuit) because of a changing current in that coil (or circuit).

Semi-log graph or **log-linear graph** A graph of the logarithm of a variable against a linear variable.

Short-term effects Radiation damage apparent in the few weeks following exposure, mainly to the skin, lungs, ovaries and testes.

Simple harmonic motion Simple harmonic motion occurs when an object moves such that its acceleration is always directed toward a fixed point and is proportional to its distance from the fixed point.

Solar constant (G) The radiative power from the Sun per unit area incident on a surface at right angles to the solar radiation. Its value at the radius of the Earth is $1.36\ \text{kW m}^{-2}$.

Specific heat capacity (c) The specific heat capacity, c, of a substance is defined by $Q = mc\Delta\theta$ where Q is the heat input, m the mass and $\Delta\theta$ the temperature change. Units are $\text{J kg}^{-1}\ \text{K}^{-1}$ (or $\text{J kg}^{-1}\ {}^{\circ}\text{C}^{-1}$).

Spontaneous fission The splitting of a nucleus without stimulation by an incident neutron.

Standard candle A standard candle in astronomy is an object of known luminosity, such as a Type 1a supernova or a Cepheid variable star. Measuring the apparent brightness of such an object in a distant galaxy allows the distance to be calculated, using the inverse square law.

Thermal equilibrium Two systems in thermal contact are in thermal equilibrium if there is no flow of heat between them. They are said to be at the same temperature.

Thermal neutrons Relatively slow-moving neutrons. They have a kinetic energy of the order of $\frac{1}{40}$ eV, equivalent to kT at room temperature.

Thermodynamic system A thermodynamic system is the collection of particles, e.g. atoms and molecules, within a volume of space.

Time constant (τ) For an RC system undergoing exponential decay, the time taken for the charge to fall to $\frac{1}{e}$ (= 0.37) of its original charge. $\tau = RC$

Torque (τ) The resultant torque, τ, on an object with moment of inertia I and angular acceleration α is defined by $\tau = I\alpha$.

Ultrasound A-scan Scan in which the strength of reflection of ultrasound is shown by the amplitude of a trace

Ultrasound B-scan Scan in which the strength of reflection of ultrasound is shown by the brightness of a trace - producing a 2D display.

U value The U value of a building element (e.g. a wall) is the rate of heat transfer per unit area per unit temperature difference (units are $\text{W m}^{-2}\ \text{K}^{-1}$).

Unified atomic mass unit (u) One twelfth of the mass of one atom of carbon-12.

Uniform electric field There is a uniform electric field in a region where the field strength is the same in magnitude and direction at all points.

Work (W) The work done by a system is $p\Delta V$ where p is the pressure exerted and ΔV is the increase in volume.

X-ray attenuation The decrease in intesity of X-rays as they pass through a material.

X-ray beam intensity The X-ray energy per unit area, per unit time, passing normally through a surface.

X-ray fluoroscopy Real-time digital images produced of soft tissues using image intesifiers.

X-rays X-rays are high energy, ionising electromagnetic radiation produced by accelerating electrons or high energy atomic transitions.

Index

O

P

Q

R

S

T

U

V

W

X

Z